ABHANDLUNGEN
DER AKADEMIE DER WISSENSCHAFTEN DER DDR
Abteilung Mathematik — Naturwissenschaften — Technik

Jahrgang 1981 · Nr. 2N

Nonlinear Analysis
Theory and Applications

Proceedings of the Seventh International Summer School
held at Berlin, GDR
from August 27 to September 1, 1979

Herausgegeben von
Professor Dr. sc. Reinhard Kluge

bearbeitet von
Dr. habil. Wolfdietrich Müller

Akademie der Wissenschaften der DDR
Institut für Mathematik

Mit 34 Abbildungen und 2 Tabellen

AKADEMIE-VERLAG · BERLIN · 1981

Herausgegeben im Auftrag des Präsidenten
der Akademie der Wissenschaften der DDR
von Vizepräsident Prof. Dr. Heinrich Scheel

Erschienen im Akademie-Verlag, DDR-1080 Berlin, Leipziger Straße 3—4

Lizenznummer: 202 · 100/422/81
Gesamtherstellung: VEB Druckhaus „Maxim Gorki", 7400 Altenburg
Bestellnummer: 762 924 4 (2001/81/2N) · LSV 1065
Printed in GDR
DDR 84,— M
ISSN 0138-1059

Preface

From August 27 to 31, 1979, the seventh international summer school "Nonlinear Analysis. Theory and Applications" was held in Berlin at the Central Institute of Mathematics and Mechanics of the Academy of Sciences of the GDR. It continued the summer schools devoted to nonlinear problems organized alternately by the Czechoslovakian Academy of Sciences and by the Central Institute in Berlin.

The first school took place at Babylon (CSSR) in 1971, the second at Neuendorf/Hiddensee (GDR) in 1972, the third at Starà Lesná (CSSR) in 1974, the fourth at Berlin (GDR) in 1975, the fifth at Berlin (GDR) in 1977, the sixth at Horní Bradlo (CSSR) in 1978.

The 105 participants of the seventh summer school came from Czechoslovakia, France, Italy, Poland, Romania, Switzerland, the U.S.S.R., the U.S.A and the German Democratic Republic.

In the 31 invited lectures and 23 communications special attention was given to qualitative aspects (direct problems, optimal control, inverse problems), to approximation methods, to numerical realizations on the computer and to applications.

It is hoped that the 52 papers contained in this volume provide an informative survey of the scientific programme of the summer school.

Berlin, December 1979

Prof. Dr. sc. Reinhard Kluge
Chairman of the summer school

Contents

Invited Lectures

Communications

Invited Lectures

Saddle points and nonlinear differential equations

Herbert Amann, Zürich

1. In this note we describe some recent results concerning the solvability of nonlinear equations in Hilbert spaces, which have partly been obtained jointly with E. Zehnder (cf. [1, 2]). The abstract results have been motivated by problems in the theory of nonlinear partial differential equations. For this reason we shall illustrate our general results by means of the following simple examples.

Let Ω be a bounded domain in $\mathbb{R}^n$ with smooth bondary $\partial\Omega$ and let $f \in C^1(\mathbb{R})$ be given. Then we are looking for solutions of the nonlinear elliptic boundary value problem (BVP)

$$\left.\begin{aligned} -\Delta u &= f(u) \quad \text{in} \quad \Omega, \\ u &= 0 \qquad \text{on} \quad \partial\Omega. \end{aligned}\right\} \tag{E}$$

For simplicity we assume that f is asymptotically linear, that is, the limit

$$f'(\infty) := \lim_{|\xi|\to\infty} f'(\xi) \tag{1.1}$$

exists.

In the second example we are looking for 2π-periodic solutions of the nonlinear wave equation

$$\left.\begin{aligned} u_{tt} - u_{xx} &= f(u) \qquad \text{for} \quad (t, x) \in \mathbb{R} \times (0, \pi), \\ u(t, 0) = u(t, \pi) &= 0 \qquad \text{for} \quad t \in \mathbb{R}. \end{aligned}\right\} \tag{W}$$

Here we assume that $f \in C^2(\mathbb{R})$, satisfies (1.1), and that

$$f'(\xi) \neq 0 \quad \forall \xi \in \mathbb{R}.$$

By standard arguments it can be shown that (E) and (W) are very special cases of the following abstract problem:

> *Find a solution $u \in D(A)$ of the equation*
>
> $$Au = F(u),$$
>
> *where A is a self-adjoint linear operator in a real Hilbert space H, and $F: H \to H$ is a continuous potential operator, that is, there exists a functional $\Phi \in C^1(H, \mathbb{R})$ satisfying $\Phi(0) = 0$ and $\Phi' = F$.* (A)

Of course, problem (A) contains much more general realizations than (E) and (W). In particular we should like to mention that our abstract results can be applied to the problem of the existence of periodic solutions to nonlinear Hamiltonian systems of ordinary differential equations, in which case we obtain interesting new results (cf. [2]).

Consider now problem (E). In this case it is well known that A has a compact resolvent and that the eigenvalues λ_k of A form an increasing sequence tending to infinity,

$$\lambda_0 < \lambda_1 < \lambda_2 < \dots,$$

each λ_k having finite multiplicity.

It is a consequence of the work of many authors, e.g. Ambrosetti, Ambrosetti-Mancini, Ambrosetti-Rabinowitz, Berger, Rabinowitz, Dancer, Amann, Hess, Thews, and many others (cf. [2] for more precise bibliographical remarks) that the solvability of problem (E) depends on the way the nonlinearity f "interacts" with the spectrum of A. To describe the most significant case, let us assume that

$$f(0) = 0, \tag{1.2}$$

so that (E) has the trivial solution. In this case, the best results known to the author, guaranteeing the existence of a nontrivial solution, are due to Thews [8] and Hess [6]. Thews proves the existence of a nontrivial solution of (E) if

$$f'(0) < \lambda_k \leqq \lambda_l < f'(\infty) < \lambda_{l+1} \tag{1.3}$$

and

$$f'(\xi) \leqq \lambda_{l+1} \qquad \forall\, \xi \in \mathbb{R} \tag{1.4}$$

whereas Hess requires (1.3) and

$$\big(f(\xi) - \lambda_{k-1}\xi\big)\, \xi > 0 \qquad \forall\, \xi \in \mathbb{R}. \tag{1.5}$$

In addition there are in both cases existence results if the roles of $f'(0)$ and $f'(\infty)$ are interchanged (roughly: if all inequality signs are reversed).

In the case of the nonlinear wave equation (W), much less seems to be known. If we presuppose (1.2), then the best result concerning the existence of a nontrivial 2π-periodic solution known to the author is due to Mancini [7] (cf. [1.2] for further bibliographical remarks). In the case of problem (W), the linear operator A has a pure point spectrum extending from $-\infty$ to $+\infty$:

$$-\infty \leftarrow \cdots < \lambda_{-2} < \lambda_{-1} < \lambda_0 = 0 < \lambda_1 < \lambda_2 < \cdots \to \infty .$$

Each nonzero eigenvalue has finite multiplicity, and $\lambda_0 = 0$ is an eigenvalue of infinite multiplicity. Given this information, Mancini proves the existence of a nontrivial solution of (W) if *either*

$$f'(0) > \lambda_k > f'(\infty),$$
$$f'(\infty) \notin \sigma(A), \quad \text{and} \quad f'(\xi) \geqq \lambda_{k-1} \geqq 0 \qquad \forall\, \xi \in \mathbb{R}$$

or

$$f'(0) < \lambda_k < f'(\infty),$$
$$f'(\infty) \notin \sigma(A), \quad \text{and} \quad f'(\xi) \leqq \lambda_{k+1} \leqq 0 \qquad \forall\, \xi \in \mathbb{R}.$$

We should like to point out that the rôles of $f'(0)$ and $f'(\infty)$ are not symmetric, since f has to be increasing if $f'(0) > f'(\infty)$ and decreasing if $f'(0) < f'(\infty)$.

The above results suggest the existence of a nontrivial solution to (E) or (W) provided the nonlinearity f "crosses" at least one eigenvalue of finite multiplicity of (A), that is, if there is at least one eigenvalue of finite multiplicity between $f'(0)$ and $f'(\infty)$. The following theorem, which is a consequence of our general abstract results shows that this is the case without any further restrictions on f, provided we consider the "non-resonance case": $f'(\infty) \notin \sigma(A)$.

Theorem. *Consider problem* (E) *or* (W) *and suppose that* $f(0) = 0$. *Then there exists at least one nontrivial solution provided*

(i) $f'(\infty) \notin \sigma(A)$,

and

(ii) *there exists an eigenvalue* λ *of* A *such that either* $f'(0) < \lambda < f'(\infty)$ *or* $f'(\infty) < \lambda < f'(0)$.

It should be observed that in the case of the wave equation the restriction $f' \neq 0$ implies $\lambda \neq 0$.

2. We turn now to the abstract problem (A) and impose throughout the following

Hypothesis. *There exist real numbers* $\alpha < \beta$ *in the resolvent set of* A *such that* $\sigma(A) \cap (\alpha, \beta)$ *contains at most finitely many eigenvalues of finite multiplicities and such that*

$$\alpha \|u - v\|^2 \leqq \langle F(u) - F(v), u - v\rangle \leqq \beta \|u - v\|^2$$

for all $u, v \in H$.

It is easily verified that this hypothesis is satisfied if α and β are arbitrary real numbers in the resolvent set of A in the case of example (E), and if either $\alpha > 0$ or $\beta > 0$ in the case of problem (W), provided $\alpha \leqq f'(\xi) \leqq \beta$ for all $\xi \in \mathbb{R}$. Intuitively speaking, the above hypothesis means that the nonlinearity interacts only with that part of the spectrum of A which lies in the interval (α, β).

Next, by means of the spectral resolution of A it can be shown that there exist an orthogonal decomposition $H = X \oplus Y \oplus Z$ and operators $R \in \mathscr{L}(H, X)$, $S \in \mathscr{L}(H, Y)$ and $T \in \mathscr{L}(H, Z)$ such that

$$-R^2 + S^2 + T^2 = (A - \alpha)^{-1}.$$

Then, letting

$$\Gamma(u) := \Phi(u) - \frac{\alpha}{2} \|u\|^2 \quad \forall\, u \in H,$$

we define a functional $f: X \times Y \times Z \to \mathbb{R}$ by

$$f(x, y, z) := \frac{1}{2} (\|x\|^2 - \|y\|^2 - \|z\|^2) + \Gamma(Rx + Sy + Tz).$$

Clearly, $f \in C^1(X \times Y \times Z, \mathbb{R})$, and it is not difficult to see that $f'(x, y, z) = 0$ iff $Rx + Sy + Tz$ is a solution of $Au = F(u)$. Hence the solutions of $Au = F(u)$ are in a one-to-one correspondence with the critical points of the functional f.

It should be noted that the obvious choice

$$\varphi(u) := \frac{1}{2} \langle Au, u\rangle - \Phi(u)$$

for a functional having as critical points the solutions of $Au = F(u)$ is not a good one, since φ is only defined on $D(A)$ and, in general, the quadratic form $\langle Au, u\rangle$ will be strongly indefinite, that is, positive definite and negative definite on infinite-dimensional subspaces.

An analysis of the functional f based upon the second part of the above hypothesis implies that, for each $(y, z) \in Y \times Z$,

$$f(\cdot, y, z): X \to \mathbb{R} \qquad \text{is strongly convex}$$

and

$$f(x, \cdot, z): Y \to \mathbb{R} \qquad \text{ist strongly concave}$$

for each fixed $(x, z) \in X \times Z$. Thus $f(\cdot, \cdot, z): X \times Y \to \mathbb{R}$ has for each fixed $z \in Z$ a unique saddle point $\big(x(z), y(z)\big) \in X \times Y$, that is,

$$D_1 f\big(x(z), y(z), z\big) = 0,$$
$$D_2 f\big(x(z), y(z), z\big) = 0$$

for each $z \in Z$. Thus it remains to find points $z \in Z$ for which $D_3 f\big(x(z), y(z), z\big) = 0$. This suggests studying the functional $g(z) := f\big(x(z), y(z), z\big)$ on the subspace Z of H, which, by the way, is spanned by the finitely many eigenfunctions of A belonging to the eigenvalues in (α, β).

For technical reasons we do not use g but the equivalent functional

$$a(z) := -g(T^{-1}z) \qquad \forall\, z \in Z,$$

where $T^{-1} := (T \mid Z)^{-1}$. Then, using the properties of a saddle point, it can be shown that $a \in C^1(Z, \mathbb{R})$, and that $a'(z) = 0$ iff $u(z) := Rx(T^{-1}z) + Sy(T^{-1}z) + z$ is a solution of $Au = F(u)$. Thus, the problem of finding solutions of the equation $Au = F(u)$ is completely equivalent to the problem of finding critical points of the smooth functional a on the finite-dimensional space Z.

The above arguments carry through also in the special case that $\sigma(A) \cap (\alpha, \beta) = \emptyset$, in which case $Z = \{0\}$. Hence we obtain the following existence and uniqueness theorem.

Theorem 1. *Suppose that $\sigma(A) \cap (\alpha, \beta) = \emptyset$. Then the equation $Au = F(u)$ has exactly one solution.*

The above theorem, which in special cases is well known, expresses the fact that our problem has a unique solution if the nonlinearity does not "interact" at all with the spectrum of A.

For a further analysis of the functional a we impose now an additional hypothesis (F_∞) concerning the asymptotic behavior of the nonlinearity.

There exists a symmetric linear operator $F'(\infty) \in \mathscr{L}(H)$ such that

$$\frac{F(u) - F'(\infty)\,u}{\|u\|} \to 0 \quad \text{as} \quad \|u\| \to \infty$$

and such that $\qquad (F_\infty)$

(i) $0 \notin \sigma\big(A - F'(\infty)\big)$

and

(ii) *$F'(\infty)$ commutes with the orthogonal projection of H onto Z.*

Clearly, (F_∞) means that F is asymptotically linear in the sense of Krasnosel'skii and that there is no "resonance at infinity". Condition (ii) is a very mild technical requirement which is satisfied in all of our applications. It allows a detailed analysis of the functional a.

As a consequence of condition (F_∞) we deduce the existence of constants $\nu > 0$ and $\delta \geqq 0$ such that

$$\|a'(z)\| \geqq \nu \|z\| - \delta \qquad \forall\, z \in Z. \tag{2.1}$$

This inequality implies that there are no ciritical points of a outside of some large ball and that there is also no ciritical point "at infinity", more precisely, that a satisfies the Palais-Smale condition.

In order to find critical points of a we consider now the flow defined by the differential equation

$$\dot{z} = a'(z) \tag{2.2}$$

on Z. It can be shown that a' is globally Lipschitz continuous so that this flow is well defined.

Let S denote the set of all bounded solutions of (2.2). If $S \neq \emptyset$, then, taking an arbitrary bounded trajectory, it must have an ω-limit point which due to the fact that we are dealing with gradient flows, has to be a critical point of a. Thus it remains to show that $S \neq \emptyset$.

Since (2.1) implies the boundedness of S, we can, following C. C. Conley [4], associate with S a generalized Morse index $h(S)$, which is the homotopy type of a pointed topological space. On the basis of hypothesis (F_∞), and using Conley's theory, this index can be computed and it turns out that $h(S)$ equals the homotopy type of a pointed m_∞-dimensional sphere, where m_∞ is the positive Morse index of $[A - F'(\infty)] \mid Z$, that is, the dimension of a maximal subspace of Z on which $A - F'(\infty)$ is positive definite. Since this generalized Morse index is distinct from the generalized Morse index of the empty set, we obtain the following existence theorem

Theorem 2. *If* (F_∞) *is satisfied, then the equation* $Au = F(u)$ *has at least one solution.*

Suppose now that $F(0) = 0$. Then the equation $Au = F(u)$ has the trivial solution and Theorem 2 does not give any information. In this case we are interested in the existence of nonzero solutions. Since it follows easily that $a'(0) = 0$, we are interested in the existence of nonzero critical points of a.

For our further considerations we have to impose an additional *regularity hypothesis* (R), which implies $a \in C^2(Z, \mathbb{R})$. This regularity condition, which is somewhat technical and cumbersome to state explicitly, is satisfied in all of our applications.

Since $a'(0) = 0$, it suffices by the above considerations to prove the existence of a bounded trajectory of (2.2), which does not contain 0 in its closure. Indeed, in this case any ω-limit point of such a trajectory is then a nonzero critical point of a. Using again Conley's generalized Morse theory, a relatively complicated theorem about certain normal forms of flows, and a certain amount of algebraic topology, the existence of a nonzero critical point of a can be shown, provided

$$m_\infty \notin \left[m^+\big(a''(0)\big),\, m^+\big(a''(0)\big) + m^0\big(a''(0)\big)\right], \tag{2.3}$$

where $m^+(a''(0))$ is the positive Morse index of $a''(0)$, and $m^0(a''(0))$ is the dimension of a maximal isotropic subspace of $a''(0)$.
Thus we obtain the following

Theorem 3. *Let* (F_∞) *and* (R) *be satisfied and suppose that* $F(0) = 0$. *Then the equation* $Au = F(u)$ *has at least one nontrivial solution, provided* (2.3) *is satisfied.*

Hence in concrete applications it remains to verify the hypotheses of the above theorems. This can be done in cases of problems (E) and (W) (and in more general situations, as well as in the case of Hamiltonian systems). By this way, we obtain the Theorem stated in Section 1.

For detailed proofs and more general applications we refer to [1, 2].

3. The previous results show that there exists a nontrivial solution if $F(0) = 0$ and "the nonlinearity crosses" at least one eigenvalue of finite multiplicity of A. This arises the question if it can be proved that the equation $Au = F(u)$ has several nonzero solutions if F "crosses" several eigenvalues of A. In general, for example in the case of problem (E), this seems to be an open question.

However we have some partial results in this direction for the case of problem (W). The reason for this is the fact that this problem possesses more structure. Indeed, if u is a solution of (W), then, for every $\tau \in [0, 2\pi]$, the function $(t, x) \mapsto u(t + \tau, x)$ is also a solution of (W). By evaluating this rather trivial fact it can be shown that in this case the equation $Au = F(u)$ is equivariant with respect to a smooth action of the circle group S^1. This, in turn, implies that the functional a is invariant with respect to this action. Now, by applying a variant of the Ljusternik-Schnirelmann theory, not based on the notion of category but on the cohomological index appropriate for transformation groups and introduced by Fadell and Rabinowitz [5], we can deduce the existence of several nonzero solutions of problem (W). A simple result in this direction is given in the following

Theorem. *Consider problem* (W) *and suppose that, for some* $k \in \mathbb{Z}$ *and* $l \in \mathbb{N}$, *either*

$$f'(0) < \lambda_{k+1} < \cdots < \lambda_{k+l} < f'(\infty) < 0$$

or

$$f'(\infty) < \lambda_{k+1} < \cdots < \lambda_{k+l} < f'(0) < 0.$$

Then problem (W) *possesses at least* l *nonzero* 2π*-periodic solutions.*

For a proof of this theorem and for more general results we refer to the forthcoming paper [3].

References

[1] H. Amann, Saddle points and multiple solutions of differential equations. Math. Zeitschr. **169** (1979), 127–166.

[2] H. Amann and E. Zehnder, Nontrivial solutions for a class of nonresonance problems and applications to nonlinear differential equations. Ann. Scuola Norm. Sup. Pisa (4) **7** (1980), 539–603.

[3] H. Amann and E. Zehnder, Multiple periodic solutions for a class of nonlinear autonomous wave equations. Houston J. Math., 1981.

[4] C. C. Conley, Isolated invariant sets and the Morse index. CBMS Regional Conference Series in Math. **38** (1978), AMS, Providence, R. I.

[5] E. R. Fadell and P. H. Rabinowitz, Generalized cohomological index theories for Lie group actions with an application to bifurcation questions for Hamiltonian systems. Invent. math. **45** (1978), 139–174.
[6] P. Hess, Solutions nontriviales d'un problème aux limites elliptique non linéaire, C. R. Acad. Sci. Paris (to appear).
[7] G. Mancini, Periodic solutions of some semilinear autonomous wave equations. Boll. U.M.I. (5), **15**–B (1978), 649–672.
[8] K. Thews, A reduction method for some nonlinear Dirichlet problems. J. Non linear Analysis, Theory, Methods & Appl. **3** (1979), 795–813.

Adress of the author:

Mathematisches Institut der Universität Zürich
Freiestr. 36
CH-8032 Zürich
Schweiz

On optimal control problems described by operator equations in Banach and Hilbert spaces

Hans Benker, Merseburg

1. Introduction

In this paper we shall consider special cases of the following general optimal control problem in Banach spaces

$$j(Q, u) \doteq \underset{u \in U \subset B_2}{\text{Minimum}}, \tag{1}$$

$$T(Q, u) = 0 \tag{2}$$

with:

B_i — Banach spaces ($i = 1, 2, 3$), $T \in (B_1 \times B_2 \to B_3)$, $j \in (B_1 \times B_2 \to R^1)$.

For the general problem (1), (2) it is only under strong assumptions possible to give existence and optimality conditions (see [1, 2, 15]). Therefore many authors consider special cases of the problem (1), (2).

Very important special cases of (1), (2) are the so called minimum norm (for $p = 1$) and quadratic (for $p = 2$) optimal control problems

$$j(Q, u) = \|Q - R\|_1^p + K\|u\|_2^p \doteq \underset{u \in U \subset B_2}{\text{Minimum}} \quad (K \geqq 0) \tag{3}$$

with the linear state equation

$$AQ + Bu + f = 0. \tag{4}$$

The problem (3), (4) has been extensively studied by many authors (see [3—14]) for special cases of p, K and the state equation (4). In order to be able to apply results of approximation theory, we write frequently the problem (3), (4) in the form

$$j(Q, u) = (\|Q - R\|_1^p + K\,\|u\|_2^p)^{1/p} \doteq \underset{(Q,u) \in V}{\text{Minimum}} \tag{5}$$

with

V $= \{(Q, u) \mid AQ + Bu + f = 0;\ u \in U \subset B_2\}$,

B_i — Banach spaces with the norm $\|\ldots\|_i$ ($i = 1, 2, 3$)[1],

K — constant $\geqq 0$, $p \in N = \{1, 2, \ldots\}$ (arbitrary but fixed),

U — convex, closed set ($\neq \emptyset$),

$A \in L(B_1 \to B_3)$, $B \in L(B_2 \to B_3)$ — linear operators (additionally we assume that A^{-1} exists, A^{-1} and B are bounded and $D(A)$ is dense in B_1),

$f \in B_3$, $R \in B_1$ are given.

---

[1] The dual spaces we denote by B_i' (with the norm $\|\cdots\|_{i'}$). If we write H_i then the spaces are Hilbert spaces.

Remarks. 1. Obviously, problem (5) has the same solution as the problem (3), (4). The form (5) was chosen in order to be able to apply results of approximation theory. Since

$$\|(Q, u)\|_P := (\|Q\|_1^p + K\,\|u\|_2^p)^{1/p} \quad (\text{for } K > 0) \tag{6}$$

is a norm on the product space $P = B_1 \times B_2$, we can consider our optimal control problem (5) as an approximation problem in the product space P, i.e.

$$\|(Q, u) - (R, 0)\|_P \doteq \underset{(Q,u)\in V}{\text{Minimum}}. \tag{5'}$$

2. Optimal control problems of the form (3), (4) (resp. (5)) practically arise for $p = 1$ and $p = 2$.

3. Since the operator A has by assumption a bounded inverse, we can write the solution of the state equation $AQ + Bu + f = 0$ in the form $Q = Lu + f'$ (with $L = -A^{-1}B$ bounded and $f' = -A^{-1}f$) For problem (5) then follows (with $r = R - f'$)

$$J(u) = (\|Lu - r\|_1^p + K\,\|u\|_2^p)^{1/p} \doteq \underset{u\in U}{\text{Minimum}}. \tag{7}$$

4. Optimal control problems of the form (5) with $K > 0$ practically arise if we wish to approximate a given state R and additionally to minimize the costs of control. Furthermore, problem (5) is considered for the regularization of the problem

$$\|Q - R\|_1 \doteq \underset{(Q,u)\in V}{\text{Minimum}}.$$

The aim of this paper is to give optimality conditions for the general problem (5) applying the conditions of approximation theory and to establish resp. to generalize properties of optimal control. Furthermore duality theorems are discussed and lower bounds for the functional from (5) are obtained.

2. Optimality conditions

Applying the Kolmogorov condition of approximation theory we obtain the following necessary and sufficient optimality condition for our optimal control problem (5).

Theorem 1. *Necessary and sufficient for the optimality of* (Q^0, u^0) *is the existence of a linear functional* $l \in (B_1 \times B_2)'$ *with the properties*

$$\|l\|_{P'} = 1, \tag{8}$$

$$l(Q^0 - R, u^0) = \|(Q^0 - R, u^0)\|_P = (\|Q^0 - R\|_1^p + K\,\|u^0\|_2^p)^{1/p} \tag{9}$$

$$l(Q^0 - Q, u^0 - u) = l_1(Q^0 - Q) + Kl_2(u^0 - u) \leqq 0 \qquad \forall\, (Q, u) \in V^1 \tag{10}$$

Proof. see [13].

---

[1] A functional $l = (l_1, l_2) \in (B_1 \times B_2)' = P'$ we will for convenience write in the form $l(Q, u) = l_1Q + Kl_2u$ ($l_1 \in B_1'$, $l_2 \in B_2'$). For the norm $\|l\|_{P'}$ of l then follows $\|l\|_{P'} = (\|l_1\|_{1'}^q + K\,\|l_2\|_{2'}^q)^{1/q}$ for $1 < p < \infty$ (with $1/p + 1/q = 1$) resp. $\|l\|_{P'} = \max\,\{\|l_1\|_{1'}, \|l_2\|_{2'}\}$ for $p = 1$.

Remarks. 1. Theorem 1 we obtain by applying the necessary and sufficient optimality condition for convex functionals having a nonempty subdifferential.

If for $B_i = H_i$ the functional from (5) is differentiable, i.e. $l_1 = j_Q^0$, $Kl_2 = j_u^0$, then the condition (10) has the well-known form

$$(j_Q^0, Q - Q^0)_1 + (j_u^0, u - u^0)_2 \geqq 0 \qquad \forall\, (Q, u) \in V. \tag{11}$$

2. The sufficiency of theorem 1 is also valid if the state equation is nonlinear (i.e. $T(Q, u) = 0$).

3. Theorem 1 can be used in order to give maximum principles for concrete problems in function spaces and to obtain properties of optimal control (see [12]).

4. Optimality conditions for special cases of the problem (5) were proved in the papers [17—22] applying generalized inverses and methods of convex analysis.

3. Properties of the optimal control

In the following we always suppose that for our control problem a solution exists. If the Banach spaces B_i are reflexive then for $K > 0$ the existence of an optimal control follows for problem (5) under the given assumption from the convex analysis. For $K = 0$ we must here additionally demand that the control set U is bounded. Now we give results on the dependence of the optimal control and the optimal state on the constant K.

Lemma 1. *If the problem*

$$j_K(Q, u) = j(Q, u) + K\|u\|_2^p \doteq \underset{u \in U \subset B_2}{\text{Minimum}}, \tag{12}$$

$$T(Q, u) = 0 \tag{13}$$

has for every $K \geqq 0$ at least a solution, denoted by $(Q(K), u(K))$, then the following properties hold:

a) $\|u(K)\|_2^p \leqq 1/K \cdot \text{const}$ (*if* $j(Q, u) \geqq 0$ $\forall\, (Q, u)$ *and* $0 \in U$), (14)

b) $j_{K_1}(Q(K_1), u(K_1)) \leqq j_{K_2}(Q(K_2), u(K_2))$ *for* $K_1 \leqq K_2$, (15)

c) $\|u(K_2)\|_2 \leqq \|u(K_1)\|_2$ *for* $K_1 \leqq K_2$, (16)

d) $j(Q(K_1), u(K_1)) \leqq j(Q(K_2), u(K_2))$ *for* $K_1 \leqq K_2$, (17)

e) $K_2(\|u(K_2)\|_2^p - \|u(K_1)\|_2^p) \leqq j(Q(K_1), u(K_1)) - j(Q(K_2), u(K_2))$
$\leqq K_1(\|u(K_2)\|_2^p - \|u(K_1)\|_2^p)$ *for* $K_1 \leqq K_2$, (18)

f) $j(Q(K_1), u(K_1)) = j(Q(K_2), u(K_2))$ *if* $\|u(K_1)\|_2 = \|u(K_2)\|_2$. (19)

Proof. Follows by simple estimation similar as in the theory of penalty functions. Therefore the proof is omitted.

Remarks. 1. The relations of monotonicity from Lemma 1 are important in order to give further properties and to solve special problems.

2. Furthermore follows from $\|u(K_1)\|_2 = \|u(K_2)\|_2$ also $u(K_1) = u(K_2)$ if j is convex, T linear and $p = 2$ (and $B_i = H_i$) or j is additionaly strongly convex for $p \neq 2$.

Lemma 2. *Let $u(K)$ be a solution of the problem*

$$j_K(Q, u) = j(Q, u) + K \|u\|_2^p \doteq \underset{u \in B_2}{\text{Minimum}}, \tag{20}$$

$$T(Q, u) = 0. \tag{21}$$

If $K_0 = \text{Min}\,\{K \geqq 0/\|u(K)\|_2 = C\}$ exists then $u(K_0)$ is also a solution of the problem ($C > 0$ is a given constant)

$$j(Q, u) \doteq \underset{\|u\|_2 \leqq C}{\text{Minimum}}, \tag{22}$$

$$T(Q, u) = 0. \tag{23}$$

Proof. Follows immediately from Lemma 1c), α) and f). q.e.d.

Lemma 3. *The solution $(Q(K), u(K))$ of the problem (12,) (13) for $K > 0$ is also a solution for $K = 0$ if the control set $\overline{U}$ has the form $\overline{U} = U \cap \{u/\|u\|_2 \leqq \|u(K)\|_2\}$.*

Proof. We have $j(Q(K), u(K)) + K \|u(K)\|_2^p \leqq j(Q, u) + K \|u\|_2^p \quad \forall\, u \in U$. From this inequality follows

$$j(Q(K), u(K)) \leqq j(Q, u) + K(\|u\|_2^p - \|u(K)\|_2^p) \leqq j(Q, u) \quad \forall\, u \in \overline{U}. \qquad \text{q.e.d.}$$

Remarks. 1. The importance of Lemma 2 lies in the fact that from the solution of problems without control constraints we can obtain solutions for problems with constraints.

2. From Lemma 3 we obtain for instance that for

$$U = U_1 = \{u/\|u\|_2 \leqq 1\} \tag{24}$$

and $\|u(K)\|_2 = 1$ $u(K)$ also solves the problem for $K = 0$.

Now are treated continuity properties of the optimal control and the optimal functional value with respect to K.

Lemma 4. *For problem (12), (13) holds*

$$\lim_{K_n \to K_0 + 0} j_{K_n}(Q(K_n), u(K_n)) = j_{K_0}(Q(K_0), u(K_0)) \quad (\textit{for } K_0 \geqq 0).$$

Proof. We have $j_{K_0}(Q(K_0), u(K_0)) = j(Q(K_0), u(K_0)) + K_0 \|u(K_0)\|_2^p \leqq j(Q(K_n), u(K_n)) + K_n \|u(K_n)\|_2^p \leqq j(Q(K_0), u(K_0)) + K_n \|u(K_0)\|_2^p$. From this inequalities the assertion immediately follows. q.e.d.

Remarks. 1. Lemma 4 gives a typical assertion of the regularization theory. It can also be generalized to

$$\lim_{K_n \to K_0 + 0,\, \varepsilon_n \to 0} j_{K_n}(Q_{\varepsilon_n}(K_n), u_{\varepsilon_n}(K_n)) = j_{K_0}(Q(K_0), u(K_0)),$$

where $(Q_\varepsilon(K), u_\varepsilon(K))$ satisfies the following inequalities

$$j_K(Q(K), u(K)) \leqq j_K(Q_\varepsilon(K), u_\varepsilon(K)) \leqq j_K(Q(K), u(K)) + \varepsilon.$$

2. For the special problem (5) we can additionally prove the convergence of the sequences $u(K_n) \to u(K_0)$, $Q(K_n) \to Q(K_0)$ for $K_n \to K_0 + 0$ (if $B_i = H_i$) (see [12]).

If furthermore is given problem (7) with $p = 2$ and $U = H_2$ then in Lemma 1 hold strong inequalities in b), c) and d) for $L^T r \neq 0$ (for $L^T r = 0$ immediately follows $u(K) = 0$) and $u(K)$ is continuous on $[\varepsilon, \infty)$ ($\varepsilon > 0$ arbitrary but fixed). This was proved by Balakrishnan [5] and Vinter [4].

3. Lemma 4 also holds if only for $K > K_0$ exists a solution $(Q(K), u(K))$ of the problem, i.e. we have

$$\lim_{K_n \to K_0 + 0} j_{K_n}(Q(K_n), u(K_n)) = \text{Infimum } j_{K_0}(Q, u).$$

For quadratic problems we now can apply lemmas 2, 3 and 4 in order to give analytic solutions too for problems with constraints.

Lemma 5. *Let $K \geqq 0$ be fixed and $u(\beta)$ be a solution of the problem ($\beta \geqq 0$)*[1]

$$J_\beta(u) = \|Lu - r\|_1^2 + (K + \beta)\|u\|_2^2 \doteq \underset{u \in H_2}{\text{Minimum}}. \tag{25}$$

Then holds for a given $M > 0$:

a) *If* $\sup_{\beta > 0} \|u(\beta)\|_2 \leqq M$ *then* $u(o) = u^0$ *is the solution (with minimal norm) of the problem*

$$J(u) = \|Lu - r\|_1^2 + K\|u\|_2^2 \doteq \underset{\|u\|_2 \leqq M}{\text{Minimum}}. \tag{26}$$

b) *If* $\sup_{\beta > 0} \|u(\beta)\|_2 > M$ *then* $u^0 = u(\beta_0)$ *with* β_0 *from* $\|u(\beta_0)\|_2 = M$ *is the solution (with minimal norm) of problem* (26).

Proof. For $K = 0$ Lemma 5 was proved by Balakrishnan and Vinter [4, 5]. For $K > 0$ the assertions a) and b) immediately follow from lemmas 2, 3 and 4 (with remark 2).

Remark. If the operator L is compact we can give the solution of problem (25) as an infinite series of the form (for $K + \beta > 0$)

$$u(\beta) = \sum_{i=1}^{\infty} (L^T r, v^i)\, v^i/(K + \beta + \lambda_i).\text{[2]} \tag{27}$$

This is the well-known expansion of the solution of the equation $(L^T L + (K + \beta) I)\, u(\beta) = L^T r$ (necessary and sufficent optimality condition for (25)) on eigenelements of the operator $L^T L$. Applying Lemma 5 we must now maximize the function

$$\|u(\beta)\|_2^2 = \sum_{i=1}^{\infty} (L^T r, v^i)^2/(K + \beta + \lambda_i)^2$$

with respect to β and we obtain for the optimal control u^0 of (26)

$$u^0 = \sum_{i=1}^{\infty} (L^T r, v^i)\, v^i/(K + \lambda_i) \text{ resp. } = \sum_{i=1}^{\infty} (L^T r, v^i)\, v^i/(K + \beta_0 + \lambda_i).$$

---

[1] For $K = 0$ L additionally must be compact.

[2] For a not compact operator L also exists an expansion of the optimal control $u(\beta)$ into an infinite series (see [16]).

In the assertion b) of Lemma 5 the optimal control is a so-called bang-bang optimal control, i.e., u^0 belongs to the boundary of the control set. In the following we shall give a definition for bang-bang controls and derive sufficient conditions for the bang-bang-ness of an optimal control.

Definition. A control u is called *weakly bang-bang* if $\lambda u \in U$ implies that $\lambda \leqq 1$ (or equivalent: $\lambda u \notin U \ \forall \ \lambda > 1$).

Now we generalize a result of Rogak/Kazarinoff [6, 7].

Lemma 6. *Let $u^0 \in U$ (U star-shaped) be an optimal control for the problem*

$$j(Q, u) = \|Q - R\|_1^2 + K\|u\|_2^2 \doteq \underset{u \in U}{\text{Minimum}} \quad \text{with} \quad Q = Lu + \varepsilon f(Q). \tag{28}$$

Furthermore we assume that the solution of the state equation from (28) *has the form $Q = Lu + \varepsilon C_\varepsilon(u)$ with $C_\varepsilon \in (H_2 \to H_1)$ bounded. If then*

$$(\|L\|^2 + K)\, M < |m|/2 \quad \text{and} \quad (Lu, R)_1 \neq 0 \quad \text{for some} \quad u \in U \tag{29}$$

hold with $M = \operatorname{Sup}_{u \in U} \|u\|_2^2$ and $m = \operatorname{Inf}_{u \in U} (-Lu, 2R) = -(Lv, 2R)$ then u^0 is weakly bang-bang for sufficiently small ε.

Lemma 7. *Under the same assumption as in Lemma 6, for $Kn > 2\,\|L^T R\|\, N$* (30) *with $n = \operatorname{Inf} \|u\|$ and $N = \operatorname{Sup} \|u\|$ (Inf and Sup over all bang-bang controls) the optimal ontrol u^0 is not bang-bang for sufficiently small ε.*

The proofs of lemmas 6 and 7 are similar to the proofs for $\varepsilon = 0$ given by the author in [14] and therefore they are omitted.

Remarks. 1. Lemmas 6 and 7 also hold for more general state equations if their solutions have the form $Q = Lu + \varepsilon C_\varepsilon(u)$.

2. For the special control set $U = U_1 = \{u / \|u\|_2 \leqq 1\}$ the conditions (29) and (30) have the form

$$(\|L\|^2 + K) < \|L^T R\|_2 \tag{29'}$$

resp.

$$K > \operatorname{Min}\{2\,\|L^T R\|_2, \|R\|_1^2\}, \tag{30'}$$

which easily follows from the proofs of the lemmas 6 and 7.

4. Duality theorems and lower bounds

Applying a general duality theorem of the approximation theory by the author [12, 13] were given the following duality conditions

$$\operatorname*{Inf}_{(Q,u) \in V} (\|Q - R\|_1^p + K\,\|u\|_2^p)^{1/p} = \operatorname*{Sup}_{\|l\|_{p'} \leqq 1} \Big(l_1 R - \operatorname*{Sup}_{(Q,u) \in V} (l_1 Q + K l_2 u)\Big), \tag{31}$$

$$\operatorname*{Inf}_{u \in U} (\|Lu - r\|_1^p + K\,\|u\|_2^p)^{1/p} = \operatorname*{Sup}_{\|l\|_{p'} \leqq 1} \Big(l_1 r - \operatorname*{Sup}_{u \in U} (L^T l_1 + K l_2)\, u\Big) \tag{32}$$

and if the $B_i = H_i$ are Hilbert spaces and $U = U_1$

$$\underset{\|u\|_2 \leqq 1}{\mathrm{Min}} \left(\|Lu - r\|_1^p + K\,\|u\|_2^p\right)^{1/p} = \underset{\|l\|_{p'} \leqq 1}{\mathrm{Max}} \left((l_1, r)_1 - \|L^T l_1 + K l_2\|_2\right). \tag{33}$$

Obviously the dual problems have not a simpler structure than the primal problems. But they are an important tool in order to compute lower bounds. Estimating the dual functionals from (33), by the author [12] were derived the following bounds (for $r \neq 0$, $K > 0$):

$$\|r\|_1 \geqq \underset{\|u\|_2 \leqq 1}{\mathrm{Min}} \left(\|Lu - r\|_1^p + K\|u\|_2^p\right)^{1/p} = \begin{cases} \|r\|_1^2/\left(\|r\|_1^q + \|L^T r\|_2^q/K^{q-1}\right)^{1/q} & \text{for } p \geqq 2 \\ K\|r\|_1^2/\|L^T r\|_2 & \text{for } p = 1 \text{ and } \|L^T r\|_2/\|r\|_1 > K \\ \|r\|_1 \ (\text{i.e. } u^0 = 0) & \text{for } p = 1 \text{ and } \|L^T r\|_2/\|r\|_1 \leqq K \end{cases} \tag{34}$$

(with $1/p + 1/q = 1$).

The disadvantages of these bounds lie in the fact that we need the inverse operator A^{-1} ($L = A^{-1}B$) and that for $K = 0$ they can not be applied. Therefore we now go another way, applying a simple inequality of the convex analysis.

Lemma 8. *Let $j(Q, u)$ be a convex and continuous functional on the Hilbert space $H = H_1 \times H_2$. Then holds for arbitrary (Q^0, u^0) with $AQ^0 + Bu^0 + f = 0$*

$$j(Q^0, u^0) \geqq j(Q, u) - (N, AQ + Bu + f)_3 \tag{35}$$

if $\partial j(Q, u) \cap \left(R(A^T) \times R(B^T)\right) \neq \emptyset$ and the triple (Q, u, N) satisfies the following equations

$$A^T N = l_Q,\ B^T N = l_u \tag{36}$$

for any $l = (l_Q, l_u) \in \partial j(Q, u) \cap \left(R(A^T) \times R(B^T)\right)$.

Proof. Follows immediately from the definition of the subdifferential for $j(Q, u)$ which is nonempty for our functional:

$$\begin{aligned} j(Q^0, u^0) - j(Q, u) &\geqq (l_Q, Q^0 - Q)_1 + (l_u, u^0 - u)_2 \\ &= (A^T N, Q^0 - Q)_1 + (B^T N, u^0 - u)_2 \\ &= (N, AQ^0 + Bu^0 + f - AQ - Bu - f)_3 \\ &= -(N, AQ + Bu + f)_3. \text{ q.e.d.} \end{aligned}$$

Remarks. 1. Obviously the equations (36) are only solvable if $D(A^T) \cap D(B^T) \neq 0$. Therefore we must demand this in the following.

2. By a convenient choice of the triple (Q, u, N) we shall obtain lower bounds for our functional.

Theorem 2. *If we assume that* a) $K > 0$, b) $R \in D(A)$, $AR + f \in D(A^T) \cap D(B^T)$ *and* c) $\|A^T(AR + f)\|_1 + \|B^T(AR + f)\|_2 > 0$ *then follows*

$$J^0 = \underset{AQ + Bu + f = 0}{\mathrm{Minimum}} \left(\|Q - R\|_1^2 + K\,\|u\|_2^2\right) \geqq \frac{\|AR + f\|_3^4}{\|A^T(AR + f)\|_1^2 + \|B^T(AR + f)\|_2^2/K} > 0. \tag{37}$$

Proof. Here we have $l_Q = 2(Q - R)$, $l_u = 2Ku$. Therefore we obtain from Lemma 8

$$\begin{aligned} j(Q^0, u^0) &\geqq \|A^T N\|_1^2/4 + \|B^T N\|_2^2/4K - (N, AQ + Bu + f)_3 \\ &= -\|A^T N\|_1^2/4 - \|B^T N\|_2^2/4K - (N, AR + f)_3 = : S(N). \end{aligned} \tag{38}$$

If we set $N = -\beta(AR + f)$ and maximize $S(N)$ with respect to $\beta > 0$ then we obtain the bound (37). q.e.d.

Remarks. 1. The bound (37) gives for $Q = Lu + \tilde{f}$ (i.e. $A = I$, $B = -L$ and $\tilde{f} = -f$) the bound from (34) so that (37) is a generalization with the advantage that we do not need the inverse operator A^{-1}. For unbounded operators A arises here the disadvantage that the assumption b) is not fulfilled in many cases (for bounded operators b) is satisfied in most cases).

2. If for unbounded operators the assumptions b) and c) are not fulfilled we can try to find an element $d \in D(A^T) \cap D(B^T) \subset H_3$ for that hold $(A^T d, R)_1 + (d, f)_3 \neq 0$ and $\|A^T d\|_1 + \|B^T d\|_2 > 0$. If we have such an element d then by setting $N = \beta d$ and maximizing (38) with respect to β we obtain the bound

$$\frac{((A^T d, R)_1 + (d, f)_3)^2}{\|A^T d\|_1^2 + \|B^T d\|_2^2/K}. \tag{37'}$$

Now we give a bound for the case $K = 0$.

Theorem 3. *If an element* $N \neq 0$ *exists with*
a) $N \in D(A^T)$, b) $N \notin \operatorname{Ker}(A^T)$, c) $N \in \operatorname{Ker}(B^T)$, d) $(A^T N, R)_1 + (N, f)_3 \neq 0$ *then we have the bound*

$$J^0 = \operatorname*{Minimum}_{AQ + Bu + f = 0} \|Q - R\|_1 \geqq \frac{|(A^T N, R)_1 + (N, f)_3|}{\|A^T N\|_1} > 0. \tag{39}$$

Proof. From Lemma 8 follows $\|Q^0 - R\|_1^2 \geqq -\|A^T N\|_1^2/4 - (A^T N, R)_1 + (N, f)_3 =: S(N)$. Replacing N by βN and maximizing S with respect to β we obtain the bound (39). q.e.d.

In order to avoid the difficult determination of an element N which satisfies the assumption a)—d) of Theorem 3 we apply the following lemma.

Lemma 9. *Let the functional* $j(Q)$ *be continuous and convex. Then holds*

$$j(Q^0) \geqq j(Q) - (N, AQ + f)_3 - \|N\|_3 \|B\| \cdot M \tag{40}$$

for all (Q^0, Q, N) *satisfying the following conditions*
a) $\exists\, u^0 \in U$ *so that* $AQ^0 + Bu^0 + f = 0$, b) $A^T N \in \partial j(Q)$, c) $M \geqq \operatorname*{Sup}_{u \in U} \|u\|_2$.

Proof. We have $j(Q^0) - j(Q) \geqq (l_Q, Q^0 - Q)_1 = (A^T N, Q^0 - Q)_1 = (N, -Bu^0 - f - AQ)_3 \geqq -(N, AQ + f)_3 - \|N\|_3 \|B\| \cdot M$. q.e.d.

Lemma 9 will now be applied for the determination of a lower bound for our problem with $K = 0$.

Theorem 4. *If* a) $\|AR + f\|_3 > \|B\|$, b) $AR + f \in D(A^T) \setminus \operatorname{Ker}(A^T)$ *then holds*

$$J^0 = \operatorname*{Minimum}_{AQ + Bu + f = 0, \|u\|_2 \leqq 1} \|Q - R\|_1 \geqq \frac{\|AR + f\|_3 (\|AR + f\|_3 - \|B\|)}{\|A^T(AR + f)\|_1} > 0. \tag{41}$$

Proof. From Lemma 9 follows

$$\begin{aligned} \|Q^0 - R\|_1^2 &\geqq \|Q - R\|_1^2 - (N, AQ + f)_3 - \|N\|_3 \|B\| \\ &= -\|A^T N\|_1^2/4 - (N, AR + f)_3 - \|N\|_3 \|B\| =: S(N). \end{aligned} \tag{42}$$

Setting $N = -\beta(AR + f)$ and maximizing S with respect to $\beta > 0$ we obtain the bound (41). q.e.d.

Remarks. 1. Obviously for $AR + f = 0$ follows the optimal solution $u^0 = 0$, $Q^0 = R$, $J^0 = 0$ for the problems from theorems 2 to 4.

2. From (41) furthermore follows (for $\|r\| > \|L\|$)

$$\operatorname*{Min}_{\|u\|_2 \leqq 1} \|Lu - r\|_1 \geqq \|r\| - \|L\|.$$

The same bound we obtain by a simple estimation of the functional below.

3. If the assumption b) of Theorem 4 is not fulfilled but an element $d \in D(A^T)$ exists with $(A^T d, R)_1 + (d, f)_3 > \|d\|_3 \|B\|$ then by setting $N = -\beta d$ and maximizing S from (42) with respect to β we can also obtain a lower bound.

References

[1] Bittner, L., On optimal control of processes governed by abstract functional, integral and hyperbolic differential equations, Math. Operationsforsch. Statist. **6** (1975), 107–134.

[2] v. Wolfersdorf, L., On a method of L. Bittner for deriving maximum principles to optimal control processes, Math. Operationsforsch. Statist. Ser. Optimization **9** (1978), 197–205.

[3] Weigand, W. A., and D'Souza, A. F., Optimal control of linear distributed parameter systems with constrained inputs, J. Basic Engrg. **6** (1969), 161–167.

[4] Vinter, R. B., A generalization to dual Banach spaces of a theorem by Balakrishnan, SIAM J. Control **12** (1974), 150–166.

[5] Balakrishnan, A. V., An operator theoretic formulation of a class of control problems and a steepest descent method of solution, J. SIAM Control **1** (1963), 109–127.

[6] Rogak, E. D., Kazarinoff, N. D., and Scott-Thomas, J. F., Sufficient conditions for bang-bang control in Hilbert space, J. Optimization Theory Appl. **5** (1970), 20–44.

[7] Rogak, E. D., and Kazarinoff, N. D., Remarks on bang-bang control in Hilbert space, J. Optimization Theory Appl. **10** (1972), 211–221.

[8] Canales, R., A lower bound on the performance of optimal regulators, IEEE Trans. Automatic Control **15** (1970), 409–415.

[9] Aronoff, E., and Leondes, C. T., Lower bounds for a quadratic cost functional, Int. J. Systems Sci. **7** (1976), 17–25.

[10] Porter, W. A., Modern foundation of systems engineering, New York 1966.

[11] Yamamoto, Y., Optimal control for nonlinear systems with quadratic performance, J. Math. Anal. Appl. **64** (1978), 348–353.

[12] Benker, H., On a general minimum norm optimal control problem in Banach and Hilbert spaces, Beiträge z. Analysis **15** (1980), 141–150.

[13] Benker, H., Anwendung der Approximationstheorie auf ein spezielles Problem der optimalen Steuerung, Beiträge z. Analysis **14** (1979), 85–91.

[14] Benker, H., und Dietrich, H., Anwendung der Dualitätstheorie auf Probleme der optimalen Steuerung, Math. Operationsforsch. Statist. Ser. Optimization **10** (1979), 131–142.

[15] Benker, H., and Dittmar, E., Ein Maximumprinzip für Probleme der optimalen Steuerung I, II, III, Beiträge z. Analysis **7** (1975), 131–141; **10** (1977), 91–100; **12** (1978) 147–154.

[16] Vidyasagar, M., Optimal control by direct inversion of a positive-definite operator in a Hilbert space, J. Optimization Theory Appl. **7** (1971), 173–177.
[17] Porter, W. A., and Williams, J. P., A note on the minimum effort control problem, J. Math. Anal. Appl. **13** (1966), 251–264.
[18] Porter, W. A., and Williams, J. P., Extension of the minimum effort control problem, J. Math. Anal. Appl. **13** (1966), 536–549.
[19] Minamide, N., Matsuo, T., and Nakamura, K., Optimal control in Hilbert and Hilbertable Banach space subject to a linear constraint, Research Rep. of Automatic Control Laboratory 15/4, 1968, 36–44.
[20] Minamide, N., and Nakamura, K., Minimum error control problem in Banach space, Research Rep. of Automatic Control Laboratory 16/4, 1969, 51–58.
[21] Minamide, N., and Nakamura, K., A minimum cost control problem in Banach space, J. Math. Anal. Appl. **36** (1971), 73–85.
[22] Minamide, M., and Nakamura, K., Linear bounded phase coordinate control problems under certain regularity and normality conditions, SIAM J. Control **10** (1972), 82–92.

Address of the author:

Technische Hochschule „Carl Schorlemmer“ Leuna-Merseburg
Sektion Mathematik
4200 Merseburg
Geusaer Straße

Regularity results for some elliptic variational inequalities with bounded measurable coefficients and applications[1]

Marco Biroli, Milano

1. Introduction and results

We will deal in this paper with regularity results for 2^0 order elliptic variational inequalities with bounded coefficients.

It is well known that the most interesting regularity results for 2^0 order elliptic equations with bounded coefficients are the De Giorgi-Nash-Moser theorem, [14, 26, 25] and the Meyer's inequality, [22]; our goal is to give an exact extension of these two results to the case of 2^0 order elliptic variational inequalities with obstacle and to sketch some applications.

Let be $\Omega \subset \mathbb{R}^N$ a bounded open set with smooth boundary $\partial\Omega$, Γ_0 a smooth open set in $\partial\Omega$ and $\Gamma = \overline{\Gamma}_0$ (we don't give here the most general hypothesis on Ω, $\partial\Omega$, Γ for sake of simplicity). We denote

$$H^{1,p}_{\Gamma}(\Omega) = \{v \in H^{1,p}(\Omega);\ v|_{\Gamma} = 0\} \tag{1.1}$$

and, if $\Gamma = \partial\Omega$, $H^{1,p}_{\Gamma}(\Omega) = H^{1,p}_0(\Omega)$; the space $(H^{1,p}_{\Gamma}(\Omega))'$ is the dual of $H^{1,p}_{\Gamma}(\Omega)$ and $\langle\cdot,\cdot\rangle$ is the duality between $(H^{1,p}_{\Gamma}(\Omega))'$ and $H^{1,p}_{\Gamma}(\Omega)$. Let be

$$K^{\psi} = \{v \in H^{1,2}_{\Gamma}(\Omega);\ v \leqq \psi \quad \text{a.e. in } \Omega\} \tag{1.2}$$

where ψ is a measurable function on Ω and suppose $K^{\psi} \neq \emptyset$. We denote by $A\colon H^{1,2}_{\Gamma}(\Omega) \to (H^{1,2}_{\Gamma}(\Omega))'$ the operator defined by

$$\langle Au, v\rangle = \int_{\Omega} \left\{\sum_{ij=1}^{N} a_{ij}(x) \frac{\partial u}{\partial x_i}(x) \frac{\partial v}{\partial x_j}(x) + \lambda u(x)\, v(x)\right\} dx \tag{1.3}$$

$\forall\, u, v \in H^{1,2}_{\Gamma}(\Omega)$, where $a_{ij} \in \mathscr{L}^{\infty}(\Omega)$, $i, j = 1, 2, \ldots, N$ with

$$\sum_{ij=1}^{N} a_{ij}(x)\, \xi_i \xi_j \geqq a\, |\xi|^2 \qquad \forall\, \xi \in \mathbb{R}^N,\ a > 0, \text{ a.e. in } \Omega \tag{1.4}$$

and $\lambda > 0$ if $\Gamma \neq \partial\Omega$, $\lambda \geqq 0$ if $\Gamma = \partial\Omega$.

Consider the problem

$$\left.\begin{aligned} &\langle Au, v - u\rangle \geqq \langle f, v - u\rangle \\ &\forall\, v \in K^{\psi},\ u \in K^{\psi} \end{aligned}\right\} \tag{1.5}$$

---

[1] Lavoro eseguito nell' ambito del gruppo G.N.A.F.A. del C.N.R.

where f is defined by

$$\langle f, v\rangle = \int_{\Omega} \left\{\sum_{i=1}^{N} f_i(x) \frac{\partial v}{\partial x_i}(x) + f_0(x)\, v(x)\right\} dx,$$

$f_i \in \mathcal{L}^2(\Omega)$, $i = 0, 1, 2, \ldots, N$ and indicate the solution u to (1.5) by $S(\psi)$.

Theorem 1. *Let be $\psi \in C^\alpha(\bar{\Omega})$, $0 < \alpha < 1$, $\psi|_\Gamma \geqq 0$, $f_i \in \mathcal{L}^p(\Omega)$, $i = 1, 2, \ldots, N$, $f_0 \in \mathcal{L}^q(\Omega)$ where $p > N$, $q > N/2$; the solution u to* (1.5) *is in $C^\beta(\Omega)$, $0 < \beta < 1$,* [8].

Theorem 2. *Let be $\Gamma = \partial\Omega$; there is a $p > 2$ such that, if $\psi \in H^{1,p}(\Omega)$, $\psi|_{\partial\Omega} \geqq 0$, $f_i \in \mathcal{L}^p(\Omega)$, $i = 1, 2, \ldots, N$, $f_0 = 0$ the solution u to* (1.5) *is in $H_0^{1,p}(\Omega)$,* [10].

In 2. we give a proof of Th. 1 and in 3. we use the result in Th. 1 to show the existence of an Hölder continuous solution to a Bensoussan-Lions type quasi variational inequality, [1, 21], with *general* boundary conditions.

In 4. we give the proof of Th. 2 and in 5. we give some results on homogeneisation estimates for variational and quasi variational inequalities with $a_{ij} \notin C^1(\bar{\Omega}_j)$, $i, j = 1, 2, \ldots, N$ (we sketch only the proof of these results).

Remark 1. We observe that recently a result on Hölder continuous solutions to (1.5) has been given by J. Frehse, U. Mosco, [16]. This result is different from Th. 1 because it has more general hypothesis on the obstacle $\left(\psi(x) - \psi(y) \geqq -c\,|x - y|^\alpha \text{ if } x = \{x_i\}_{i=1}^N,\ y = \{y_i\}_{i=1}^N,\ x_i \geqq y_i,\ i = 1, 2, \ldots, N\right)$, but supposes $\Gamma = \partial\Omega$, $f_i = 0$, $i = 1, 2, \ldots, N$ and $K^\psi \cap H_0^{1,\infty}(\Omega) \neq \emptyset$.

We observe that Frehse-Mosco's result implies the existence of an Hölder continuous solution to a Bensoussan-Lions type quasi variational inequality in the case $\Gamma = \partial\Omega$, $f_i = 0$, $i = 1, 2, \ldots, N$.

Remark 2. We can extend the results of Th. 1, 2 to

(a) the case of general coercive 2^0 order elliptic operators in divergence form,
(b) the parabolic case, (c) the case of convex

$$K^{\psi_1\psi_2} = \{v \in H_\Gamma^{1,2}(\Omega),\ \psi_1 \leqq v \leqq \psi_2 \text{ a.e. in } \Omega\}$$

with $\operatorname{Inf}_{x\in\Omega} \{\psi_2(x) - \psi_1(x)\} > 0$ if $-\psi_1$, ψ_2 verify the same hypothesis which holds for ψ in Th. 1, 2.

2. Proof of Theorem 1

We observe that we can suppose $\psi|_\Gamma = 0$, $0 \in \Omega$.

The proof uses a regularisation method and is divided into three steps.

(1) The case $A\psi \in \mathcal{L}^\infty(\Omega)$.

From the Lewy-Stampacchia-Mosco's inequality, [24], we have for the solution u of (1.5) $Au \in \mathcal{L}^\infty(\Omega)$, then from De Giorgi-Nash-Moser theorem we have the result.

(2) The case $\psi \in H_\Gamma^{1,p}(\Omega)$, $p > N$.

This step uses the following lemma:

Lemma 1. *Let be* $\psi \in H_\Gamma^{1,p}(\Omega)$, $P > N$, *and consider the sequence* $\{\psi_n\}$ *defined by*

$$n^{-1}A\psi_n + \psi_n = \psi; \tag{2.1}$$

we have

$$\|\psi_n - \psi\|_{\mathscr{L}^\infty(\Omega)} \leqq C \|\psi\|_{H_\Gamma^{1,p}(\Omega)} n^{-1/2}, \tag{2.2}$$

$$\|A\psi_n\|_{\mathscr{L}^\infty(\Omega)} \leqq C \|\psi\|_{H_\Gamma^{1,p}(\Omega)} n^{1/2}. \tag{2.3}$$

Let be ψ_n defined by (2.1); we have

$$n^{-1}A(\psi_n - \psi) + \psi_n - \psi = -n^{-1}A\psi. \tag{2.4}$$

Let be $\omega = n(\psi_n - \psi)$, $y = n^{-1/2}x$ and $\Omega_y(\Gamma_y)$ the image of $\Omega(\Gamma)$ for the map $x \mapsto y$. From (2.4) by the map $x \mapsto y$ we have the following equation in y

$$A\omega + \omega = -A\psi, \tag{2.5}$$

then repeating the proof of $\mathscr{L}^\infty$-estimates, [19, 29, 30],

$$\|\omega\|_{\mathscr{L}^\infty(\Omega_y)} \leqq C \|\psi\|_{H_{\Gamma_y}^{1,p}(\Omega_y)}$$

where C does not depend on n; by the map $x = n^{1/2}y$ we have

$$\|\omega\|_{\mathscr{L}^\infty(\Omega)} \leqq Cn^{1/2} \|\psi\|_{H_\Gamma^{1,p}(\Omega)},$$

then the first inequality follows.

The inequality (2.3) follows from the inequality (2.2) and the equation (2.1). ■

Let be $\{\psi_n\}$ the sequence in the lemma 1; from [23] we have

$$\|S(\psi_n) - S(\psi)\|_{\mathscr{L}^\infty(\Omega)} \leqq Cn^{-1/2} \|\psi\|_{H_\Gamma^{1,p}(\Omega)}. \tag{2.6}$$

From the first step we have

$$|S(\psi_n)(x + h) - S(\psi_n)(x)| \leqq Cn^{1/2} \|\psi\|_{H_\Gamma^{1,p}(\Omega)} |h|^\gamma, \tag{2.7}$$

$x, x + h \in \bar{\Omega}$, $0 < \gamma < 1$.

From (2.6) (2.7) we have for $u = S(\psi)$

$$|u(x + h) - u(x)| \leqq Cn^{1/2} \|\psi\|_{H_\Gamma^{1,p}(\Omega)} |h|^\gamma + 2Cn^{-1/2} \|\psi\|_{H_\Gamma^{1,p}(\Omega)};$$

choosing n as the first integer greater than $|h|^{-\gamma}$, we have

$$|u(x + h) - u(x)| \leqq C' |h|^{\gamma/2},$$

then $u \in C^{\gamma/2}(\bar{\Omega})$.

(3) The case $\psi \in C^\alpha(\bar{\Omega})$, $\psi|_\Gamma = 0$.

Let be $\varrho \in C_0^\infty(\mathbb{R}^N)$, supp $(\varrho) \subset B(0; 1)$, such that $\int \varrho(x)\, dx = 1$; we denote $\varrho_n(x) = n^N \varrho(nx)$.

The proof of this step uses the following lemmas:

Lemma 2. *Let be $f \in C^\alpha(\mathbb{R}^N)$; we have*

$$\|\varrho_n * f\|_{H^{1,\infty}(\mathbb{R}^N)} \leqq C n^{1-\alpha} \|f\|_{C^\alpha(\mathbb{R}^N)}.$$

It is easy to verify that $\varrho_n *: H^{1,\infty}(\mathbb{R}^N) \to H^{1,\infty}(\mathbb{R}^N)$ is a contraction. We have also

$$\left|\frac{\partial}{\partial x_i} \varrho_n * f(x)\right| \leqq \left|\int f(y) \frac{\partial}{\partial x_i} \varrho_n(x-y)\, dy\right|$$

$$= \left|\int f(y)\, n^{N+1} \frac{\partial}{\partial x_i} \varrho\big(n(x-y)\big)\, dy\right|$$

$$= n\left|\int f(z/n) \frac{\partial}{\partial x_i} \varrho(nx - z)\, dz\right| \leqq K_1 n \|f\|_{\mathscr{L}^\infty(\mathbb{R}^N)},$$

then $\|\varrho_n *\|_{\mathscr{L}(\mathscr{L}^\infty(\mathbb{R}^N);\, H^{1,\infty}(\mathbb{R}^N))} \leqq K_1 n$; we have now the result by interpolation [3]. ■

Lemma 3. *Let be $f \in C^\alpha(\mathbb{R}^N)$; we have*

$$\|\varrho_n * f - f\|_{\mathscr{L}^\infty(\mathbb{R}^N)} \leqq n^{-\alpha} \|f\|_{C^\alpha(\mathbb{R}^N)}.$$

We have

$$|\varrho_n * f(x) - f(x)| \leqq \left|\int_{B(0,1/n)} \varrho_n(y) \big(f(x-y) - f(x)\big)\, dy\right|$$

$$\leqq \int_{B(0,1/n)} \varrho_n(y) \|f\|_{C^\alpha(\mathbb{R}^N)} n^{-\alpha}\, dy = \|f\|_{C^\alpha(\mathbb{R}^N)} n^{-\alpha}. \;■$$

We prolongate the function ψ to a function on $\mathbb{R}^N$ in $C^\alpha(\mathbb{R}^N)$ which we indicate again by ψ.

Consider the sequence $\{\tilde{\psi}_n\}$, $\tilde{\psi}_n = \varrho_n * \psi$; we indicate again by $\tilde{\psi}_n$ the restriction of functions $\tilde{\psi}_n$ to Ω.

Let be $\psi_n = \zeta_n \tilde{\psi}_n$, where $\zeta_n(x)$ is defined as

$$\zeta_n(x) = \begin{cases} 1 & \text{if dist}\,(x, \Gamma) \geqq 1/n, \\ 0 & \text{if } x \in \Gamma, \end{cases}$$

$$0 \leqq \zeta_n(x) \leqq 1 \quad \text{if dist}\,(x, \Gamma) < 1/n,$$

$$\left|\frac{\partial \zeta_n}{\partial x_i}(x)\right| \leqq K n^{-1} \quad \text{if dist}\,(x, \Gamma) \leqq 1/n,\ i = 1, \ldots, N.$$

We have

$$\left\|\frac{\partial(\zeta_n \tilde{\psi}_n)}{\partial x_i}\right\|_{\mathscr{L}^\infty(\Omega)} \leqq \left\|\frac{\partial \zeta_n}{\partial x_i} \tilde{\psi}_n\right\|_{\mathscr{L}^\infty(\Omega)} + \left\|\zeta_n \frac{\partial \tilde{\psi}_n}{\partial x_i}\right\|_{\mathscr{L}^\infty(\Omega)} \leqq K n^{-\alpha},\ i = 1, \ldots, N,$$

$$\|\zeta_n \tilde{\psi}_n - \psi\|_{\mathscr{L}^\infty(\Omega)} \leqq \|(1 - \zeta_n)\, \psi\|_{\mathscr{L}^\infty(\Omega)} + \|\zeta_n(\tilde{\psi}_n - \psi)\|_{\mathscr{L}^\infty(\Omega)} \leqq K n^{-\alpha}.$$

Then

$$\|S(\psi_n) - S(\psi)\|_{\mathscr{L}^\infty(\Omega)} \leqq K n^{-\alpha};\ |S(\psi_n)\,(x+h) - S(\psi_n)\,(x)| \leqq K n^{1-\alpha} |h|^{\gamma/2}. \tag{2.8}$$

From (2.8) we have for $u = S(\psi)$

$$|u(x+h) - u(x)| \leqq Kn^{1-\alpha} |h|^{\gamma/2} + Kn^{-\alpha}.$$

Choosing n as the first integer greater than $|h|^{-\gamma/2}$ we have

$$|u(x+h) - u(x)| \leqq C_1 |h|^{\alpha\gamma/2},$$

then $u \in C^{\alpha\gamma/2}(\bar{\Omega})$.

3. An application of Theorem 1

Let be

$$Mv(x) = 1 + \operatorname{Inf}_{x+\xi\in\Omega, \xi\geqq 0} v(x+\xi).$$

We consider the problem

$$\left.\begin{array}{l} \langle Au, v-u\rangle \geqq \langle f, v-u\rangle \\ \forall\, v \in K^{Mu},\ u \in K^{Mu} \end{array}\right\} \tag{3.1}$$

where A, f are as in Th. 1.

We suppose

$$M\colon C^\alpha(\bar{\Omega}) \to C^\beta(\bar{\Omega}), \quad 0 < \alpha, \beta < 1 \quad \text{or} \quad \alpha = \beta = 0. \tag{3.2}$$

Theorem 3. *Let be $u \in C(\bar{\Omega})$ a solution to* (3.1); *u is Hölder continuous*, [9].

Let be $x_0 \in \bar{\Omega}$; if $u(x_0) < Mu(x_0)$ there is a neighbourhood N of x_0 such that

$$u(x) < Mu(x), \quad x \in N \cap \bar{\Omega}, \tag{3.3}$$

then

$$\langle Au, v\rangle = \langle f, v\rangle \tag{3.4}$$

$\forall\, v \in H^{1,2}_\Gamma(\Omega)$, supp $(v) \subset N \cap \bar{\Omega}$.

From (3.4) we have $u \in C^{\alpha_1}(N' \cap \bar{\Omega})$, $0 < \alpha_1 < 1$, where $N' \subset\subset N$ is a neighbourhood of x_0.

Suppose now $u(x_0) = Mu(x_0)$ and denote by Σ_0 the set of the points $x_0 + \bar{\xi}$, $\bar{\xi} \geqq 0$ such that

$$u(x_0 + \bar{\xi}) = \operatorname{Min}_{\xi\geqq 0, x_0+\xi\in\bar{\Omega}} u(x_0+\xi).$$

We claim that $u(y) < Mu(y)$, $y \in \Sigma_0$; if $u(y) = Mu(y)$ we have

$$u(x_0) = 1 + u(y) = 1 + Mu(y) \geqq 1 + Mu(x_0),$$

then we have a contradiction.

Being u continuous we have a $\bar{\Omega}$-neighbourhood N_0 of Σ_0 such that

$$u(y) < Mu(y) \quad \text{in} \quad N_0.$$

We have

$$\langle Au, v\rangle = \langle f, v\rangle$$

$\forall\, v \in H^{1,2}_{\Gamma}(\Omega)$, supp $(v) \subset N_0$, then $u \in C^{\alpha_2}(N_1 \cap \bar{\Omega})$ where $N_1 \subset\subset N_0$ is a neighbourhood of Σ_0 in $\bar{\Omega}$.

If $y = x_0 + \xi$, $\xi \geqq 0$ we have for dist $(y, \Sigma_0) > \delta > 0$

$$u(y) > Mu(x_0) + \delta_1, \quad \delta_1 > 0,$$

then there is a $\bar{\Omega}$-neighbourhood of the set $W^{\delta'} = \{y \mid \text{dist } (y, \Sigma_0) > \delta'\}$, $\delta' > 0$, such that in this neighbourhood

$$u(x) > Mu(x_0) + \delta_2, \ \delta_2 > 0.$$

We can now choose $\delta_3 > 0$ such that if $x \in V_{\delta_3} = \{x \mid (x - x_0) < \delta_3, x \in \bar{\Omega}\}$,

$$Mu(x) \leqq Mu(x_0) + \delta_2,$$

then there is a $\bar{\Omega}$-neighbourhood N_2 of Σ_0 such that if $x \in V_{\delta_3'}$, $\delta_3' > 0$ suitable,

$$Mu(x) = 1 + \text{Inf}_{\xi \geqq 0, x+\xi \in N_2} u(x + \xi).$$

We can now prolongate $u|_{N_2}$ to a function $\tilde{u} \in C^{\beta}(\bar{\Omega})$ such that

$$\text{Inf}_{\xi \geqq 0, x+\xi \in \Omega}\, \tilde{u}(x + \xi) = \text{Inf}_{\xi \geqq 0, x+\xi \in N_2} u(x + \xi), \ x \in V_{\delta_3'}.$$

From (3.2) we have $M\tilde{u} \in C^{\beta'}(\bar{\Omega})$, $0 < \beta' < 1$, then $M\tilde{u}|_{V_{\delta_3'}} = Mu|_{V_{\delta_3'}} \in C^{\beta'}(\bar{\Omega})$. By a localization of Th. 1 we have $u \in C^{\beta''}(V_{\delta_4})$, $0 < \beta'' < 1$, $0 < \delta_4 < \delta_3'$. Being u Hölder continuous in a neighbourhood of each point $x \in \bar{\Omega}$ and Ω bounded, u is Hölder continuous in $\bar{\Omega}$.

Remark 3. The method used in the proof of Th. 3 is due to Caffarelli, Friedman [13].

Remark 4. For the existence of solutions to (3.1) in $C(\bar{\Omega})$ see e.g. [5].

4. Proof of Theorem 2

The proof of Th. 2 uses the penalisation method and the Schauder's fixed point theorem.

We can suppose $\psi|_{\partial\Omega} = 0$.

Consider the penalised problem

$$-\Delta u_\lambda + \frac{1}{\lambda}(u_\lambda - \psi)^+ = 0 \tag{4.1}$$

where $\psi \in H_0^{1,2}(\Omega)$ and denote $u_\lambda = S_\lambda(\psi)$.

We have easily the following lemma

Lemma 4. *Let be $\psi_1, \psi_2 \in H_0^{1,2}(\Omega)$; we have*

$$\|\text{grad } S_\lambda(\psi_1) - \text{grad } S_\lambda(\psi_2)\|_{(\mathscr{L}^2(\Omega))^N} \leqq \|\text{grad } \psi_1 - \text{grad } \psi_2\|_{(\mathscr{L}^2(\Omega))^N}.$$

Using essentially the methods given in [7], [20] we have also

Lemma 5. *Let be Ω diffeomorphic to a smooth convex set, $\psi \in H_0^{1,\infty}(\Omega)$; we have*

$$\|\text{grad } S_\lambda(\psi)\|_{(\mathscr{L}^\infty(\Omega))^N} \leqq \|\text{grad } \psi\|_{(\mathscr{L}^\infty(\Omega))^N}.$$

From the lemmas 4, 5 and the Tartar's results on nonlinear interpolation, [31], we have

Lemma 6. $S_\lambda : H_0^{1,p}(\Omega) \to H_0^{1,p}(\Omega)$, $2 \leqq p < +\infty$, *is bounded and*

$$\|S_\lambda(\psi)\|_{H_0^{1,p}(\Omega)} \leqq C \|\psi\|_{H_0^{1,p}(\Omega)} \tag{4.2}$$

where C is independent on λ.

Consider now the problem

$$-\Delta u_\lambda + \frac{1}{\lambda} u_\lambda^+ = f \tag{4.3}$$

where $f = \operatorname{div} g$ and denote grad $u_\lambda = T_\lambda g$, where u_λ is the solution to (4.3); we have easily

Lemma 7. $T_\lambda : (\mathscr{L}^2(\Omega))^N \to (\mathscr{L}^2(\Omega))^N$ *is a contraction.*

We suppose that the hypothesis on Ω of lemma 5 holds again, we have

Lemma 8. $T_\lambda : (\mathscr{L}^p(\Omega))^N \to (\mathscr{L}^p(\Omega))^N$, $2 \leqq p < +\infty$, is bounded and

$$\|T_\lambda g\|_{(\mathscr{L}^p(\Omega))^N} \leqq C \|g\|_{(\mathscr{L}^p(\Omega))^N}$$

where C is independent on λ.

We fix now $p_0 > 2$ and for $2 \leqq p \leqq p_0$ $\dfrac{1}{p} = \dfrac{\theta(p)}{p_0} + \dfrac{1-\theta(p)}{2}$; we denote

$$\tilde{\mathscr{L}}^p(\Omega) = [\mathscr{L}^2(\Omega), \mathscr{L}^{p_0}(\Omega)]_{\theta(p),p}$$

(by the K-real method).

We observe that $\tilde{\mathscr{L}}^p(\Omega)$ is the space $\mathscr{L}^p(\Omega)$ with an equivalent norm, [3]. We denote also by $\tilde{H}_0^{1,p}(\Omega)$ the space $H_0^{1,p}(\Omega)$ with the norm

$$\|\varphi\|_{\tilde{H}_0^{1,p}(\Omega)} = \|\operatorname{grad} \varphi\|_{(\tilde{\mathscr{L}}^p(\Omega))^N}$$

and by $\tilde{H}^{-1,p}(\Omega)$ the space $H^{-1,p}(\Omega)$ with the norm

$$\|\varphi\|_{\tilde{H}^{-1,p}(\Omega)} = \inf_{\operatorname{div} g = \varphi} \|g\|_{(\tilde{\mathscr{L}}^p(\Omega))^N}.$$

We observe that $\tilde{H}_0^{1,p}(\Omega)$ and $\tilde{H}^{-1,p}(\Omega)$ are uniformly convex spaces. From the lemmas 7, 8 and the Tartar's results on nonlinear interpolation, [31], we have

Lemma 9. $T_\lambda : (\tilde{\mathscr{L}}^p(\Omega))^N \to (\tilde{\mathscr{L}}^p(\Omega))^N$ *is bounded and*

$$\|T_\lambda g\|_{(\tilde{\mathscr{L}}^p(\Omega))^N} \leqq C^{\theta(p)} \|g\|_{(\tilde{\mathscr{L}}^p(\Omega))^N}, \quad 2 \leqq p \leqq p_0. \tag{4.4}$$

Lemma 10. *Let be* $g_1, g_2 \in \mathscr{L}^p(\Omega)$, $0 \leqq g_1(x) \leqq c g_2(x)$ *a.e. in* Ω; *we have*

$$\|g_1\|_{\tilde{\mathscr{L}}^p(\Omega)} \leqq C \|g_2\|_{\tilde{\mathscr{L}}^p(\Omega)}.$$

It is enough to recall the definition of the norm of $\tilde{\mathscr{L}}^p(\Omega)$ in the K-real method.

Lemma 11. *There is* $c > 0$ *such that*

$$\left\| \Delta + \frac{1}{C} A \right\|_{\mathscr{L}(\tilde{H}_0^{1,p}(\Omega), \tilde{H}^{-1,p}(\Omega))} \leqq \delta < 1 \tag{4.5}$$

where C, δ *do not depend on* p.

The proof is the same of (4.35) (4.36), p. 41, [2]. using the lemma 10. We give now the proof of Th. 2 and we specify that in the following we suppose Ω *diffeomorphic to a smooth convex set.*

We observe that we can suppose $\psi = 0$ and $C = 1$ in the lemma 11. Consider the penalised problem

$$Au_\lambda + \frac{1}{\lambda} u_\lambda^+ = f. \tag{4.6}$$

We can write (4.6) as

$$-\Delta u_\lambda + \frac{1}{\lambda} u_\lambda^+ = f - (\Delta + A) u_\lambda.$$

We denote now $u_\lambda = G(Z)$ if

$$-\Delta u_\lambda + \frac{1}{\lambda} u_\lambda^+ = f - (\Delta + A) Z.$$

From the lemmas 9, 11 we have, if $f \in H^{-1,p}(\Omega)$,

$$\begin{aligned} \|G(Z)\|_{\tilde{H}_0^{1,p}(\Omega)} &\leqq C^{\theta(p)} \|f - (\Delta + A) Z\|_{\tilde{H}^{-1,p}(\Omega)} \\ &\leqq C^{\theta(p)}(K_1 + \|(\Delta + A) Z\|_{\tilde{H}^{-1,p}(\Omega)}) \\ &\leqq C^{\theta(p)}(K_1 + \delta \|Z\|_{\tilde{H}_0^{1,p}(\Omega)}). \end{aligned}$$

From (4.7) there is $\bar{p} > 2$ such that $G(B_{R_p}) \subset B_{R_p}$ where $B_{R_p} \subset \tilde{H}_0^{1,p}(\Omega)$ is the ball in $\tilde{H}_0^{1,p}(\Omega)$ of radius R_p, R_p suitable, $2 \leqq p \leqq \bar{p}$. We observe that R_p depends on p but does not depend on λ. We show now that G is continuous for the weak topology of $\tilde{H}_0^{1,p}(\Omega)$. Let be

$$\lim_{n\to+\infty}{}^* Z_n = Z \quad \text{in} \quad \tilde{H}_0^{1,p}(\Omega),$$

we have

$$\|G(Z_n)\|_{\tilde{H}_0^{1,p}(\Omega)} \leqq \text{const},$$

then we can suppose

$$\lim_{n\to\infty}{}^* G(Z_n) = \chi \quad \text{in} \quad \tilde{H}_0^{1,p}(\Omega)$$

and we can easily show $\chi = G(Z)$.

From Schauder's fixed point theorem G has a fixed point u_λ in $\tilde{H}_0^{1,p}(\Omega)$ and

$$\|u_\lambda\|_{\tilde{H}_0^{1,p}(\Omega)} \leqq R_p$$

where R_p does not depend on λ.

It is easy to verify that u_λ is the solution to (4.6), then from (4.6) (4.8) passing to the limit for $\lambda \to 0$ we have the result.

5. An application of Theorem 2

We give now an application of Th. 2 to homogeneisation estimates for variational and quasi variational inequalities in the case of irregular coefficients.

Let be $b_{ij} \in \mathscr{L}^\infty(P)$, $P = \prod_{i=1}^{N} [0, y_i]$, $i, j = 1, \ldots, N$, with

$$\sum_{ij=1}^{N} b_{ij}(x)\, \xi_i \xi_j \geqq b\, |\xi|^2 \qquad \forall\, \xi \in \mathbb{R}^N,\ b > 0 \qquad \text{a.e. in } P.$$

Let be

$$a_{ij}^\varepsilon(x) = b_{ij}(x/\varepsilon), \quad i, j = 1, \ldots, N$$

and $A_\varepsilon \colon H_0^{1,2}(\Omega) \to H_0^{-1,2}(\Omega)$ defined by

$$\langle A_\varepsilon u, v\rangle = \int_\Omega \sum_{ij=1}^{N} a_{ij}^\varepsilon(x) \frac{\partial u}{\partial x_i}(x) \frac{\partial v}{\partial x_j}(x)\, dx.$$

Let be also

$$\langle A u, v\rangle = \int_\Omega \sum_{ij=1}^{N} q_{ij} \frac{\partial u}{\partial x_i}(x) \frac{\partial v}{\partial x_j}(x)\, dx$$

the homogeneised operator of A_ε (the q_{ij} are constants in x, [1, 2]). Let be p the Meyer's index relative to the b_{ij} and suppose

$$\|b_{ij}(y + h) - b_{ij}(y)\|_{\mathscr{L}^{p^*}(P)} \leqq K\, |h|^\beta, \quad p^* = \frac{2p}{p-2}, \qquad 0 < \beta \leqq 1 \tag{5.1}$$

(we recall that p depends only on b and $\operatorname{Sup}_{y \in P} b_{ij}(y)$).

Consider the problems

$$\left.\begin{aligned} &\langle A_\varepsilon u^\varepsilon, v - u^\varepsilon\rangle \geqq \langle f, v - u^\varepsilon\rangle \\ &\forall\, v \in K^\psi, \quad u^\varepsilon \in K^\psi, \end{aligned}\right\} \tag{5.2$_\varepsilon$}$$

$$\left.\begin{aligned} &\langle A u, v - u\rangle \geqq \langle f, v - u\rangle \\ &\forall\, v \in K^\psi,\ u \in K^\psi, \end{aligned}\right\} \tag{5.2}$$

where $K^\psi \subset H_0^{1,2}(\Omega)$.

Theorem 4. *Let be $f \in \mathscr{L}^q(\Omega)$, $q > N$, $\psi \in C^\alpha(\bar{\Omega})$, $0 < \alpha < 1$, $\psi|_{\partial\Omega} \geqq 0$; we have*

$$\|u^\varepsilon - u\|_{\mathscr{L}^r(\Omega)} \leqq \varepsilon^{s\beta/(1+\beta)} \tag{5.3}$$

where $r = \dfrac{2N}{N-2}$ *if* $N > 2$, $r < +\infty$ *if* $N = 2$ *and*

$$s = \frac{\sigma\alpha}{2(N - 2 + 3\sigma)}$$

(σ is the Hölder exponent relative to the b and $b_{ij}(y)$ in the De Giorgi-Nash-Moser theorem).

Moreover

$$\|u^\varepsilon\|_{C^\gamma(\bar{\Omega})} \leqq C,\ 0 < \gamma < 1,$$

then

$$\|u^\varepsilon - u\|_{\mathscr{L}^\infty(\Omega)} \leqq \varepsilon^\tau \tag{5.4}$$

where $\tau = 2s\beta\gamma[(1+\beta)(N-2)(1-\gamma)]^{-1}$.

Suppose at first $b_{ij} \in C^1(\bar{\Omega})$ and $\psi \in W^{1,q}(\Omega)$; from [4] we have

$$\|u^\varepsilon - u\|_{\mathscr{L}^\infty(\Omega)} \leqq K\varepsilon^{\sigma/2(N-2+3\sigma)}.$$

Suppose now $\psi \in C^\alpha(\bar{\Omega})$; by the regularisation method used in the step 3 in the proof of Th. 1 we have

$$\|u^\varepsilon - u\|_{\mathscr{L}^\infty(\Omega)} \leqq K\varepsilon^{\sigma\alpha/2(N-2+3\sigma)}. \tag{5.5}$$

By the regularisation method on the coefficients developed for the equations in [11] we have (5.3) from (5.5).

By the same methods used in the lemma 2 [4] we have easily:

Lemma 12. *Let be* $\{v_n\}$ *a sequence such that*

$$\|v_n - v\|_{\mathscr{L}^{2N/N-2}(\Omega)} \leqq Kn^{-\varrho},\ \varrho \geqq 0,$$

$$\|v_n\|_{C^\gamma(\bar{\Omega})} \leqq K,$$

then

$$\|v_n - v\|_{\mathscr{L}^\infty(\Omega)} \leqq K' n^{-2\varrho\gamma/N-2(1-\gamma)}.$$

From Th. 1 we have $\|u^\varepsilon\|_{C^\gamma(\bar{\Omega})} \leqq K$, $0 < \gamma < 1$; then from (5.3) and the lemma 12 we have (5.4). ∎

Let be now $M\colon \mathscr{L}^\infty(\Omega) \to \mathscr{L}^\infty(\Omega)$ as in 3. and consider the quasi variational inequalities

$$\left.\begin{array}{l} \langle A_\varepsilon u^\varepsilon, v - u^\varepsilon\rangle \geqq \langle f, v - u^\varepsilon\rangle \\ \forall\, v \in K^{Mu^\varepsilon},\ u^\varepsilon \in K^{Mu^\varepsilon}, \end{array}\right\} \tag{5.6$_\varepsilon$}$$

$$\left.\begin{array}{l} \langle Au, v - u\rangle \geqq \langle f, v - u\rangle \\ \forall\, v \in K^{Mu},\ u \in K^{Mu}. \end{array}\right\} \tag{5.6}$$

We suppose that there is a subsolution u_0^ε (u_0) of (5.6$_\varepsilon$) ((5.6)), [21], such that $u_0^\varepsilon, u_0 \geqq -1 + \delta$, $\delta > 0$, a.e. in Ω.

From [21] we have existence and uniqueness of the solution to (5.6) (5.6$_\varepsilon$). The following homogeneisation estimate can be obtained using Th. 4, the approximation by iterations to u^ε (u) and the estimate on this approximation given in [5, 18].

Theorem 5. *Let be* $f \in \mathscr{L}^\infty(\Omega)$, u^ε *and* u *the solutions to* (5.6$_\varepsilon$) *and* (5.6); *we have*

$$\|u^\varepsilon - u\|_{\mathscr{L}^\infty(\Omega)} \leqq K\varepsilon^{\tau\eta},\quad 0 < \eta < 1$$

where τ *is as in Th.* 4 *and* γ *is the Hölder exponent given in* [16] *for a monotone obstacle.*

Remark 5. The hypothesis (5.1) holds in the case

$$b_{ij}(y) = \begin{cases} \alpha^1_{ij} \text{ in } P_1 = \prod_{i=1}^{N-1} [0, y_i] \times [0, y_{N/2}], \\ \alpha^2_{ij} \text{ in } P_2 = \prod_{i=1}^{N-1} [0, y_i] \times [y_{N/2}, y_N]. \end{cases}$$

Remark 6. The detailled proofs of Th. 5, 6 are given in [12].

References

[1] A. Bensoussan, J. L. Lions, Nouvelle formulation de problèmes de contrôle impulsionel et applications, C.R. Acad. Sci. Paris **A276** (1973), 1189–1192.

[2] A. Bensoussan, J. L. Lions, G. Papanicolau, Asymptotic analysis for periodic structures, North Holland 1978.

[3] J. Bergh, J. Lofstrom, Interpolation spaces; an introduction, Springer 1976.

[4] M. Biroli, An estimate on homogeneisation for variational inequalities, Rend. Sem. Mat. Padova **58** (1977), 9–24.

[5] M. Biroli, An estimate on convergence of approximation by iterations of solution to a quasi-variational inequality and some consequences in continuous dependence and in G-convergence, Ann. Mat. Pura Appl. (4) **102** (1979), 79–101.

[6) M. Biroli, G-convergence for elliptic variational and quasi variational inequalities, Recent Methods in Nonlinear Analysis, Rome May 1978, Pitagora 1978.

[7] M. Biroli, A regularity result for the solution to a quasi variational inequality with variable coefficients, Ric. Mat., **23** (1979), 259–267.

[8] M. Biroli, A De Giorgi-Nash-Moser result for a variational inequality, Boll. U.M.I. **16** (1979), 598–606.

[9] M. Biroli, Un risultato di holderianita per la soluzione di certe disequazioni quasi variazionali, Rend. Sem. Mat. Fis. Milano **48** (1978), 191–195.

[10] M. Biroli, Une estimation dans $\mathscr{L}^p(\Omega)$ du gradient de la solution d'une inéquation elliptique du second ordre, C.R. Acad. Sci. Paris **A288** (1979), 453–455.

[11] M. Biroli, S. Marchi, An estimate in homogeneisation problem for elliptic equations with bounded discontinuous coefficients, Quaderni I.A.C. III, **92** (1978) 1–12.

[12] M. Biroli, S. Marchi, T. Norando, in preparation.

[13] L. A. Caffarelli, A. Friedman, Regularity of the solution of the quasi variational inequality for the impulse control problem, Comm. in P.D.E. **3**, 8 (1978), 745–753.

[14] E. De Giorgi, Sulla differenziabilità- ed analiticità delle estremali degli integrali multipli regolari, Mem. Acc. Sc. Torino **3** (1957), 1–19.

[15] F. De Giorgi, S. Spagnolo, Sulla convergenza di integrali della energia per operatori ellittici del 2° ordine, Boll. U.M.I. **8** (1973), 391–411.

[16] J. Frehse, U. Mosco, Sur la regularité des solutions faibles de certaines inéquations variationnelles et quasi variationnelles, C.R. Acad. Sci. Paris **A288** (1979), 627–630.

[17] B. Hanouzet, J. L. Joly, Méthodes d'ordre dans l'interpretation de certaines inéquations variationelles, C.R. Acad Sci. Paris **A281** (1975), 373–376.

[18] B. Hanouzet, J. L. Joly, Convergence uniforme des itérées definissant la solution d'une inéquation quasi variationnelle, C.R. Acad. Sci. Paris **A286** (1978), 735–738.

[19] O. A. Ladyzenskaya, N. N. Ural'ceva, Equations aux derivées partielles de type elliptique, Dunod, 1968

[20] H. Lewy, G. Stampacchia, On existence and smoothness of some non coercive inequalities, Arch. Rat. Mech. Anal. **41** (1971), 241–253.

[21] J. L. Lions, Sur quelques questions d'analyse de mecanique et de contrôle optimal, Les presses de l'Université de Montreal 1976.
[22] N. G. Meyers, An estimate for the gradient of solutions of second order elliptic equations, Ann. Sc. Norm Sup. Pisa **17** (1963), 189–206.
[23] F. Mignot, J. P. Puel, Inéquations d'évolution paraboliques avec convexe dépendant du temps: applications aux inéquations quasi variationnelles d'évolution, Arch. Rat. Mech. Anal. **64** (1977), 59–91.
[24] U. Mosco, Implicit variational problems and quasi variational problems, Nonlinear operators and calculus of variations, Lecture Notes in Mathematics, vol. 543, Springer 1976.
[25] J. Moser, A new proof of De Giorgi's theorem concerning the regularity problem for elliptic differential equations, Comm. Pure Appl. Math. **13** (1960), 457–468.
[26] J. Nash, Continuity of solutions to parabolic and elliptic equations, Comm. Pure Appl. Math. **80** (1958), 931–954.
[27] G. Pulvirenti, Sulla sommabilità ..., Le Matematiche **22** (1968), 250–265.
[28] G. Pulvirenti, Ancora sulla sommabilità ..., Le Matematiche **23** (1971), 160–165.
[29] G. Stampacchia, Problemi al contorno ellittici con dati discontinui dotati di soluzioni holderiane, Ann. Mat. Pura Appl. **51** (1960), 1–32.
[30] G. Stampacchia, Equations elliptiques du second ordre à coefficients discontinus, Les presses de l'Université de Motreal, 1965.
[31] L. Tartar, Interpolation non linéaire et regularité, J. Functional Anal. **9** (1972), 469–489.

Address of the author:

Istituto di Matematica del Politecnico di Milano (Analisi)
Via Bonardi, 9
20 133 Milano
Italy

Nonlinear equality constraints in optimization problems with set restrictions

Leonard Bittner, Greifswald

1.

As far as we can see, equalities as side conditions, defined by mappings from one space to another space of infinite dimension, cause difficulties in deriving optimality conditions even today, if among the feasibility conditions there occurs a constraint set Ω eventually not open. As is well known, in such a case a certain neighbourhood of a hypothetical solution x_0 of the optimization problem under debate is usually approximated by some convex set, for example by the so called cone of feasible displacements, the tangent cone or some other "regular" set. The well known optimization theories of Dubovitski-Miljutin, Pshenični, Neustadt, Halkin and Boltjanski make use of such an approximation of the constraint set Ω in the vicinity of x_0 and are capable of taking account of a finite number of (real) scalar equations $g_k(x) = 0$ for a feasible solution. As is well known too, an implicit function theorem, Brouwer's fixed point theorem or degree theory (among other things) is employed in order to establish multiplier rules. This suggests to look for cases where a more general fixed point theorem can and must be employed for a similar purpose. In this paper we are going to take account of such equality constraints, which allow to apply Schauder's fixed point theorem. Boltjanski introduced a special kind of approximative cones (for a constraint set Ω in the vicinity of a point $x_0 \in \Omega$), which he called tents. The notion of a tent, somewhat specified, turns out to be most convenient for our aim, hence we shall adopt it. From the following considerations there emerges an interesting analytical question to which we want to attract the attention of the reader (see p. 43).

At the end of this paper we show that under certain elementary conditions a vector minimum problem has special efficient solutions, which satisfy the so called Neumann-Morgenstern property.

2. A traditional optimization problem revised with respect to equality constraints

For the sake of simplicity let us explain the indicated views by the hand of the following problem, which encloses a lot of concrete problems of optimal control. Let

X, U be real Banach spaces;
V, Z real locally convex topological vector spaces;
K and L convex cones with vertex 0 in Z or V, distinct

from Z or V respectively and equipped with a nonvoid interior, if the dimension of the corresponding space is infinite;

$z_1 \geqq z_2$, $v_1 \geqq v_2$ denote $z_1 - z_2 \in K$, $v_1 - v_2 \in L$;
z_1, z_2 (similarly v_1, v_2) be called *comparable* if either

$z_1 \leqq z_2$ or $z_2 \leqq z_1$ or both holds;

Ω be a subset of X;
f, g, h mappings from Ω into Z, U, V respectively.

The **optimization problem** now reads: *Find a point $x_0 \in \Omega$, which satisfies $g(x_0) = 0$, $h(x_0) \geqq 0$, such that for every other point $x \in \Omega$, which satisfies $g(x) = 0$, $h(x) \geqq 0$, either $f(x_0)$ and $f(x)$ are not comparable or $f(x_0) \leqq f(x)$.* As usually such a point x_0 is called an efficient solution. Due to Boltjanski, a convex cone $\Delta\Omega$ with vertex 0 in X is called *a tent for Ω at $x_0 \in \Omega$*, if to each point $\Delta x \in \Delta\Omega$ of sufficiently small norm we can assign a unique point $x_{\Delta x} = x_0 + \Delta x + \omega(\Delta x)$ such that

(i) $x_{\Delta x} \in \Omega$,

(ii) $\dfrac{\|\omega(\Delta x)\|}{\|\Delta x\|} \to 0$ for $\Delta x \to 0$,

(iii) ω be continuous.

If the mapping ω, defined by $\omega(\Delta x) = x_{\Delta x} - x_0 - \Delta x$, appears to be completely continuous, then let us call $\Delta\Omega$ a *c.c.tent*. A tent is not uniquely determined and turns out to be a part of the tangent cone or the cone of feasible displacements.

Assumptions. In the sequel we assume a convex cone $\widetilde{\Delta\Omega}$ with vertex 0 in X as an approximation for Ω in the vicinity of x_0, where $\widetilde{\Delta\Omega}$ need not be a tent for Ω at x_0, but be the union of tents. Further we assume certain linear mappings

$$F: X \to Z,\ G: X \to U,\ H: X \to V,$$

which play the role of derivatives for f, g, h at x_0 and may be identified f.e. with the Fréchet-derivatives, provided those exist. In detail we claim:

1) If int $G\widetilde{\Delta\Omega} = \emptyset$, then $G\widetilde{\Delta\Omega}$ be not dense in U; hence there exists a convex body U_1 disjoint to $G\widetilde{\Delta\Omega}$.

2) If int $G\widetilde{\Delta\Omega} \neq \emptyset$ and $0 \in$ int $G\widetilde{\Delta\Omega}$, f.e. $0 = G\Delta x_0$ with $\Delta x_0 \in \widetilde{\Delta\Omega}$ (and $\|\Delta x_0\| \ll 1$), then there be a tent $\Delta\Omega$ containing Δx_0 and contained in $\widetilde{\Delta\Omega}$; further there be a bounded closed convex neighbourhood U_0 of the origin in U and a mapping $\gamma: U_0 \to \Delta\Omega$ so that

(i) $$f(x_{\Delta x}) = f(x_0 + \Delta x + \omega(\Delta x)) = f(x_0) + F\Delta x + o_f(\Delta x)$$
$$g(x_{\Delta x}) = g(x_0) + G\Delta x + o_g(\Delta x),$$
$$h(x_{\Delta x}) = h(x_0) + H\Delta x + o_h(\Delta x),$$
$$\frac{o_f(\Delta x)}{\|\Delta x\|} \to 0, \quad \frac{o_g(\Delta x)}{\|\Delta x\|} \to 0, \quad \frac{o_h(\Delta x)}{\|\Delta x\|} \to 0 \text{ for } \Delta x \to 0, \qquad \Delta x \in \Delta\Omega,$$

where ω is the mapping connected with the tent $\Delta\Omega$,

(ii) $\gamma(0) = \Delta x_0$,

(iii) $G\gamma(u) = u \; \forall \, u \in U_0$,

(iv) For each sufficiently small positive ε the composed mapping $u \in U_0 \to o_g(\varepsilon\gamma(u)) \in U$ is completely continuous.

(v) The composed mappings $F\gamma$, $H\gamma$ are continuous at 0.

The only uncommon feature among these assumptions is the requirement concerning γ. According to (2.iii) γ denotes a local right-hand inverse of G in the neighbourhood of 0, according to (2.ii) a single-valued section of G^{-1} through Δx_0. This gives rise to the question, whether there is such a γ. If U is of finite dimension, such a γ can easily be defined (cf. the literature mentioned above), but if U is of infinite dimension, the construction of such a mapping γ seems to be an open question, although we are able to contribute some ideas below. Precisely the question turns to the

Inversion problem. Given a convex cone $\Delta\Omega$ with vertex 0, a point $\Delta x_0 \in \Delta\Omega$, a linear (continuous) mapping $G: X \to U$ such that $G\Delta x_0 = 0$ and $G\Delta\Omega = U$, find a single-valued (continuous) mapping γ, defined in a certain neighbourhood U_0 of the origin so that (2.ii) and (2.iii) holds.

Theorem 1 (multiplier rule). *If x_0 denotes an efficient solution for the optimization problem and if there is an approximative cone $\widetilde{\Delta\Omega}$ (a circumtent) for Ω in the vicinity of x_0, which satisfies all assumptions* (1)—(2.v) *stated above, then there exist linear continuous functionals z^*, u^*, v^*, not all zero, over Z, U, V such that*

$$(z^*F + u^*G - v^*H)\,\Delta x \geqq 0 \;\forall \Delta x \in \widetilde{\Delta\Omega},\quad v^*h\,(x_0) = 0,$$

z^*, v^* are isotonic (order preserving).

Proof. First case: int $G\widetilde{\Delta\Omega} = \emptyset$. Due to assumption (1) there is a convex body U_1 disjoint to the convex set $G\widetilde{\Delta\Omega}$. Thus a linear continuous functional $u^* \neq 0$ and a real number α exist so that

$$\inf u^*G\widetilde{\Delta\Omega} \geqq \alpha \geqq \sup u^*U_1.$$

Since $\widetilde{\Delta\Omega}$ is a cone, we obtain $\inf u^*G\widetilde{\Delta\Omega} \geqq 0$. $v^* = 0$, $z^* = 0$ together with u^* trivially satisfy the assertion.

Second case: int $G\widetilde{\Delta\Omega} \neq \emptyset$, but $0 \notin$ int $G\widetilde{\Delta\Omega}$. Then the point $\{0\}$ can be separated from the convex body $G\widetilde{\Delta\Omega}$ by a hyperplane $u^*u = \alpha$ such that as above $\inf u^*G\widetilde{\Delta\Omega} \geqq 0$.

Third case: int $G\widetilde{\Delta\Omega} \neq \emptyset$ and $0 \in$ int $G\widetilde{\Delta\Omega}$, say $0 = G\Delta x_0$ with $\Delta x_0 \in \widetilde{\Delta\Omega}$ and sufficiently small norm. According to assumption (2) there is a tent $\Delta\Omega \subset \widetilde{\Delta\Omega}$ with a mapping ω, a neighbourhood U_0 of the origin in U and a mapping $\gamma: U_0 \to \Delta\Omega$ satisfying (2.i) up to (2.v). Without loss of generality we may assume that both Δx_0 and $\gamma(U_0)$ belong to the domain of ω and that $\gamma(U_0)$ is bounded. The proof relies on the following

Proposition. *For each sufficiently small positive ε there is a corresponding point $u_\varepsilon \in U^0$ such that*

$$x(\varepsilon) := x_0 + \varepsilon\gamma(u_\varepsilon) + \omega(\varepsilon\gamma(u_\varepsilon)) \in \Omega,\; g(x(\varepsilon)) = 0,\; u_\varepsilon \to 0$$

for $\varepsilon \to 0$.

In order to verify this, consider the equation

$$g(x_0 + \varepsilon\gamma(u) + \omega(\varepsilon\gamma(u))) = 0, \; u \in U_0$$

for $0 < \varepsilon \ll 1$. Due to assumption (2.i), (2.iii) this is equivalent to

$$g(x_0) + \varepsilon G\gamma(u) + o_g(\varepsilon\gamma(u)) = 0$$

or

$$u = -\frac{1}{\varepsilon}\, o_g(\varepsilon\gamma(u)) =: T_\varepsilon(u).$$

Since $\gamma(U_0)$ is supposed to be bounded, $\|\gamma(u)\| \leqq c \;\forall\; u \in U_0$ with a certain number c, thus for $\varepsilon \to 0$ uniformly with respect to $u \in U_0$

$$T_\varepsilon(u) = -\frac{1}{\varepsilon\|\gamma(u)\|}\, o_g(\varepsilon\gamma(u)) \cdot \|\gamma(u)\| \to 0.$$

Consequently, there is a positive ε_0 so that for $0 < \varepsilon < \varepsilon_0$

$$T_\varepsilon(u) \in U_0 \qquad \forall\; u \in U_0, \quad \text{i.e.} \quad T_\varepsilon\colon U_0 \to U_0.$$

Due to assumption (2.iv) there is a positive $\varepsilon_1 \leqq \varepsilon_0$ so that T_ε is completely continuous. Now Schauder's fixed point theorem yields a solution $u_\varepsilon \in U_0$ for $u = T_\varepsilon(u)$. Obviously $u_\varepsilon = T_\varepsilon(u_\varepsilon) \to 0$ for $\varepsilon \to 0$. This proves the proposition. Now let us consider the subsets

$$A = \{(f(x_0) + F\Delta x,\, g(x_0) + G\Delta x,\, h(x_0) + H\Delta x) \mid \Delta x \in \widetilde{\Delta\Omega}\},$$

$$B = (f(x_0) - \operatorname{int} K) \times \{0\} \times \operatorname{int} L$$

in $Z \times U \times V$. Evidently they are convex. Further A, B have no point in common. Suppose the contrary, that there is a $\Delta x_0 \in \widetilde{\Delta\Omega}$ satisfying

$$F\Delta x_0 \in -\operatorname{int} K,\; G\Delta x_0 = 0,\; h(x_0) + H\Delta x_0 \in \operatorname{int} L.$$

Note that $\lambda\Delta x_0$ also satisfies these conditions for each positive $\lambda < 1$, since $h(x_0) \in L$. Hence, if A, B have a point in common, they have even a common point generated by a Δx_0 of arbitrarily small norm. Suppose that the point Δx_0 mentioned above at the beginning of the third case coincides precisely with that point. Then — as was shown in the proposition — for $0 < \varepsilon \leqq \varepsilon_1$ there is a $u_\varepsilon \in U_0$ so that $g(x(\varepsilon)) = 0$. It is easy to convince oneself that for a certain positive $\varepsilon_2 \leqq \varepsilon_1$ we have also $h(x(\varepsilon)) \geqq 0$, if $0 < \varepsilon \leqq \varepsilon_2$. Indeed, $l_0 := h(x_0) + H\Delta x_0 \in \operatorname{int} L$. Hence a convex neighbourhood V_0 of the origin in V exists so that $l_0 + V_0 \subset \operatorname{int} L$. For sufficiently small positive ε, say for $0 < \varepsilon \leqq \varepsilon_2$ with $0 < \varepsilon_2 \leqq \varepsilon_1$

$$H(\gamma(u_\varepsilon) - \Delta x_0) + \frac{1}{\varepsilon}\, o_h(\varepsilon\gamma(u_\varepsilon)) \in V_0$$

as a result of assumption (2.v), (2.i) and the proposition. This implies

$$l_0 + H(\gamma(u_\varepsilon) - \Delta x_0) + \frac{1}{\varepsilon}\, o_h(\varepsilon\gamma(u_\varepsilon)) \in \operatorname{int} L,$$

$$h(x(\varepsilon)) = (1 - \varepsilon)\, h(x_0) + \varepsilon l_0 + \varepsilon H(\gamma(u_\varepsilon) - \Delta x_0) + o_h(\varepsilon\gamma(u_\varepsilon)) \in \operatorname{int} L.$$

Similarily from $F\Delta x_0 \in -\operatorname{int} K$ we obtain

$$f(x(\varepsilon)) = f(x_0) + \varepsilon F\Delta x_0 + \varepsilon F(\gamma(u_\varepsilon) - \Delta x_0) + o_f(\varepsilon\gamma(u_\varepsilon)) \in f(x_0) - \operatorname{int} K$$

for sufficiently small positive ε, say $0 < \varepsilon \leqq \varepsilon_3$ with $\varepsilon_3 \leqq \varepsilon_2$.

Thus $f(x_0) - f(x(\varepsilon)) \in \operatorname{int} K$, $x(\varepsilon)$ satisfies all constraints, consequently $f(x(\varepsilon)) - f(x_0) \in K$ due to the optimality of x_0, which entails $0 \in \operatorname{int} K$, $K = Z$ in contrast to the supposition $K \neq Z$. Therefore $A \cap B = \emptyset$, $0 \notin A - B$ and $A - B$ is a convex subset. Now we prove that $\operatorname{int}(A - B) \neq \emptyset$. Obviously

$$A - B = \{(F\Delta x + k,\ G\Delta x,\ h(x_0) + H\Delta x - l) \mid \Delta x \in \widetilde{\Delta\Omega},\ k \in \operatorname{int} K,\ l \in \operatorname{int} L\}.$$

Consider any point $\Delta x_0 \in \widetilde{\Delta\Omega}$ with (sufficiently small norm and) $G\Delta x_0 = 0$, a tent $\Delta\Omega$ containing Δx_0 and contained in $\widetilde{\Delta\Omega}$, a neighbourhood U_0 of the origin in U, a mapping $\gamma\colon U_0 \to \Delta\Omega$ with $\gamma(0) = \Delta x_0$ just as was required in assumptions (1), (2). Further take any point $k_0 \in \operatorname{int} K$, $l_0 \in \operatorname{int} L$. Let Z_0, Z_0' be neighbourhoods of the origin in Z so that

$$k_0 + Z_0 \subset \operatorname{int} K,\quad Z_0' + Z_0' \subset Z_0.$$

Analogously let V_0, V_0' be neighbourhoods of the origin in V so that

$$l_0 + V_0 \subset \operatorname{int} L,\quad V_0' - V_0' \subset V_0.$$

At last let U_0' be such a neighbourhood of the origin in U that $U_0' \subset U_0$ and

$$F(\gamma(u) - \Delta x_0) \in -Z_0',\quad H(\gamma(u) - \Delta x_0) \in V_0'$$

for all $u \in U_0'$, which is possible because of the continuity of $F\gamma$, $H\gamma$ at $u = 0$. Then the set

$$(F\Delta x_0 + k_0 + Z_0') \times U_0' \times (h(x_0) + H\Delta x_0 - l_0 + V_0')$$

turns out to be a neighbourhood of the point $(F\Delta x_0 + k_0, 0, h(x_0) + H\Delta x_0 - l_0)$ and a subset of $A - B$. In order to verify this, assume

$$z = F\Delta x_0 + k_0 + k_0' \text{ with arbitrary } k_0' \in Z_0',\ u \in U_0',$$

$$v = h(x_0) + H\Delta x_0 - l_0 + v_0' \text{ with arbitrary } v_0' \in V_0'.$$

Pose

$$\Delta x = \gamma(u),\ k = k_0 + F(\Delta x_0 - \gamma(u)) + k_0',$$

$$l = l_0 + H(\gamma(u) - \Delta x_0) - v_0'.$$

Then

$$u = G\Delta x,\ k \in k_0 + Z_0' + Z_0' \subset k_0 + Z_0 \subset \operatorname{int} K,$$

$$z = F\Delta x + k,\ l \in l_0 + V_0' - V_0' \subset l_0 + V_0 \subset \operatorname{int} L,$$

$$v = h(x_0) + H\Delta x - l,$$

hence $(z, u, v) \in A - B$.

Therefore 0 can be separated from $A - B$ by a hyperplane, i.e. there are continuous linear functionals z^*, u^*, v^* over Z, U, V, not all zero, so that

$$z^*(F\Delta x + k) + u^*G\Delta x - v^*(h(x_0) + H\Delta x - l) \geqq 0$$

holds for all $\Delta x \in \widetilde{\Delta\Omega}$, $k \in \operatorname{int} K$, $l \in \operatorname{int} L$, which implies the assertion of the theorem in a well known way.

3. Examples for the existence of a local right hand inverse

In this section we shall deal with the inversion problem formulated above (p. 43) and shall give some examples for the existence of a mapping γ.

a) At first let us consider a finite dimensional, say m-dimensional space U. Since $0 = G\Delta x_0 \in \operatorname{int} G\Delta\Omega$, there is a (nondegenerate) simplex $\sigma(u_0, u_1, \ldots, u_m)$, which contains 0 in its interior and is contained in $G\Delta\Omega$ together with the points $-u_1, \ldots, -u_m$. Suppose

$$\pm u_i = G\Delta x_i^\pm,\ \Delta x_i^\pm \in \Delta\Omega.$$

Denote max $\{0, \pm\lambda\}$ by $\lambda^\pm$ for each real λ. Every point $u \in U$ can be written as $\sum_{k=0}^{m} \lambda_k u_k$ with uniquely determined $\lambda_k = \lambda_k(u)$ satisfying $\sum \lambda_k = 1$, the points $u \in \sigma(u_0, \ldots, u_m)$ being distinguished by $\lambda_i \geqq 0$. Let be

$$0 = \sum_{k=0}^{m} \lambda_k^0 u_i,\ \sum_{k=0}^{m} \lambda_k^0 = 1,$$

thus $\lambda_k^0 > 0$. $\lambda_k = \lambda_k(u)$ depends continuously on u. We obtain

$$\begin{aligned} u &= \sum \lambda_k u_k = \sum (\lambda_k - \lambda_k^0)\, u_k = \sum [(\lambda_k - \lambda_k^0)^+ - (\lambda_k - \lambda_k^0)^-]\, u_k \\ &= \sum [(\lambda_k - \lambda_k^0)^+\, G\Delta x_k^+ + (\lambda_k - \lambda_k^0)^-\, G\Delta x_k^-] \\ &= G[(1 - \sum |\lambda_k - \lambda_k^0|)\, \Delta x_0 + \sum (\lambda_k - \lambda_0^k)^+\, \Delta x_k^+ + \sum (\lambda_k - \lambda_k^0)^-\, \Delta x_k^-] \end{aligned}$$

for each u. Let ϱ_0 be an arbitrary positive number < 1 and

$$\begin{aligned} U_0 &= \{u = \sum \lambda_k u_k \mid \lambda_k \geqq 0, \sum \lambda_k = 1, \sum |\lambda_k - \lambda_k^0| \leqq \varrho_0\}, \\ \gamma(u) &= \left(1 - \sum |\lambda_k(u) - \lambda_k^0|\right) \Delta x_0 + \sum \left(\lambda_k(u) - \lambda_k^0\right)^+ \Delta x_k^+ + \sum \left(\lambda_k(u) - \lambda_k^0\right)^- \Delta x_k^- \end{aligned}$$

for each $u \in U_0$. Then U_0 is a closed, convex neighbourhood of $0, \subset \sigma(u_0, \ldots, u_m) \subset G\Delta\Omega$ and γ a continuous mapping of U_0 into conv $(\Delta x_0, \Delta x_1^\pm, \ldots, \Delta x_m^\pm)$ and $G\gamma(u) = u$.

b) Now suppose that U is a Banach space with a Schauder base $u_1, u_2, \ldots$ so that each $u \in U$ may be written as $u = \sum_{i=1}^{\infty} \xi_i u_i$ with uniquely determined $\xi_i = \xi_i(u)$. Moreover suppose that $\|u\|$ and $\sum |\xi_i(u)|$ are equivalent norms (as for instance in $U = l_1$). Let $\Delta\Omega$ be a closed, convex cone with vertex 0, $G: X \to U$ be a linear bounded mapping with $G\Delta\Omega = U$. Suppose either that $\Delta\Omega$ has a nonempty interior and a $\Delta x_1 \in \operatorname{int} \Delta\Omega$ satisfies $G\Delta x_1 = 0$ or that there exists a neighbourhood U_1 of the origin in U and a corresponding bounded set $X_1 \subset \Delta\Omega$ with $GX_1 = U_1$ at once.

If any $\Delta x_1 \in \operatorname{int} \Delta\Omega$ satisfies $G\Delta x_1 = 0$, then such a set X_1 always exists. Namely, let X_1' be any open bounded neighbourhood of Δx_1 completely contained in $\Delta\Omega$. G maps X_1' into an open neighbourhood U_1' of 0. Let U_1 be a neighbourhood of 0 and a subset of U_1'. Then there is a certain subset X_1 of the bounded set $X_1' \subset \Delta\Omega$ so that $GX_1 = U_1$.

Evidently the norms $\|u_i\|$ are bounded, i.e. $\pm u_i$ are contained in a certain positive multiple αU_1 of U_1, hence $\pm u_i$ have inverse images $\Delta x_i^\pm \in G^{-1}(\pm u_i)$, contained in the same multiple αX_1 of the set X_1. Therefore

$$\pm u_i = G\Delta x_i^\pm ; \ \Delta x_i^\pm \in \Delta\Omega, \ \sup \|\Delta x_i^\pm\| < \infty.$$

Let Δx_0 be in $\Delta\Omega$ and satisfy $G\Delta x_0 = 0$, as required in the inversion problem. Assume

$$r \geqq \|\Delta x_0\|, \|\Delta x_i^\pm\|; \ \sum_{i=1}^{\infty} |\xi_i(u)| \leqq d\|u\| \qquad \forall\, u; \ 1 > \varrho_0 > 0$$

so that

$$U_0 := \left\{ u = \sum_{i=1}^{\infty} \xi_i u_i \,\middle|\, \sum_{i=1}^{\infty} |\xi_i| \leqq \varrho_0 \right\} \subset U_1.$$

Put

$$\gamma_n(u) = \left(1 - \sum_{i=1}^{n} |\xi_i(u)|\right) \Delta x_0 + \sum_{i=1}^{n} [\xi_i^+(u)\, \Delta x_i^+ + \xi_i^-(u)\, \Delta x_i^-],$$

$$\gamma(u) = \left(1 - \sum_{i=1}^{\infty} |\xi_i(u)|\right) \Delta x_0 + \sum_{i=1}^{\infty} [\xi_i^+(u)\, \Delta x_i^+ + \xi_i^-(u)\, \Delta x_i^-] \qquad \forall\, u \in U_0.$$

Obviously, $\gamma(u)$ is well defined, since $\sum [\xi_i^+(u) + \xi_i^-(u)] = \sum |\xi_i(u)| \leqq \varrho_0$ and $\|\Delta x_i^\pm\| \leqq r$. Further $\gamma(u) = \lim \gamma_n(u)$, $\gamma_n(u) \in \Delta\Omega$ for $u \in U_0$, hence $\gamma(u) \in \Delta\Omega$, each $\xi_i(\cdot)$ is a linear continuous functional. Let u' be any other point of U_0 like u, then

$$\|\gamma(u) - \gamma(u')\| \leqq \sum \big||\xi_i(u)| - |\xi_i(u')|\big| \, \|\Delta x_0\| + \sum |\xi_i^+(u) - \xi_i^+(u')| \, \|\Delta x_i^+\|$$
$$+ \sum |\xi_i^-(u) - \xi_i^-(u')| \, \|\Delta x_i^-\| \leqq 3r \sum |\xi_i(u) - \xi_i(u')| \leqq 3rd\, \|u - u'\|,$$

i.e. γ is continuous. Since G is continuous. we obtain

$$G\gamma(u) = \sum [\xi_i^+ u_i + \xi_i^-(-u_i)] = \sum \xi_i u_i \quad \text{with} \quad \xi_i = \xi_i(u) \text{ for } u \in U_0.$$

May be, in certain Banach spaces U a Hamel base $\{u_i\}_{i \in I}$ can be used in a similar way.

c) Thirdly let X be a Hilbert space and G as previously a linear continuous mapping from X onto U such that already $G\Delta\Omega = U$. $\Delta\Omega$ be a closed convex cone, $\Delta x_0 \in \Delta\Omega$ and $G\Delta x_0 = 0$. Then $G^{-1}u \cap \Delta\Omega \neq \emptyset$ for each $u \in U$. Moreover, $G^{-1}u \cap \Delta\Omega$ is convex and closed. The required mapping γ must have values $\gamma(u) \in G^{-1}u \cap \Delta\Omega$ for all u of some neighbourhood U_0 of 0, especially $\gamma(0)$ must be equal to Δx_0. Hence it seems to be reasonable to define $\gamma(u)$ as the projection of Δx_0 into $G^{-1}u \cap \Delta\Omega$, i.e. $\gamma(u)$ be this element of $G^{-1}u \cap \Delta\Omega$ that satisfies

$$\|\gamma(u) - \Delta x_0\| \leqq \|x - \Delta x_0\| \qquad \forall\, x \in G^{-1}u \cap \Delta\Omega$$

for all u of some neighbourhood U_0 of 0.

Evidently $\gamma(u)$ is well defined, $\gamma(0) = \Delta x_0$. γ appears to be continuous, if the set $G^{-1}u \cap \Delta\Omega$ depends on u continuously with respect to the Hausdorff set metric.

Denote the kernel of G by N and its orthogonal complement by M so that $X = N \oplus M$. Let G_M be the restriction of G to M. G_M is one-to-one and onto so that G_M^{-1} is a continuous linear mapping from U to M and $G^{-1}u = G_M^{-1}u + N$. $n \in N$ and $G_M^{-1}u + n \in \Delta\Omega$, if and only if $n \in (\Delta\Omega - G_M^{-1}u) \cap N$. The point $\gamma(u)$, defined above, now can be written as $\gamma(u) = G_M^{-1}u + n(u)$, where $n(u)$ is the unique point in $(\Delta\Omega - G_M^{-1}u) \cap N$ satifsying

$$\|n(u) - (\Delta x_0 - G_M^{-1}u)\| \leqq \|n - (\Delta x_0 - G_M^{-1}u)\|$$

with respect to all other n of this set.

4. Conditions guaranteeing the complete continuity of $o_g(\varepsilon\gamma(\cdot))$

Suppose that $\Delta\Omega$ is a c.c.tent for a subset Ω of X at x_0, thus the corresponding ω is a completely continuous mapping. Suppose that $g(x) = Ax + C(x)\ \forall\, x \in X$, where A denotes a linear bounded and C a completely continuous Fréchet-differentiable mapping from X to U. Put $G = g'(x_0) = A + C'(x_0)$. According to Krasnoselski $C'(x_0)$ is completely continuous too. Since g is differentiable at x_0, we have

$$g(x_0 + \Delta x) = g(x_0) + G\Delta x + \bar{o}_g(\Delta x),$$

where $\|\bar{o}_g(\Delta x)\|/\|\Delta x\| \to 0$ for $\Delta x \to 0$. Hence for each $\Delta x \in \Delta\Omega$

$$g(x_{\Delta x}) = g\big(x_0 + \Delta x + \omega(\Delta x)\big) = g(x_0) + G\Delta x + o_g(\Delta x),$$

where

$$o_g(\Delta x) = G\omega(\Delta x) + \bar{o}_g\big(\Delta x + \omega(\Delta x)\big).$$

Lemma. *If ω and C are completely continuous, γ continuous (from a neighbourhood of 0 in U to X), then $o_g(\varepsilon\gamma(\cdot))$ is completely continuous for each sufficiently small positive ε (in a certain neighbourhood of 0).*

Proof. Since γ is continuous at 0, there is a bounded neighbourhood U_0 of 0 such that $\|\gamma(u) - \gamma(0)\| < 1$, i.e. γ is bounded over U_0. Obviously $o_g\big(\varepsilon\gamma(\cdot)\big)$ is continuous over U_0 as the composition of continuous mappings.

$$\bar{o}_g(\Delta x) = g(x_0 + \Delta x) - G\Delta x = C(x_0 + \Delta x) - C(x_0) - C'(x_0)\,\Delta x,$$

which shows that $\bar{o}_g$ is completely continuous. If the sequence $\{u_n\}$ belongs to U_0, then $\{\varepsilon\gamma(u_n)\}$ is bounded, there is a subsequence $\{u_{n_k}\}$ so that $\big\{\omega\big(\varepsilon\gamma(u_{n_k})\big)\big\}$ is converging, and there is a further subsequence $\big\{u_{n_{k_l}}\big\}$ so that $\big\{\bar{o}_g\big(\varepsilon\gamma(u_{n_{k_l}}) + \omega\big(\varepsilon\gamma(u_{n_{k_l}})\big)\big)\big\}$, hence $\big\{o_g\big(\varepsilon\gamma(u_{n_{k_l}})\big)\big\}$ is converging.

This implies the complete continuity of $o_g\big(\varepsilon\gamma(\cdot)\big)$ over U_0. Similarily it can be shown that $o_g\big(\varepsilon\gamma(\cdot)\big)$ is completely continuous, if g is continuously differentiable, γ completely continuous (in a neighbourhood of $0 \in U$) and ω (only) continuous.

5. Examples for tents with completely continuous ω

a) Let Y be another Banach space, B a linear, bounded mapping from X to Y, C a completely continuous mapping from X to Y and

$$\varphi(x) = Bx + C(x) \qquad \forall\, x \in X.$$

Define

$$\Omega = \{x \in X \mid \varphi(x) = 0\}.$$

Suppose $x_0 \in \Omega$, C continuously Fréchet-differentiable (in the vicinity of x_0) and $\operatorname{im} \varphi'(x_0) = \operatorname{im} \big(B + C'(x_0)\big) = Y$. Define

$$\Delta\Omega = N := \ker \varphi'(x_0) = \{\Delta x \in X \mid \varphi'(x_0)\, \Delta x = 0\}.$$

Suppose N has a topological complement M so that $X = N + M$. Pose $\psi(n, m) = \varphi(x_0 + n + m)$ for $n \in N$, $m \in M$. Evidently the implicit function theorem guarantees the existence of a differentiable function ω, defined in a vicinity of 0 for $n = \Delta x \in \Delta\Omega$ with values in M and satisfying $\varphi\big(x_0 + \Delta x + \omega(\Delta x)\big) = 0$. We claim that ω is completely continuous in a small neighbourhood of 0 and that $x_{\Delta x} = x_0 + \Delta x + \omega(\Delta x) \in \Omega$ for $\Delta x \in \Delta\Omega$. For there is such a small neighbourhood of 0 that $\|\omega(\Delta x) - \omega(0)\| < 1$ for its elements Δx in $\Delta\Omega$. Let $\{\Delta x_k\}$ be a bounded sequence belonging to $\Delta\Omega$ and this neighbourhood, and ω_k be equal to $\omega(\Delta x_k)$. Denote the restriction of $\varphi'(x_0)$ to M by $\varphi'_M(x_0)$. Then

$$\varphi(x_0 + \Delta x_k + \omega_k) - \varphi(x_0) - \varphi'(x_0)\,(\Delta x_k + \omega_k) = -\varphi'(x_0)\,(\Delta x_k + \omega_k) = -\varphi'_M(x_0)\,\omega_k,$$

hence $\omega_k = -[\varphi'_M(x_0)]^{-1}\big(C(x_0 + \Delta x_k + \omega_k) - C(x_0) - C'(x_0)\,(\Delta x_k + \omega_k)\big)$. $\{\omega_k\}$ is bounded like $\{\Delta x_k\}$, thus there is a subsequence $\{\Delta x_{k_l} + \omega_{k_l}\}$ so that $C(x_0 + \Delta x_{k_l} + \omega_{k_l}) - C(x_0) - C'(x_0)\,(\Delta x_{k_l} + \omega_{k_l})$ converges to a certain element y and ω_{k_l} to $-[\varphi'_M(x_0)]^{-1}\, y$. A special case is at hand, if Y denotes a finite dimensional space, say R^m, φ an m-dimensional function with components φ_i ($i = 1, \ldots, m$) over X, if the φ_i are continuously differentiable and its gradients $\varphi_i'(x_0)$ are linearly independent. Put $B = 0$, $C(x) = \varphi(x)$ and $M = \operatorname{span}\{x_1, \ldots, x_m\}$, where $x_1, \ldots, x_m$ are such elements that $\varphi'(x_0)\, x_1, \ldots, \varphi'(x_0)\, x_m$ form a base for R^m.

b) Assume the same specification with respect to Y, B, C, φ, $N = \ker \varphi'(x_0)$, M as in a). Additionally assume that Y is partially ordered by means of a convex cone Y_+ (i.e. $y \geqq 0 \Leftrightarrow y \in Y_+$) and define

$$\Omega = \{x \in X \mid \varphi(x) \geqq 0\}.$$

Suppose $x_0 \in \Omega$. Define

$$\Delta\Omega = \{\Delta x \in X \mid \varphi(x_0) + \varphi'(x_0)\,\Delta x \geqq 0\}.$$

Pose $\psi(\Delta x, m) = \varphi(x_0 + \Delta x + m) - \varphi(x_0) - \varphi'(x_0)\,\Delta x$ for $\Delta x \in X$, $m \in M$. The implicit function theorem again guarantees the existence of a differentiable bounded function ω, defined in a small vicinity of 0 for $\Delta x \in X$ with values in M and satisfying $\psi\big(\Delta x, \omega(\Delta x)\big) = 0$. We claim that ω is completely continuous and that $x_{\Delta x} = x_0 + \Delta x + \omega(\Delta x) \in \Omega$ for $\Delta x \in \Delta\Omega$. In order to verify this, we consider a bounded $\{\Delta x_k\}$ and

$$\varphi(x_0 + \Delta x_k + \omega_k) - \varphi(x_0) - \varphi'(x_0)\,\Delta x_k - \varphi'(x_0)\,\omega_k = -\varphi'_M(x_0)\,\omega_k,$$

which implies

$$\omega_k = -[\varphi'_M(x_0)]^{-1}\big(C(x_0 + \Delta x_k + \omega_k) - C(x_0) - C'(x_0)\,\Delta x_k - C'(x_0)\,\omega_k\big).$$

Obviously there is a (common) subsequence $\{\Delta x_{k_l}\}$, $\{\omega_{k_l}\}$ so that $C(x_0 + \Delta x_{k_l} + \omega_{k_l})$, $C'(x_0)\,\Delta x_{k_l}$, $C'(x_0)\,\omega_{k_l}$ converge simultaneously. $x_{\Delta x} \in \Omega$, since $\varphi(x_{\Delta x}) = \varphi(x_0) + \varphi'(x_0)\,\Delta x$.

6. Existence of efficient points with the Neumann-Morgenstern property

Let M denote a partially ordered set and Min M the subset of minimal points of M. It is said that the Neumann-Morgenstern property holds, if for every $m \in M$ there is a $m' \in \operatorname{Min} M$ so that $m' \leqq m$. This property has proved to be of extreme importance for the derivation of optimality conditions of vector optimization problems. We want to point out the fact that the range $f(\Omega_a)$ or Min $f(\Omega_a)$ of a vector-valued cost function f has this property under certain elementary conditions.

Let Ω_a be a set or a topological space, Z a partially ordered topological space, f a mapping from Ω_a to Z. We call f lower (upper) semicontinuous if for each $z \in Z$ and each sequence $\{x_i\}_{i\in I} \subset \Omega_a$ converging to some $x \in \Omega_a$, the inequalities $f(x_i) \leqq z$ $\forall\, i$ $\big(f(x_i) \geqq z\ \forall\, i\big)$ imply $f(x) \leqq z$ $\big(f(x) \geqq z\big)$.[1] We call the ordering in Z *closed downward* (*upward*) if for each $z \in Z$ the segments $\{w \in Z \mid z \leqq w\}$ ($\{w \in Z \mid w \leqq z\}$) are closed subsets of Z. If f is a real-valued function, then the semicontinuity defined above, coincides with the common notion. If Z is a topological vector space and the ordering is compatible with the linear structure, then the ordering is both closed upward and downward, if the cone of nonnegative elements is closed. A continuous mapping f into Z is lower (upper) semicontinuous, if the ordering is closed upward (downward).

Theorem 2. *If* (i) $f(\Omega_a)$ *is a compact subset of* Z *and the ordering in* Z *is closed upward or* (ii) Ω_a *is a compact space and* f *lower semicontinuous, then there is always an efficient solution* x_e *for the vector minimum problem; moreover, for each nonefficient* $x_a \in \Omega_a$ *there is always such an efficient solution* x_e *that satisfies* $f(x_e) \leqq f(x_a)$*, but not* $f(x_a) \leqq f(x_e)$.

Proof. We present only the proof for (i), since it is similar for (ii). x_e is efficient if and only if $f(x_e)$ is minimal in $f(\Omega_a)$. Let Z_0 be a totally ordered subset of $f(\Omega_a)$. Take Z_0 as directed (index) set I by setting $i \triangleq i'$ for $i, i' \in Z_0 = I$, if $i' \leqq i$, where $\leqq$ denotes the original ordering in Z. Pose $z_i = i\ \forall\ i \in I$. Then $\{z_i\}_{i\in I}$ is a generalized sequence, which has a generalized subsequence $\{z_{i(j)}\}_{j\in J}$ converging to a certain $z_0 \in f(\Omega_a)$. Let i_0 be any index of I. There is a corresponding $j_0 \in J$ so that $i_0 \triangleq i(j)$ for all j with $j_0 \triangleq j$ (the ordering in J is denoted by the same symbol). Hence for $j_0 \triangleq j$ we obtain $z_{i(j)} \leqq z_{i_0}$, hence $z_0 \leqq z_{i_0}\ \forall\ i_0 \in I$, i.e. the chain $Z_0 = I$ has a lower bound in $f(\Omega_a)$. Due to a version of Zorn's lemma, for each given $z_a = f(x_a)$ there is a minimal point $z_e = f(x_e) \leqq z_a$. If z_a is not minimal, there is a $z \in f(\Omega_a)$ with $z \leqq z_a$, but not $z_a \leqq z$. With this z we may associate a minimal point z_e such that $z_e \leqq z$. Then $z_e \leqq z_a$, but not $z_a \leqq z_e$, otherwise $z_a \leqq z_e \leqq z$, which contradicts the choice of z. For (ii) one has to argue with the original images x_i of $z_i = f(x_i)$ in the same way as we did with the z_i (noticing that in a metric space there is an ordinary subsequence $\{x_{i(j_\nu)}\}_{\nu=1,2,\ldots}$ of $\{x_{i(j)}\}_{j\in J}$, which converges to the same x_0 as $\{x_{i(j)}\}_{j\in J}$).

References

[1] Boltjanski, W. G., Optimale Steuerung diskreter Systeme (translation from russ.). Akademische Verlagsgesellschaft, Leipzig 1976.

[2] Dubovitskii, A. J., A. A. Miljutin, Extremum problems under constraints (in russ.). Ž. Vyčisl. Mat. i Mat. Fiz. 5, (1965), 395–453.

[1] If Ω_a is a metric space, then $\{x_i\}$ may be an ordinary sequence.

[3] Dunford, N., J. T. Schwartz, Linear Operators, I. Interscience, New York 1958.
[4] Gähler S., Optimality conditions in polyoptimization. *In:* Theory of Nonlinear Operators (Proc. Internt. Summer School Berlin 1977), Abh. Akad. Wiss. DDR Jahrg. 1978 Nr. 6N, Akademie-Verlag, Berlin 1978, 353–355.
[5] Halkin, H., L. Neustadt, General necessary conditions for optimization problems. Proc. Nat. Acad. Sci. USA **56** (1966), 1066–1071.
[6] Krasnoselski, M. A., Topological methods in the theory of nonlinear integral equations (in russ.). Moscow 1956.
[7] Neustadt, L. V., An abstract variational theory with applications to a broad class of optimization problems. I. II. SIAM J. Control **4** (1966), 505–527 and **5** (1967), 90–137.
[8] Pšeničnyi, B. N., Notwendige Optimalitätsbedingungen (transl.). Akademische Verlagsgesellschaft, Leipzig 1972.
[9] Gähler, S., m-Verfeinerungen von Topologien. Math. Nachr. **73** (1976), 225–241.
[10] Kirsch, A., W. Warth, J. Werner, Notwendige Optimalitätsbedingungen. Springer-Verlag, Berlin 1978.

Address of the author:

Ernst-Moritz-Arndt-Universität Greifswald
Sektion Mathematik
22 Greifswald
Friedrich-Ludwig-Jahn-Str. 15a

On the speed of c_0-sequences in connection with a lemma of Toeplitz

Gottfried Bruckner, Berlin

The convergence of iterated processes has been considered for a long time. Since the last years also the speed of convergence is investigated, e.g. by Kluge and Telschow ([4, 5]) and Poljak [6].

Continuing the paper [1] we consider here the speed of convergence in an abstract context using a "classical" lemma of Toeplitz (cf. [2] or [3]) as a main tool. This lemma characterizes linear bounded transformations of the space c_0 of real sequences converging to zero. So far as I know this lemma has been used in [7] for the first time to obtain convergence assertions on stochastic iteration processes.

While in [4, 5] and [6] the question of the speed is seen under the aspect of evaluation in terms of standard sequences, we here (as well as in [1]) mainly ask for the dependence of the speed on the original process and the parameters of the iteration method (§ 3). In § 1 we give an introduction and in § 2 the main notions and definitions.

§ 1. To declare our approach we consider a simple example. Let be

H a real Hilbert space, $\|\cdot\|$ its norm, $(\cdot, \cdot)$ the scalar product,

T, S operators from H into H, gradients, Lipschitz continuous, T strongly monotone, S monotone,

u^* the (unique) solution of the variational inequality

$$(Tu, u - v) \leq 0, \quad v \in \{y \in H, Sy = 0\}.$$

To approximate u^* we consider the penalty method

$$(T + e_n S)\, u_n = 0, \quad n = 1, 2, \ldots, \quad e_n \to \infty,$$

and the iterated penalty method

$$v_0 \in H,\ v_n = A_n v_{n-1},\ n = 1, 2, \ldots,$$

where

$$A_n = I - \lambda_n (T + e_n S), \quad \lambda_n = \frac{2}{L(T) + e_n L(S) + c(T)}.$$

Let be

$$e_n = \frac{2c(T)\, f_n - L(T) - c(T)}{L(S)}.$$

Then A_n is strictly contractive with the contraction constant

$$k_n = 1 - 1/f_n,$$

u_n is the fixed point of A_n, and by induction we get

$$\|v_n - u_n\| \leqq \sum_{1 \leqq i \leqq n} \|u_i - u_{i-1}\| \, k_i \cdots k_n =_{\text{def.}} x_n .$$

We have

$$\|v_n - u^*\| \leqq \|v_n - u_n\| + \|u_n - u^*\| \leqq x_n + \|u_n - u^*\|.$$

If we assume that convergence and speed of the convergence of the original process are known we can consider these questions concerning $v_n \to u^*$ as reduced to the same questions concerning the sequence $x_n \to 0$. Therefore, in what follows sequences of the form

$$x_n = \sum_{1 \leqq i \leqq n} a_i k_i \cdots k_n$$

will be investigated.

Remark. Our example stands for a lot of other iteration procedures such as iterated projection, penalty or regularization methods for stationary and instationary problems (cf. [4, 5]) or iterated stochastic procedures (cf. [6]).

The following questions concerning the sequence x_n may be asked:

(1) Convergence. (Cf. [1, 4, 5, 6])
(2) Calculation of the speed s. (Cf. [1, 4, 5, 6])
(3) Dependence of $s(x_n)$ on a_n, k_n fixed. (Cf. [1] and § 3)
(4) Dependence of $s(x_n)$ on k_n. (§ 3)
(5) Determination of k_n if $s(x_n)$ and $s(a_n)$ are given. (The "inverse problem", communicated by R. Kluge.)

§ 2. Using "classical" theorems on bounded linear mappings in sequence spaces we first regive known convergence criteria. After this the speed of convergence will be characterized as an order relation in these spaces.

We consider the Banach spaces

$$c_0 = \{a = [a_1, a_2, \ldots], a_n \to 0\}, \ \|a\| = \text{l.u.b. } |a_i|,$$

$$cs = \left\{a = [a_1, a_2, \ldots], \sum_{1 \leqq i < \infty} a_i < \infty\right\}, \ \|a\| = \sup_n \left|\sum_{1 \leqq i < \infty} a_i\right|.$$

Let $T = (t_{ij})$ be a matrix of real numbers and

$$T[a_1, a_2, \ldots] = [x_1, x_2, \ldots]$$

where

$$x_n = \sum_{1 \leqq i < \infty} t_{in} a_i . \tag{1}$$

Theorem 1 (Toeplitz, *cf.* [2]). *T is a bounded linear mapping of c_0 into c_0 if and only if*

$$\sum_{1 \leqq i < \infty} |t_{in}| \leqq K, \quad n = 1, 2, \ldots, \tag{2}$$

$$t_{in} \to 0 \ (n \to \infty), \ i = 1, 2, \ldots \tag{3}$$

Theorem 2. *T is a bounded linear mapping from cs into* c_0 *if and only if* (3) *and*

$$|t_{ij}| \leqq K_1 \quad \textit{for all } i \textit{ and } j. \tag{4}$$

In the sequel we will only use the fact that (2) and (3) are sufficient conditions for T to map from c_0 into c_0.

For the application to iteration procedures the following special case is interesting: Let k_n be numbers with the property $0 < k \leqq k_n < 1$, $k_1 \cdots k_n \to 0$ $(n \to \infty)$. If

$$t_{in} = \begin{cases} (1 - k_i)\ k_i \cdots k_n, & i \leqq n, \\ 0, & i > n, \end{cases} \tag{6}$$

then (2) and (3) are fulfilled; if

$$t_{in} = \begin{cases} k_i \cdots k_n, & i \leqq n, \\ 0, & i > n, \end{cases} \tag{7}$$

then (3) and (4) are fulfilled.

Then we obtain the following known criteria (cf. e.g. [4]):

Corollary 1. *If T is of the form* (6) *then T maps* c_0 *into* c_0.

Corollary 2. *If T is of the form* (7) *then T maps cs into* c_0.

Speed of convergence of positive sequences of c_0

Let be

$$d = [d_1, d_2, \ldots] \ \in c_0,$$

$$e = [e_1, e_2, \ldots] \ \in c_0.$$

We say "d and e are of the same speed", i.e.

$$d_n \sim e_n$$

if

$$c_1 d_n \leqq e_n \leqq c_2 d_n, \quad c_1 > 0,\ c_2 > 0,$$

holds. (If d_n is "known" such as n^a or p^n then in the same way the speed of an arbitrary sequence e_n is defined.)

We say "d is faster than e", i.e.

$$d_n \succ e_n$$

if

$$d_n = o(e_n) \qquad (\text{i.e. } d_n/e_n \to 0).$$

We say "d is at least as fast as e", i.e.

$$d_n \succsim e_n$$

if $d_n \sim e_n$ or $d_n \succ e_n$ holds.

We call d and e comparable if a $\succsim$ relation holds. "$\succsim$" is an order relation and c_0 a partially ordered set, "$\sim$" is an equivalence relation.

Remark. Also another "weaker" order relation could be considered: $d \geqq e$ if $d_n = O(e_n)$. However we will not deal with it here.

§ 3. For the sake of simplicity we only consider here the special case (6) important for applications to iteration procedures. We will give qualitative answers to the questions (3) and (4) of § 1.

Let k_n be numbers with the property

$$0 < k \leqq k_n < 1, \quad k_1 \cdots k_n \to 0 \; (n \to \infty). \tag{8}$$

Then we have (cf. [1])

$$1 - k_n \sim \ln (1/k_n). \tag{9}$$

Let also k'_n be numbers with the property (8). We set

$$k'_n = k_n^{\lambda_n}, \tag{10}$$

then clearly

$$\ln (1/k'_n) = \lambda_n \ln (1/k_n) \tag{11}$$

and vice versa.

Lemma 1. (i) $c_1 \leqq \lambda_n \leqq c_2$, $c_1, c_2 > 0$, *if and only if* $\ln (1/k_n) \sim \ln (1/k'_n)$.

(ii) $\lambda_n \to 0$ *if and only if* $\ln (1/k_n) \succ \ln (1/k'_n)$.

(iii) $\lambda_n \to \infty$ *if and only if* $\ln (1/k_n) \prec \ln (1/k'_n)$.

Let $a_n \geqq 0$ be such that $a_n/(1 - k_n) \to 0$. Then by Corollary 1

$$x_n = \sum_{1 \leqq i \leqq n} a_i k_i \cdots k_n \to 0 \; (n \to \infty).$$

Proposition 1. (i) *If* $a_n \sim a'_n$, $k_1 \cdots k_n \sim k'_1 \cdots k'_n$ *then* $x_n \sim x'_n$. (ii) *Let* $1 - k_n$ *and* $1 - k'_n$ *be comparable. If* $a_n/(1 - k_n) \sim a'_n/(1 - k'_n)$, $k_1 \cdots k_n \sim k'_1 \cdots k'_n$ *then* $x_n \sim x'_n$.

Proof. From $c_1 k_1 \cdots k_n \leqq k'_1 \cdots k'_n \leqq c_2 k_1 \cdots k_n$ we find $(c_1/c_2)\, k_i \cdots k_n \leqq k'_i \cdots k'_n \leqq (c_2/c_1)\, k_i \cdots k_n$, and (i) follows immediately. If $1 - k_n$ and $1 - k'_n$ are comparable then via (9) one of the cases (i), (ii) or (iii) of Lemma 1 will be true. Because of $k_1 \cdots k_n \sim k'_1 \cdots k'_n$ this has to be case (i), i.e. $1 - k_n \sim 1 - k'_n$. Then $a_n \sim a'_n$, and the assertion follows from Proposition 1 (i).

Dependence of the speed of x_n on the speed of a_n

Proposition 2. *Let be* $x_n = \sum_{1 \leqq i < \infty} t_{in} a_i$, $x'_n = \sum_{1 \leqq i < \infty} t_{in} a'_i$. *If* $a'_n/a_n \to 0$, $t_{in}/x_n \to 0$ $(n \to \infty)$, $i = 1, 2, \ldots$, *then* $x'_n/x_n \to 0$.

Proof. We have

$$x'_n/x_n = (1/x_n) \sum_{1 \leqq i < \infty} t_{in} a'_i = \sum_{1 \leqq i < \infty} (t_{in} a_i/x_n)\, (a'_i/a_i) = \sum_{1 \leqq i < \infty} t^*_{in} a^*_i,$$

and

$$\sum_{1 \leqq i < \infty} t^*_{in} = \sum_{1 \leqq i < \infty} t_{in} a_i/x_n = 1, \; t^*_{in} = a_i (t_{in}/x_n) \to 0 \; (n \to \infty).$$

Using Theorem 1 we get the assertion.

If $a_n \geqq 0$, $n = 1, 2, \ldots$, $a_i > 0$ then generally $t_{in} \leqq x_n/a_i$, i.e. $t_{in} = O(x_n)$ holds. As we have in our special case (6)

$$x_n = \sum_{1 \leqq i < \infty} a_i k_i \cdots k_n = k_1 \cdots k_n \sum_{1 \leqq i < \infty} a_i / k_1 \cdots k_{i-1} = k_1 \cdots k_n d_n,$$

where d_n is monotonely increasing, we obtain the sharper relation

$$x_n \lesssim k_1 \cdots k_n =_{\text{def.}} g_n. \tag{12}$$

From this either

$$x_n \prec k_i \cdots k_n, \; i = 1, 2, \ldots, \tag{13}$$

or

$$x_n \sim k_i \cdots k_n, \; i = 1, 2, \ldots \tag{14}$$

In case (14) we have also $x'_n \sim k_i \cdots k_n$, as $x'_n \prec k_i \cdots k_n$ cannot be. Therefore $x_n \sim x'_n$. And in case (13) Proposition 2 gives $x_n \prec x'_n$.

Corollary. *A transformation T of the kind* (6) *is chain preserving, i.e.* $[a] \lesssim [a']$ *implies* $T[a] \lesssim T[a']$.

If $h_n =_{\text{def.}} a_n/(1 - k_n) \sim d_n$, d_n monotone, then

$$h_n = O(x_n) \tag{15}$$

(cf. [1]). We will see that under additional assumptions the stronger relation $x_n \lesssim h_n$ holds.

Let k_n, h_n, g_n be given as

$$k_n = k(n), \; g_n = g(n), \; h_n = h(n)$$

where k, g, h are monotone differentiable functions. Let further be

$$K(x) = \ln\left(1/k(x)\right), \quad f(x) = \int_{c_1}^{x} K(y)\, dy,$$

c_1 a suitable positive constant. Then

$$f'(x) = K(x).$$

We have

$$\frac{1}{g_n} = e^{\ln \frac{1}{g_n}} = e^{\sum_{1 \leqq i \leqq n} \ln \frac{1}{k_i}} \sim e^{\int_{c_1}^{n} \ln \frac{1}{k(x)}} = e^{f(n)}$$

and more general

$$g(x) \sim e^{-f(x)} \quad (x \to \infty).$$

We assume that h_n and g_n are comparable, i.e.

$$h(x) = H\left(f(x)\right) \tag{16}$$

where

$$H(z) = e^{-z} d(z). \tag{17}$$

We assume further that d is strictly monotone and differentiable converging

$$\text{(i) to } \infty, \quad \text{(ii) to } 0 \quad (\text{if } z \to \infty).$$

(The case $0 < d_1 \leqq d(z) \leqq d_2 < \infty$ implies $h_n \sim g_n$ and $x_n \sim h_n \ln(1/h_n)$). Let $H(z)$ be strictly monotone ($\to 0$ if $z \to \infty$). Then

$$H'(z) = e^{-z}\big(d'(z) - d(z)\big) \tag{18}$$

is negative, i.e.

$$d(z) > d'(z). \tag{19}$$

Let us consider

$$b(z) = \frac{|H'(z)|}{H(z)} = 1 - \frac{d'(z)}{d(z)}. \tag{20}$$

In the case (i) $d'(z)$ is positive, i.e. $0 < b(z) < 1$.
In the case (ii) $d'(z)$ is negative, i.e. $b(z) > 1$.
Furthermore, the involved functions are supposed to be such that the construction (21) is possible. Then we have the

Proposition 3. *Let be either $b(z) \leqq b < 1$ (if $z > z_0$) or $b(z)$ monotone. Then $x_n \lesssim h_n$ holds.*

Proof. We have

$$x_n = \sum_{1 \leqq i \leqq n} a_i k_i \cdots k_n = g_n \sum_{1 \leqq i \leqq n} \frac{a_i}{g_{i-1}} \sim g_n \sum_{1 \leqq i \leqq n} h_i \ln \frac{1}{k_i} \frac{1}{g_i}$$

and write it as

$$x_n \sim e^{-f(n)} \int_{c_2}^{n} H\big(f(y)\big) \cdot f'(y) \cdot e^{f(y)}\, dy, \tag{21}$$

c_2 a suitable positive constant. Substituting $z = f(y)$ we obtain

$$x_n \sim e^{-f(n)} \int_{f(c_2)}^{f(n)} H(z)\, e^z\, dz.$$

We get by partial integration

$$\int_c^w H(z)\, e^z\, dz = H(z)\, e^z\big|_c^w + \int_c^w |H'(z)|\, e^z\, dz$$

and by the mean value theorem

$$\int_c^w |H'(z)|\, e^z\, dz = \int_c^w b(z)\, H(z)\, e^z\, dz = l(w) \int_c^w H(z)\, e^z\, dz$$

where either $l(w) \leqq l < 1$ or $l(w)$ monotone, and $0 < l(w) < 1$ in the case (i) and $l(w) > 1$ in the case (ii). Consequently,

$$\int_c^w H(z)\, e^z\, dz = \frac{H(w)\, e^w - C_1}{1 - l(w)}.$$

Hence

$$x_n \sim e^{-f(n)} \frac{Hf((n))\, e^{f(n)} - C_1}{1 - l(f(n))} = H(f(n)) \frac{1 - \dfrac{C_1}{e^{f(n)} H(f(n))}}{1 - l(f(n))} \sim h_n \frac{1}{1 - l(f(n))} \sim h_n l_n,$$

l_n monotonely non-decreasing, in the case where $l(x) < 1$. In the case (ii) we have

$$x_n \sim e^{-f(n)} \cdot \frac{C_1 - H(f(n)) \cdot e^{f(n)}}{1 - l(f(n))} = H(f(n))\, d_n,$$

where

$$d_n = \frac{\dfrac{C_1}{d(f(n))} - 1}{l(f(n)) - 1} \geqq \frac{1}{-d'(f(n))} \left(C_1 - d(f(n))\right) \geqq \frac{C_2}{-d'(f(n))} \to \infty \ (n \to \infty),$$

as $-d'(s) \to 0$ $(x \to \infty)$, $d(x) \to 0$ $(x \to \infty)$, if b is monotonely increasing. In the case where b is monotonely decreasing, also $d(x)\,(l(x) - 1) \to 0$, i.e. $d_n \to \infty$. Qed.

Remark. We obtain from the proof together with (19) and (20) under the assumptions of Proposition 3:

$$x_n \sim h_n$$

holds if and only if there is a c such that

$$d(z) > d'(z) \geqq c d(z) \quad (z \geqq z_0),\ 0 < c < 1,$$

holds. Examples are $d(x) = e^x/x^a$, $a > 0$; e^{bx}/x^a, $0 < b < 1$, $a \geqq 0$.

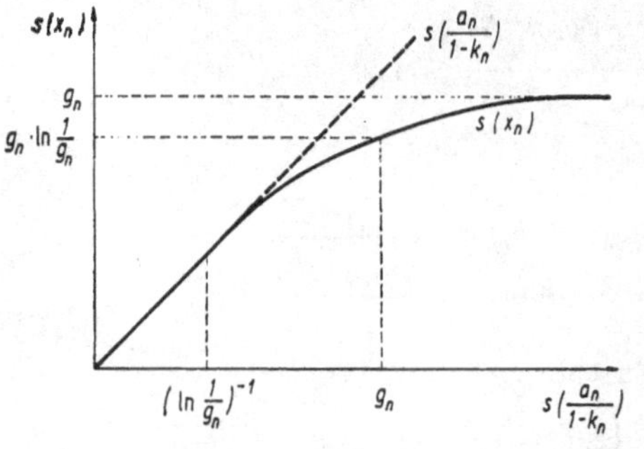

Fig. 1

If the assumptions of proposition 3 are fulfilled we may draw $s(x_n)$ (the speed of x_n) as a monotone curve over $s(h_n,)$ k_n being fixed. This is done in Fig. 1.

Dependence of $s(x_n)$ on k_n

We replace k_n by

$$k'_n = k_n^{\lambda_n}$$

and generally assume

$$g'_n \succ g_n, \quad \text{i.e. } \frac{k'_1 \cdots k'_n}{k_1 \cdots k_n} \to 0. \tag{22}$$

We consider the cases

(i) $\lambda_n \to \infty$ (i.e. $1 - k_n \succ 1 - k'_n$),

(ii) $1 + \lambda \leqq \lambda_n \leqq \lambda_1,\ \lambda > 0$

(iii) $1 < \lambda_n \leqq \lambda_1$

(ii), (iii): (special cases of $1 - k_n \sim 1 - k'_n$).

We put

$$x_n = \sum_{1 \leqq i \leqq n} a_i k_i \cdots k_n,$$

$$x'_n = \sum_{1 \leqq i \leqq n} a_i k'_i \cdots k'_n,$$

$$y_n = \sum_{1 \leqq i \leqq n} a_i [(1 - k'_i)/(1 - k_i)]\, k'_i \cdots k'_n.$$

Case (i)

At first we remark that by Proposition 2

$$y_n \prec x'_n \quad \text{if} \quad y_n \prec g'_n, \tag{23}$$

$$y_n \sim x'_n \quad \text{if} \quad y_n \sim g'_n. \tag{24}$$

Proposition 4. *If $x_n \prec a_n/(1 - k_n)$ then $x_n \prec y_n$.*

If $x_n \sim a_n/(1 - k_n)$ then $x_n \sim y_n$.

Proof.

$$\frac{y_n}{x_n} = \sum_{1 \leqq i \leqq n} a_i \frac{1 - k'_i}{1 - k_i} \cdot \frac{1}{x_n} \cdot k'_i \cdots k'_n = \sum_{1 \leqq i \leqq n} \frac{a_i}{(1 - k_i)\, x_i} \cdot (1 - k_i) \frac{x_i k_{i+1} \cdots k_n}{x_n} \cdot k'_i \cdot \frac{k'_{i+1} \cdots k'_n}{k_{i+1} \cdots k_n}.$$

We have

$$\frac{x_i}{x_n} k_{i+1} \cdots k_n = \frac{\sum\limits_{1 \leqq \nu \leqq n} a_\nu k_\nu \cdots k_i k_{i+1} \cdots k_n}{\sum\limits_{1 \leqq \nu \leqq n} a_\nu k_\nu \cdots k_n} \leqq 1.$$

Hence

$$\frac{y_n}{x_n} \leqq c \cdot \sum_{1 \leqq i \leqq n} \underbrace{\frac{a_i}{(1 - k_i)\, x_i}}_{a'_i} \cdot \underbrace{(1 - k'_i) \cdot k'_i \cdot \frac{k'_{i+1} \cdots k'_n}{k_{i+1} \cdots k_n}}_{t'_{in}}. \tag{25}$$

By (22) we have $t'_{in} \to 0$ $(n \to \infty)$. Further

$$\sum_{1 \leqq i \leqq n} t'_{in} = \sum_{1 \leqq i \leqq n} (1 - k_i^{\lambda_i})\, k_i^{\lambda_i} k_{i+1}^{\lambda_{i+1} - 1} \cdots k_n^{\lambda_n - 1}.$$

Let n_0 be such that

$$\lambda_n > 1,\ n \geqq n_0, \text{ and } \lambda_n \leqq (\lambda_n - 1) \cdot \varkappa,\ n \geqq n_0.$$

(E.g., if $\varkappa = 2$ then $\lambda_n \geqq 2$ for $n \geqq n_0$.) Then

$$\sum_{1 \leqq i \leqq n} t'_{in} = \sum_{1 \leqq i \leqq n_0} t'_{in} + \sum_{n_0 \leqq i \leqq n} t'_{in} \leqq C + \sum_{n_0 \leqq i \leqq n} (1 - k_i^{(\lambda_i - 1)\varkappa})\, k_i^{\lambda_i - 1} \cdots k_n^{\lambda_n - 1} \leqq C_1 + \sum_{1 \leqq i \leqq n} (1 - k_i^{*\varkappa})\, k_i^* \cdots k_n^*$$

where $k_i^* \leqq 1$. We have for $k \leqq 1$

$$1 - k^\varkappa \leqq \begin{cases} \varkappa(1 - k) & \text{if} \quad \varkappa \geqq 1 \\ (1 - k) & \text{if} \quad 0 \leqq \varkappa \leqq 1. \end{cases} \tag{26}$$

Hence

$$\sum_{1\leqq i\leqq n} t'_{in} \leqq C_1 + C_2 \sum_{1\leqq i\leqq n} (1 - k_i^*)\, k_i^* \cdots k_n^* \leqq C\,. \tag{27}$$

As $a'_n \to 0$ Theorem 1 gives the assertion.

If $x_n \sim a_n/(1 - k_n)$, then from (25) and (27) we get $y_n/x_n \leqq c_1$. Otherwise, if $a_n/(1 - k_n) \sim d_n$, d_n monotone, then $y_n \geqq c_2 a_n/(1 - k_n)$ by (15). This means together $y_n \sim x_n$. Qed.

In the cases (ii) and (iii) clearly $y_n \sim x'_n$ holds.

Case (ii)

Proposition 5. *If $x_n \prec a_n/(1 - k_n)$ then $x_n \prec x'_n$.*
If $x_n \sim a_n/(1 - k_n)$ then $x_n \sim x'_n$.

Proof. We put $\lambda_n = 1 + \varepsilon_n$, then $\varepsilon_n \geqq \varepsilon > 0$.

$$\frac{x'_n}{x_n} = \sum_{1\leqq i\leqq n} \frac{a_i}{(1 - k_i)\, x_i} \cdot (1 - k_i) \cdot \frac{x_i}{x_n} \cdot k'_i \cdots k'_n$$

$$= \sum_{1\leqq i\leqq n} \frac{a_i}{(1 - k_i)\, x_i} (1 - k_i) \frac{\sum\limits_{1\leqq \nu\leqq i} a_\nu k_\nu \cdots k_i k_{i+1} \cdots k_n}{\sum\limits_{1\leqq \nu\leqq i} a_\nu k_\nu \cdots k_n} \cdot k_i^{1+\varepsilon_i} \cdot k_{i+1}^{\varepsilon_{i+1}} \cdots k_n^{\varepsilon_n}$$

$$\leqq \sum_{1\leqq i\leqq n} \underbrace{\frac{a_i}{(1 - k_i)\, x_i}}_{a'_i} \underbrace{(1 - k_i)\, k_i^{\varepsilon_i} \cdots k_n^{\varepsilon_n}}_{t'_{in}}.$$

Clearly, $t'_{in} \to 0$ $(n \to \infty)$ as $\varepsilon_n \geqq \varepsilon > 0$.

$$\sum_{1\leqq i\leqq n} t'_{in} = \sum_{1\leqq i\leqq n} (1 - k_i)\, k_i^{\varepsilon_i} \cdots k_n^{\varepsilon_n} \leqq \sum_{1\leqq i\leqq n} (1 - k_i)\, k_i^{\varepsilon} \cdots k_n^{\varepsilon} \leqq c_1$$

(cf. the proof of Proposition 4). As $a'_n \to 0$ the first assertion follows by Theorem 1. If $x_n \sim a_n/(1 - k_n)$ we see in the same way that $x'_n/x_n \leqq c$. If $a_n/(1 - k_n)$ is monotone, $a_n/(1 - k_n) \leqq c_1 x'_n$. From both we obtain the second assertion. Qed.

Remark. The assumption $\varepsilon_n \geqq \varepsilon > 0$ is only needed for $n \geqq n_0$.

Case (iii)

We put again $\lambda_n = 1 + \varepsilon_n$, $\varepsilon_n > 0$. To guarantee that $g'_n/g_n \to 0$ we have to assume

$$k_1^{\varepsilon_1} \cdots k_n^{\varepsilon_n} \to 0 \quad (n \to \infty)\,. \tag{28}$$

This is true if and only if

$$\sum_{1\leqq n\leqq \infty} \varepsilon_n (1 - k_n) = \infty\,. \tag{29}$$

Furthermore, we assume

$$\varepsilon_n \to 0\,. \tag{30}$$

We have

$$\varepsilon_n (1 - k_n) \sim \varepsilon_n \ln \frac{1}{k_n} = \ln \frac{k_n}{k'_n} \sim 1 - \frac{k'_n}{k_n} \tag{31}$$

as $k'_n/k_n = k_n^{\varepsilon_n} \geqq k_n \geqq \varepsilon_1 > 0$ for $n \geqq n_0$. By (30) and (31)

$$1 - k'_n/k_n \succ 1 - k_n \quad \text{and} \quad a_n/(1 - k'_n/k_n) \prec a_n/(1 - k_n).$$

Proposition 6. *Let be* $k'_n = k_n^{1+\varepsilon_n}$, $\varepsilon_n > 0$, $\varepsilon_n \to 0$, $\sum_{1 \leqq n \leqq \infty} \varepsilon_n(1 - k_n) = \infty$. *If* $x_n \prec a_n/\varepsilon_n(1 - k_n)$ *then* $x_n \prec x'_n$. *If* $x_n \sim a_n/(1 - k_n)$ *then* $x_n \sim x'_n$.

Proof.

$$\frac{x'_n}{x_n} = \sum_{1 \leqq i \leqq n} \underbrace{\frac{a_i}{(1 - k_i^{\varepsilon_i})\, x_i}}_{a'_i} \underbrace{(1 - k_i^{\varepsilon_i}) \frac{\sum_{1 \leqq \nu \leqq i} a_\nu k_\nu \cdots k_i k_{i+1} \cdots k_n}{\sum_{1 \leqq \nu \leqq n} a_\nu k_\nu \cdots k_n} k_i^{1+\varepsilon_i} \cdot k_{i+1}^{\varepsilon_{i+1}} \cdots k_n^{\varepsilon_n}}_{t'_{in}}.$$

We have $t'_{in} \leqq (1 - k_i^{\varepsilon_i})\, k_i^{\varepsilon_i} \cdots k_n^{\varepsilon_n} \to 0$ $(n \to \infty)$, $\sum_{1 \leqq i \leqq n} t'_{in} \leqq \sum_{1 \leqq i \leqq n} (1 - k_i^*)\, k_i^* \cdots k_n^* \leqq 1$ and $a'_n \to 0$ by the assumption. Then Theorem 1 implies $x'_n/x_n \to 0$. Clearly we have $x'_n \leqq x_n$. The second assertion follows as in Proposition 5.

Proposition 6 says nothing about the behaviour of x'_n in the segment

$$a_n/\varepsilon_n(1 - k_n) \precsim x_n \prec a_n/(1 - k_n). \tag{32}$$

We consider the following

Example. *Let be* $k_n = k < 1$, $\varepsilon_n = 1/n$, $a_n = k^n$. *Then* $x_n \sim a_n/\varepsilon_n(1 - k_n)$ *and* $x'_n \sim x_n$.

Proof. Clearly (29) and (30) are fulfilled, so we are in the case (iii). We have $a_n/(1 - k_n) = k^n/(1 - k) \sim g_n$, hence $x_n \sim nk^n$, and $a_n/\varepsilon_n(1 - k_n) = nk^n/(1 - k) \sim x_n$. We have further

$$x'_n = \sum_{1 \leqq i \leqq n} a_i k'_i \cdots k_n = \sum_{1 \leqq i \leqq n} k^i \cdot k^{n-i+1+\sum_{i \leqq \nu \leqq n} \frac{1}{\nu}} = k^{n+1} \sum_{1 \leqq i \leqq n} k^{\left(\sum_{i \leqq \nu \leqq n} \frac{1}{\nu}\right)}$$

and

$$\frac{x'_n}{x_n} \sim \frac{1}{n} \sum_{1 \leqq i \leqq n} k^{\left(\sum_{i \leqq \nu \leqq n} \frac{1}{\nu}\right)}.$$

As

$$\frac{1}{n} + \ln \frac{n}{i} \leqq \sum_{i \leqq \nu \leqq n} \frac{1}{\nu} \leqq \frac{1}{i} + \ln \frac{n}{i}$$

we get

$$c_1 \cdot \frac{1}{n} \sum_{1 \leqq i \leqq n} k^{\frac{1}{i} + \ln \frac{n}{i}} \leqq \frac{x'_n}{x_n} \leqq c_2 \cdot \frac{1}{n} \sum_{1 \leqq i \leqq n} k^{\frac{1}{n} + \ln \frac{n}{i}}$$

or (as $k^{\frac{1}{i}} \geqq k$, $k^{\frac{1}{n}} \leqq 1$)

$$c_1 \frac{k}{n} \sum_{1 \leqq i \leqq n} k^{\ln \frac{n}{i}} \leqq \frac{x'_n}{x_n} \leqq c_2 \cdot \frac{1}{n} \sum_{1 \leqq i \leqq n} k^{\ln \frac{n}{i}}.$$

Further

$$k^{\ln \frac{n}{i}} = \left(\frac{n}{i}\right)^{\ln k} = \left(\frac{i}{n}\right)^{\ln \frac{1}{k}}.$$

Hence

$$\frac{1}{n} \sum_{1 \leq i \leq n} k^{\ln \frac{n}{i}} = \frac{1}{n \cdot n^{\ln \frac{1}{k}}} \sum_{1 \leq i \leq n} i^{\ln \frac{1}{k}} = \frac{1}{n \cdot n^a} \sum_{1 \leq i \leq n} i^a, \ a > 0.$$

As

$$\sum_{1 \leq i \leq n} i^a \geq \int\limits_0^n x^a \, dx = \frac{n^{a+1}}{a+1},$$

we obtain

$$\frac{1}{n} \sum_{1 \leq i \leq n} k^{\ln \frac{n}{i}} \geq \frac{1}{n^{a+1}} \frac{n^{a+i}}{a+1} = \frac{1}{a+1}.$$

Consequently

$$\frac{x_n'}{x_n} \geq c \cdot \frac{1}{a+1} > 0, \quad \text{Qed.}$$

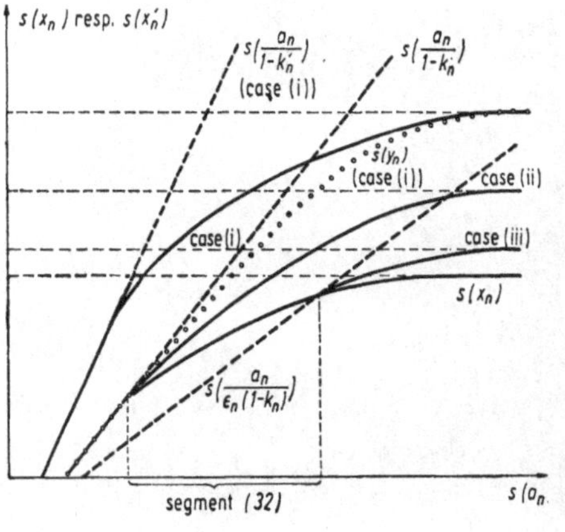

Fig. 2

To illustrate the Propositions 4, 5, 6 in Fig. 2 we draw $s(x_n')$ over $s(a_n)$ in the cases (i), (ii) and (iii). We see that for slow a_n $s(x_n')$ is identical with $s(x_n)$ and then somewhere "branches" from it. If g_n' becomes slower this "branching point" goes to the right.

Besides, the reflections show that the "inverse problem" (5) will in general have no unique solution.

References

[1] G. Bruckner, On the speed of convergence of iteration methods, Abh. Akad. Wiss. DDR 1978 Nr. 6N, 309–314.

[2] N. Dunford and J. T. Schwartz, Linear Operators, Part I, New York 1958.

[3] G. M. Fichtenholz, Differential- und Integralrechnung II, Deutscher Verlag der Wissenschaften, Berlin 1964.

[4] R. Kluge and G. Telschow, On the convergence and speed of convergence of some iteration methods for variational inequalities I, Abh. Akad. Wiss. DDR 1977, Nr. 1N, 135–163.

[5] — II, Math. Nachr. 75, 73–89 (1976).

[6] Б. Т. Поляк, Сходимость и скорость сходимости итеративных стохастических алгоритмов I, Автомат. и Телемех. **12**, 83—94 (1976).
[7] Я. З. Цыпкин, А. Й. Калинский, К. А. Ларионов, Алгоритмы адаптации и обучения в нестационарных условиях, Изв. АН. СССР, Техническая кибернетика **5**, 9—21 (1970).

Address of the author:

Akademie der Wissenschaften der DDR
Institut für Mathematik
1080 Berlin
Mohrenstr. 39
DDR

On some optimal control problems for elliptic systems in the plane

Lothar v. Wolfersdorf, Freiberg

Abstract. Necessary optimality conditions are derived for control problems of distributed and boundary control with semilinear first order elliptic systems in the plane under additional constraints insuring the correctness of the corresponding linearized boundary value problem.

1. Introduction

In our papers [1, 2] we derived necessary optimality conditions for control processes with nonlinear singular integral equations and with semilinear first order elliptic systems written in I. N. Vekua's complex normal form, where in general the corresponding linearized problems are not uniquely solvable but are problems of Noetherian type. Now we consider related optimal control problems for such elliptic systems with additional constraints insuring the correctness of the linearized boundary value problems important from the point of view of applications. There are several possibilities for obtaining a correctly posed linearized problem by adding integral or point conditions to the boundary condition with nonnegative index (cf. the discussion in [3], Chap. IV, § 6). In the sequel we restrict ourselves to two such possibilities often used in recent papers on this subject, the first one going back to the work of W. Haack, G. Hellwig, and W. Wendland (cf. [4, 5]) and the second one to that of I. N. Vekua (cf. [3], loc. cit.) on the linear Riemann-Hilbert boundary value problems for generalized analytic functions (cf. also J. Nitsche [6]). The derivation of necessary optimality conditions for these problems is again based on the theory in [1]. But in contrast to the approach there, which uses the reduction of the boundary value problems to singular integral equations, we directly employ suitably defined adjoint boundary value problems as we have done in the second method of [2] for problems without additional constraints.

By means of other methods necessary optimality conditions for some problems of distributed control with systems of complex differential equations in case of Dirichlet boundary conditions with additional point condition on the boundary of the domain have been derived by W. Tutschke [7] and H.-K. Klink [8] without introducing a corresponding adjoint boundary value problem.

2. Statement of problem

Let G be a bounded simply connected region in complex z-plane of class C_μ^1, $0 < \mu < 1$, with boundary Γ, i.e. for Γ the representation $t = t(s)$, s arc length, holds with $t'(s) \in C_\mu(\Gamma)$. (In the sequel we denote complex Hölder spaces by C_μ and real ones by H_μ, analogously we distinguish between spaces L_∞ and M_∞ of bounded measurable functions over the field of real and complex numbers, respectively). The complex state function $W(z) = w(x, y) + iv(x, y) \in C_\mu(\overline{G}) \cap D_{\bar{z},\infty}(G)$, where $D_{\bar{z},\infty}(G)$ means the class of complex-valued functions $W(z)$ in G with generalized derivative $\partial_{\bar{z}} W \in M_\infty(G)$ (cf. [3], Chap. I), satisfies the *semilinear first order differential equation*

$$\partial_{\bar{z}} W = \Psi(z, \bar{z}, W, \overline{W}, U) \quad \text{in} \quad G \tag{1}$$

with a (for simplicity) real-valued scalar distributed control function $U(z) \in L_\infty(G)$ and the *linear boundary condition*

$$\alpha(t)\, w(t) + \beta(t)\, v(t) \equiv \operatorname{Re}\left[\overline{\lambda(t)}\, W\right] = h(t) \quad \text{on} \quad \Gamma. \tag{2}$$

The given complex-valued function $\Psi(z, \bar{z}, W, \overline{W}, U)$ is supposed to be continuous in its arguments possessing continuous partial derivatives of first order Ψ_W, $\Psi_{\overline{W}}$, and Ψ_U. The given complex-valued function $\lambda(t) = \alpha(t) + i\beta(t) \in C_\mu(\Gamma)$ satisfies the normality condition $\lambda(t) \neq 0$ on Γ and has nonnegative index

$$n = \frac{1}{2\pi} [\arg \lambda(t)]_\Gamma \geqq 0. \tag{3}$$

The real-valued function $h(t) \in H_\mu(\Gamma)$ has the form

$$h(t) = f(t) + g(t)\, u(t) \tag{4}$$

with given functions $f(t)$, $g(t) \in H_\mu(\Gamma)$ and an additional real-valued scalar boundary control function $u(t) \in H_\mu(\Gamma)$.

The *admissible control functions U and u* are elements of arbitrary given convex subsets V_1 and V_2 of $L_\infty(G)$ and $H_\mu(\Gamma)$, respectively. In particular the values of U and u may be taken from prescribed (non-degenerate) finite or infinite intervals of real axis I_k, $k = 1, 2$, respectively:

$$U(z) \in I_1 \quad \text{for a.a.} \quad z \in G, \qquad u(t) \in I_2 \quad \text{for all} \quad t \in \Gamma. \tag{5}$$

Remark. In analogous way the case of *vector control functions U and u* with values in prescribed convex domains of corresponding Euclidean spaces can be dealt with.

Moreover we pose the *additional conditions* on the state function W:

$$W(z_i) = \varkappa_i, \qquad i = 1, \ldots, n, \tag{6}$$

where z_i, $i = 1, \ldots, n$, are given fixed points in (the interior of) G and $\varkappa_i$, $i = 1, \ldots, n$, are given complex numbers, and either in case I

$$\int_\Gamma \operatorname{Im}\left[\overline{\lambda(t)}\, W\right] \varrho(t)\, ds = \mu, \tag{7}$$

where $\varrho(t) \in H_\mu(\Gamma)$ is a given positive function on Γ and μ is a given real number, or in case II

$$\operatorname{Im}\left[\overline{\lambda(t_0)}\, W(t_0)\right] = \mu_0, \tag{8}$$

where t_0 is a prescribed point on Γ and again μ_0 is a given real number. In case II there may be given also $r \leqq n$ point conditions of form (6) and $2(n - r) + 1$ ones of form (8). (For a generalization of case I to $2n + 1$ prescribed integral conditions of form (7) see J. Nitsche [6]).

Aim of control is the minimization of a *cost functional* $J = J_1 + J_2$ with

$$J_1 = \iint_G F_1(z, \bar{z}, W, \overline{W}, U)\, dx\, dy, \tag{9}$$

where the given real-valued function $F_1(z, \bar{z}, W, \overline{W}, U)$ is continuous in its arguments possessing continuous partial derivatives of first order with respect to W, $\overline{W}$, and U[1] and

$$J_2 = \int_\Gamma F_2\big(t, w(t), v(t), u(t)\big)\, ds, \tag{10}$$

where the given real-valued function $F_2(t, w, v, u)$ is continuous in its arguments and possesses continuous partial derivatives of first order with respect to w, v, and u, the first two satisfying Hölder-Lipschitz conditions of the form

$$\begin{aligned} &|G(t, w_1, v_1, u_1) - G(\tau, w_2, v_2, u_2)| \\ &\leqq A[|t - \tau|^\mu + |w_1 - w_2| + |v_1 - v_2| + |u_1 - u_2|], \qquad A = \text{const}, \end{aligned} \tag{11}$$

in any bounded domain of variables.

Remark. In the same manner analogous control problems for *multiply connected regions* G of class C^1_μ with boundary $\Gamma = \bigcup_{k=0}^{m} \Gamma_k$ in case $n \geqq m$ can be dealt with, where Γ_k, $k = 1, \ldots, m$ are simple non-intersecting closed contours surrounded by the contour Γ_0 of same type. As additional conditions one may choose e.g. the point conditions (6) in $n - m$ points $z_i \in G$ and besides $m + 1$ integral conditions of form (7) with given positive functions $\varrho_k(t) \in H_\mu(\Gamma_k)$ on each Γ_k or point conditions of form (8) in prescribed points $t_k \in \Gamma_k$, respectively. (For further generalization of conditions see again [3], Chap, IV, § 6).

3. Adjoint boundary value problems

At first we consider a *linearized version of boundary value problem* (1), (2) with (3) and (6), (7) or (8), respectively, where the differential operator in (1) is substituted by a linear one:

$$\mathfrak{C}(W) = \partial_{\bar{z}} W + aW + b\overline{W} \tag{12}$$

with given functions $a(z), b(z) \in M_\infty(G)$. We introduce the Banach spaces $X = C_\mu(\overline{G}) \cap D_{\bar{z},\infty}(G)$ with elements $\xi = W(z)$ and equipped with the norm

$$\|\xi\|_X = \|W\|_{C_\mu(\overline{G})} + \|\partial W/\partial \bar{z}\|_{M_\infty(G)}$$

and $Y = M_\infty(G) \times H_\mu(\Gamma) \times C^n \times R$ with elements $\eta = \big(Z_1(z), Z_2(t), d_1, \ldots d_n, D_0\big)$ and norm

$$\|\eta\|_Y = \|Z_1\|_{M_\infty(G)} + \|Z_2\|_{H_\mu(\Gamma)} + \|d\|_{C^n} + |D_0|,$$

[1] With respect to z, $\bar{z}$ the function F_1 and its derivatives only need be bounded and measurable.

where $D_0 \in R$ and $d = (d_1, \ldots, d_n) \in C^n$, the complex Euclidean space of dimension n. The *operator* $A : X \to Y$, defined by

$$AW = \begin{cases} \mathfrak{C}(W) \quad \text{in } G, \\ \operatorname{Re}[\bar{\lambda} W] \quad \text{on } \Gamma, \\ W(z_i), \qquad i = 1, \ldots, n, \\ \int\limits_\Gamma \operatorname{Im}[\bar{\lambda} W](t)\, \varrho(t)\, ds \quad \text{or} \quad \operatorname{Im}[\bar{\lambda} W](t_0), \quad \text{respectively,} \end{cases} \tag{13}$$

is a linear bounded invertibel one. This follows in case II directly from theorem 4.13 in [3], Chap. IV, § 6. But the proof of this theorem can be easily extended to case I, too, because in account of the positiveness of $\varrho(t)$ from the homogeneous integral condition

$$\int\limits_\Gamma \operatorname{Im}[\bar{\lambda} W](t)\, \varrho(t)\, ds = 0 \tag{14}$$

there follows a homogeneous point condition

$$\operatorname{Im}[\bar{\lambda} W](t_0) = 0 \quad \text{for some} \quad t_0 \in \Gamma.^1 \tag{15}$$

We further introduce *conjugate spaces* X^+ *and* Y^+ to X and Y, respectively. Setting

$$Q(z) = \prod_{i=1}^{n} (z - z_i)^{-1}, \tag{16}$$

we choose in case I

$$X^+ = \hat{M}_\infty(G) \times H_\mu(\Gamma) \tag{17}$$

with elements $\xi^+ = \left(W_1^+(z), W_2^+(t)\right)$, where $W_1^+(z) \in \hat{M}_\infty(G)$ has the representation

$$W_1^+(z) = Q(z)\, W_0(z), \qquad W_0(z) \in M_\infty(G), \tag{18}$$

and norm

$$\|\xi^+\|_{X^+} = \|W_0\|_{M_\infty(G)} + \|W_2^+\|_{H_\mu(\Gamma)}.$$

Analogously the space Y^+ consists of the elements $\eta^+ = \left(Z^+(z), C_0\right)$, where $C_0 \in R$ and $Z^+(z)$ in $\bar{G}$ has the representation

$$Z^+(z) = Q(z)\, Z_0(z), \qquad Z_0(z) \in C_\mu(\bar{G}) \cap D_{\bar{z},\infty}(G) = X. \tag{19}$$

From (19) follows, that there exist uniquely determined complex numbers c_i, $i = 1, \ldots, n$, for each $Z^+(z)$ with

$$Z^+(z) - \sum_{i=1}^{n} \frac{c_i}{z - z_i} \in C(\bar{G}). \tag{20}$$

As norm in Y^+ we may take $\|\eta^+\|_{Y^+} = \|Z_0\|_X + |C_0|$.

---

[1] Another proof in case I is given in [5], Chap. 1, but only for the more restricted space $X = C_\mu^1(\bar{G})$ of classical solutions of (12).

In case II we choose for simplicity

$$X^+ = M_p(G) \times H_\mu(\Gamma), \qquad 1 < p < 2, \tag{21}$$

with elements $\xi^+ = \big(W_1^+(z), W_2^+(t)\big)$ and norm

$$\|\xi^+\|_{X^+} = \|W_1^+\|_{M_p(G)} + \|W_2^+\|_{H_\mu(\Gamma)},$$

where $M_p(G)$ means the space of complex functions in G summabel to power p. As Y^+ we take the space with elements $\eta^+ = Z^+(z)$ in $\bar{G}$ such that

$$Z^+(z) - \frac{1}{\pi i} \frac{C_0 \overline{\lambda(t_0)}}{z - t_0} = Q(z)\, Z_0(z), \qquad Z_0(z) \in D_{1,p}(G), \tag{22}$$

where $C_0 \in R$ and $D_{1,p}(G)$, $1 < p < 2$, is the complex analogue of the Sobolev space $W_1^p(G)$ (cf. [3], Chap. I). From (22) follows that there exist besides of the real number C_0 uniquely determined complex numbers c_i, $i = 1, \ldots, n$, for each $Z^+(z)$ with

$$Z^+(z) - \frac{1}{\pi i} \frac{C_0 \overline{\lambda(t_0)}}{z - t_0} - \sum_{i=1}^{n} \frac{c_i}{z - z_i} \in M_r(G), \tag{23}$$

where the exponent r, $2 < r < 2p/(2-p)$, is chosen according to the well-known imbedding property of the space $W_p^1(G)$, $1 < p < 2$.[1] The norm in Y^+ is now defined by $\|\eta^+\|_{Y^+} = \|Z_0\|_{D_{1,p}(G)} + |C_0|$.

On the product spaces $X \times X^+$ and $Y \times Y^+$ we define the continuous *bilinear forms*

$$\langle \xi, \xi^+ \rangle_{X,X^+} = \operatorname{Re} \iint_G W(z)\, W_1^+(z)\, dx dy + \frac{1}{2} \int_\Gamma \frac{1}{|\lambda|^2} \operatorname{Im}\,[\bar{\lambda} W(t)]\, W_2^+(t)\, ds \tag{24}$$

and

$$\langle \eta, \eta^+ \rangle_{Y,Y^+} = \pi \sum_{i=1}^{n} \operatorname{Re}\,[d_i c_i] + \frac{1}{2} D_0 C_0$$
$$+ \operatorname{Re} \iint_G Z_1(z)\, Z^+(z)\, dx\, dy - \frac{1}{2} \int_\Gamma \frac{1}{|\lambda|^2} Z_2(t) \operatorname{Im}\,[\lambda t'(s)\, Z^+(t)]\, ds, \tag{25}$$

where in case II the second integral in (25) is understood as Cauchy principal value (Note for the existence of this integral in case II also that $Z_0(t) \in M_s(\Gamma)$, $1 < s < p/(2-p)$ for the boundary values $Z_0(t)$ of $Z_0(z) \in D_{1,p}(G)$).

For functions $W(z) \in X$ and $Z^+(z)$ with (19), (20) there holds the following *generalization of complex Green's formula* (cf. [3], Chap. III, § 9):

$$\operatorname{Re} \iint_G [W \mathfrak{C}'(Z^+) + Z^+ \mathfrak{C}(W)]\, dx\, dy$$
$$= \operatorname{Re} \left[\frac{1}{2i} \int_\Gamma W(t)\, Z^+(t)\, dt \right] - \pi \operatorname{Re} \sum_{i=1}^{n} c_i W(z_i), \tag{26}$$

---

[27]) Note that the functions $(z - z_i)^{-1}$ and $(z - t_0)^{-1}$ do not lie in $D_{1,p}(G)$ for $p \geq 1$ and in $M_r(G)$ for $r \geq 2$.

analogously for $W(z) \in X$ and $Z^+(z)$ with (22), (23)

$$\operatorname{Re} \iint_G [W\mathfrak{C}'(Z^+) + Z^+\mathfrak{C}(W)]\, dxdy = -\frac{C_0}{2} \operatorname{Im} [\overline{\lambda(t_0)}\, W(t_0)]$$

$$+ \operatorname{Re} \left[\frac{1}{2i} \oint_\Gamma W(t)\, Z^+(t)\, dt\right] - \pi \operatorname{Re} \sum_{i=1}^{n} c_i W(z_i), \tag{27}$$

where $\mathfrak{C}'$ denotes the adjoint differential operator to $\mathfrak{C}$:

$$\mathfrak{C}'(Z) = \partial_{\bar z} Z - aZ - \bar b \bar Z. \tag{28}$$

Therefore, the bilinear forms (24) and (25) by means of the defining relation

$$\langle \xi, A^+\eta^+\rangle_{X,X^+} = \langle A\xi, \eta^+\rangle_{Y,Y^+} \tag{29}$$

generate the following *conjugate operators* $A^+ : Y^+ \to X^+$ to $A : X \to Y$:

$$A^+\eta^+ = \begin{cases} -\mathfrak{C}'(Z^+) & \text{in } G \\ \operatorname{Re} [\lambda t'(s)\, Z^+(t)] + C_0 |\lambda(t)|^2 \varrho(t) & \text{on } \Gamma \end{cases} \tag{30}$$

in case I and

$$A^+\eta^+ = \begin{cases} -\mathfrak{C}'(Z^+) & \text{in } G \\ \operatorname{Re} [\lambda t'(s)\, Z^+(t)] & \text{on } \Gamma \end{cases} \tag{31}$$

in case II, where in last case we suppose Γ possessing curvature of Hölder class C_μ and $\lambda(t)$ being of class C_μ^1 in a neighbourhood of t_0 to insure

$$\operatorname{Re} \left[\lambda(t)\, t'(s) \frac{1}{\pi i} \frac{\overline{\lambda(t_0)}}{t - t_0}\right] \in H_\mu(\Gamma) \tag{32}$$

(cf. [9], § 7 and § 13).

Also the so defined conjugate operators $A^+ : Y^+ \to X^+$ are linear bounded an *invertibel*. To show the last assertion we consider the boundary value problem for $Z_0(z)$ corresponding to equation $A^+\eta^+ = \xi^+$ in X^+ (cf. [3], Chap. IV, § 6 and in case II additionally § 7). This problem has index

$$\varkappa_0' = n' + n = -1, \qquad n' = \frac{1}{2\pi} [\arg \lambda(t)\, t'(s)]_\Gamma = -n - 1 \tag{33}$$

eing the index of the adjoint boundary value problem with differential operator $\mathfrak{C}'(Z)$ ind boundary operator $\operatorname{Re} [\lambda t' Z]$. Besides there is a free constant in an additive term in the boundary condition of (30) in case I and in the relation (22) in case II leading to corresponding free constants in the boundary value problem for function $Z_0(z)$. The operator A^+ therefore has index $\varkappa' = 0$ (cf. [10]), i.e. it is $k' = l'$ for the nullity k' and the deficiency l' of A^+ (cf. e.g. [11], A, Chap. I and B, Chap III). If $k' > 0$ there would be also $l \geqq k' > 0$ for the deficiency l of A because of the necessary solvability condition $\langle A\xi, \eta^+\rangle_{Y,Y^+} = 0$ for any η^+ with $A^+\eta^+ = \Theta$ following from (29). But this contradicts the above stated existence of an inverse to A. Therefore, $l' = k' = 0$, i.e. A^+ has a (bounded) inverse, too.

Remarks. 1. For simplicity we have chosen all appearing Hölder exponents equal to another. This allowed us to work with the same space $H_\mu(\Gamma)$ in X and X^+, and also in Y^+ in case I. If e.g. in case I we change the above assumptions about region G and function $\varrho(t)$ to $G \in C_\nu^1$ and $\varrho(t) \in H_\sigma(\Gamma)$ with $0 < \nu, \sigma < 1$, in X^+ and Y^+ we have to use the space $H_\tau(\Gamma)$ with $\tau = \min(\mu, \nu, \sigma)$ and analogously in case II (cf. [3], Chap. IV, § 1).

2. The definition of operators A and A^+ can be easily generalized to the above mentioned more general problems in simply and multiply connected regions G. Also in these problems A and A^+ have bounded inverses with analogously defined bilinear forms.

4. Necessary optimality conditions

Differential equation (1), boundary condition (2), and additional conditions (6) with (7) in case I and with (8) in case II are considered as components of the *operator equation*

$$T(\xi, \omega) = \Theta, \tag{34}$$

where $T : X \times U \to Y$ with

$$\xi = W(z) \in X = C_\mu(\overline{G}) \cap D_{\bar{z},\infty}(G),$$

$$\omega = \big(U(z), u(t)\big) \in U = L_\infty(G) \times H_\mu(\Gamma),$$

and Θ the zero element of $Y = M_\infty(G) \times H_\mu(\Gamma) \times C^n \times R^1$ as in point 3. Under the above assumptions on the data of (1), (2) the operator T possesses continuous Fréchet derivatives $T_\xi : X \to Y$ and $T_\omega : U \to Y$ given by

$$T_\xi(\xi, \omega)\, W = \begin{cases} \partial_{\bar{z}} W - [\Psi_W W + \Psi_{\overline{W}} \overline{W}] & \text{in } G \\ \operatorname{Re}[\bar{\lambda} W](t) & \text{on } \Gamma \\ W(z_i), \qquad i = 1, \dots, n, \\ \int_\Gamma \operatorname{Im}[\bar{\lambda} W](t)\, \varrho(t)\, ds \quad \text{or} \quad \operatorname{Im}[\bar{\lambda} W](t)_0, \end{cases} \tag{35}$$

respectively, and

$$T_\omega(\xi, \omega)\,(U, u) = \begin{cases} -\Psi_U(z)\, U(z) & \text{in } G \\ -g(t)\, u(t) & \text{on } \Gamma \end{cases} \tag{36}$$

with zero components otherwise.

Now let $\omega^0 = \big(U^0(z), u^0(t)\big)$ be a pair of *optimal control functions* with relating optimal state function $\xi^0 = W^0(z)$. We use the abbrevations $A = T_\xi^0$ and $B = T_\omega^0$, where generally an upper index 0 means the value of the corresponding quantity for the optimal process (ξ^0, ω^0), e.g. $T_\xi^0 = T_\xi(\xi^0, \omega^0)$. Then the operator $A : X \to Y$ has the form (13) with coefficients $a(z) = -\Psi_W^0(z)$, $b(z) = -\Psi_{\overline{W}}^0(z)$, and we can define a conjugate operator $A^+ : Y^+ \to X^+$ to A by means of formula (30) and (31) of point 3, respectively.

As in [2] we start from the *general necessary optimality condition*

$$J^0_\omega(\omega - \omega^0) + J^0_\xi(\xi - \xi^0) \geqq 0 \tag{37}$$

for any admissibel $\omega = \big(U(z), u(t)\big) \in V_1 \times V_2$ and relating uniquely determined $\xi = W(z) \in X$ from

$$A(\xi - \xi^0) + B(\omega - \omega^0) = \Theta. \tag{38}$$

The Fréchet derivatives J^0_ω and J^0_ξ, which exist under the above assumptions on F_1 and F_2, are given by

$$J^0_\omega \omega = \iint\limits_G \frac{\partial F_1^0}{\partial U}(z)\, U(z)\, dx\, dy + \int\limits_\Gamma \frac{\partial F_2^0}{\partial u}(t)\, u(t)\, ds, \tag{39}$$

$$J^0_\xi \xi = 2 \operatorname{Re} \iint\limits_G \frac{\partial F_1^0}{\partial W}(z)\, W(z)\, dx\, dy + \int\limits_\Gamma \operatorname{Re}\left[\overline{\sigma(t)}\, W(t)\right] ds \tag{40}$$

with

$$\sigma(t) = \frac{\partial F_2^0}{\partial w}(t) + i \frac{\partial F_2^0}{\partial v}(t), \qquad t \in \Gamma. \tag{41}$$

Further we split up $J^0_\xi \xi$ into the following two parts $K_1\xi$ and $K_2\xi$:

$$K_1\xi = 2 \operatorname{Re} \iint\limits_G \frac{\partial F_1^0}{\partial W}(z)\, W(z)\, dx\, dy - \int\limits_\Gamma \frac{1}{|\lambda|^2} \operatorname{Im}[\bar\lambda W(t)] \operatorname{Im}[\lambda\bar\sigma]\, ds, \tag{42}$$

which is equal to $\langle \xi, \xi^+ \rangle_{X,X^+}$ with

$$\xi^+ = \big(W_1^+(z), W_2^+(t)\big) = (2\partial F_1^0/\partial W, -2 \operatorname{Im}[\lambda\bar\sigma]) \in X^+, \tag{43}$$

and

$$K_2\xi = \int\limits_\Gamma \frac{1}{|\lambda|^2} \operatorname{Re}[\bar\lambda W(t)] \operatorname{Re}[\lambda\bar\sigma]\, ds. \tag{44}$$

Introducing an adjoint state $\eta^+ = \big(Z^+(z), C_0\big)$ with (19) in case I and $\eta^+ = Z^+(z)$ with (22) in case II, respectively, as uniquely determined solution of the conjugate equation

$$A^+\eta^+ = \xi^+ \quad \text{in } X^+, \qquad \xi^+ \text{ defined by (43)}, \tag{45}$$

we get using relations (29) and (38)

$$\begin{aligned} K_1(\xi - \xi^0) &= -\langle B(\omega - \omega^0), Z^+ \rangle_{Y,Y^+} \\ &= \int\limits_G \operatorname{Re}[\Psi_U^0(z)\, Z^+(z)]\,[U(z) - U^0(z)]\, dx\, dy \\ &\quad - \int\limits_\Gamma g(t)\, q_1(t)\,[u(t) - u^0(t)]\, ds, \end{aligned} \tag{46}$$

where we have put

$$q_1(t) = \frac{1}{2} \frac{1}{|\lambda(t)|^2} \operatorname{Im}[\lambda(t)\, t'(s)\, Z^+(t)]. \tag{47}$$

Finally, on account of equation

$$\operatorname{Re}\left[\bar{\lambda}(W - W^0)\right] = g(t)\,[u(t) - u^0(t)] \quad \text{on } \Gamma \tag{48}$$

following from (38) we have

$$K_2(\xi - \xi^0) = \int_\Gamma g(t)\, q_2(t)\, [u(t) - u^0(t)]\, ds \tag{49}$$

with

$$q_2(t) = \frac{1}{|\lambda(t)|^2} \operatorname{Re}\left[\lambda(t)\, \overline{\sigma(t)}\right]. \tag{50}$$

Summing up, from (37), (39), (46), and (49), we obtain the necessary optimality conditions

$$\iint_G \left[\frac{\partial F_1^0}{\partial U}(z) + \operatorname{Re}\{\Psi_U^{0\prime}(z)\, Z^+(z)\}\right] [U(z) - U^0(z)]\, dx\, dy \geqq 0 \tag{51}$$

for any admissible distributed control function $U(z) \in V_1$, and

$$\int_\Gamma \left[\frac{\partial F_2^0}{\partial u}(t) - g(t)\, q(t)\right] [u(t) - u^0(t)]\, ds \geqq 0 \tag{52}$$

for any admissible boundary control function $u(t) \in V_2$, where $Z^+(z)$ is (the main part of) the adjoint state $\eta^+ \in Y^+$ defined by (45) and $q(t) = q_1(t) - q_2(t)$ with (47), (50), and (43). In case of controls $U(z)$ and $u(t)$ with point constraints (5) the integral inequalities (51) and (52) are equivalent to corresponding pointwise inequalities or minimum principles for $U^0(z)$ and $u^0(t)$, respectively.

Remarks. 1. As in [1, 2] one can construct a conjugate space U^+ to U and a corresponding conjugate operator $B^+ : Y^+ \to U^+$ to B, too. But transforming $K_1\xi$ in form (46) is most easily done in direct manner here.

2. If the controls $U(z)$ and $u(t)$ also occur in the additional conditions (6), (7) or (8), respectively, the optimality conditions (51) and (52) contain the parameters c_i, $i = 1, \ldots, n$ and C_0 of adjoint state in explicit form, too. E.g. let an additional expression

$$\int_\Gamma \nu(t)\, u(t)\, ds, \qquad \nu(t) \in L_1(\Gamma) \quad \text{or} \quad \nu_0 u(t_0), \qquad \nu_0 = \text{const} \tag{53}$$

be on the left side of condition (7) or (8), respectively. This expression then also occurs as last component of the operator B leading to the additional term

$$-\frac{C_0}{2} \int_\Gamma \nu(t)\, [u(t) - u^0(t)]\, ds \quad \text{or} \quad -\frac{C_0}{2}\, \nu_0[u(t_0) - u^0(t_0)] \tag{54}$$

in the optimality condition (52), respectively.

3. Analogous optimality conditions to (51), (52) hold in problems with multiply connected regions G and vector-valued controls $U(z)$ and $u(t)$, respectively.

References

[1] L. v. Wolfersdorf, Necessary optimality conditions for control processes with singular integral equations and elliptic equations. *In:* Theory of Nonlinear Operators (Proc. Internat. Summer School Berlin 1977), Abh. Akad. Wiss. DDR, Jahrg. 1978, Nr. 6N, Akademie-Verlag, Berlin 1978, 279—292.
[2] L. v. Wolfersdorf, Optimal control with semilinear first order complex differential equation of I. N. Vekua's type. Applicable Analysis 1981.
[3] I. N. Vekua, Verallgemeinerte analytische Funktionen. Akademie-Verlag, Berlin 1963.
[4] W. Haack, W. Wendland, Vorlesungen über Partielle und Pfaffsche Differentialgleichungen. Birkhäuser Verlag, Basel und Stuttgart 1969.
[5] W. Wendland, Elliptic systems in the plane. Pitman Publ. Comp., London 1979.
[6] J. Nitsche, Untersuchungen über die linearen Randwertaufgaben linearer und quasilinearer elliptischer Differentialgleichungssysteme I, II. Math. Nachr. **14** (1955), 75—127, 157—182.
[7] W. Tutschke, Ein Optimalitätskriterium für Lösungen des Dirichletproblems bei elliptischen Systemen erster Ordnung mit Parameterfunktion. Beiträge zur Analysis **12** (1978), 19—27.
[8] H.-K. Klink, Komplexe Methoden in der Kontrolltheorie. Untersuchungen optimaler Lösungen des Dirichletproblems elliptischer Systeme erster Ordnung in der Ebene mit Steuerfunktion. Diss. A, TH Magdeburg 1978.
[9] N. I. Muschelischwili, Singuläre Integralgleichungen. Akademie-Verlag, Berlin 1974.
[10] S. Prößdorf, L. v. Wolfersdorf, Zur Theorie der Noetherschen Operatoren und einiger singulärer Integral- und Integrodifferentialgleichungen. Wiss. Z. Techn. Hochsch. Karl-Marx-Stadt **13** (1971) **1**, 103—116.
[11] D. Przeworska-Rolewicz, S. Rolewicz, Equations in linear spaces. PWN, Warsaw 1968.

Address of the author:

Bergakademie Freiberg
Sektion Mathematik
9200 Freiberg
Bernhard-v.-Cotta-Str. 2
DDR

On an initial-boundary value problem for the nonlinear Schrödinger equation with selfconsistent potential

Herbert Gajewski, Berlin

1. Introduction

We shall study the following initial-boundary value problem

$$iu_t + (\beta - i\gamma)\, u_{xx} + (k + v_x)\, u = 0,\ i^2 = -1, \tag{1.1}$$

$$v_{tt} - v_{xx} + (|u|^2)_x = 0,\ (t, x) \in (0, T) \times (0, 1),\ 0 < T < \infty, \tag{1.2}$$

$$u(0, x) = u_0(x),\ v(0, x) = v_0(x),\ v_t(0, x) = v_1(x), \tag{1.3}$$

$$u_x(t, 0) = i\alpha_0\big(2a_0 - u(t, 0)\big),\ u_x(t, 1) = i\alpha_1\big(u(t, 1) - 2a_1\big), \tag{1.4}$$

$$v(t, 0) = v(t, 1) = 0, \tag{1.5}$$

where α_0, α_1, β, γ are positive real constants and a_0, a_1, k are (in general) complex constants. The subscripts t and x indicate partial differentiating.

With the new function $n = -v_x$ we can rewrite (1.1), (1.2) as follows

$$\begin{aligned} &iu_t + (\beta - i\gamma)\, u_{xx} + (k - n)\, u = 0, \\ &n_{tt} - n_{xx} - (|u|^2)_{xx} = 0. \end{aligned} \tag{1.6}$$

This system of equations (for $\gamma = 0$) was firstly derived by Karpman [7] to describe mathematically transverse plasma waves and by Zakharov [8] for describing longitudinal (Langmuir) plasma waves. The physical meaning of the function $u = u(t, x)$ is that of the temporally slowly varying complex amplitude (envelope) of the high-frequency electric wave field $E = \operatorname{Re}\big(u \exp(-i\omega_0 t)\big)$. The other function $n = n(t, x)$ in (1.6) may be interpreted physically as low-frequency ion density perturbation. The equations (1.1), (1.2) along with the boundary conditions (1.4) model the interaction between a plasma layer localized in $[0, 1]$ and external electric fields $E_0 = \operatorname{Re}\big(a_0 \exp\big(i(\alpha_0 x - \omega_0 t)\big)\big)$, $x < 0$, $E_1 = \operatorname{Re}\big(a_1 \exp\big(i(\alpha_1(1 - x) - \omega_0 t)\big)\big)$, $x > 1$. The unknown boundary values $u(t, 0)$ and $u(t, 1)$ represent the reflexion and transmission properties of the plasma layer [1. 4].

Recently, the initial value problem (1.1)—(1.3) has been solved for solutions which tend to zero as $|x| \to \infty$ as well as for spatially periodic solutions [2]. The boundary conditions (1.4), (1.5) turn out to be more difficult. As an initial step we studied in a previous paper [4] the nonlinear Schrödinger equation

$$iu_t + (\beta - i\gamma)\, u_{xx} + (k + |u|^2)\, u = 0$$

along with (1.4). This equation results from (1.1), (1.2) by neglecting the term v_{tt}, as usual in the so-called quasistationary theory. In [4] we could include the "undamped" (cf. [9]) case $\gamma = 0$. In this paper we shall essentially need the damping parameter γ to be strictly positive. The existence of a global solution of (1.1)—(1.5) if $\gamma = 0$ remains open.

Section 2 contains notations and preliminaries. Our main result, a global existence and uniqueness theorem for (1.1)—(1.5) is proved in Section 3. In Section 4 we establish Galerkin's method as an appropriate procedure to solve (1.1)—(1.5) numerically.

2. Preliminaries

Throughout this paper the letter c stands for various constants. For a complex number z we denote by $\bar{z}$, $|z|$, $\operatorname{Re} z$ and $\operatorname{Im} z$ conjugate complex number, modulus, real and imaginary part, respectively. C^l, H^l, H_0^1, H^{-1} and L^q are the usual spaces of complex-valued functions defined on the interval (0,1) ([11]). We use the following norms

$$\|u\|_{C^l} = \sum_{j=0}^{l} \max \left|\frac{d^j u}{dx^j}\right|, \quad \|u\|_{H^l} = \left(\sum_{j=0}^{l} \int_0^1 \left|\frac{d^j u}{dx^j}\right|^2 dx\right)^{1/2},$$

$$\|u\|_q = \left(\int_0^1 |u|^q dx\right)^{1/q}, \quad 1 \leqq q < \infty, \quad \|u\|_\infty = \operatorname*{ess\,sup}_{x \in [0,1]} |u(x)|.$$

To abbreviate we write

$$C = C^0, \ \|u\|_{C^0} = \|u\|_C, \ H = H^0 = L^2, \ \|u\| = \|u\|_2, \ (u, v) = \int_0^1 u\bar{v}\, dx.$$

The symbol $(\cdot, \cdot)$ we also use to denote the pairing between H^{-1} and H_0^1. The space H^1 is continuously embedded into C and it holds [2]

$$\|u\|_C^2 \leqq \|u\| \left(\|u\| + 2\, \|u_x\|\right), \qquad \forall\, u \in H^1. \tag{2.1}$$

In what follows we need the operator A defined by

$$Au = -u_{xx} + 2ipu_x + (ip_x + p^2)\, u, \tag{2.2}$$

$$D(A) = \{u \in H^2 \mid u_x(0) = -i\alpha_0 u(0),\ u_x(1) = i\alpha_1 u(1)\}.$$

Here $p \in H^3$ is a real function satisfying

$$p(0) = -\alpha_0, \quad p(1) = \alpha_1. \tag{2.3}$$

The function $p = p(x) = (\alpha_0 + \alpha_1)\, x - \alpha_0$ may serve as a simple example. Using the identity

$$(Au, u) = \|u_x\|^2 + \|pu\|^2 - 2 \operatorname{Im} (pu_x, u), \quad u \in D(A), \tag{2.4}$$

it is easy to see (cf. [4]) that the following lemma holds.

Lemma 2.1. *The operator $A \in (D(A) \to H)$ is selfadjoint and nonnegative. Its energetic space is H^1. A has a pure point spectrum. Its eigenvalues are $\lambda_n = n^2\pi^2$, $n = 0, 1, \ldots$ Each eigenvalue is single. The corresponding orthonormal eigenfunctions are*

$$h_n = r_n \, e^{iP(x)} \cos n\pi x, \quad P(x) = \int_0^x p(s) \, ds, \quad r_n = \begin{cases} 1 \text{ if } n = 0, \\ \sqrt{2} \text{ if } n = 1, 2, \ldots \end{cases} \tag{2.5}$$

Provided with the norm

$$\|u\|_V^2 = \|u + Au\|^2,$$

and the corresponding scalar product

$$((u, v)) = (u + Au, v + Av))$$

$D(A)$ becomes a Hilbert space V. A simple calculation shows (cf. [4]) that the V-norm and the H^2-norm are equivalent on V. Let $\langle \cdot, \cdot \rangle$ denote the pairing between V and its dual space V'. Because of Riesz' representation theorem the mapping $E \in (H \to V')$ defined by

$$\langle Ef, u \rangle = (f, u + Au), \quad u \in V$$

is one-to-one and isometric. Thus we can identify V' and H. As a consequence of Lemma 2.1 we have

Lemma 2.2. *Let $g \in H$ and $g_n = \sum_{l=0}^{n} (g, h_l) \, h_l$. Then $g_n \to g$ in H. Moreover, if $g \in V$, then $g_n \to g$ in H^2.*

In what follows $S = [0, T]$ is a bounded time interval. For a Banach space B we denote by

- $C^l(S; B)$ the Banach space of l-times continuously differentiable (B-valued) functions provided with the norm

$$\|u\|_{C^l(S;B)} = \sum_{j=0}^{l} \max_{t \in S} \left\| \frac{d^j u}{dt^j} \right\|_B,$$

- $L^2(S; B)$ the Banach space of quadratically Bochner-integrable functions provided with the norm

$$\|u\|_{L^2(S;B)} = \int_S \|u(t)\|_B^2 \, dt,$$

- $H^l(S; B)$ the Banach space of functions $u \in L^2(S; B)$ having derivatives $u^{(j)} \in L^2(S; B)$, $0 \leqq j \leqq l$, taken in the sense of distributions on $(0, T)$ with values in B,
- $C_w(S; B)$ the space of all functions $u \in H^l(S; B)$ such that $u^{(j)}$, $0 \leqq j \leqq l$, is weakly continuous.

Remark 2.1. Clearly, the relation $L^2(S; H) = L^2((0, T) \times (0, 1))$ holds. Accordingly we shall consider H-valued functions also as ordinary ones and vice-versa.

3. Existence and uniqueness

In order to replace (1.4) by corresponding homogeneous boundary conditions we need a function u_a satisfying (1.4). Let $u_a \in H^3$ be a such function, for example

$$u_a = -i\left(\alpha_0 a_0 (1-x)^2 e^{iP(x)} + \alpha_1 a_1 x^2 e^{i(P(x)-P(1))}\right). \tag{3.1}$$

Now we can rewrite problem (1.1)—(1.5) in the more condensed form

$$\begin{aligned} &iu_t + (\beta - i\gamma)\, u_{xx} + (k + v_x)\, u = 0, \\ &u(0) = u_0,\ u - u_a \in C(S; V) \cap C^1(S; H) \cap H^1(S; H^1), \end{aligned} \tag{3.2}$$

$$v_{tt} - v_{xx} + (|u|^2)_x = 0,\ v(0) = v_0,\ v_t(0) = v_1,\ v \in C^1(S; H_0^1) \cap C(S; H^2) \cap C^2(S; H). \tag{3.3}$$

Our main result is

Theorem 3.1. *Suppose* $\alpha_0, \alpha_1, \beta \geqq 0, j > 0, u_0 - u_a \in V, v_0 \in H_0^1 \cap H^2, v_1 \in H_0^1$. *Then the problem* (3.2), (3.3) *has a unique solution.*

The idea of the existence proof is to approximate problem (3.2), (3.3) by regularized problems like

$$iu_t + (\beta - i\gamma)\, u_{xx} + (k + v_x)\, u = 0,\ u(0) = u_{0\varepsilon},\ u - u_a \in H^1(S; V) \cap H^2(S; H), \tag{3.4}$$

$$v_{tt} - (v + \varepsilon v_t)_{xx} + (|u|^2)_x = 0,\ v(0) = v_0,\ v_t(0) = v_{1\varepsilon},\ v \in H^2(S; H_0^1) \cap H^1(S; H^2) \tag{3.5}$$

and then to pass to the limit $\varepsilon \to 0$. Here $u_{0\varepsilon}$ and $v_{1\varepsilon}$ are regularized initial values satisfying

$$\begin{aligned} &u_{0\varepsilon} - u_a \in V \cap H^3,\ u_{0\varepsilon} \to u_0 \quad \text{in} \quad H^2, \\ &v_{1\varepsilon} \in H_0^1 \cap H^2,\ v_{1\varepsilon} \to v_1 \quad \text{in} \quad H^1 \quad \text{as} \quad \varepsilon \to 0. \end{aligned} \tag{3.6}$$

The existence of such initial values is guaranteed by Lemma 2.2

Before we can supply the details of the proof of Theorem 3.1 we need some results concerning the regularized problems. We start with the following lemma which can be easily verified.

Lemma 3.1. *Let* (u, v) *be solution of* (3.4), (3.5). *Then* (w, y, z) *with* $w = u - u_a$, $y = v_t$, $z = \int_0^1 v_0\, dx - v_x$ *is solution of the problem*

$$\left.\begin{aligned} &w_t - (i\beta + \gamma)(w_{xx} + u_{axx}) - i(k' - z)(w + u_a) = 0,\ w(0) = w_{0\varepsilon}, \\ &w \in H^1(S; V) \cap H^2(S; H), \\ &y_t + (z + |w + u_a|^2 - \varepsilon y_x)_x = 0,\ y(0) = y_{0\varepsilon},\ y \in H^1(S; H_0^1) \cap L^2(S; H^2), \\ &z_t + y_x = 0,\ z(0) = z_0,\ z \in H^2(S; H) \cap H^1(S; H^1), \end{aligned}\right\} \tag{3.7}$$

where $k' = k + \int_0^1 v_0\, dx$ *and*

$$w_{0\varepsilon} = u_{0\varepsilon} - u_a \in V \cap H^3,\ y_{0\varepsilon} = v_{1\varepsilon} \in H_0^1 \cap H^2,\ z_0 = \int_0^1 v_0\, dx - v_{0x} \in H^1.$$

Conversely, let (w, y, z) *be solution of* (3.7). *Then* (u, v) *with* $u = w + u_a$, $v = x \int_0^1 z_0\,dx - \int_0^x z\,ds$ *is solution of* (3.4), (3.5), *where*

$$k = k' - \int_0^1 z_0\,dx,\quad u_{0\varepsilon} = w_{0\varepsilon} + u_a,\quad v_0 = x \int_0^1 z_0\,dx - \int_0^x z_0\,ds,\quad v_{1\varepsilon} = y_{0\varepsilon}.$$

Now we want to solve the problem (3.7). For this purpose we study the auxiliary problems

$$\left.\begin{array}{l} w_t - (i\beta + \gamma)(w_{xx} + u_{axx}) - i(k' - Q_r z)\,P_r(w + u_a) = 0, \\ w \in H^1(S; V) \cap H^2(S; H), \\ y_t + \big(z + |P_r(w + u_a)|^2 - \varepsilon y_x\big)_x = 0,\ y \in H^1(S; H_0^1) \cap L^2(S; H^2), \\ z_t + y_x = 0,\ z \in H^2(S; H) \cap H^1(S; H^1),\ \big(w(0), y(0), z(0)\big) = (w_{0\varepsilon}, y_{0\varepsilon}, z_0). \end{array}\right\} \tag{3.8}$$

Here for $r > 0$ the operators $P_r \in (C \to C)$ and $Q_r \in (H \to H)$ are defined by

$$(P_r u)(x) = \begin{cases} u(x) & \text{if } |u(x)| \leqq r \\ r\dfrac{u(x)}{|u(x)|} & \text{if } |u(x)| > r, \end{cases} \qquad u \in C,\ x \in [0, 1], \tag{3.9}$$

$$Q_r u = \begin{cases} u & \text{if } \|u\| \leqq r \\ r\dfrac{u}{\|u\|} & \text{if } \|u\| > r. \end{cases} \qquad u \in H \tag{3.10}$$

Lemma 3.2. *The problem* (3.8) *has a unique solution* $(w_{r\varepsilon}, v_{r\varepsilon}, z_{r\varepsilon})$.

Proof. For the time being we introduce the Hilbert spaces

$$W = V \times H_0^1 \times H,\ Y = H^1 \times H \times H,\ W' = H \times H^{-1} \times H$$

and denote by $\langle\cdot,\cdot\rangle$ the pairing between W' and W. Let $B \in (W \to W')$ be the operator defined by

$$B_r(w, y, z) = \big(-(i\beta + \gamma)(w_{xx} + u_{axx}) - i(k' - Q_r z)\,P_r(w + u_a),\ (z + |P_r(w + u_a)|^2 - \varepsilon y_x)_x,\ y_x\big).$$

Now we can write (3.8) as an (equivalent) standard evolution equation

$$f_t + B_r f = 0,\ f(0) = f_{0\varepsilon},\ f \in H^1(S; W) \cap H^2(S; W'), \tag{3.11}$$

where $f = (w, y, z)$ and $f_{0\varepsilon} = (w_{0\varepsilon}, y_{0\varepsilon}, z_0)$. (That (3.8) and (3.11) are in fact equivalent is easily to be seen. Especially, the inclusion $y \in L^2(S; H^2)$ follows from $f \in H^1(S: W)$ and from the equation

$$\varepsilon y_{xx} + \int_0^t y_{xx}\,ds = z_{0x} + y_t + (|P_r(w + u_a|^2)_x \in L^2(S; H)$$

which results from $(3.8)_{2,3}$.)

We wish to apply known existence and uniqueness results to (3.11). To this end we have to verify some properties of $f_{0\varepsilon}$ and B_r. Firstly we note that

$$f_{0\varepsilon} \in W, \quad B_r f_{0\varepsilon} \in Y. \tag{3.12}$$

Next, using (2.1) and the simple inequalities

$$\|P_r u\|_C \leqq r, \ \|P_r u_1 - P_r u_2\|_C \leqq \|u_1 - u_2\|_C, \ u_1, u_2, u \in C, \tag{3.13}$$

$$\|Q_r u\| \leqq r, \ \|Q_r u_1 - Q_r u_2\| \leqq \|u_1 - u_2\|, \ u_1, u_2, u \in H, \tag{3.14}$$

we find for $f_j = (w_j, y_j, z_j)$, $g = (g_1, g_2, g_3) \in W$, $j = 1, 2$, setting $f = f_1 - f_2$

$$\begin{aligned}\langle B_r f_1 - B_r f_2, g\rangle &= \big(-(i\beta + \gamma)\, w_{xx} - i((k' - Q_r z_1)\, P_r(w_1 + u_a) - (k' - Q_r z_2)\\ &\quad \times P_r(w_2 + u_a)),\ g_1 + A g_1\big) + \big(\varepsilon y_x - z - |P_r(w_1 + u_a)|^2\\ &\quad + |P_r(w_2 + u_a)|^2, g_{2x}\big) + (y_x, g_3)\\ &\leqq c(r)\, \|f\|_W\, \|g\|_W. \end{aligned}\tag{3.15}$$

Thus $B_r \in (W \to W')$ is Lipschitzian. Finally we have

$$\begin{aligned}2\,\mathrm{Re}\,\langle B_r f_1 - B_r f_2, f\rangle &= 2\,\mathrm{Re}\,\big(-(i\beta + \gamma)\, w_{xx} - i\big((k' - Q_r z_1)\, P_r(w_1 + u_a) - (k' - Q_r z_2)\\ &\quad \times P_r(w_2 + u_a)\big),\ w + Aw\big) + 2\big(\varepsilon y_x - z - |P_r(w_1 + u_a)|^2\\ &\quad + |P_r(w_2 + u_a)|^2, y_x\big) + 2(y_x, z)\\ &\geqq \min(\varepsilon, \gamma)\, \|f\|_W^2 - c(r)\, \|f\|_Y^2. \end{aligned}\tag{3.16}$$

Now the relations (3.12), (3.15) and (3.16) guarantee the existence of a unique solution $f_{r\varepsilon} = (w_{r\varepsilon}, y_{r\varepsilon}, z_{r\varepsilon})$ of (3.11) that is of (3.8). (Cf. [6], Satz 3, Bemerkung 5 and [5], Satz 3.)

Lemma 3.3. *With a constant c independent of r and ε it holds* (3.17)

$$\|w_{r\varepsilon}\|_{C(S;H^1)} + \|w_{r\varepsilon}\|_{L^2(S;H^2)} + \|y_{r\varepsilon}\|_{C(S;H)} + \sqrt{\varepsilon}\, \|y_{r\varepsilon}\|_{L^2(S;H^1)} + \|z_{r\varepsilon}\|_{C(S;H)} \leqq c. \tag{3.17}$$

Proof. Setting $w = w_{r\varepsilon}$, $u = w_{r\varepsilon} + u_a$, $y = y_{r\varepsilon}$, $z = z_{r\varepsilon}$ we get from $(3.8)_1$

$$\begin{aligned} 0 &= 2\,\mathrm{Re}\,\big(u_t - (i\beta + \gamma)\, u_{xx} - i(k' - Q_r z)\, P_r u, u\big) = (\|u\|^2)_t - 2\,\mathrm{Re}\,\big((i\beta + \gamma)\, i\\ &\quad \times \sum_{j=0}^{1} \alpha_j (u(t, j) - 2a_j)\, \bar{u}(t, j)\big) + 2\gamma\, \|u_x\|^2 + 2\,\mathrm{Im}\, k'(P_r u, u)\\ &\geqq (\|u\|^2)_t + \beta \sum_{j=0}^{1} \alpha_j\, |u(t, j)|^2 + 2\gamma\, \|u_x\|^2 - c(1 + \|u\|^2). \end{aligned}$$

Gronwall's lemma yields because of (3.6)

$$\|u\|_{C(S;H)} + \|u\|_{L^2(S;H^1)} \leqq c. \tag{3.18}$$

Next we deduce from $(3.8)_1$ using (2.4)

$$\begin{aligned} 0 &= 2\,\mathrm{Re}\,\big(w_t - (i\beta + \gamma)\, u_{xx} - i(k' - Q_r z)\, P_r u, Aw\big) = (w, Aw)_t + 2\,\mathrm{Re}\,\big((i\beta + \gamma)\\ &\quad \times (Aw - u_{axx} - 2ipw_x - (ip_x + p^2)\, w) - i(k' - Q_r z)\, P_r u, Aw\big)\\ &\geqq (w, Aw)_t + 2\gamma\, \|Aw\|^2 - c(\|u_{axx}\| + \|w\|_{H^1} + \|u\|_{H^1}\, \|z\|)\, \|Aw\| \end{aligned}$$

and hence by (3.6)

$$\|w(t)\|_{H^1}^2 + \gamma \int_0^t \|Aw\|^2\, ds \leqq c\left(1 + \int_0^t \|u\|_{H^1}^2 \|z^2\|\, ds\right). \tag{3.19}$$

On the other hand $(3.8)_2$ implies, taking into account $(3.8)_3$,

$$0 = \left(y_t + (z + |P_r u|^2 - \varepsilon y_x)_x, y\right) = \frac{1}{2}\left(\|y\|^2 + \|z\|^2\right)_t + \varepsilon \|y_x\|^2 + \left((|P_r u|^2)_x, y\right)$$
$$\geqq \frac{1}{2}\left(\|y\|^2 + \|z\|^2\right)_t + \varepsilon \|y_x\|^2 - c\,\|u\|_{H^1}^2 \|y\|$$

or, using (3.6),

$$\|y(t)\|^2 + \|z(t)\|^2 + \varepsilon \int_0^t \|y_x\|^2\, ds \leqq c\left(1 + \int_0^t \|u\|_{H^1}^2 \|y\|^2\, ds\right). \tag{3.20}$$

Adding (3.19) and (3.20), the desired estimate (3.17) follows by (3.18) and Gronwall's lemma.

Lemma 3.4. *The problem* (3.4), (3.5) *has a unique solution* $(u_\varepsilon, v_\varepsilon)$. *Moreover, the following a priori estimate holds*

$$\|u_\varepsilon\|_{C(S;H^2)} + \|u_{\varepsilon t}\|_{C(S;H)} + \|u_{\varepsilon t}\|_{L^2(S;H^1)} + \|v_\varepsilon\|_{C^1(S;H^1)} + \|v_\varepsilon\|_{C(S;H^2)}$$
$$+ \|v_\varepsilon\|_{C^2(S;H)} + \sqrt{\varepsilon}\, \|v_{\varepsilon t}\|_{L^2(S;H^2)} \leqq c. \tag{3.21}$$

Proof. From (3.17) and the definition of the projectors P_r and Q_r (cf. (3.9), (3.10)) we see that the unique solution $(w_{r\varepsilon}, y_{r\varepsilon}, z_{r\varepsilon})$ of (3.8) is also the unique solution of (3.7) provided the parameter r is chosen sufficiently large. Therefore by Lemma 3.1 $(u_\varepsilon, v_\varepsilon)$ with

$$u_\varepsilon = w_{r\varepsilon} + u_a, \quad v_\varepsilon = x \int_0^1 z_0\, dx - \int_0^x z_{r\varepsilon}\, ds$$

is the unique solution of (3.4), (3.5). The estimate (3.17) expressed in terms of u_ε and v_ε is

$$\|u_\varepsilon\|_{C(S;H^1)} + \|u_\varepsilon\|_{L^2(S;H^2)} + \|v_\varepsilon\|_{C(S;H^1)} + \|v_\varepsilon\|_{C^1(S;H)} + \sqrt{\varepsilon}\, \|v_{\varepsilon t}\|_{L^2(S:H^1)} \leqq c. \tag{3.22}$$

Now we want to prove (3.21). Suppressing the subscript ε, we get from (3.4)

$$0 = 2\,\mathrm{Re}\left(iu_t + (\beta - i\gamma)\, u_{xx} + (k + v_x)\, u, (\beta - i\gamma)\, u_{xxt} + v_x u_t\right) = 2\beta \sum_{j=0}^{1} \alpha_j\, |u_t(t, j)|^2$$
$$+ 2\gamma\, \|u_{xt}\|^2 + (\beta^2 + \gamma^2)\left(\|u_{xx}\|^2\right)_t + \left(v_x^2, (|u|^2)_t\right) + 2\mathrm{Re}\,\big(\beta(v_x, \bar{u} u_{xxt} + \bar{u}_{xx} u_t)$$
$$+ i\gamma(v_x, \bar{u} u_{xxt} - \bar{u}_{xx} u_t) + k\left(u, (\beta - i\gamma)\, u_{xxt} + v_x u_t\right)\big)$$
$$= 2\beta \sum_{j=0}^{1} \alpha_j\, |u_t(t, j)|^2 + 2\gamma\, \|u_{xt}\|^2 + (\beta^2 + \gamma^2)\left(\|u_{xx}\|^2\right)_t + \left(v_x^2, (|u|^2)_t\right)$$
$$- 2\beta\, \mathrm{Re}\,\Big((v_{xx}, \bar{u} u_{xt} + \bar{u}_x u_t) + (v_x, \bar{u}_x u_{xt} + \bar{u}_x u_{xt}) + i \sum_{j=0}^{1} \alpha_j v_x(t, j)\, \big(u(t, j)\, \bar{u}_t(t, j)\big)$$

$$- \big(u(t, j) - 2a_j)\, \overline{u}_t(t, j)\big)\big) - 2\gamma \operatorname{Im} \big((v_x, \overline{u}u_{xx})_t - (v_{xt}, \overline{u}u_{xx}) - (v_x, \overline{u}_t u_{xx} + \overline{u}_{xx} u_t)\big)$$
$$+ 2\operatorname{Re} \big(k(\beta + i\gamma)\, ((u, u_{xx})_t - (u_t, u_{xx})) + (u, v_x u_t)\big)\big)$$
$$= 2\beta \sum_{j=0}^{1} \alpha_j\, |u_t(t, j)|^2 + 2\gamma\, \|u_{xt}\|^2 + (\beta^2 + \gamma^2)\, (\|u_{xx}\|^2)_t + \big(v_x^2, (|u|^2)_t$$
$$- \beta\Big(\big(v_{xx}(|u|^2)_{xt}\big) + 2\big(v_x(|u_x|^2)_t\big) - 4 \sum_{j=0}^{1} \alpha_j v_x(t, j) \operatorname{Im} a_j \overline{u}(t, j)\Big) - 2\gamma \operatorname{Im} \big((v_x, \overline{u}u_{xx})_t$$
$$- (v_{xt}, \overline{u}u_{xx})\big) + 2 \operatorname{Re} \big(k((\beta + i\gamma)\, ((u, u_{xx})_t - (u_t, u_{xx})) + (u, v_x u_t))\big).$$

On the other hand it follows from (3.5) that

$$0 = \big(v_{tt} - (v + \varepsilon v_t)_{xx} + (|u|^2)_x, \beta v_{txx} + 2v_t v_x\big) = \frac{\beta}{2} (\|v_{tx}\|^2 + \|v_{xx}\|^2)_t - \beta\big(v_{txx}, (|u|^2)_x\big)$$
$$+ \varepsilon\beta\, \|v_{txx}\|^2 - \left(v_t^2 + \frac{1}{3} v_x^2, v_x\right)_t + \big(|u|^2, (v_x^2)_t + 2v_t v_{xx}\big) + 2\varepsilon(v_{txx}, v_t v_x).$$

Adding these equations, integrating with respect to t and using (3.4), (3.21), we get for arbitrary small $\delta > 0$

$$\int_0^t \left(2\beta \sum_{j=0}^{1} \alpha_j\, |u_t(s, j)|^2 + 2\gamma\, \|u_{xt}\|^2 + \varepsilon\beta\, \|v_{txx}\|^2\right) ds + \Big[(\beta^2 + \gamma^2)\, \|u_{xx}\|^2 + \frac{\beta}{2} (\|v_{tx}\|^2$$
$$+ \|v_{xx}\|^2) + (|u|^2, v_x^2)\Big]_0^t = \Big[\beta\big(v_{xx}, (|u|^2)_x\big) + 2\beta(v_x, |u_x|^2) + 2\gamma \operatorname{Im} (v_x, \overline{u}u_{xx})$$
$$- 2 \operatorname{Re} \big(k(\beta + i\gamma)\, (u, u_{xx})\big) + \left(v_t^2 + \frac{1}{3} v_x^2, v_x\right)\Big]_0^t - \int_0^t \Big\{2\beta(v_{xt}, |u_x|^2) + 4\beta \sum_{j=0}^{1} \alpha_j$$
$$\times v_x(s, j) \operatorname{Im} a_j \overline{u}(t, j) + 2\gamma \operatorname{Im} (v_{xt}, u\overline{u}_{xx}) - 2 \operatorname{Re} \big(k((\beta + i\gamma)\, (u_t, u_{xx}) - (u, v_x u_t))\big)$$
$$+ 2(|u|^2, v_t v_{xx}) + 2\varepsilon(v_{txx}, v_t v_x)\Big\}\, ds \leqq \delta(\|v_{xx}\|^2 + \|u_{xx}\|^2)$$
$$+ c\left(1 + \int_0^t (\|u_{xx}\|^2 + \|v_{tx}\|^2 + \|v_{xx}\|^2)\, ds\right) + \frac{\varepsilon}{2} \int_0^t \|v_{txx}\|^2\, ds.$$

Hence, using Gronwall's lemma and taking into account (3.4), (3.5), (3.22), we deduce (3.21).

Proof of Theorem 3.1. (existence) By a well-known compactness lemma (cf. [8]) there exist a sequence (ε_n) tending to zero, a sequence (u_j, v_j) of corresponding solutions of (3.4), (3.5) as well as functions $u \in C_w(S; H^2) \cap C(S; H^1) \cap H^1(S; H^1)$ and $v \in C_w^1(S; H_0^1) \cap C_w(S; H^2) \cap H^2(S; H)$ such that

$$u_j \to u \text{ (strongly) in } L^2(S; H^1),\ u_j \rightharpoonup u \text{ (weakly) in } L^2(S; H^2),$$
$$u_{jt} \rightharpoonup u_t \text{ in } L^2(S; H^1),\ v_j \rightharpoonup v \text{ in } L^2(S; H^2),$$
$$v_{jtt} \rightharpoonup v_{tt} \text{ in } L^2(S; H),\ \sqrt{\varepsilon_j} v_{jtxx} \to 0 \text{ in } L^2(S; H).$$

Therefore we can pass to the limit $\varepsilon \to 0$ in (3.4) and (3.5). We get

$$iu_t + (\beta - i\gamma)\, u_{xx} + (k + v_x)\, u = 0,\ u(0) = u_0,\ u - u_a \in C_w(S; V) \cap C(S; H^1) \cap H^1(S; H^1),$$
$$v_{tt} - v_{xx} + (|u|^2)_x = 0,\ v(0) = v_0,\ v_t(0) = v_1,\ v \in C^1_w(S; H^1_0) \cap C_w(S; H^2) \cap H^2(S; H). \quad (3.23)$$

These equations imply $u \in C^1_w(S; H)$ and $v \in C^2(S; H)$. The inclusion $u - u_a \in C(S; V) \cap C^1(S; H)$ can be proved essentially in the same way as the corresponding statement in [4]. (Cf. the last part of the proof of Theorem 3.1 in [4].)

In order to show that $v \in C(S; H^2) \cap C^1(S; H^1_0) \cap C^2(S; H)$ we use a technique of Strauss [10]. Let $r = r_\varepsilon = r_\varepsilon(t)$ be an even smoothing kernel (cf. [10]). We define

$$\big(r * v(t)\big) = \int r_\varepsilon(t - s)\, v(s)\, ds, \qquad \int = \int_{-\infty}^{\infty}.$$

Further let $h = h_\delta = h_\delta(s)$ be $\{1$ for $s \in [\delta, t - \delta]$, 0 for $s \notin [0, t]$ and linear in the intervals $[0, \delta]$ and $[t - \delta, t]\}$. We have

$$\begin{aligned} 0 &= \int \big(r * (hv_{tx}), r * (hv_{tx})\big)_t\, ds = \int \big(r' * (hv_{tx}), r * (hv_{tx})\big)\, ds \\ &= -\int \big(r' * (hv_t), r * (hv_{txx})\big)\, ds = -\int \big(r * (h'v_t + hv_{tt}), r * (hv_{txx})\big)\, ds \\ &= \int \Big\{\big(r * (h'v_{tx}), r * (hv_{tx})\big) - \big(r * (hv_{tt}), r * hv_{xxt})\big)\Big\}\, ds \end{aligned}$$

and

$$\begin{aligned} 0 &= \int \big(r * (hv_{xx}), r(hv_{xx})\big)_t\, ds = \int \big(r' * (hv_{xx}), r * (hv_{xx})\big)\, ds \\ &= \int \big(r * (h'v_{xx} + hv_{xxt}), r * (hv_{xx})\big)\, ds. \end{aligned}$$

Adding these equations we get

$$\begin{aligned} 0 = \int \Big\{&\big(r * (h'v_{tx}), r * (hv_{tx})\big) + \big(r * (h'v_{xx}), r * (hv_{xx})\big) + \big(r * (h(v_{xx} - v_{tt})), \\ &r * (hv_{xxt})\big)\Big\}\, ds. \end{aligned}$$

Using (2.3) we can rewrite the third term as follows

$$\begin{aligned} &\int \big(r * (h(v_{xx} - v_{tt})), r * (hv_{xxt})\big)\, ds = \int \big(r * (h(v|u|^2)_x), r * (hv_{xxt})\big)\, ds \\ &= \int \big(r * (h(|u|^2)_x), r' * (hv_{xx}) - r * (h'v_{xx})\big)\, ds = \int \Big\{\big(-r * (h(|u|^2)_{xt}) \\ &\quad - r * (h'(|u|^2)_x), r * (hv_{xx})\big) - \big(r * (h(|u|^2)_x), r * (h'v_{xx})\big)\Big\}\, ds. \end{aligned}$$

Thus it follows

$$\begin{aligned} 0 &= \int \Big\{\big(r * (h'v_{tx}), r * (hv_{tx})\big) + \big(r * (h'v_{xx}), r * (hv_{xx})\big) - \big(r * (h(|u|^2)_{xt}) \\ &\quad - r * (h'(|u|^2)_x), r * (hv_{xx})\big) - \big(r * (h(|u|^2)_x), r * (h'v_{xx})\big)\Big\}\, ds \\ &= \int \Big\{\big(h'_\delta v_{tx}, r * r * (h_\delta v_{tx})\big) + \big(h'_\delta v_{xx}, r * r * (h_\delta v_{xx})\big) - \big(h'_\delta(|u|^2)_x, r * r * (h_\delta v_{xx})\big) \\ &\quad - \big(r * r * (h_\delta(|u|^2)_x), h'_\delta v_{xx}\big) - \big(r * (h_\delta v_{xx}), r * (h_\delta(|u|^2)_{xt})\big)\Big\}\, ds. \end{aligned}$$

Letting $\delta \to 0$ and taking into account that $(|u|^2)_x \in C(S; H)$ and $v \in C^2_w(S; H) \cap C^1_w(S; H^1_0) \cap C_w(S; H^2)$, we find (cf. [10])

$$\begin{aligned} 0 &= \Big[\big(v_{tx}, r_\varepsilon * r_\varepsilon * (h_0 v_{tx})\big) + \big(v_{xx}, r_\varepsilon * r_\varepsilon * (h_0 v_{xx})\big) - \big((|u|^2)_x, r_\varepsilon * r_\varepsilon * (h_0 v_{xx})\big) \\ &\quad - \big(r_\varepsilon * r_\varepsilon * (h_0(|u|^2)_x), v_{xx}\big)\Big]_0^t - \int \big(r_\varepsilon * (h_0(|u|^2)_{xt}), r_\varepsilon * (h_0 v_{xx})\big)\, ds. \end{aligned}$$

Now we can take the limit $\varepsilon \to 0$ and obtain

$$\varphi(t) := \frac{1}{2}\left(\|v_{tx}(t)\|^2 + \|v_{tx}(t)\|^2\right) = \frac{1}{2}\left(\|v_{tx}(0)\|^2 + \|v_{xx}(0)\|^2\right)$$

$$+ \left((|u|^2)_x, v_{xx}\right)\Big]_0^t - \int_0^t \left((|u|^2)_{xt}, v_{xx}\right) ds \in C(S).$$

This equation implies $v \in C^1(S; H_0^1) \cap C(S; H^2)$. In fact, let $t_n \to t$. Since $v \in C_w^1(S; H_0^1) \cap C_w(S; H^2)$, we have

$$\frac{1}{2}\left(\|v_{tx}(t_n) - v_{tx}(t)\|^2 + \|v_{xx}(t_n) - v_{xx}(t)\|^2\right) = \varphi(t_n) + \varphi(t) - \left(v_{tx}(t_n)\, v_{tx}(t)\right)$$

$$- \left(v_{xx}(t_n), v_{xx}(t)\right) \to 2\varphi(t) + 2\varphi(t) = 0.$$

Finally, the remaining inclusion $v \in C^2(S; H)$ follows from (3.23).

(uniqueness) The proof of the uniqueness assertion is standard.

4. Galerkin's method

According to Galerkin's method we look for approximative solutions of (3.2), (3.3) having the form

$$u_n = \sum_{j=1}^{n} d_j(t)\, h_j + u_a, \quad d_j(0) = (u_0 - u_a, h_j), \quad j = 0, 1, \ldots, n, \tag{4.1}$$

$$v_n = \sum_{j=1}^{n} e_j(t)\, g_j, \quad e_j(0) = (v_0, g_j), \quad e_j'(0) = (v_1, g_j), \quad j = 1, \ldots, n,$$

where the h_j's are the eigenfunctions of the operator A defined by (2.5) while $g_j = \sqrt{2}\sin j\pi x$, $j = 1, 2, \ldots$

We call functions (u_n, v_n) of the form (4.1) the n-th Galerkin approximation of the solution (u, v) of (3.2), (3.3) if

$$\left(iu_{nt} + (\beta - i\gamma)\, u_{nxx} + (k + Q_r v_{nx})\, u_n, h_j\right) = 0, \quad j = 0, 1, \ldots, n, \tag{4.2}$$

$$\left(v_{ntt} - v_{nxx} + (|u_n|^2)_x, g_j\right) = 0, \; j = 1, \ldots, n. \tag{4.3}$$

Here Q_r is the projector of H onto the r-ball in H (cf. (3.10)) and r is an a priori bound for $\|v_x\|_{C(S;H)}$. Such a bound can be found by calculating explicitly the constant c in (3.20).

Remark 4.1. The equations (4.2) differ from the usual Galerkin equations as far as the projector Q_r is involved. We introduced Q_r in (4.2) in order to be able to show that the u_n's are uniformly bounded (cf. (4.4)).

Lemma 4.1. *Under the hypotheses of Theorem 3.1 a unique Galerkin approximation $(u_n, v_n) \in C^1(S; H^2) \times C^2(S; H^2)$ exists for $n = 1, 2, \ldots$ Moreover, the following a priori estimate holds*

$$\|u_n\|_{C(S;H)} + \|u_n\|_{L^2(S;H^1)} \leqq c, \quad n = 1, 2, \ldots \tag{4.4}$$

Proof. The first assertion follows from classical results on ordinary differential equations. To prove (4.4) we set $U_{nj} = u_n(\cdot, j) - a_j$, $j = 0, 1$, and get from (4.2)

$$0 = 2\,\mathrm{Im}\left(iu_{nt} + (\beta - i\gamma)\, u_{nxx} + (k + Q_r v_{nx})\, u_n,\, u_n - u_a\right) = (\|u_n\|^2)_t + 2\sum_{j=0}^{1} \alpha_j$$
$$\times \mathrm{Re}\left((\beta - i\gamma)\,(U_{nj} - a_j)\,(\overline{U}_{nj} + \bar{a}_j)\right) + 2\gamma\,\|u_{nx}\|^2 + 2\,\mathrm{Im}\, k\,\|u_n\|^2 - 2\,\mathrm{Im}\,(iu_n, u_a)_t$$
$$2\sum_{j=0}^{1} \alpha_j\, \mathrm{Re}\left((\beta - i\gamma)\,(U_{nj} - a_j)\,\overline{u}_a(j)\right) + 2\,\mathrm{Im}\left((\beta - i\gamma)\,(u_{nx}, u_{ax}) - (k + Q_r v_{nx})\right.$$
$$\left.\times\, u_n, u_a)\right).$$

Integrating with respect to t yields after simple calculations

$$\|u_n(t)\|^2 + \gamma \int_0^t \|u_{nx}\|^2\, ds = c\left(1 + \int_0^t \|u_n\|^2\, ds\right).$$

Now Gronwall's lemma proves (4.4).

Theorem 4.1. *Under the hypotheses of Theorem* 3.1 *the Galerkin approximations* (u_n, v_n) *converge to the solution* (u, v) *of* (3.2), (3.3). *More precisely we have*

$$\begin{aligned} &u_n \to u \ \text{in}\ C(S; H)\ \text{and}\ L^2(S; H^1),\\ &v_n \to v \ \text{in}\ C(S; H^1)\ \text{and}\ C^1(S; H). \end{aligned} \tag{4.5}$$

Proof. Let $y_n = u_n - u$, $z_n = v_n - v$, $R_n u = \sum_{j=n+1}^{\infty} (u - u_a, h_j)\, h_{\cdot j}$, $R_n v = \sum_{j=n+1}^{\infty} (v, g_j)\, g_j$. We see from Lemma 2.2 and well known results on Fourier series that

$$\left.\begin{aligned} &u_n(0) \to u_0 \text{ in } H^2,\ v_n(0) \to v_0 \text{ in } H^2,\ v_{nt}(0) \to v_1 \text{ in } H^1,\\ &R_n u \to 0 \text{ in } L^2(S; H^1),\ R_n u_t \to 0 \text{ in } L^2(S; H),\\ &R_n v \to 0 \text{ in } H^1(S; H^1) \text{ and } H^2(S; H). \end{aligned}\right\} \tag{4.6}$$

Using (3.2) and (4.3), we get

$$0 = 2\,\mathrm{Im}\left(iy_{nt} + (\beta - i\gamma)\, y_{nxx} + (k + Q_r v_{nx})\, u_n - (k + Q_r v_x)\, u,\, y_n + R_n u\right) = (\|y_n\|^2)_t$$
$$+ 2\beta \sum_{j=0}^{1} \alpha_j\, |y_n(t, j)|^2 + 2\gamma\, \|y_{nx}\|^2 + 2\,\mathrm{Im}\left[\left(ky_n + Q_r v_{nx} y_n + (Q_r v_{nx} - Q_r v_x)\, u, y_n + R_n u\right)\right.$$
$$\left. + (iy_n, R_n u)_t - (iy_n, R_n u_t) + i\sum_{j=0}^{1} \alpha_j (\beta - i\gamma)\, y_n(t, j)\, R_n \overline{u}(t, j) - \left((\beta - i\gamma)\, y_{nx}, (R_n u)_x\right)\right].$$

Taking into account (3.14) and (4.6), we obtain

$$\|y_n(t)\|^2 + \int_0^t \left(\beta \sum_{j=0}^{1} \alpha_j |y_n(s, j)|^2 + \gamma \|y_{nx}\|^2\right) ds \leqq \varepsilon_n + c \int_0^t (\|y_n\|^2 + \|z_{nx}\|^2)\, ds, \tag{4.7}$$

where $\varepsilon_n \to 0$ as $n \to \infty$.

On the other hand it follows from (3.3), (4.3) that

$$0 = \left(z_{ntt} - z_{nxx} + (|u_n|^2 - |u|^2)_x, z_{nt} + R_n v_t\right) = \frac{1}{2}\left(\|z_n(t)\|^2 + \|z_{nx}(t)\|^2\right)_t$$

$$+ 2\operatorname{Re}\,(y_n\bar{u}_{nx} + u\bar{y}_{nx}, z_{nt} + R_n v_t) + (z_{nt}, R_n v_t)_t - (z_{nt}, R_n v_{tt}) + \left(z_{nx}, (R_n v_t)_x\right).$$

By (4.4) and (4.6) we find

$$\|z_{nt}(t)\|^2 + \|z_{nx}(t)\|^2 \leqq \varepsilon_n + \int_0^t \Big(\frac{\gamma}{2}\|y_{nx}\|^2 + c\big(\|y_n\|^2 + \|z_{nx}\|^2$$

$$+ (1 + \|u_{nx}\|^2)\|z_{nt}\|^2\big)\Big)\,ds.$$

Adding (4.7) and (4.8), we obtain

$$\|y_n(t)\|^2 + \|z_{nt}(t)\|^2 + \|z_{nx}(t)\|^2 + \int_0^t \|y_{nx}\|^2\,ds$$

$$\leqq \varepsilon_n + c\int_0^t \left(\|y_n\|^2 + \|z_{nx}\|^2 + (1 + \|u_{nx}\|^2)\|z_{nt}\|^2\right) ds,$$

where $\varepsilon_n \to 0$ as $n \to \infty$. Because of Gronwall's lemma this inequality implies our convergence assertions (4.5).

Remark 4.2. From (2.1) and $u_n \to u$ in $L^2(S; H^1)$ it follows that $u_n(\cdot, 0) \to u(\cdot, 0)$ and $u_n(\cdot, 1) \to u(\cdot, 1)$ in $L^2(S)$. This fact is of practical interest because the boundary values of u represent the physically relevant reflexion and transmission properties of the plasma layer described by (3.2), (3.3).

Remark 4.3. Using the orthonormality of the base functions g_j and h_j, we can rewrite (4.2), (4.3) as follows

$$id_j' + (\beta + i\gamma)\, j^2\pi^2 d_j = \left((\beta - i\gamma)\,(A(u_n - u_a) - u_{nxx}) + (k + Q_r v_{nx})\, u_n, h_j\right) = 0,$$

$$e_j'' + j^2\pi^2 e_j = \sqrt{2}\, j\pi(|u_n|^2, \cos j\pi x).$$

This form suggests a natural way to calculate the coefficients d_j, e_j iteratively. As to an appropriate time discrete variant of such an iteration procedure we refer to [12].

References

[1] Andreev, N. E., K. Sauer, Zeitliche Entwicklung des Reflexionskoeffizienten einer Plasmaschicht bei starker elektromagnetischer Einstrahlung, Akad. Wiss. DDR, Zentralinst. f. Elektronenphysik Preprint 77-4.

[2] Gajewski, H., Über Näherungsverfahren zur Lösung der nichtlinearen Schrödingergleichung, Math. Nachr. **85** (1978), 283–302.

[3] Gajewski, H., On Cauchy's problem for the spatially one-dimensional Zakharov system, Theory of Nonlinear Operators (Proc. Fifth Internat. Summer School Berlin 1977), Abh. Akad. Wiss. DDR Jahrg. 1978 Nr. 6 N, Akademie Verlag, Berlin 1978, 79–96.

[4] Gajewski, H., On an initial-boundary value problem for the nonlinear Schrödlinger equation, Internat. J. Math. & Math. Sci. **2** (1979), 503–522.
[5] Gajewski, H., K. Gröger, Ein Projektions-Iterationsverfahren für Evolutionsgleichungen, Math. Nachr. **72** (1976), 119–136.
[6] Gröger, K., Regularitätsaussagen für Evolutionsgleichungen mit stark monotonen Operatoren, Math. Nachr. **67** (1975), 21–34.
[7] Karpman, V. I., Plasma Physics **13** (1971), 477.
[8] Lions, J. L., Quelques méthodes de résolution des problèmes aux limites non linéaires, Dunod Gauthier-Villars, Paris 1969.
[9] Pereira, N. R., L. Stenflo, Nonlinear Schrödinger equation including growth and damping, Phys. Fluids **20** (1977), 1733–1735.
[10] Strauss, W. A., On continuity of functions with values in various Banach spaces, Pacific J. Math. **19** (1966), 543–551.
[11] Triebel, H., Höhere Analysis, Deutscher Verlag der Wissenschaften, Berlin 1972.
[12] Zacharias, K., On the numerical treatment of wave-envelope equations, Theory of Nonlinear Operators (Proc. Fifth Internat. Summer School Berlin 1977), Abh. Akad. Wiss. DDR Jahrg. 1978 Nr. 6N, Akademie Verlag, Berlin 1978, 293–300.
[13] Zakharov, V. E., Ž. Exp. Theor. Fiz. (JETP) **62** (1972), 1745.

Address of the author:

Akademie der Wissenschaften der DDR
Institut für Mathematik
1080 Berlin
Mohrenstraße 39
DDR

On initial value problems in thermodynamics

Konrad Gröger and Rolf Hünlich, Berlin

Introduction

The methods and results of convex analysis and the theory of evolution equations in Hilbert spaces have been useful tools for the formulation and the theoretical and numerical solution of many problems arising in physics and especially in mechanics (see e.g. Duvaut-Lions [4], Moreau [15], Halphen-Nguyen [9], Gröger [6], Hünlich [11]). Using the same tools in this paper we try to give a unified treatment of arbitrary thermodynamic processes (such as thermomechanic processes, heat conduction, diffusion, chemical reactions etc.). To this end it is necessary to find out what is common for all thermodynamic systems and what is moreover essential for the formulation of laws governing the behavior of such systems. In other words, one needs an abstract idea of a thermodynamic system and its laws. Therefore in Section 1 of this paper we make an attempt to define a notion of a thermodynamic system. In fact we shall restrict our considerations to processes that are in a certain sense small, and we use this restriction already in the definition of a thermodynamic system, since it leads to a remarkable simplification. Thus our definition of a thermodynamic system is reasonable only in the context of this paper. In the main we can represent a thermodynamic system by a linear space (the state space) and two functionals defined on this space which are interpreted as internal energy and dissipation potential of the system, respectively. From these two functionals one can derive the laws characterizing the behavior of the system. These laws guarantee that for any process of the system the Fundamental Laws of Thermodynamics are satisfied.

That the behavior of a thermodynamic system can be characterized by two functions was pointed out by Ziegler [18]. Although we use functions that are somewhat different from those proposed by Ziegler we follow his intentions.

In the second section we show that in view of the laws introduced in Section 1 the evolution of a thermodynamic system is described by an initial value problem of the type

$$Au + Bu' \ni f,\ u(0) = u_0,$$

where A and B are in general nonlinear (multivalued) subdifferential mappings. Since there exists no general theory for problems of that type we discuss special cases leading to satisfactory results. In this connection we introduce the notion of the serial composition of two thermodynamic systems. We show that for serial compositions one can obtain results under assumptions that are verified without difficulty for interesting concrete systems.

The third section of the paper is devoted to approximation methods. For a large class of systems we prove the convergence of a discretization method with respect to time. Moreover, we discuss the problems arising after the discretization for each time step. Again it turns out to be useful to consider systems that are serial compositions. We show that under natural additional assumptions for each time step one has to solve an equation with a strongly monotone and Lipschitzian operator. Hence one can use standard projection and iteration methods to do this. We illustrate our abstract results by a model of a thermoplastic body.

Throughout this paper we use the basic notions and the usual notation of convex analysis without further explanation (see e.g. Barbu [1]). In general our notation is the same as that of Gröger [6]. If H is a Hilbert space we denote by $\Gamma(H)$ the set of convex, proper, lower semicontinuous functionals $\varphi : H \to]-\infty, +\infty]$.

1. Thermodynamic systems

As announced in the introduction we start this section with a definition of the notion of a thermodynamic system that suites for our purposes. After that we comment the items of this definition.

Definition 1. We speak of a *thermodynamic system* if we are given an object $\mathcal{T} = (Y, E, F, U, K)$ consisting of two Hilbert spaces Y, U, two functionals $E, F \in \Gamma(Y)$ and an operator $K \in \mathcal{L}(U; Y)$ such that

$$E(0) < \infty, \; 0 \in \partial F(0), \; \operatorname{Ker} K = \{0\}, \; \operatorname{Im} K = \overline{\operatorname{Im} K}. \tag{1.1}$$

The space Y is called the *state space* of the system $\mathcal{T}$, E its *internal energy* and F its *dissipation potential*. The space U is called the space of *generalized displacements* of $\mathcal{T}$.

1.1. Comments on the state space Y and the internal energy E. We have assumed Y to be a linear space (and not a manifold) because we want to consider "small" state deviations only. The point 0 is seen as the "natural state" of the system. Thus $y \in Y$ is a state deviation. The elements of Y' (dual to Y) are called *thermodynamic forces* or simply *forces*. For $z \in Y'$, $y \in Y$ the dual pairing $\langle z, y \rangle$ denotes the work of the force z for the state deviation y. The assumption that Y is a *Hilbert* space is made for the sake of simplicity; more general spaces are needed only for exceptional cases.

The meaning of the functional E can be explained as follows. For $y \in Y$ the set $\partial E(y)$ is a subset of Y', i.e. a set of thermodynamic forces. Each force $z^s \in \partial E(y)$ is interpreted as a *static* force corresponding to the state y. Therefore, following Moreau [15], we shall say that ∂E defines the *statical law* of the system $\mathcal{T}$. If E is differentiable in a neighbourhood of the natural state then it can be approximated by a quadratic functional since we consider only small state deviations. In that case the statical law becomes a linear one (generalizing Hooke's law in elasticity). If E is not differentiable (for example if there exist unilateral constraints for the state deviations) then an approximation of E by a quadratic functional does not make sense.

The existence of E as a state function along with its interpretation is one way of expressing the First Law of Thermodynamics.

Many authors instead of the internal energy prefer the so called free energy in order to derive the statical law of a thermodynamic system. However, in general the free energy is not a convex functional and it is not so easy to apply methods of convex analysis.

The internal energy of a real thermodynamic system can depend on time explicitely. For the sake of simplicity we have ignored that possibility.

Example 1. Let $Y = L^2(G; \mathscr{S} \times \mathbb{R}^{n+1}) = Y'$ (where $\mathscr{S}$ is the space of symmetric 3×3 matrices). Choosing Y in this way we think of a porous body (occupying the domain $G \subset \mathbb{R}^3$) subject to deformation, heat conduction, diffusion and chemical reactions (of n substances). If $y = (\varepsilon, \eta, c) \in Y$, $c = (c_1, \ldots, c_n)$, $z = (\sigma, \vartheta, \mu) \in Y'$, $\mu = (\mu_1, \ldots, \mu_n)$, the components of y and z are to be interpreted as follows:
ε deformation, η entropy (deviation), c concentration (deviation), σ stress, ϑ temperature (deviation), μ chemical potential. The components of y are called *state variables*. Furthermore, let

$$E(y) = \int_G e(y(x))\, dx \quad \text{for} \quad y = (\varepsilon, \eta, c) \in Y,$$

where $e: \mathscr{S} \times \mathbb{R}^{n+1} \to \mathbb{R}$ is a quadratic (strongly convex) function, the *density* of the internal energy. If

$$(\sigma^s, \vartheta^s, \mu^s) = E'(\varepsilon, \eta, c)$$

then σ^s, ϑ^s, μ^s are the static stress, the temperature and the chemical potential of the state $y = (\varepsilon, \eta, c)$, respectively. The discussion of the dissipation potential will show that in general the choice of the entropy as a state variable is quite natural.

1.2. Comments on the dissipation potential F. The meaning of the functional F can be explained only if processes $y: S \to Y$ are considered ($S = [0, T]$ denotes a fixed interval of time). If $y: S \to Y$ is a process such that $y'(t)$ exists then $\partial F(y'(t)) \in Y'$ is interpreted as the set of possible *dissipative thermodynamic forces* at the time t for the process y. In other words, if for a system the rate of its state deviation is y' then there acts one of the dissipative forces $z^d \in \partial F(y')$. We shall say that ∂F defines the *dissipation law* of the system.

For a real thermodynamic system the dissipation potential would depend not only on y' but also explicitely on y. We have neglected this dependence since we restrict ourselves to small state deviations.

If $z^d \in \partial F(y')$ then $\langle z^d, y' \rangle \geqq 0$ since ∂F is monotone and $0 \in \partial F(0)$ by assumption (cf. (1.1)). Hence the *dissipative power* of any process is not less than zero. This is one of the possible formulations of the Second Law of Thermodynamics. Moreover, the existence of a potential of the dissipative forces generalizes Onsager's reciprocity relations (cf. Ziegler [18]).

If $\langle z^d, y' \rangle$ has the same value for all $z^d \in \partial F(y')$ (what can happen even if $\partial F(y')$ contains more than one element) then $\tilde{F}(y') := \langle z^d, y' \rangle$ is called the *dissipation function* of the system. In his approach to thermodynamic systems Ziegler [18] assumes $\tilde{F}$ to be given instead of F. The introduction of F is due to Moreau (cf. [15]).

If the entropy is a state variable, i.e. one of the components of y, then the dissipation law $z^d \in \partial F(y')$ which can be written also as $y' \in \partial F^*(z^d)$ includes a relation for the entropy rate. Considering concrete systems one can see that this relation is valid only if the so called internal entropy production of the system is neglected. We don't want to go into details about internal entropy production. Let us mention only that it is of the same order of smallness as the dissipative power $\langle z^d, y' \rangle$. Therefore, our interpretation of F is justified if both z^d and y' are "small".

For many thermodynamic systems it is convenient to formulate the dissipation law by means of additional variables called "driving forces" and "flows". We illustrate this by an example.

Example 2. Let Y be given and interpreted as in Example 1. We are interested in a reasonable dissipation law

$$(\varepsilon', \eta', c') \in \partial F^*(\sigma^d, \vartheta^d, \mu^d).$$

Usually the entropy rate does not depend on the temperature directly but via its gradient. Similarly the concentration rate depends on the gradients of the chemical potentials and, if reactions take place, on the so called affinities of the system with respect to the possible reactions. This motivates the following definitions:

$$D := L^2(G; \mathscr{S} \times \mathbb{R}^{3(n+1)+m}), \; D(L) := L^2(G; \mathscr{S}) \times H^1(G; \mathbb{R}^{n+1}) \subset Y',$$

$$Lz^d = L(\sigma^d, \vartheta^d, \mu^d) := (\sigma^d, -\operatorname{grad} \vartheta^d, -\operatorname{grad} \mu^d, a^d). \qquad \text{where}$$

$$a^d = (a_1^d, \ldots, a_m^d), \; a_k^d = c_{0k} \vartheta^d + \sum_{i=1}^{n} c_{ik} \mu_i^d, \; k = 1, \ldots, m.$$

The elements of D are called *driving forces*. The linear operator $L: D(L) \to D$ assigns to "regular" (dissipative) forces a driving force. The variables a_k^d, $k = 1, \ldots, m$, denote the affinities with respect to the various possible chemical reactions. They could depend also on σ^d (cf. Biot [2]). The elements of the space D' dual to D are interpreted as *flows*. More precisely, if $q = (p, q_\eta, q_c, r) \in D'$ then p is the deformation rate, q_η the entropy flow, q_c the mass flow and r the reaction rate. Let

$$\Phi(d) := \int_G \varphi(d(x)) \, dx \quad \text{for} \quad d \in D,$$

where $\varphi: \mathscr{S} \times \mathbb{R}^{3(n+1)+m} \to \mathbb{R}$ is a given quadratic (strongly convex) functional. We define now $F \in \Gamma(Y)$ by

$$F^*(z^d) := \begin{cases} \Phi(Lz^d) & \text{if} \quad z^d \in D(L) \\ +\infty & \text{if} \quad z^d \in Y' \setminus D(L) \end{cases}, \quad z^d \in Y'.$$

Then $F^* \in \Gamma(Y')$, $\partial F^* = L^* \, \partial \Phi L$ and

$$z^d \in \partial F(y') \Leftrightarrow y' = L^* q, \; q \in \partial \Phi(d), \; d = Lz^d.$$

The relation $y' = L^* q$ can be written detailed as follows:

$$(\varepsilon', \eta', c') = (p, \operatorname{div} q_\eta + s_\eta, \operatorname{div} q_c + s_c), \; q_{\eta n}|_{\partial G} = 0, \; q_{cn}|_{\partial G} = 0,$$

where $s_\eta = \sum_{k=1}^{m} c_{0k} r_k$, $s_c = \left(\sum_{k=1}^{m} c_{1k} r_k, \ldots, \sum_{k=1}^{m} c_{nk} r_k \right)$.

Since r_k, $k = 1, \ldots, m$, are the rates of the various reactions these relations show clearly the meaning of the constants c_{ik}. In particular, mass conservation is guaranteed if

$$\sum_{i=1}^{n} c_{ik} = 0, \quad k = 1, \ldots, m.$$

Remark 1. We may assume that for every thermodynamic system there exist a Hilbert space D, an operator L and a functional Φ such that

$$L : D(L) \subset Y' \to D \text{ is linear and closed, } \overline{D(L)} = Y',$$

$$\Phi \in \Gamma(D), \ 0 \in \partial\Phi(0),$$

$$F^*(z) = \begin{cases} \Phi(Lz) & \text{if } z \in D(L) \\ +\infty & \text{if } z \in Y' \setminus D(L) \end{cases}, \qquad \partial F^* = L^* \, \partial\Phi L.$$

Indeed, if there is no other "natural" choice one can choose $D = Y'$, $L =$ identity, $\Phi = F^*$.

Example 3. Let $Y = L^2(G) = Y'$ and let us interprete the elements of Y and Y' as entropy fields and temperature fields, respectively. Furthermore, let C be a nonempty closed convex subset of Y' and let $F^*(\vartheta) := I_C(\vartheta)$. Then the dissipation law is $\eta' \in \partial I_C(\vartheta^d)$. This law corresponds to an (ideal) control system assuring that the temperatur ϑ^d remains always in C. Clearly there is an analogy to the constitutive relation of a rigid-plastic body.

Remark 2. If it is said that for a thermodynamic system there exists an "external" source this means that the dissipation law $y' \in \partial F^*(z^d)$ is to be replaced by $y' \in \partial F^*(z^d) + s$, where $s : S \to Y$ is a given "source" function.

Remark 3. We have formulated the laws of a thermodynamic system without making use of heat supply (using instead of this entropy supply). This seems to be natural from the physical point of view and it is also convenient from the mathematical point of view, since for typical systems the entropy supply is an element of a suitable L^2-space whereas the heat supply being an energy rate belongs to an L^1-space.

1.3. Comments on the space U and the operator K. If the state space Y is choosen and interpreted as in Example 1 then by means of the elements of Y and Y' we can deal with deformations and stresses but not with displacements and mechanical forces (at least not in a simple way). Therefore into our definition of a thermodynamic system we have included another space U the elements of which are to be regarded as *generalized displacements*. The elements of the dual space U' will be called *generalized forces*. The operator K assigns to every generalized displacement $u \in U$ a state $y \in Y$. We shall say that a state y satisfies the compatibility condition if $y = Ku$ for a suitable $u \in U$. Furthermore, we shall say that there is equilibrium between a thermodynamic force $z \in Y'$ and a generalized force $f \in U'$ if

$$\forall\, u \in U: \langle f, u \rangle = \langle z, Ku \rangle$$

(principle of virtual work), i.e. if $f = K^*z$. We shall need this relation later on for $z = z^s + z^d$, i.e. for the sum of a static and a dissipative thermodynamic force.

Example 4. Let $U = H_0^1(G; \mathbb{R}^3) \times L^2(G; \mathbb{R}^{n+1})$. If $u = (v, \eta, c) \in U$ and $f = (g, \vartheta, \mu) \in U' = H^{-1}(G; \mathbb{R}^3) \times L^2(G; \mathbb{R}^{n+1})$ then v is a displacement, g a body force and η, c, ϑ, μ have the same meaning as in Example 1. Let $Y = L^2(G; \mathcal{S} \times \mathbb{R}^{n+1})$ and

$$Ku = K(v, \eta, c) := \big(\varepsilon(v), \eta, c\big),$$

where $\varepsilon(v)$ is the infinitesimal strain associated to v, i.e.

$$\varepsilon_{ij}(v) := \frac{1}{2}\,(v_{i,j} + v_{j,i}).$$

Then the adjoint operator $K^* \in \mathcal{L}(U'; Y')$ is given by

$$K^*z = K^*(\sigma, \vartheta, \mu) = (-\operatorname{div}\sigma, \vartheta, \mu).$$

If there act only body forces on the system, i.e., if $f = (g, 0, 0)$, $g \in H^{-1}(G; \mathbb{R}^3)$, we have equilibrium between f and $z^s + z^d$ if

$$-\operatorname{div}(\sigma^s + \sigma^d) = g,\ \vartheta^s + \vartheta^d = 0,\ \mu^s + \mu^d = 0.$$

Since ϑ^s and μ^s are the temperature and the chemical potential of the body this leads to an interpretation of ϑ^d and μ^d.

2. Initial value problems

We want to consider quasi-static processes of a thermodynamic system $\mathcal{T} = (Y, E, F, U, K)$ during a fixed period of time $S = [0, T]$. We assume that we are given an initial state $y_0 \in Y$ such that $y_0 = Ku_0$ and a function $f \in L^2(S; U')$ representing the "external" generalized forces. The process we are looking for is described by functions u, y, z^s, z^d satisfying the following conditions:

$$\left.\begin{array}{lll} y = Ku & \text{compatibility,} & \\ K^*(z^s + z^d) = f & \text{quasi-static equilibrium,} & \\ z^s \in \partial E(y) & \text{statical law,} & \\ z^d \in \partial F(y') & \text{dissipation law,} & \\ y(0) = y_0 & \text{initial condition,} & \\ (u, y) \in H^1(S; U \times Y),\ (z^s, z^d) \in L^2(S; Y' \times Y') & & \text{regularity.} \end{array}\right\} \tag{2.1}$$

The regularity requirement is made in order to obtain a precise formulation of the problem. The multivalued mappings ∂E and ∂F are considered in the usual way as mappings from $L^2(S; Y)$ into $L^2(S; Y')$. Analogously, K is considered as a mapping from $L^2(S; U)$ into $L^2(S; Y)$. The problem (2.1) can be written shortly as follows:

$$K^*\big(\partial E(Ku) + \partial F(Ku')\big) \ni f,\ u(0) = u_0,\ u \in H^1(S; U). \tag{2.2}$$

As a matter of fact (2.1) and (2.2) are equivalent only with respect to existence but not necessarily with respect to uniqueness: It can happen that z^s, z^d are not unique even if (2.2) has a unique solution u.

2.1. Existence and uniqueness results. For our discussion of the problem (2.2) the following definition turns out to be useful.

Definition 2. We shall call a thermodynamic system $\mathcal{T} = (Y, E, F, U, K)$ *normal* if

$$\partial(E \circ K) = K^* \, \partial E K, \quad \partial(F \circ K) = K^* \, \partial F K. \tag{2.3}$$

The relation (2.3) holds for example if E and F are continuous for at least one state from $\operatorname{Im} K$ (see e.g. Joffe-Tichomirov [14]). However this standard result from convex analysis is not applicable to various interesting thermodynamic systems. In some cases the normality of a thermodynamic system can be proved by means of the following lemma.

Lemma 1. *Let* $K \in \mathcal{L}(U; Y)$ *be such that* $\operatorname{Ker} K = \{0\}$, $\operatorname{Im} K = \overline{\operatorname{Im} K}$, *and let* $\varphi \in \Gamma(Y)$ *satisfy the condition*

$$z \in \partial\varphi(y) \Rightarrow \|z\|_{Y'} \leqq \gamma(\|y\|_Y + \|K^* z\|_{U'}), \tag{2.4}$$

where $\gamma : \mathbb{R}^+ \to \mathbb{R}^+$ *is an increasing function. Then* $\partial(\varphi \circ K) = K^* \, \partial\varphi K$.

Proof. It is easy to see that $K^* \, \partial\varphi K \subset \partial(\varphi \circ K)$. Hence it is sufficient to show that $K^* \, \partial\varphi K$ is maximal monotone. To this end we prove for arbitrary $f \in U'$ the solvability of

$$K^* \, \partial\varphi K u + K^* J K u \ni f \qquad (J \text{ duality map of } Y). \tag{2.5}$$

Let $z := JK(K^*JK)^{-1} f$. Then $K^* z = f$ and $Pz = 0$, where P denotes the orthogonal projection of Y' onto $\operatorname{Ker} K^*$. Let (c_n) be a sequence of positive numbers such that $c_n \uparrow +\infty$. Then for every $n \in \mathbb{N}$ there exists $y_n \in Y$ such that

$$\partial\varphi(y_n) + c_n P J y_n + J y_n \ni z. \tag{2.6}$$

Since $\partial\varphi + c_n PJ + J$ is strongly monotone uniformly with respect to n the sequence (y_n) is bounded in Y. In view of (2.4) this implies that $(c_n P J y_n)$ is bounded too. Hence we may assume without loss of generality that $y_n \rightharpoonup y$ in Y, $c_n P J y_n \rightharpoonup \zeta$ in Y', where $PJy = \lim\limits_{n\to\infty} PJy_n = 0$ and $\zeta \in \operatorname{Im} P$. Hence

$$\begin{aligned} &y_n \rightharpoonup y \text{ in } Y, \; z - c_n P J y_n - J y_n \rightharpoonup z - \zeta - Jy \text{ in } Y', \\ &\varlimsup_{n\to\infty} \langle z - c_n P J y_n - J y_n, y_n \rangle \leqq \langle z, y \rangle - 0 - \langle Jy, y \rangle = \langle z - \zeta - Jy, y \rangle. \end{aligned} \tag{2.7}$$

In view of the maximal monotonicity of $\partial\varphi$ from (2.6), (2.7) it follows $z - \zeta - Jy \in \partial\varphi(y)$ (cf. Brézis [3], Prop. 2.5). Since $PJy = 0$ we have $y = Ku$. Thus

$$K^* \, \partial\varphi K u + K^* J K u \ni K^*(z - \zeta) = K^* z = f,$$

i.e., the problem (2.5) is solvable. This completes the proof.

Let $\mathcal{T}$ be a normal thermodynamic system. Then we can rewrite problem (2.2) as follows:

$$\partial(E \circ K)\, u + \partial(F \circ K)\, u' \ni f, \; u(0) = u_0, \; u \in H^1(S; U). \tag{2.8}$$

Moreover, it is easy to see that there exist also the following two possibilities to formulate this problem:

$$\partial(E + I_{\mathrm{Im}K})\, y + \partial F(y') \ni g,\ y(0) = y_0,\ y \in H^1(S; Y), \tag{2.9}$$

$$\partial E(y) + \partial(F + I_{\mathrm{Im}K})\, y' \ni g,\ y(0) = y_0,\ y \in H^1(S; Y); \tag{2.10}$$

here g denotes any function from $L^2(S; Y')$ such that $K^*g = f$. To our knowledge there does not exist a general (unified) theory of initial value problems of type (2.8). However, the existing results on different types of evolution equations in Hilbert spaces can be applied to such problems if only E and F are specialized appropriately. In what follows we deal shortly with some important special cases. After that we discuss in the next subsection more detailed so called serial compositions of thermodynamic systems.

In the remaining part of this subsection $\mathcal{T} = (Y, E, F, U, K)$ is assumed to be a normal thermodynamic system.

Theorem 1. *Suppose that $\partial E : Y \to Y'$ is Lipschitzian and that $\partial F \subset Y \times Y'$ is strongly monotone. Then for arbitrary $f \in L^2(S; U')$ and $u_0 \in U$ there exists a unique solution u to problem* (2.8).

Proof. The problem (2.8) is easily reduced to an initial value problem for an ordinary differential equation in the space U, and the assertion follows from standard results on such problems (see e.g. Gajewski-Gröger-Zacharias [5], Chap. V).

Remark 4. The assumptions of Theorem 1 are satisfied for example for viscoelastic media.

Theorem 2. *Suppose that ∂E belongs to $\mathcal{L}(Y; Y')$ and is positive definite. Then for $f \in H^1(S; U')$ and $u_0 \in U$ such that*

$$f_0^d := f(0) - K^*\, \partial E K u_0 \in D\big((F \circ K)^*\big) \tag{2.11}$$

there exists a unique solution u to problem (2.8).

Proof. Elementary calculations show that u is a solution to (2.8) if and only if $f^d := f - K^*\, \partial E K u$ is a solution to

$$(f^d)' + \partial(E \circ K)\, \partial(F \circ K)^* f^d \ni f',\quad f^d(0) = f_0^d,\quad f^d \in H^1(S; U'). \tag{2.12}$$

Since $\partial(E \circ K)$ can be considered as the duality map of U the unique solvability of (2.12) follows from well known results on evolution equations with subdifferential operators (see e.g. Brézis [3]).

Remark 5. The unknown f^d of problem (2.12) has the meaning of a "dissipative generalized force". Using the relation $\partial(F \circ K) = K^*\, \partial F K$ one can prove that (2.11) holds if and only if there exists $z_0^d \in D(F^*)$ such that $K^* z_0^d = f_0^d$.

Remark 6. The assumptions of Theorem 2 are satisfied for various processes (deformation of elastoplastic materials, heat conduction, diffusion etc.). Let us mention that using a result of Gröger-Nečas [8] one could weaken somewhat the assumption that ∂E is linear (cf. also Trávniček [17]). Below we shalls show that the definiteness assumption with respect to ∂E can also be weakened (see Theorem 4).

Theorem 3. *Suppose that ∂F belongs to $\mathscr{L}(Y; Y')$ and is positive definite. Then for arbitrary $f \in L^2(S; U')$ and $u_0 \in U$ such that $Ku_0 \in D(E)$ there exists a unique solution u to problem* (2.8).

Proof. Since $\partial(F \circ K)$ can be considered as the duality map of U the theorem follows from the same result on evolution equations as Theorem 2.

Remark 7. The assumptions of Theorem 3 are satisfied for viscoelastic bodies with a linear viscosity law the motion of which can be restricted by obstacles.

Now we want to generalize Theorem 2 as announced in Remark 6. To this end we prove a lemma which is of interest for its own.

Let H be a Hilbert space and let A be such that

$$A \in \mathscr{L}(H; H') \text{ is symmetric and positive,} \quad \operatorname{Im} A = \overline{\operatorname{Im} A}. \tag{2.13}$$

Furthermore, let $\varphi \in \Gamma(H)$ satisfy the conditions

$$v \in \partial\varphi^*(g) \Rightarrow \|v\|_H \leqq \text{const}\,(\|g\|_{H'} + \|Av\|_{H'} + 1), \tag{2.14}$$

$$v_i \in \partial\varphi^*(g), \quad i = 1, 2, \quad Av_1 = Av_2 \Rightarrow v_1 = v_2. \tag{2.15}$$

Lemma 2. *Suppose that* (2.13)—(2.15) *hold. Then for every $f \in H^1(S; H')$ and $v_0 \in H$ such that $f(0) - Av_0 \in D(\varphi^*)$ there exists a unique solution v to the problem*

$$Av + \partial\varphi(v') \ni f, \quad v(0) = v_0, \quad v \in H^1(S; H). \tag{2.16}$$

Proof. (i) The existence of a solution to (2.16) will be proved later via discretization of time (see the proof of Lemma 3). One could prove the existence also by a regularization of $\partial\varphi$ or A. (ii) Let v_1, v_2 be two solutions to (2.16). Then due to the monotonicity of $\partial\varphi$ we have for every $t \in S$

$$\frac{1}{2}\langle A(v_1 - v_2)(t), (v_1 - v_2)(t)\rangle = \int_0^t \langle A(v_1 - v_2)(s), (v_1' - v_2')(s)\rangle\, ds \leqq 0.$$

Hence $Av_1 = Av_2$ and $Av_1' = Av_2'$. In view of (2.15) this result along with $v_i' \in \partial\varphi^*(f - Av_i)$ shows that $v_1 = v_2$.

Now we return to the investigation of a normal thermodynamic system $\mathscr{T} = (Y, E, F, U, K)$. From Lemma 2 it follows immediately

Theorem 4. *Suppose that $A := \partial E$ belongs to $\mathscr{L}(Y; Y')$, that $\operatorname{Im} A$ is closed and that*

$$y \in \partial(F + I_{\operatorname{Im} K})^* z \Rightarrow \|y\|_Y \leqq \text{const}\,(\|z\|_{Y'} + \|Ay\|_{Y'} + 1). \tag{2.17}$$

Then for $g \in H^1(S; Y')$ and $y_0 \in \operatorname{Im} K$ such that $g(0) - Ay_0 \in D(F^*)$ there exists a solution y to problem (2.10). *If in addition*

$$y_i \in \partial(F + I_{\operatorname{Im} K})^* z, \quad Ay_1 = Ay_2 \Rightarrow y_1 = y_2, \tag{2.18}$$

then the solution to problem (2.10) *is unique.*

Remark 8. At the first glance it seems difficult to verify the crucial condition (2.17) in applications. However we shall see in the following subsection that for so called serial compositions of thermodynamic systems one can reduce (2.17) to simpler conditions that are easy to check in applications. If A is positive definite then (2.17), (2.18) are trivially satisfied and we come back to Theorem 2 since (2.10) is equivalent to (2.8) (cf. also Remark 6).

2.2. Serial compositions of thermodynamic systems

Definition 3. If $\mathcal{T}_i = (Y, E_i, F_i, U, K)$, $i = 1, 2$, are two thermodynamic systems, the serial composition $\overline{\mathcal{T}}$ of $\mathcal{T}_1$ and $\mathcal{T}_2$ is introduced as follows:

$$\overline{\mathcal{T}} = (\overline{Y}, \overline{E}, \overline{F}, \overline{U}, \overline{K}), \quad \text{where} \quad \overline{Y} := Y \times Y, \quad \overline{E}(y_1, y_2) := E_1(y_1) + E_2(y_2),$$
$$\overline{F}(y_1, y_2) := F_1(y_1) + F_2(y_2), \quad \overline{U} := U \times Y, \quad \overline{K}(u, y) := (Ku - y, y).$$

Remark 9. It is easy to see that $\overline{\mathcal{T}}$ is a thermodynamic system in the sense of Definition 1 and that the formal definition of $\overline{\mathcal{T}}$ corresponds to the intuitive idea of a serial composition. We are interested in conditions on $\mathcal{T}_1$ and $\mathcal{T}_2$ which guarantee that $\overline{\mathcal{T}}$ satisfies the assumptions of one of the preceding theorems. The following theorem yields a sufficient condition for the normality of $\overline{\mathcal{T}}$.

Theorem 5. *Suppose that ∂E_1 is bounded and that for $i = 1$ or $i = 2$*

$$z \in \partial F_i(y) \Rightarrow \|z\|_{Y'} \leqq \gamma(\|y\|_Y + \|K^*z\|_{U'}), \tag{2.19}$$

where $\gamma : \mathbb{R}^+ \to \mathbb{R}^+$ is an increasing function. Then $\overline{\mathcal{T}} = (\overline{Y}, \overline{E}, \overline{F}, \overline{U}, \overline{K})$ is normal.

Proof. In order to prove $\partial(\overline{F} \circ \overline{K}) = \overline{K}^* \partial\overline{F}\overline{K}$ we use Lemma 1. Let $\bar{z} \in \partial\overline{F}(\bar{y})$ for $\bar{z} = (z_1, z_2)$, $\bar{y} = (y_1, y_2)$, i.e. $z_i \in \partial F_i(y_i)$, $i = 1, 2$. Then $\overline{K}^*\bar{z} = (K^*z_1, z_2 - z_1)$. If ∂F_1 satisfies (2.19) then

$$\begin{aligned}\|\bar{z}\|_{\overline{Y}'} &\leqq \|z_1\|_{Y'} + \|z_2\|_{Y'} \leqq 2\,\|z_1\|_{Y'} + \|z_2 - z_1\|_{Y'}\\ &\leqq 2\gamma(\|y_1\|_Y + \|K^*z_1\|_{U'}) + \|z_2 - z_1\|_{Y'} \leqq 2\gamma(\|\bar{y}\|_{\overline{Y}} + \|\overline{K}^*\bar{z}\|_{\overline{U}'}) + \|\overline{K}^*\bar{z}\|_{\overline{U}'}\\ &\leqq \bar{\gamma}(\|\bar{y}\|_{\overline{Y}} + \|\overline{K}^*\bar{z}\|_{\overline{U}'}),\end{aligned}$$

where $\bar{\gamma}(r) := 2\gamma(r) + r$. If ∂F_2 satisfies (2.19) one can prove a similar result. Thus for $\overline{F}$ an analogue of (2.4) is valid and Lemma 1 yields the relation $\partial(\overline{F} \circ \overline{K}) = \overline{K}^* \partial\overline{F}\overline{K}$. The operator ∂E_1 being bounded satisfies a condition of the type (2.19). Therefore one can prove in the same way that $\partial(\overline{E} \circ \overline{K}) = \overline{K}^* \partial\overline{E}\overline{K}$.

Remark 10. It is now easy to formulate conditions for $\mathcal{T}_1$, $\mathcal{T}_2$ that guarantee the applicability of Theorem 1 — Theorem 3 to the system $\overline{\mathcal{T}}$. For example, if $\partial E_i \in \mathcal{L}(Y; Y')$, $i = 1, 2$, are positive definite and ∂F_1 is bounded then $\overline{\mathcal{T}}$ satisfies the conditions of Theorem 2. In the following theorem we present sufficient conditions for the applicability of Theorem 4 to a serial composition $\overline{\mathcal{T}}$.

Theorem 6. *Suppose that $A_i := \partial E_i \in \mathcal{L}(Y;\ Y')$, that A_1 is positive definite and that* Im A_2 *is closed. Furthermore let the following conditions be satisfied:*

$$z \in \partial F_1(y) \Rightarrow \|z\|_{Y'} \leqq \text{const}\ (\|y\|_Y + \|K^*z\|_{U'} + 1), \tag{2.20}$$
$$y \in \partial F_2^*(z) \Rightarrow \|y\|_Y \leqq \text{const}\ (\|z\|_{Y'} + \|A_2 y\|_{Y'} + 1). \tag{2.21}$$

Then for the system $\bar{\mathcal{T}}$ there holds an analogue of condition (2.17), *i.e., we have with* $\bar{A} := \partial\bar{E}$

$$\bar{y} \in \partial(\bar{F} + I_{\mathrm{Im}\bar{K}})^* \bar{z} \Rightarrow \|\bar{y}\|_{\bar{Y}} \leqq \text{const}\, (\|\bar{z}\|_{\bar{Y}'} + \|\bar{A}\bar{y}\|_{\bar{Y}'} + 1), \tag{2.22}$$

If

$$\begin{aligned} &z_i \in \partial F_1(y), \quad i = 1, 2, \quad K^* z_1 = K^* z_2 \Rightarrow z_1 = z_2, \\ &y_i \in \partial F_2^*(z), \quad i = 1, 2, \quad A_2 y_1 = A_2 y_1 \Rightarrow y_1 = y_2, \end{aligned} \tag{2.23}$$

then an analogue of condition (2.18) *holds for* $\bar{\mathcal{T}}$, *i.e.*

$$\bar{y}_i \in \partial(\bar{F} + I_{\mathrm{Im}\bar{K}})^* \bar{z}, \quad \bar{A}\bar{y}_1 = \bar{A}\bar{y}_2 \Rightarrow \bar{y}_1 = \bar{y}_2. \tag{2.24}$$

Proof. The assertion (2.22) can be written as follows:

$$\left.\begin{aligned} &\bar{z} \in \partial\bar{F}(\bar{y}) + \bar{\xi}, \quad \bar{y} \in \mathrm{Im}\,\bar{K}, \quad \bar{\xi} \in \mathrm{Ker}\,\bar{K}^* \Rightarrow \\ &\|\bar{y}\|_{\bar{Y}} \leqq \text{const}\, (\|\bar{z}\|_{\bar{Y}'} + \|\bar{A}\bar{y}\|_{\bar{Y}'} + 1). \end{aligned}\right\} \tag{2.25}$$

Let $\bar{y} = (y_1, y_2)$, $\bar{z} = (z_1, z_2)$. We have $\bar{\xi} = (\zeta, \zeta)$, where $\zeta \in \mathrm{Ker}\, K^*$, since $\bar{K}^*(\zeta_1, \zeta_2) = (K^*\zeta_1, \zeta_2 - \zeta_1)$. Clearly, (2.25) holds if

$$\left.\begin{aligned} &z_i \in \partial F_i(y_i) + \zeta, \quad \zeta \in \mathrm{Ker}\, K^*, \quad i = 1, 2 \Rightarrow \\ &\|y_i\|_Y \leqq \text{const}\, (\|z_1\|_{Y'} + \|z_2\|_{Y'} + \|A_1 y_1\|_{Y'} + \|A_2 y_2\|_{Y'} + 1), \quad i = 1, 2. \end{aligned}\right\} \tag{2.26}$$

Since A_1 is positive definite, the condition for $\|y_1\|_Y$ is trivially satisfied. In view of (2.21) under the premises of (2.26) the norm of y_2 can be estimated by $\|z_2 - \zeta\|_{Y'}$ and $\|A_2 y_2\|_{Y'}$. Because $\|\zeta\|_{Y'} \leqq \text{const}\, (\|y_1\|_Y + \|z_1\|_{Y'} + 1)$ (cf. (2.20)) finally we obtain that (2.26) is valid. This proves the first assertion of the theorem. The proof of the second assertion uses the same idea and can therefore be omitted.

3. Approximation methods

In this section we shall discuss some approximation methods for solving certain initial value problems considered above. The first one is an implicit discretization with respect to time. In the next stage we have to handle the problems that arise in each time step. Using a more detailed structure of the thermodynamic system (serial composition, product structure of the state space) we can reduce these problems to equations with continuous monotone potential operators. In some special cases those operators are even Lipschitzian and strongly monotone. Then further standard approximation methods can be applied. This will be illustrated by an example of a thermoplastic system.

3.1. Discretization of time. In the preceding section we have considered thermodynamic systems the corresponding initial value problem of which could be reduced to the form

$$Av + \partial\varphi(v') \ni f, \quad v(0) = v_0, \quad v \in H^1(S; H) \tag{3.1}$$

(cf. (2.16)). Now we shall approximate such problems using an implicit difference scheme with respect to time (cf. Moreau [15], Gröger [6], Nguyen [16]). Let $n \in \mathbb{N}$

be fixed, $h_n = T/n$, $S_n^k =](k-1)\,h_n, kh_n]$, $k = 1, \ldots, n$, and let $L_n^2(S; H)$ denote the space of functions $v \in L^2(S; H)$ which are constant on each intervall S_n^k,

$$v(t) = v^k \qquad \forall\, t \in S_n^k, \quad k = 1, \ldots, n.$$

Furthermore, for fixed $w \in H$ by the expression

$$(\Delta_n^w v)\,(t) = (v^k - v^{k-1})/h_n \qquad \forall\, t \in S_n^k, \quad k = 1, \ldots, n; \quad v^0 = w$$

an operator $\Delta_n^w : L_n^2(S; H) \to L_n^2(S; H)$ will be defined. Then the discrete problem corresponding to (3.1) reads as

$$Av_n + \partial\varphi(\Delta_n^{v_0} v_n) \ni f_n, \quad v_n \in L_n^2(S; H), \tag{3.2}$$

where $f_n \in L_n^2(S; H')$ is given by $f_n^k = f(kh_n)$, $k = 1, \ldots, n$. For fixed $w \in H$ we set

$$\varphi_n^w(v) = h_n \varphi\big((v - w)/h_n\big), \quad v \in H,$$

and rewrite problem (3.2) in the following form

$$Av_n^k + \partial\varphi_n^{v_n^{k-1}}(v_n^k) \ni f_n^k, \quad v_n^k \in H, \quad k = 1, \ldots, n; \quad v_n^0 = v_0. \tag{3.3}$$

We shall see that the conditions (2.13)—(2.15) imply not only the existence of a uniquely determined solution to (3.1) as well as to (3.2) but also the convergence of the approximation procedure in some weak sense. In order to obtain a result on strong convergence we replace condition (2.15) by the more restrictive

$$\begin{aligned} & v_i \in \partial\varphi^*(g_i), \quad (Av_i, g_i) \to (h, g) \quad \text{in} \quad H' \times H' \\ & \Rightarrow v_i \to v \quad \text{in} \quad H. \end{aligned} \tag{3.4}$$

Lemma 3. *Suppose that* (2.13), (2.14) *and* (3.4) *hold. Let $f \in H^1(S; H')$ and $v_0 \in H$ be given such that $f(0) - Av_0 \in D(\varphi^*)$ and let $f_n \in L_n^2(S; H')$ be defined by $f_n^k = f(kh_n)$, $k = 1, \ldots, n$. Then for each $n \in \mathbb{N}$ there exists a unique solution v_n to* (3.2) *and it holds*

$$(v_n, \Delta_n^{v_0} v_n) \to (v, v') \quad \text{in} \quad L^2(S; H \times H)$$

where v is the unique solution to (3.1).

Proof. Let H be identified with its dual. (i) For fixed $n \in \mathbb{N}$, $w \in H$ the mapping $A + \partial\varphi_n^w$ is maximal monotone. From (2.14) it follows that $(A + \partial\varphi_n^w)^{-1}$ is bounded. This implies the surjectivity of $A + \partial\varphi_n^w$. Hence for each $n \in \mathbb{N}$ there exists a solution to (3.2). Since condition (2.15) is a consequence of (3.4) the unicity can be proved as in Lemma 2. (ii) Let (v_n), $n = 1, 2, \ldots$, be the sequence of solutions to (3.2). Then it holds

$$(A\Delta_n^{v_0} v_n - \Delta_n^{f(0)} f_n, \Delta_n^{v_0} v_n)_{L^2(S;H)} + \varphi^*\big(f_n(T) - Av_n(T)\big) - \varphi^*\big(f(0) - Av_0\big) \leqq 0. \tag{3.5}$$

This inequality shows that

$$\|A\Delta_n^{v_0} v_n\|^2_{L^2(S;H)} \leqq \text{const}\,(\|\Delta_n^{v_0} v_n\|_{L^2(S;H)} + 1).$$

On the other hand, from (2.14) we get the estimation

$$\|\Delta_n^{v_0} v_n\|_{L^2(S;H)} \leqq \text{const}\,(\|A\Delta_n^{v_0} v_n\|_{L^2(S;H)} + 1).$$

Hence the sequence $(\Delta_n^{v_0} v_n)$ is bounded in $L^2(S; H)$. (iii) There exists an element $v \in H^1(S; H)$ and a subsequence (v_{n_i}) such that $v(0) = v_0$ and

$$(v_{n_i}, \Delta_{n_i}^{v_0} v_{n_i}) \rightharpoonup (v, v') \quad \text{in} \quad L^2(S; H \times H) \tag{3.6}$$

as well as $v_{n_i}(T) \rightharpoonup v(T)$ in H. The relation

$$\limsup \, (\Delta_{n_i}^{v_0} v_{n_i}, f_{n_i} - A v_{n_i})_{L^2(S;H)} \leqq (v', f - Av)_{L^2(S;H)}$$

yields that v is a solution to the initial value problem (3.1). From the unicity result proved in Lemma 2 it follows that the convergence assertion (3.6) is valid with respect to the sequence (v_n). (iv) The solution v satisfies the equation

$$(Av' - f', v')_{;H)} + \varphi^*\big(f(T) - Av(T)\big) - \varphi^*\big(f(0) - Av_0\big) = 0$$

and along with (3.5) this implies

$$\limsup \, (A \Delta_n^{v_0} v_n, \Delta_n^{v_0} v_n)_{L^2(S;H)} \leqq (Av', v')_{L^2(S;H)}.$$

Thus we obtain

$$(Av_n, A\Delta_n^{v_0} v_n) \to (Av, Av') \quad \text{in} \quad L^2(S; H \times H).$$

(v) Now the assertion $(v_n, \Delta_n^{v_0} v_n) \to (v, v')$ in $L^2(S; H \times H)$ follows immediately from assumption (3.4).

Remark 11. The unique solution to the discrete problem (3.2) is given by the expression

$$v_n^k = \big(A + \partial\varphi_n^{v_n^{k-1}}\big)^{-1} (f_n^k), \quad k = 1, \dots, n; \quad v_n^0 = v_0$$

(cf. (3.3)) where $(A + \partial\varphi_n^w)^{-1}$ for fixed $n \in \mathbb{N}$, $w \in H$ is a single-valued monotone bounded potential operator from H' into H. We are going to show that this mapping is continuous. Let be $v_i = (A + \partial\varphi_n^w)^{-1} (g_i)$ and $g_i \to g$ in H'. Since $(A + \partial\varphi_n^w)^{-1}$ is hemicontinuous we obtain $v_i \rightharpoonup v$ in H. Then the inequality

$$\langle Av_i, v_i \rangle \leqq \langle g_i, v_i - v \rangle + \langle Av_i, v \rangle + \varphi_n^w(v) - \varphi_n^w(v_i)$$

implies $Av_i \to Av$ in H' and the assertion follows once more from (3.4).

The discretization method described above can be applied to the initial value problem (2.10) if assumption (2.18) of Theorem 4 is replaced by

$$\begin{aligned} & y_i \in \partial(F + I_{\operatorname{Im} K})^* z_i, \quad (Ay_i, z_i) \to (h, z) \quad \text{in} \quad Y' \times Y' \\ & \Rightarrow y_i \to y \quad \text{in} \quad Y. \end{aligned}$$

The following theorem shows that in the case of a serial composition of two thermodynamic systems this condition can be reduced to a more practicable form.

Theorem 7. *Let $\mathcal{T}_i = (Y, E_i, F_i, U, K)$, $i = 1, 2$, be two thermodynamic systems. Suppose that $A_i := \partial E_i \in \mathcal{L}(Y; Y')$, that A_1 is positive definite and that* Im A_2 *is closed. Furthermore let the following conditions be satisfied:*

$$\begin{aligned} & z_i \in \partial F_1(y_i), \quad (K^* z_i, y_i) \to (f. y) \quad \textit{in} \quad U' \times Y \\ & \Rightarrow z_i \to z \quad \textit{in} \quad Y', \end{aligned} \tag{3.7}$$

$$\begin{aligned} & y_i \in \partial F_2^*(z_i), \quad (A_2 y_i, z_i) \to (h, z) \quad \textit{in} \quad Y' \times Y' \\ & \Rightarrow y_i \to y \quad \textit{in} \quad Y. \end{aligned} \tag{3.8}$$

Then for the serial composition $\bar{\mathcal{T}}$ it holds

$$\bar{y}_i \in \partial(\bar{F} + I_{\mathrm{Im}\bar{K}})^* \bar{z}_i, \quad (\bar{A}\bar{y}_i, \bar{z}_i) \to (\bar{h}, \bar{z}) \quad in \quad \bar{Y}' \times \bar{Y}'$$
$$\Rightarrow \bar{y}_i \to \bar{y} \quad in \quad \bar{Y}.$$

We omit the proof of this theorem, since it is analogous to that of Theorem 6. Note that the conditions (3.7) and (3.8) are trivially satisfied if ∂F_1 is continuous and A_2 is positive definite. For details and further comments see Hünlich [13].

3.2. Approximation methods for solving the problems arising after discretization of time

Let us consider the serial composition of two thermodynamic systems satisfying the assumptions of Theorem 6 and 7. Then the corresponding initial value problem can be solved by means of the discretization procedure described above. Now we are interested in approximation methods for solving the discrete problem for fixed n.

Let us suppose that the external generalized force is given by $\bar{f} = (f, 0)$ where $f \in H^1(S; U')$. Taking into account the definition of a serial composition, setting $y = y_1 + y_2$, $z = z_1^s + z_1^d = z_2^s + z_2^d$, $F_{in}^{k-1} = F^{y_{in}^{k-1}}_{in}$, $i = 1, 2$, and omitting the index n if there is no misunderstanding we may write the discrete problem in the form (cf. (2.1) and Remark 11)

$$\left.\begin{aligned} &Ku^k = y^k, \quad K^*z^k = f^k, \\ &y^k = \partial\big((E_1 + F_{1n}^{k-1})^* + (E_2 + F_{2n}^{k-1})^*\big)(z^k), \end{aligned}\right\} \tag{3.9}$$

$$y_i^k = \partial(E_i + F_{in}^{k-1})^*(z^k), \quad y_i^0 = y_{i0}, \quad i = 1, 2. \tag{3.10}$$

For fixed k the relations (3.9) describe a situation analogous to that arising in physically nonlinear elasticity. The third equation (3.9_3) represents a constitutive relation between the state variables y^k and the thermodynamic forces z^k which is expressed by a continuous monotone potential operator depending on the state variables y_1^{k-1}, y_2^{k-1} at the preceding time step (cf. Remark 11).

This analogy becomes yet clearer if we consider a thermodynamic system with a more detailed structure of its state space. Let us suppose that $Y = Y_0 \times Y_{\mathrm{int}}$ where Y_0 denotes the space of mechanical deformations and Y_{int} the space of so called *internal state variables*. Then we have $Y' = Y_0' \times Y_{\mathrm{int}}'$ where Y_0' will be the space of usual stresses. The conditions of compatibility and equilibrium are expressed only by the mechanical part of the variables. Therefore we set $U = U_0 \times Y_{\mathrm{int}}$, $Ku = (K_0 v, \alpha)$ for $u = (v, \alpha) \in U$ and $f = (g, 0)$ where $g \in H^1(S; U_0')$. Finally using the notation $y = (P_0 y, P_{\mathrm{int}} y) = (\varepsilon, \alpha)$, $z = (\sigma, \zeta)$ we reduce the relations (3.9), (3.10) to

$$\left.\begin{aligned} &K_0 v^k = \varepsilon^k, \quad K_0^* \sigma^k = g^k, \\ &\varepsilon^k = P_0\, \partial\big((E_1 + F_{1n}^{k-1})^* + (E_2 + F_{2n}^{k-1})^*\big)(\sigma^k, 0), \end{aligned}\right\} \tag{3.11}$$

$$\left.\begin{aligned} &\varepsilon_1^k = P_0\, \partial(E_1 + F_{1n}^{k-1})^*(\sigma^k, 0), \quad \varepsilon_0^1 = \varepsilon_{10}, \\ &\alpha_1^k = P_{\mathrm{int}}\, \partial(E_1 + F_{1n}^{k-1})^*(\sigma^k, 0), \quad \alpha_1^0 = \alpha_{10}, \\ &(\varepsilon_2^k, \alpha_2^k) = \partial(E_2 + F_{2n}^{k-1})^*(\sigma^k, 0), \quad (\varepsilon_2^0, \alpha_2^0) = (\varepsilon_{20}, \alpha_{20}). \end{aligned}\right\} \tag{3.12}$$

Now we see that for fixed k corresponding to the product structure of the state space the problem has been split into two sets of relations. The first one (3.11) represents the basic problem of nonlinear elasticity. It contains a constitutive relation between the deformations ε^k and the stresses σ^k described by a continuous monotone potential operator, too. The other relations (3.12) determine the values of the remaining variables.

In some practically important cases the operator at the right hand side of (3.11_3) will be Lipschitzian and strongly monotone even if the corresponding operator in (3.9_3) does not have one of these properties. Then the problem (3.11) (formulated as an equation for the displacements or the stresses) can be solved numerically by means of standard iteration, projection and projection-iteration methods as well as by means of finite element methods (cf. Gajewski-Gröger-Zacharias [5], Hlaváček [10]).

The situation just described can be illustrated by various examples. In the following let us consider some models of thermoplasticity (cf. also Gröger [6], Hünlich [11], Gröger-Hünlich [7]).

3.3 Thermoplastic systems

Let $G \subset \mathbb{R}^m$ be a bounded domain and let $\mathcal{S}$ be a finite dimensional linear space. Further we introduce the spaces $Y_0 = L^2(G; \mathcal{S}) = Y_0'$, $Y_{\text{int}} = L^2(G; \mathbb{R}^2) = Y_{\text{int}}'$, $Y = Y_0 \times Y_{\text{int}}$, $U = U_0 \times Y_{\text{int}}$ where U_0 denotes a function space over G. The variables $u = (v, \alpha, \eta) \in U$, $y = (\varepsilon, \alpha, \eta) \in Y$, $z = (\sigma, \zeta, \vartheta) \in Y' = Y$ except (α, ζ) shall be interpreted as inExample 1 and 4. By (α, ζ) we denote additional internal variables for describing the plastic behavior.

Next, let us define the mappings

$$E_1(y) = \frac{1}{2}(A_1 y, y)_Y, \quad F_1(y) = \psi(\eta), \quad y \in Y,$$

$$E_2(y) = \frac{1}{2}(A_2 y, y)_Y \quad F_2(y) = I_C^*(\varepsilon, \alpha) + I_{\{0\}}(\eta), \quad y \in Y,$$

$$Ku = (K_0 v, \alpha, \eta), \quad u \in U,$$

where the following conditions are assumed to be fulfilled:

$$\left.\begin{aligned} &A_1(\varepsilon, \alpha, \eta) = \big(A_0\varepsilon - D\eta, a\alpha, -D^*\varepsilon + (D^*A_0^{-1}D + b)\,\eta\big), \\ &A_2(\varepsilon, \alpha, \eta) = (B_0\varepsilon, c\alpha, d\eta), \\ &A_0, B_0 \in \mathcal{L}(\mathcal{S}, \mathcal{S}) \text{ are symmetric and positive}, \\ &\operatorname{Ker} A_0 = \{0\}, \quad D \in \mathcal{L}(\mathbb{R}, \mathcal{S}), \quad a, b, c, d > 0; \end{aligned}\right\} \tag{3.13}$$

$$\psi \in \Gamma\big(L^2(G)\big), \quad 0 \in \partial\psi(0); \tag{3.14}$$

$$\left.\begin{aligned} &C = \big\{(\sigma, \zeta) \in Y_0 \times L^2(G) \mid f\big(\sigma(x)\big) \le \zeta(x) \quad \text{for a.a.} \quad x \in G\big\}, \\ &f: \mathcal{S} \to \mathbb{R} \text{ is convex and continuously differentiable} \\ &\text{such that } f' \text{ is bounded}, \quad f(0) \le 0; \end{aligned}\right\} \tag{3.15}$$

$$K_0 \in \mathcal{L}(U_0, Y_0), \quad \operatorname{Ker} K_0 = \{0\}, \quad \operatorname{Im} K_0 = \overline{\operatorname{Im} K_0}. \tag{3.16}$$

It is easy to see that by $\mathcal{T}_i = (Y, E_i, F_i, U, K)$, $i = 1, 2$, there are given two thermodynamic systems in the sense of Definition 1. The first one represents a thermoelastic system. Its dissipation law is equivalent to $\sigma^d = \zeta^d = 0$, $\eta' \in \partial\psi^*(\vartheta^d)$ where the last relation describes the entropy transport (cf. Example 2 for further details). The system $\mathcal{T}_2$ is a plastic system including certain hardening effects. The statical law and the dissipation law give $(\varepsilon', \alpha') \in \partial I_C(\sigma - B_0\varepsilon, \zeta - c\alpha)$, $\eta' = 0$, where $(\sigma, \zeta) = (\sigma^s, \zeta^s) + (\sigma^d, \zeta^d)$. For a fixed ζ the relation $(\sigma - B_0\varepsilon, \zeta - c\alpha) \in C$ characterizes the set of admissible stresses σ. This set depends on the state variables (ε, α). The terms $-B_0\varepsilon$ and $-c\alpha$ describe the so called kinematic and isotropic hardening, respectively. Condition (3.15) implies that the functional I_C has the following properties (cf. Gröger [6]):

$$(\varepsilon, \alpha) \in \partial I_C(\sigma, \zeta) \Rightarrow \|\varepsilon\|_{Y_0} \leqq \text{const}\, \|\alpha\|_{L^2(G)},$$
$$(\varepsilon_i, \alpha_i) \in \partial I_C(\sigma_i, \zeta_i), \quad (\alpha_i, \sigma_i, \zeta_i) \to (\alpha, \sigma, \zeta) \tag{3.17}$$
$$\text{in} \quad L^2(G) \times Y_0 \times L^2(G) \Rightarrow \varepsilon_i \to \varepsilon \quad \text{in} \quad Y_0.$$

Now, let us consider the serial composition $\bar{\mathcal{T}}$ of both the systems $\mathcal{T}_1$ and $\mathcal{T}_2$ which describes a simple model of thermoplasticity. We shall prove that the assumptions of Theorem 6 and 7 are satisfied. Firstly, from (3.13) it follows that the conditions imposed on A_i, $i = 1, 2$, are valid. Next, let be $z \in \partial F_1(y)$, i.e. $\sigma = \zeta = 0$, $\vartheta \in \partial\psi(\eta)$. Then $K^*z = (0, 0, \vartheta)$ and the conditions (2.20) and (3.7) are trivially fulfilled. Now let be $y \in \partial F_2^*(z)$, i.e. $(\varepsilon, \alpha) \in \partial I_C(\sigma, \zeta)$, $\eta = 0$. From (3.17) it follows that
$\|y\|_Y \leqq \text{const}\, \|\alpha\|_{L^2(G)} \leqq \text{const}\, \|A_2 y\|_Y$ and (2.21) is true. In the same way (3.8) is seen to be valid.

Thus we can gather the following results. The thermoplastic system $\bar{\mathcal{T}}$ is normal (note that $\mathcal{T}_2$ will not be normal). For suitably chosen external generalized forces and initial values the corresponding initial value problem has a unique solution where this solution can be approximated by the uniquely determined solutions to the time discretized problems.

From physical reasons the external generalized force $\bar{f} \in H'(S; \bar{U}')$ has to be given by $\bar{f} = (f, 0)$, $f = (g, 0, 0)$ where $g \in H^1(S; U_0')$. Then the discrete problems take the form (3.11), (3.12). For fixed $n \in \mathbb{N}$, $\bar{y} \in Y$ and $z = (\sigma, 0, 0) \in Y$ it holds

$$y = \partial(E_1 + F_{1n}^{\bar{y}})^* z \Leftrightarrow$$
$$\varepsilon = A_0^{-1}\sigma + A_0^{-1}D\eta, \quad \alpha = 0, \quad \eta = (bI + \partial\psi_n^{\bar{\eta}})^{-1} D^* A_0^{-1}\sigma$$

as well as

$$y = \partial(E_2 + F_{2n}^{\bar{y}})^* z \Leftrightarrow$$
$$\varepsilon = \bar{\varepsilon} + \big(B_0 + (\partial\varphi^{\bar{\alpha}})^{-1}\big)^{-1} (\sigma - B_0\bar{\varepsilon}), \quad \alpha = \bar{\alpha} + P^{\bar{\alpha}}(\sigma - B_0\varepsilon), \quad \eta = \bar{\eta}$$

where for $\bar{\alpha} \in L^2(G)$, $\sigma \in Y_0$

$$\varphi^{\bar{\alpha}}(\sigma) = \frac{1}{2c} \int_G \big([f(\sigma(x)) + c\bar{\alpha}(x)]^+\big)^2 \, dx,$$

$$P^{\bar{\alpha}}(\sigma)\,(x) = -\frac{1}{c} \, [f(\sigma(x)) + c\bar{\alpha}(x)]^+ \quad \text{for a.a.} \quad x \in G$$

(cf. Hünlich [13]). Let us note that the mappings $\varphi^{\bar{\alpha}}$ and $P^{\bar{\alpha}}$ do not depend on n according to the fact that plastic behavior is rate independent. Finally setting $\beta = \varepsilon_2$, $\alpha = \alpha_2$, $\eta = \eta_1$, $\varphi^{k-1} = \varphi^{\alpha^{k-1}}$, $P^{k-1} = P^{\alpha^{k-1}}$ and $\psi_n^{k-1} = \psi_n^{\eta^{k-1}}$ we can write the relations (3.11), (3.12) as follows:

$$\left.\begin{aligned} &K_0 v^k = \varepsilon^k, \quad K_0^* \sigma^k = g^k, \\ &\varepsilon^k = A_0^{-1}\sigma^k + A_0^{-1}D(bI + \partial\psi_n^{k-1})^{-1} D^* A_0^{-1}\sigma^k \\ &\qquad + \beta^{k-1} + \left(B_0 + (\partial\varphi^{k-1})^{-1}\right)^{-1} (\sigma^k - B_0\beta^{k-1}), \end{aligned}\right\} \tag{3.18}$$

$$\alpha^k = \alpha^{k-1} + P^{k-1}(\sigma^k - B_0\beta^k), \quad \alpha^0 = \alpha_0,$$

$$\eta^k = (bI + \partial\psi_n^{k-1})^{-1} D^* A_0^{-1}\sigma^k, \quad \eta_0 = \eta^0,$$

$$\beta^k = \beta^{k-1} + \left(B_0 + (\partial\varphi^{k-1})^{-1}\right)^{-1} (\sigma^k - B_0\beta^{k-1}), \quad \beta^0 = \beta_0.$$

The operator A_0^{-1} is positive definite. Since $b > 0$ the mapping $(bI + \partial\psi_n^{\eta})^{-1}$ will be Lipschitzian. Finally, under additional assumptions on the function f which can be formulated easily the operator $\partial\varphi^{\alpha}$ becomes Lipschitzian. Then the mapping $\left(B_0 + (\partial\varphi^{\alpha})^{-1}\right)^{-1}$ will be Lipschitzian, too. Hence for fixed k the relation (3.18_3) contains a strongly monotone Lipschitzian potential operator from Y_0 into itself such that (3.18) can be solved by standard approximation methods mentioned above. Numerical results for a special elastoplastic system are given in Hünlich [12].

At the end let us note that more general thermodynamic systems including viscous and diffusion processes as well as chemical reactions can be handled in a similar way.

References

[1] V. Barbu, Nonlinear semigroups and differential equations in Banach spaces. Bucureşti/Leyden 1976.

[2] M. A. Biot, Variational Lagrangian irreversible thermodynamics of initially stressed solids with thermomolecular diffusion and chemical reactions. J. Mech. Phys. Solids **25** (1977), 289–307.

[3] H. Brézis, Operateurs maximaux monotones et semigroupes de contractions dans les espaces de Hilbert. North-Holland Mathematics Studies **5** (1973).

[4] G. Duvaut, J.-L. Lions, Les inéquations en mécanique et en physique. Paris 1972.

[5] H. Gajewski, K. Gröger, K. Zacharias, Nichtlineare Operatorgleichungen und Operatordifferentialgleichungen. Berlin 1974.

[6] K. Gröger, Initial value problems for elastoplastic and elastoviscoplastic systems. *In:* Nonlinear analysis, function spaces and applications, Proceedings of a spring school held in Horní Bradlo 1978. Ed. S. Fučik, A. Kufner, Leipzig 1979, 95–127.

[7] K. Gröger, R. Hünlich, An existence-uniqueness result for some models of thermoplasticity. Z. Angew. Math. Mech. **60** (1980), 169–171.

[8] K. Gröger, J. Nečas, On a class of nonlinear initial value problems in Hilbert spaces. Math. Nachr. **93** (1979), 21–31.

[9] B. Halphen, Nguyen Q. S., Sur les matériaux standards généralisés. J. Mécanique **14** (1975), 39–63.

[10] I. Hlaváček, Convergence of an equilibrium finite element model for plane elastostatics. Apl. Mat. **24** (1979), 427–457.

[11] R. Hünlich, Quasistatische Anfangsrandwertprobleme für elastisch-plastische Materialien mit linearer Verfestigung. AdW der DDR, Zentralinstitut für Mathematik und Mechanik, Report R-02/79.
[12] R. Hünlich, On simultaneous torsion and tension of a circular cylindrical bar consisting of an elastoplastic material with linear hardening. Z. Angew. Math. Mech. **59** (1979), 509–516.
[13] R. Hünlich, Prozesse in mechanischen Systemen mit internen Parametern. Thesis, Berlin 1980.
[14] A. D. Joffe, V. M. Tichomirov, Theory of extremal problems. (Russian) Moscow 1974.
[15] J. J. Moreau, On unilateral constraints, friction and plasticity. *In:* New variational techniques in mathematical physics, CIME, II ciclo 1973, Ed. G. Capriz, G. Stampacchia, 1974, 175–322.
[16] Nguyen Q. S., On the elastic plastic initial-boundary value problem and its numerical integration. Internat. J. Num. Meth. Engng. **11** (1977), 817–832.
[17] L. Trávniček, Existence and uniqueness of solutions to the boundary value problem for elastoplastic materials. (Czech) Thesis, Prague 1977.
[18] H. Ziegler, An introduction to thermomechanics. Amsterdam 1977.

Address of the author:

Akademie der Wissenschaften der DDR
Institut für Mathematik
1080 Berlin
Mohrenstr. 39
DDR

Common properties of nonlinear programming algorithms basing on sequential unconstrained minimizations

Christian Großmann, Dresden

1. Sequential unconstrained minimization techniques and duality

With the development of powerful unconstrained minization methods the importance of nonlinear programming algorithms transforming a given constrained optimization problem into a sequence of unconstrained ones has grown up. This resulted in an extensive investigation of various types of nonlinear programming algorithms basing on sequential unconstrained minimizations. The aim of this paper is to present a unified approach to these methods involving barrier-penalty-methods and methods of centers as well as augmented Lagrangian and shifting methods. Using this approach a general duality theory can be derived, the convergence theory of sequential unconstrained minimization techniques can be partially unified, the rates of convergence can be estimated and effective modifications of well known methods can be proposed.

Let be given the following nonlinear programming problem

$$f_0(x) \to \min! \quad \textit{subject to} \quad x \in G := \{x \in X \mid f_i(x) \leqq 0, \quad i = 1, \ldots, m\} \tag{1.1}$$

with a closed subset $X \subseteq R^n$ and a continuous mapping $f \mid X \to R^{m+1}$, where $f(x) = \big(f_0(x), f_1(x), \ldots, f_m(x)\big)^T$.

Let be selected an arbitrary set Y of parameters and a generating function $E \mid Y \times R^{m+1} \to \tilde{R} := R^1 \cup \{+\infty\}$. Now, we define a related function $T \mid X \times Y \to \bar{R} := \tilde{R} \cup \{-\infty\}$ by

$$T(x, y) = \inf\{E(y, v) \mid v \geqq f(x)\} \quad \text{for any} \quad x \in X, \quad y \in Y. \tag{1.2}$$

Most of the methods transforming the constrained optimization problem (1.1) into a sequence of unconstrained minimization problems are involved in the following general scheme.

Sequential unconstrained minimization technique (SUMT)

Step 1: Select a starting parameter $y^1 \in Y$. Set $k := 1$.

Step 2: Determine a solution x^k of the auxiliary problem

$$T(x, y^k) \to \min! \quad \textit{subject to} \quad x \in X. \tag{1.3}$$

Step 3: Compute a new parameter $y^{k+1} \in Y$ and set $k := k + 1$. Go to step 2.

If $X = R^n$ then the auxiliary problems (1.3) are unconstrained. The various special algorithms we get from the general scheme by specifying the parameter set Y, the generating function E and by appropriate selection of the updating rule defining the sequence $\{y^k\} \subset Y$ of parameters in step 3 of SUMT. Let especially $Y = \text{int } R^1_+$,

$$E(y, v) = v_0 + \sum_{i=1}^{m} y v_i^2$$

and let $\{y^k\}$ denote an arbitrary sequence of positive numbers tending to infinity. Then we get from SUMT the well-known quadratic penalty method with

$$T(x, y) = f_0(x) + y \sum_{i=1}^{m} \max^2 \{0, f_i(x)\}, \quad x \in X, \quad y > 0.$$

Up to now a unified theory only for each of the main classes of sequential unconstrained minimization techniques (penalty methods, methods of centers, methods of exterior centers, augmented Lagrangian methods) has been developed. The aim of this paper is to derive some common aspects of all these methods by investigating related problems

$$E(y^k, v) \to \min! \quad \textit{subject to} \quad v \in Q.$$

Thereby the set Q is defined by $Q = \{v \in R^{m+1} \mid \exists\, x \in X \text{ with } f(x) \leqq v\}$. To shorten our notation we set $v = \begin{pmatrix} v_0 \\ \underline{v} \end{pmatrix}$ with $\underline{v} = (v_1, \ldots, v_m)^T$ for any $v \in R^{m+1}$. Especially $\underline{f}(x)$ denotes $\underline{f}(x) = \big(f_1(x), \ldots, f_m(x)\big)^T$.

Let $\chi \mid R^m \to \overline{R}$ be the optimal value function related to the problem (1.1), that means

$$\chi(u) = \inf \{f_0(x) \mid x \in X,\ \underline{f}(x) \leqq u\} \quad \text{for any} \quad u \in R^m.$$

Immediately from the definitions of the function χ and of the set Q we get

$$Q \subset \text{epi } \chi \quad \text{and} \quad \text{cl } Q = \text{cl epi } \chi. \tag{1.4}$$

A point $v \in R^{m+1}$ is called *Q-regular* if $v \in Q$ and $v - \varepsilon e^0 \notin Q$ for any $\varepsilon > 0$, where $e^0 = (1, 0, \ldots, 0)^T \in R^{m+1}$.

Now, for fixed parameters $y \in Y$ let us investigate the relation between the problems

$$T(x, y) \to \min! \quad \textit{subject to} \quad x \in X \tag{1.5}$$

and

$$E(y, v) \to \min! \quad \textit{subject to} \quad v \in Q. \tag{1.6}$$

Theorem 1.1. *For fixed $y \in Y$ let $x(y)$ be a solution of* (1.5) *and let exist a solution $v(y)$ of the related problem*

$$E(y, v) \to \min! \quad \textit{subject to} \quad v \geqq f\big(x(y)\big). \tag{1.7}$$

Then $v(y)$ also solves the problem (1.6). *If furthermore $v(y)$ is Q-regular then $x(y)$ forms a solution of the perturbed problem*

$$f_0(x) \to \min! \quad \textit{subject to} \quad x \in X, \quad \underline{f}(x) \leqq \underline{v}(y). \tag{1.8}$$

Proof. Because $v(y)$ is feasible for (1.7) it is also feasible for (1.6). Let us assume a point $\tilde{v} \in Q$ to exist with $E(y, \tilde{v}) < E\big(y, v(y)\big)$. From $\tilde{v} \in Q$ we get an $\tilde{x} \in X$ satisfying $f(\tilde{x}) \leqq \tilde{v}$. Using (1.2) and the optimality of $v(y)$ for the problem (1.7) this results in

$$T(\tilde{x}, y) \leqq E(y, \tilde{v}) < E\big(y, v(y)\big) = T\big(x(y), y\big),$$

which contradicts the assumption that $x(y)$ solves (1.5).

Now we prove the second part of the theorem. If this is not true then a point $\hat{x} \in X$ exists such that $\underline{f}(\hat{x}) < \underline{v}(y)$ and $f_0(\hat{x}) < f_0(x(y))$. Then $\hat{v} = \begin{pmatrix} f_0(\hat{x}) \\ \underline{v}(y) \end{pmatrix}$ belongs to Q. This contradicts the Q-regularity of $v(y)$. ∎

The constraints in (1.7) are only lower bounds subject to each of the variables v_i. If the function $E(y, \cdot)$ has, as usually in practice, a simple separable structure then the problems (1.7) can be explicitly solved. Thus solutions of (1.6) can be directly computed if solutions of the auxiliary problems (1.5) are available.

It is to remark that the second part of theorem 1.1 generalizes the well-known theorem of Everett [6] about the solution of perturbed problems established for the Lagrangian related to (1.1). Indeed, if we select $Y = R_+^m$ and set $E(y, v) = v_0 + y^T \underline{v}$ then

$$T(x, y) = f_0(x) + y^T \underline{f}(x) \quad \text{for any} \quad x \in X, \quad y \in Y$$

holds. Furthermore we get $v(y) = f(x(y))$ and $v(y)$ is Q-regular for each fixed $y \in Y$.

Using the generating function E a general duality theory can be developed. For $y \in Y$ let denote

$$S(y) = \{t \in R^1 \mid E(y, (t, 0, \ldots, 0)^T) \leqq \inf_{v \in Q} E(y, v)\}$$

and

$$\tau(y) = \sup \{t \mid t \in S(y)\}. \tag{1.9}$$

Theorem 1.2. *Let be* $E(y, v) > \inf_{w \in Q} E(y, w)$ *for any* $y \in Y$ *and arbitrary* $v \in Q$ *not being* Q*-regular. Then*

$$\sup_{y \in Y} \tau(y) \leqq \inf_{x \in G} f_0(x) \tag{1.10}$$

holds.

Proof. If the feasible set G is empty then by definition $\inf_{x \in G} f_0(x) = +\infty$ and the inequality (1.10) trivially holds. Now, let be $G \neq \emptyset$. For any $a > \inf_{x \in G} f_0(x)$ the point $v_a = (a, 0, \ldots, 0)^T$ belongs to Q and is not Q-regular. From the property assumed in the theorem we get $E(y, v_a) > \inf_{w \in Q} E(y, w)$. This results in $a \notin S(y)$ for any $y \in Y$.

Therefore $\sup_{y \in Y} \tau(y) \leqq a$ holds. Tending $a \to \chi(0) + 0$ we get the proposition. ∎

On the base of theorem 1.2 the optimization problem

$$\tau(y) \to \sup! \quad \textit{subject to} \quad y \in Y \tag{1.11}$$

can be considered as a dual one to (1.1). Fig. 1 interprets the relation (1.10) in a simple geometric way.

Now, let the generating function E possess the structure

$$E(y, v) = v_0 + e(y, \underline{v}) \quad \text{for any} \quad y \in Y, \quad v \in R^{m+1}, \tag{1.12}$$

where e denotes a function from $Y \times R^m$ into $\bar{R}$. Then the assumptions of theorem 1.2 are fulfilled and the objective function τ of the dual problem (1.10) can be expressed by

$$\tau(y) = \inf_{x \in X} T(x, y) - e(y, 0) \quad \text{for any} \quad y \in Y.$$

A point $(x^*, y^*) \in X \times Y$ is called a saddle point of the auxiliary function T if

$$T(x^*, y) \leqq T(x^*, y^*) \leqq T(x, y^*) \quad \text{for any} \quad x \in X, \quad y \in Y$$

holds. As a generalization of the well known saddle point theorem to a more general function T we get the following theorem.

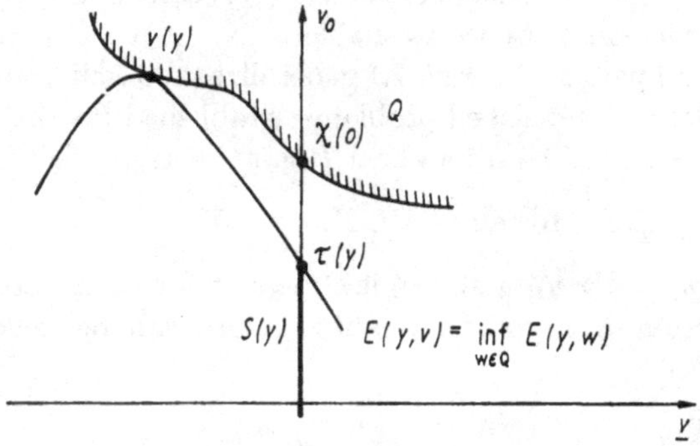

Fig. 1

Theorem 1.3. *Let be given the generating function E by (1.12) with a function $e \mid Y \times R^m \to \bar{R}$ satisfying $e(y, 0) = 0$ for any $y \in Y$. Furthermore let*

$$\sup_{y \in Y} T(x, y) = \begin{cases} f_0(x) & \text{if} \quad x \in G, \\ +\infty & \text{if} \quad x \notin G. \end{cases} \tag{1.13}$$

If (x^*, y^*) forms a saddle point of T then x^* solves the primal problem (1.1) *and y^* solves the dual problem* (1.11).

Let Φ denote the following family of functions

$$\Phi = \{e(y, \cdot) \mid y \in Y\}.$$

If the conditions of theorem 1.3 are fulfilled then the function T can be proved to possess a saddle point if and only if the optimal value function χ is Φ-subdifferentiable at the point 0, that means there is an element $y^* \in Y$ such that

$$\chi(u) + e(y^*, u) \geqq \chi(0) + e(y^*, 0) \quad \text{for any} \quad u \in R^m.$$

In the convex case the dual bound $\tau(y)$ of the optimal value of the primal problem (1.1) can be sharped in the following way.

Theorem 1.4. *Let hold the property assumed in theorem 1.3. Furthermore let the set X be convex and the functions f, $E(y, \cdot)$ be convex. If $v(y)$ forms a solution of the problem* (1.6) *and $E(y, \cdot)$ is differentiable at $v(y)$ with $\frac{\partial}{\partial v_0} E(y, v(y)) \neq 0$ then the value $\sigma(y)$ defined by*

$$\frac{\partial}{\partial v_0} E(y, v(y))\, \sigma(y) = \nabla_v E(y, v(y))^T\, v(y) \tag{1.14}$$

can be estimated by

$$\tau(y) \leqq \sigma(y) \leqq \chi(0). \tag{1.15}$$

For the proofs of the theorems 1.3, 1.4 the reader is referred to [10].

2. Theory of convergence

The convergence of sequential unconstrained minimization methods can be established on the base of the relation between solutions of the problems (1.5), (1.6) and the following two lemmas.

Lemma 2.1. *Let* $\{z^k\} \subset X$, $\{v^k\} \subset R^{m+1}$ *be sequences such that* $f(z^k) \leqq v^k$, $k = 1, 2, \ldots$, $\overline{\lim\limits_{k\to\infty}}\, \underline{v}^k \leqq 0$ *and* $\overline{\lim\limits_{k\to\infty}}\, v_0^k \leqq \chi(0)$.
Then any accumulation point of the sequence $\{z^k\}$ *solves the optimization problem* (1.1).

Proof. From the closedness of X we get $z^* \in X$ for any accumulation point z^* of the sequence $\{z^k\}$. The continuity of f and the proposed inequalities result in $\underline{f}(z^*) \leqq 0$ and $f_0(z^*) \leqq \chi(0)$. Thus z^* solves (1.1). ■

Lemma 2.2. *Let be given a sequence of parameters* $y^k \in Y$, $k = 1, 2, \ldots$, *and let exist a sequence* $\{x^k\}$ *of solutions of the related auxiliary problems* (1.3). *Furthermore let exist a sequence* $\{v^k\} \subset R^{m+1}$ *of Q-regular points solving the problems*

$$E(y^k, v) \to \min! \quad \textit{subject to} \quad v \geqq f(x^k), \tag{2.1}$$

$k = 1, 2, \ldots$ *If the optimal value function* χ *at the point* 0 *is upper semi-continuous with respect to* dom χ *and* $\lim\limits_{k\to\infty} \underline{v}^k = 0$ *holds then any accumulation point of* $\{x^k\}$ *solves the problem* (1.1).

Proof. From theorem 1.1 follows v^k to solve (1.6) for $y = y^k$. Using (1.4) and the Q-regularity of v^k we get $v_0^k = \chi(\underline{v}^k)$, $k = 1, 2, \ldots$ This results in $\{\underline{v}^k\} \subset \operatorname{dom} \chi$. With $\lim\limits_{k\to\infty} \underline{v}^k = 0$ and the u.s.c. of χ at 0 the inequality $\overline{\lim\limits_{k\to\infty}}\, v_0^k \leqq \chi(0)$ holds. Lemma 2.1 completes the proof. ■

Penalty methods we get from the general scheme of SUMT by means of generating functions E according to (1.12) and given parameter sequences $\{y^k\}$ such that the following properties hold:

(i) *The functions* $e(y^k, \cdot) \mid R^m \to \tilde{R}$ *are continuous.*

(ii) *There exists a set* $P \subset R^m_-$ *such that* $\inf\limits_{u \in P} \chi(u) = \chi(0)$ *and* $\lim\limits_{k\to\infty} \left[\inf\limits_{w \geqq u} e(y^k, w)\right] = 0$ *for any* $u \in P$ *hold.*

(iii) *For any* $\bar{u} \in R^m$ *and arbitrary sequences* $\{u^k\}$ *converging to* $\bar{u}$ *holds*

$$\underline{\lim}_{k\to\infty} \left[\inf_{w \geqq u^k} e(y^k, w)\right] \begin{cases} = +\infty & \textit{if} \quad \bar{u} \notin R^m_- \\ \geqq 0 & \textit{if} \quad \bar{u} \in R^m_-. \end{cases}$$

(iv) *For fixed* $y \in Y$, $u \in R^m$ *there is a point* $\tilde{u} \in R^m$, $\tilde{u} \geqq u$ *such that*

$$e(y, \tilde{u}) = \inf_{w \geqq u} e(y, w).$$

The properties (i)—(iv) we get by applying the general properties of penalty methods introduced by Fiacco/McCormick [7] to the concept of generating functions used here.

Theorem 2.1. *Let be* $\{y^k\}$ *and* $e \mid Y \times R^m \to \tilde{R}$ *selected such that* (i) — (iv) *hold. Then any accumulation point of a sequence* $\{x^k\}$ *generated by the related* SUMT *solves* (1.1) *if* $G \neq \emptyset$.

Proof. Using (1.12) and (iv) there exists a sequence $\{v^k\}$ of solutions of the problems (2.1). From (1.4), (1.12) and theorem 1.1 we get

$$\chi(\underline{v}^k) + e(y^k, \underline{v}^k) \leqq \chi(u) + e(y^k, u) \quad \text{for any} \quad u \in R^m \tag{2.2}$$

and

$$f_0(x^k) = v_0^k = \chi(\underline{v}^k), \quad k = 1, 2, \ldots$$

Now, (ii) leads to

$$\overline{\lim_{k\to\infty}}\, [v_0^k + e(y^k, \underline{v}^k)] \leqq \chi(0). \tag{2.3}$$

Let $\{x^k\}_{\mathscr{K}}$ denote a subsequence of $\{x^k\}$ converging to x^*. From the continuity of f_0 and (2.3) we get $\underline{\lim}_{k\in\mathscr{K}}\, e(y^k, \underline{v}^k) < +\infty$. Property (iii) results in $\overline{\lim}_{k\in\mathscr{K}}\, \underline{v}^k \leqq 0$. Combining (iii) and (2.3) this leads to $\overline{\lim}_{k\in\mathscr{K}}\, v_0^k \leqq \chi(0)$. Lemma 1.2 completes the proof. ■

We remark that the essential steps of the convergence proof given in [7] are preserved. Our proof however shows the close relation between the solvability of the problem (1.1) by a penalty method and the behaviour of the function χ. Indeed, if additionally

$$e(y, u) \geqq e(y, 0) = 0 \quad \text{for any} \quad y \in Y, \quad u \in R^m \tag{2.4}$$

holds then (2.2) results in

$$\chi(\underline{v}^k) \leqq \chi(\underline{v}^k) + e(y^k, \underline{v}^k) \leqq \underline{\lim}_{u\to 0}\, \chi(u), \quad k = 1, 2, \ldots$$

With $f_0(x^k) = \chi(\underline{v}^k)$, $k = 1, 2, \ldots$, we get $\overline{\lim}_{k\to\infty}\, f_0(x^k) \leqq \underline{\lim}_{u\to 0}\, \chi(u)$ for any sequence $\{x^k\}$ generated by the related penalty method. Thus the optimal value function χ has to be lower semi-continuous at the point 0 for a problem to be solvable by a penalty method with (2.4). For example the well-known quadratic penalty function generated by $e(y, u) = y\, \|u\|^2$, $y > 0$, satisfies (2.4).

The methods of centers and exterior centers can be founded in a similar way by transcription of the usual properties proposed in publications (see e.g. [3, 11]) to the generating functions used here. In the methods of centers e.g. we select $Y = \{y \in R^1 \mid y > \chi(0)\}$. Let us denote

$$B(y) = \{v \in R^{m+1} \mid \underline{v} < 0, v_0 < y\}, \quad y \in Y.$$

Then a function $E \mid Y \times R^{m+1} \to \tilde{R}$ can be used as a generated function in the method of centers if the following properties hold:

(i) *There is a constant* $\gamma \in \tilde{R}$ *such that* $E(y, v) < \gamma \Leftrightarrow v \in B(y)$.

(ii) *For any fixed* $v \in R^{m+1}$ *the function* $E(\cdot, v)$ *is upper semi-continuous on* Y.

(iii) *For any* $y \in Y$, $v \in B(y)$ *there is a* $\bar{v} \geqq v$ *such that* $E(y, \bar{v}) = \min_{w \geqq v} E(y, w)$.

In the method of centers the sequence $\{y^k\}$ of parameters is generated by $y^{k+1} = f_0(x^k)$, $k = 1, 2, \ldots$ More details of the method of centers and exterior centers can be found e.g. in [11].

Now, let us consider a method generalizing the augmented Lagrangian method introduced by Hestenes [12]. The optimization problem (1.1) is assumed to be convex, that means the set $X \subseteq R^n$ and the functions $f_i \mid X \to R^1$, $i = 0, 1, \ldots, m$, are supposed to be convex. Furthermore we assume that (1.1) is solvable and the generating function fulfils the following conditions:

(i) *For any $y \in Y$ the function $E(y, \cdot) \mid R^{m+1} \to \tilde{R}$ is continuous and convex. Additionally $E(y, \cdot)$ is differentiable on its effective domain.*

(ii) *For any $y \in Y$ and any related solution $v(y)$ of (1.6) there holds $\frac{\partial}{\partial v_0} E(y, v(y)) \neq 0$ if $v(y) \neq (\chi(0), 0, \ldots, 0)^T$.*

Method of sequential supporting halfspaces (SSH)

Step 1: } like general SUMT
Step 2: }

Step 3a: Determine a solution v^k of the problem

$$E(y^k, v) \to \min! \quad \textit{subject to} \quad v \geqq f(x^k). \tag{2.5}$$

If $\underline{v}^k = 0$ then stop. Otherwise we define $w^k \in R^{m+1}$ by

$$\nabla_v E(y^k, v^k)^T (w^k - v^k) = 0 \quad \text{and} \quad \underline{w}^k = 0. \tag{2.6}$$

Step 3b: Select a parameter $y^{k+1} \in Y$ such that $\operatorname{dom} E(y^{k+1}, \cdot) \cap Q \neq \emptyset$ and w^k solves the problem

$$E(y^{k+1}, v) \to \min! \quad \textit{subject to} \quad v \in Q_k \tag{2.7}$$

with $Q_k = \{v \in R^{m+1} \mid \nabla_v E(y^k, v^k)^T (v - v^k) \geqq 0\}$.
Set $k := k + 1$ and go to step 2.

As earlier we denote by τ the dual function defined in (1.9).

Theorem 2.2. *Let be $\{(x^k, v^k, w^k, y^k)\}$ a sequence generated by the SSH method and let exist a forcing function*[1] *$\delta \mid R^1_+ \to R^1_+$ such that*

$$w_0^k \geqq \tau(y^k) + \delta(\|\underline{v}^k\|), \quad k = 1, 2, \ldots \tag{2.8}$$

If a point $\bar{x} \in X$ satisfying $\underline{f}(\bar{x}) < 0$ exists and the points v^k, $k = 1, 2, \ldots$, are Q-regular then any accumulation point of the sequence $\{x^k\}$ forms a solution of the constrained optimization problem (1.1).

Proof. If the SSH method terminates after a finite number k' of steps then the optimality of $x^{k'}$ is a direct consequence of the theorem 1.1 since $v^{k'}$ is Q-regular and $\underline{v}^{k'} = 0$ holds.

---

[1] A function $\delta \mid R^1_+ \to R^1_+$ is called *forcing* if $\delta(s) \to 0$ implies $s \to 0$.

Now, let be $\underline{v}^k \neq 0$, $k = 1, 2, \ldots$ Then property (ii) leads to

$$\frac{\partial}{\partial v_0} E(y^k, v^k) \neq 0, \quad k = 1, 2, \ldots$$

Thus the sequence $\{w^k\}$ is well defined by (2.6) and we have $\sigma(y^k) = w_0^k$, $k = 1, 2, \ldots$, with $\sigma(y^k)$ according to (1.14). Using theorem 1.4 and the proposed inequalities (2.8) we get

$$\chi(0) \geqq w_0^k \geqq \tau(y^k) + \delta(\|\underline{v}^k\|) \geqq \tau(y^k), \quad k = 1, 2, \ldots \tag{2.9}$$

Moreover the set Q can be shown to be contained in the sets Q_k, $k = 1, 2, \ldots$, in step 3b of the SSH method. By selection of the parameter $y^{k+1} \in Y$ the point w^k solves the problem (2.7). On the base of theorem 1.1 this results in the following estimation

$$E(y^{k+1}, w^k) = \min_{v \in Q_k} E(y^{k+1}, v) \leqq \min_{v \in Q} E(y^{k+1}, v) = E(y^{k+1}, v^{k+1}).$$

Using $\underline{w}^k = 0$ we get $w_0^k \in S(y^{k+1})$ and consequently $\tau(y^{k+1}) \geqq w_0^k$ holds. With (2.9) this leads to

$$\chi(0) \geqq \tau(y^{k+1}) \geqq w_0^k \geqq \tau(y^k) + \delta(\|\underline{v}^k\|), \quad k = 1, 2, \ldots$$

Thus the sequence $\{\tau(y^k)\}$ is non-decreasing and bounded from above. Therefore the sequence $\{\tau(y^k)\}$ converges. This results in

$$\lim_{k \to \infty} \delta(\|\underline{v}^k\|) = 0.$$

Since δ is a forcing function we get $\lim_{k \to \infty} \underline{v}^k = 0$. The optimal value function χ is guaranteed to be continuous at 0 by the convexity of (1.1) and by the existence of a point $\tilde{x} \in X$ with $\underline{f}(\tilde{x}) < 0$. Lemma 2.2 completes the proof. ∎

Now, let us consider the special case $Y = R^m$ and $E(y, v) = v_0 + y^T \underline{v} + \frac{r}{2} \|\underline{v}\|^2$, $r > 0$ fixed. Then we get the auxiliary function

$$T(x, y) = f_0(x) + \frac{1}{2r} \sum_{i=1}^{m} [\max^2 \{0, y_i + r f_i(x)\} - y_i^2]$$

for any $x \in X$, $y \in Y$. Furthermore there holds

$$v_i^k = \begin{cases} f_0(x^k) & \text{if } i = 0, \\ \max \{f_i(x^k), -y_i^k/r\} & \text{if } i = 1, \ldots, m \end{cases} \tag{2.10}$$

for the solutions v^k of the problems (2.5). Using this generating function the inequalities (2.8) are satisfied with $\delta(s) = \frac{r}{2} s^2$. The rule for parameter selection given in step 3b of the SSH method leads to

$$y^{k+1} = y^k + r\underline{v}^k, \quad k = 1, 2, \ldots$$

With (2.10) we get

$$y_i^{k+1} = \max \{0, y_i^k + r f_i(x^k)\} \quad i = 1, \ldots, m; \quad k = 1, 2, \ldots$$

Thus the augmented Lagrangian method proposed by Hestenes [12] is involved in the SSH method.

Now, we investigate a modification of a method of exterior centers. We define

$$Y = \{y \in R^{m+1} \mid \underline{y} \leqq 0,\ y_0 \leqq \chi(0)\}$$

and

$$E(y, v) = \|v - y\|^2 \quad \text{for any} \quad y \in Y, \quad v \in R^{m+1}.$$

Then the sequence $\{y^k\}$ of parameters in the related SSH method is given by

$$y^{k+1} = w^k - \lambda_k(v^k - y^k), \quad k = 1, 2, \ldots,$$

with an arbitrary sequence of nonnegative numbers λ_k. If especially $\lambda_k = 0$, $k = 1, 2, \ldots$, then we get a method of exterior centers with the tangent parameter sequence (see e.g. [9]).

Here we propose to select λ_k according to

$$\lambda_k = \frac{(w^k - z^k)\,(v^k - y^k)}{E(y^k, v^k)}, \quad k = 1, 2, \ldots,$$

where $z^k = \left(\tau(y^k), 0, \ldots, 0\right)^T$. Then we get from the SSH method a shifted exterior centers method being numerically more stable than the related original method of exterior centers. In this case a constant $c > 0$ can be shown to exist such that

$$E(y^k, v^k) \leqq c, \quad k = 1, 2, \ldots$$

Using this bound we get

$$w_0^k \geqq \tau(y^k) + \frac{1}{2\sqrt{c}}\,\|\underline{v}^k\|^2, \quad k = 1, 2, \ldots,$$

and theorem 2.2 can be applied to prove the convergence of this new realization of the SSH method.

3. Rate of convergence

The close relation between the auxiliary problems (1.3) in SUMT and problems

$$E(y^k, v) \to \min! \quad \textit{subject to} \quad v \in Q$$

can be effectively used to estimate the rate of convergence of sequential unconstrained minimization algorithms. One possible way is to substitute the implicitly given set Q by a more simple set $\tilde{Q}$ approximating Q. Especially if optimal Lagrange multipliers λ_i^*, $i = 1, \ldots, m$, of the optimization problem (1.1) are available or can be approximated then we can use $\tilde{Q} \supset Q$ with

$$\tilde{Q} = \{v \in R^{m+1} \mid v_0 + \lambda^{*T}\underline{v} \geqq \chi(0)\}. \tag{3.1}$$

Similar to $S(y)$ and $\tau(y)$ let us define

$$\tilde{S}(y) = \left\{t \in R^1 \mid E\left(y, (t, 0, \ldots, 0)^T\right) \leqq \inf_{v \in \tilde{Q}} E(y, v)\right\}$$

and

$$\tilde{\tau}(y) = \sup\,\{t \mid t \in \tilde{S}(y)\}.$$

If $\tilde{Q} \supset Q$ holds then $\tilde{\tau}(y)$ underestimates $\tau(y)$. Using special generating functions E the problem

$$E(y, v) \to \min! \quad \textit{subject to} \quad v \in \tilde{Q} \tag{3.2}$$

can be solved explicitly. Thus $\tilde{\tau}(y)$ can be computed.

For example, let us consider the parameter set $Y = \operatorname{int} R_+^1$ and the generating function $E(y, v) = v_0 + y \sum_{i=1}^{m} |v_i|^p$ with a fixed $p > 1$. Then the point $\tilde{v}(y)$ defined by

$$\tilde{v}_0(y) = \chi(0) - \chi^{*T} \tilde{\underline{v}}(y)$$

and

$$\tilde{v}_i(y) = \left(\frac{\lambda_i^*}{py}\right)^{1/(p-1)}, \quad i = 1, \ldots, m,$$

solves the problem (3.2) if the set $\tilde{Q}$ is described by (3.1). Thus we get

$$\tilde{\tau}(y) = \chi(0) - (p-1)\, y^{\frac{1}{1-p}} \sum_{i=1}^{m} \left(\frac{\lambda_i^*}{p}\right)^{\frac{p}{p-1}}.$$

In the case $p = 2$ this leads to the estimation

$$\chi(0) - \frac{1}{4y} \|\lambda^*\|^2 \leqq \inf_{x \in X} T(x, y) \leqq \chi(0)$$

first given by Eremin [5].

If we select $Y = \operatorname{int} R_+^m$ and

$$E(y, v) = v_0 + \sum_{i=1}^{m} y_i \left(v_i + \sqrt{v_i^2 + y_i^{-2-\theta}}\right), \quad \theta \geqq 0,$$

then we get the penalty method proposed by Kaplan [13] and the related estimation

$$\chi(0) + \sum_{i=1}^{m} \left[\sqrt{\lambda_i^* (2y_i - \lambda_i^*)}\, y_i^{-1-\theta/2} - y_i^{-\theta/2}\right] \leqq \tau(y) \leqq \chi(0)$$

for sufficiently large parameters y_i, $i = 1, \ldots, m$. Under additional conditions this estimation can shown to be asymptotically sharp.

The approach described here can be applied also to get bounds for the rate of convergence of other sequential unconstrained minimization techniques as well. In [9] this has been done for methods of exterior centers.

References

[1] Антипин, А. С., Методы нелинейного программирования основанные на прямой и двойственной модификации функции Лагранжа. Препринт БНИИ системных исследований, Москва 1979.

[2] Bertsekas, D. P., Combined primal-dual and penalty methods for constrained minimization. SIAM J. Control **13** (1975) 3, 521—545.

[3] Bui-Trong-Lieu, Huard, P., La méthode des centres dans un espace topologique. Numer. Math **8** (1966), 56–67.

[4] Dolecki, S., Kurcyusz, S., On Φ-convexity in extremal problems. Tech. Report., Institute of Automatic Control, Warszawa 1976.

[5] Еремин, И. И., Метод штрафов в выпуклом программировании. Докл. АН СССР **173** (1967), 748–751.

[6] Everett, H. D., Generalized Lagrange multiplier method for solving problem of optimum allocation of resources. Operations Res. **11** (1963), 399–417.

[7] Fiacco, A. V., McCormick, G. P., Nonlinear programming: sequential unconstrained minimization techniques. Wiley, New York 1968.

[8] Гольштейн, Е. Г., Третьяков, Н. В., Модифицированные функции Лагранжа. Экон. Мат. Методы **10** (1974), 568–591.

[9] Großmann, Ch., Rates of convergence in methods of exterior centers. MOS Ser. Optimization **9** (1978) 3, 373–388.

[10] Großmann, Ch., Die Sensitivitätsanalyse als einheitlicher Zugang zu den Verfahren der sukzessiven unrestringierten Minimierung. Dissertation. B, TU Dresden 1979.

[11] Großmann, Ch. Kaplan, A. A., Strafmethoden und modifizierte Lagrangefunktionen in der nichtlinearen Optimierung. Teubner-Text, Leipzig 1979.

[12] Hestenes, M. R., Multiplier and gradient methods. JOTA **4** (1969), 303–320.

[13] Каплан, А. А., О скорости сходимости метода штарфа. Докл. АН СССР **229** (1976), 288–291.

[14] Kort, B. W., Bertsekas, D. P., Combined primal-dual methods for convex programming. SIAM J. Control Optimization **14** (1976), 268–294.

[15] Mangasarian, O. L., Unconstrained Lagrangians in nonlinear programming. SIAM J. Control **13** (1975) 4, 772–791.

[16] Mifflin, R., Convergence bounds for nonlinear programming algorithms. Math. Programming **8** (1975) 3, 251–271.

[17] Поляк, Б. Т., Третьяков, Н. В., Методы штрафных оценок для задач на условный экстремум. Ж. Вычисл. Мат. и Мат. Физ. **13** (1973), 34–46.

[18] Rockafellar, R. T., A dual approach to solving nonlinear programming problems by unconstrained optimization. Math. Programming **5** (1973) 3, 354–373.

[19] Rockafellar, R. T., The multiplier method of Hestenes and Powell applied to convex programming. JOTA **12** (1973), 555–562.

[20] Wierzbicki, A. P., Kurcyusz, S., Projection on a cone, penalty functionals and duality for problems with inequality constraints in Hilbert space. SIAM J. Control Optimization **15** (1977), 25–56.

Address of the author:

Technische Universität Dresden
Sektion Mathematik
8027 Dresden
Mommsenstr. 13
DDR

The finite-difference method for a mildly nonlinear elliptic problem

Bernd Heinrich, Karl-Marx-Stadt

Abstract. A discrete analogue $A_h u_h = G_h$ is constructed for a plane boundary value problem $Au = G$ "of monotonic type" (cf. [4]) with a mildly nonlinear elliptic differential equation of second order and nonlinear mixed boundary conditions. The properties of A_h are investigated and used to prove an a priori estimate, the existence of a solution u_h, the convergence $u_h \to u$ with respect to the C-norm and the convergence of some standard iterative methods for solving $A_h u_h = G_h$.

1. Introduction

Maximum principles and definiteness in spaces with a scalar product are essential tools to investigate finite-difference approximations for elliptic boundary value problems (B.V.P.) $Au = G$. In the nonlinear case, we use the corresponding generalizations, i.e. A is an operator "of monotonic type" (cf. [4]) or a special monotone operator (cf. [5]). The problem discussed here in general gives rise to the property of being "of monotonic type" only.

We consider the following class of B.V.P.

$$Ru \equiv -\sum_{i,j=1}^{2} a_{ij}(x) \frac{\partial^2 u}{\partial x_i \partial x_j} + c\left(x, u, \frac{\partial u}{\partial x_1}, \frac{\partial u}{\partial x_2}\right) = g(x), \quad x \in \Omega, \tag{1}$$

$$ru \equiv \alpha(x, u) + \beta(x) \frac{\partial u}{\partial \mu} = \varphi(x), \quad x \in \Gamma, \tag{2}$$

briefly denoted by

$$Au = G \quad \text{with} \quad A = (R, r)^T, \quad G = (g, \varphi)^T. \tag{3}$$

The parameters of (1), (2) are characterized by the following assumption

$\mathrm{P}(A, \bar{\Omega})$: *The plane domain Ω is bounded, Γ is the boundary of Ω,*

$$\left.\begin{aligned} &a_{ij} \in C(\bar{\Omega}), a_{12} = a_{21}, a_0 \sum_{i=1}^{2} \xi_i^2 \leqq \sum_{i,j=1}^{2} a_{ij}(x)\, \xi_i \xi_j \leqq a_1 \sum_{i=1}^{2} \xi_i^2 \quad \text{for all } \xi \in R^2 \\ &\text{and } x \in \Omega,\ a_0 > 0, \text{ i.e. } R \text{ is uniformly elliptic.}\end{aligned}\right\} \tag{4a}$$

$$\left.\begin{aligned} &\text{The function } c(x, z, p_1, p_2) \text{ from (1) is such that } c \in C(\bar{\Omega} \times R^3), c(x, \cdot) \in C^1(R^3) \text{ and} \\ &\frac{\partial c}{\partial z} \geqq 0, \left|\frac{\partial c}{\partial p_i}\right| \leqq b_0 < \infty \ (i = 1, 2) \text{ for all } (x, z, p_1, p_2) \in \bar{\Omega} \times R^3.\end{aligned}\right\} \tag{4b}$$

$$\left.\begin{array}{l}\textit{The function } \beta \textit{ from (2) is piecewise continuous and bounded on } \Gamma \textit{ and} \\ \alpha(x, z) \textit{ is such that } \alpha \in C(\Gamma \times R^1),\ \alpha(x, \cdot) \in C^1(R^1) \textit{ with } \beta \geqq 0 \textit{ on } \Gamma, \\ \qquad \dfrac{\partial \alpha}{\partial z} \geqq \alpha_0 > 0 \textit{ for all } (x, z) \in \Gamma \times R^1. \end{array}\right\} \tag{5a}$$

$$\left.\begin{array}{l}\textit{The outward direction } \mu \textit{ and the outward normal } n \textit{ on } \Gamma_{\beta>0} \subseteq \Gamma \\ \textit{satisfy } (\mu, n) \geqq \mu_0 > 0, \textit{ with } (\mu, \mu) = 1. \end{array}\right\} \tag{5b}$$

Furthermore, let the assumptions (with respect to smoothness, corners of Γ) quoted in [8] and excluding certain singular cases of Γ be satisfied.

Let g, φ from (1), (2) be continuous, but in general φ only piecewise so as not to exclude classical solutions of the B.V.P. (think of corners).

The condition (5a) may be weakened by

$$\begin{array}{l}\beta \geqq 0 \quad \text{on} \quad \Gamma_0, \dfrac{\partial \alpha}{\partial z} \geqq \alpha_0 > 0 \quad \text{on} \quad \Gamma_0 \times R^1 \quad \text{and} \quad \beta > 0,\ \alpha \equiv 0 \quad \text{on} \\ \Gamma \setminus \Gamma_0,\ \Gamma_0 \subseteq \Gamma,\ \Gamma_0 \neq \emptyset, \end{array} \tag{5'a}$$

where Γ_0 has a positive $d\Gamma$-measure. The methods and results for the B.V.P. with (5'a) are the same if we have a suitable set of "auxiliary functions" w_i. For $\Gamma = \Gamma_0$, such functions w_i (here: $i = 1, 2$) are given in (8).

The operators R, r act on functions $u \in D = C(\bar{\Omega}) \cap C^2(\Omega) \cap C^1(\Omega \cup \Gamma_{\beta>0})$. Together with the B.V.P. (1), (2) we consider the corresponding linear B.V.P.

$$\begin{array}{l} Lu = g \text{ in } \Omega,\ Lu := Ru \text{ with } c\left(x, u, \dfrac{\partial u}{\partial x_1}, \dfrac{\partial u}{\partial x_2}\right) \equiv -\sum_{i=1}^{2} b_i(x) \dfrac{\partial u}{\partial x_i} + c^*(x)\, u, \\ lu = \varphi \text{ on } \Gamma, \quad lu := ru \quad \text{with} \quad \alpha(x, u) \equiv \alpha^*(x)\, u. \end{array} \tag{6}$$

We extensively use the results contained in [16, 17] for the linear case (6).

Now we shall state two assertions for B.V.P. (3), which have discrete analogues (cf. section 3).

Theorem 1. *Let* P$(A, \bar{\Omega})$ *with* (4), (5) *be satisfied. For any $u_1, u_2 \in D$, from $Au_1 \geqq Au_2$ on $\bar{\Omega}$, i.e. $Ru_1 \geqq Ru_2$ in Ω and $ru_1 \geqq ru_2$ on Γ, the inequality $u_1 \geqq u_2$ on $\bar{\Omega}$ follows, i.e. A is an operator "of monotonic type".*

For the proof cf. [8, p. 288]. Due to $\dfrac{\partial \alpha}{\partial z} \geqq \alpha_0 > 0$ we can drop the additional assumption $u_1 - u_2 \not\equiv$ const quoted in [8]. Theorem 1 is the base for deriving a priori estimates with respect to the C-norm $\left(\|\cdot\|_{C(X)} = \max_{P \in X} |\cdot (P)|\right)$.

Theorem 2. *Let* P$(A, \bar{\Omega})$ *with* (4), (5) *be satisfied. Then the inequality*

$$\|u_1 - u_2\|_{C(\bar{\Omega})} \leqq N_0 \sup_{\Omega} |Ru_1 - Ru_2| + N_1 \sup_{\Gamma} |ru_1 - ru_2| \tag{7}$$

holds for any u_1, $u_2 \in D$ with positive constants N_0, N_1 not depending on u_1, u_2.

Proof. Cf. e.g. [17] for the linear case (6). From the properties of differentiability of c, α the representation $Ru_1 - Ru_2 = \tilde{L}(u_1 - u_2)$, $ru_1 - ru_2 = \tilde{l}(u_1 - u_2)$ follows, where $\tilde{L}, \tilde{l}$ are linear operators of type (6) depending on u_1, u_2 and acting on $u_1 - u_2$. The coefficients of $\tilde{L}, \tilde{l}$ satisfy uniformly with respect to x, u_1, u_2 the inequalities $|\tilde{b}_i| \leqq b_0$ $(i = 1, 2)$, b_0 from (4b), $\tilde{c}^* \geqq 0$, $\tilde{\alpha}^* \geqq \alpha_0 > 0$, α_0 from (5a), and $\tilde{c}^*$, $\tilde{\alpha}^*$ are just like $\tilde{b}_i$ at least bounded functions of x. Such operators $\tilde{A} = (\tilde{L}, \tilde{l})^T$ are also "of monotonic type" (cf. [8]). For the following "auxiliary functions"

$$w_0(x) = c_0 \left[c_1 - \exp \left\{ c_2 \sum_{i=1}^{2} (x_i - e_i)^2 \right\} \right], \quad w_1(x) = 1/\alpha_0 \tag{8}$$

uniformly for any $u_1, u_2 \in D$ the relations

$$\tilde{L}w_0 \geqq 1, \quad \tilde{L}w_1 \geqq 0 \quad \text{in} \quad \Omega, \quad \tilde{l}w_0 \geqq 0, \quad \tilde{l}w_1 \geqq 1 \text{ on } \Gamma \tag{9}$$

hold provided that the constants c_i, e_i $(i = 1, 2)$ are suitably chosen (here: $c_0 = 1$). Finally, (7) follows from (9) (cf. [17]). Assertions in the sense of Theorem 1, 2 are also mentioned in [4, 8, 11, 12] and in other papers by the author of [4].

Obviously from (7) the uniqueness of the solution of B.V.P. (1), (2) may be derived.

2. Finite-difference approximation

1⁰ Linear case. In [16] a discrete analogue to $Au = G$, the finite-difference scheme (F.D.S.)

$$A_h u_h = G_h \quad \text{on a grid} \quad \bar{\Omega}_h = \Omega_h + \Gamma_h, \; A_h = (L_h, l_h)^T, \; G_h = (g_h, \varphi_h)^T, \tag{10}$$

is constructed for the linear case (6). The approximation of the nonlinear problem (1), (2) (cf. 2⁰, 3⁰, 4⁰ of this section) yields the F.D.S. (10) for the linear case at the same time. Each linear finite-difference operator A_h admits a canonical representation

$$(A_h u_h)(P_0) = \sum_{P \in S(P_0)} A(P_0, P)\, u_h(P), \quad P_0 \in \bar{\Omega}_h, \quad S(P_0) = \{P_0\} + N(P_0), \tag{11}$$

$N(P_0)$: the set of "neighbours" of P_0 in the "star" of points $S(P_0)$, P_0: "center" of $S(P_0)$, $N(P_0) = S(P_0) \setminus \{P_0\}$. In [16] it was proved that the linear operator A_h from (10) has the following property $\mathrm{M}(A_h, \bar{\Omega}_h)$: *The real coefficients* $A(P_0, P)$ *from* (11) *satisfy for* $h \in (0, h_0]$ *the conditions*

(i) $$A(P_0, P) \begin{cases} > 0 & \text{at} \quad P = P_0 \\ < 0 & \text{at} \quad P \in N(P_0) = S(P_0) \setminus \{P_0\} \end{cases}, \quad P_0 \in \bar{\Omega}_h, \tag{12}$$

(ii) $$A_h(1)|_{P_0} \geqq 0 \quad \Big(A_h(1)|_{P_0} := \sum_{P \in S(P_0)} A(P_0, P)\Big), \quad P_0 \in \bar{\Omega}_h. \tag{13}$$

(iii) *At least at one point* $P_0 \in \bar{\Omega}_h$ *we have* $A_h(1)|_{P_0} > 0$. (14)

(iv) *If* $A_h(1)|_{P_0} = 0$ *at a point* $P_0 \in \bar{\Omega}_h$, *then there is a finite sequence of grid points* $Q_0, Q_1, \ldots, Q_n$ *taken from* $\bar{\Omega}_h$ *with* $Q_0 = P_0$, $Q_{j+1} \in N(Q_j)$, $A_h(1)|_{Q_j} = 0$ *for* $j = 0, 1, \ldots, n-1$ *and* $A_h(1)|_{Q_n} > 0$.

Lemma 1. *The property* $\mathrm{M}(A_h, \bar{\Omega}_h)$ *is sufficient for* A_h *to be an operator "of monotonic type", i.e. for any* u_h *with* $A_h u_h \geqq 0$ *on* $\bar{\Omega}_h$ *the inequality* $u_h \geqq 0$ *on* $\bar{\Omega}_h$ *holds* $(h \leqq h_0)$.

The well-known assertion of Lemma 1, in this or a similar fashion, is noted e.g. in [3, 4, 9] and [16, 17]. The linear operator A_h constructed in [16] fulfils $\mathrm{M}(A_h, \bar{\Omega}_h)$. Therefore A_h from (10) is "of monotonic type".

For the sake of brevity, the construction now to be given of the grid $\bar{\Omega}_h$ and the approximation of the linear main part of A are only roughly described (for details cf. [16]).

2⁰ The grid $\bar{\Omega}_h$. The "closed" grid $\bar{\Omega}_h$ is defined by $\bar{\Omega}_h = \Omega_h + \Gamma_h$. The set Ω_h consists of the points of intersection of Ω with an infinite square grid arising from mesh lines, which are parallel to the axes in a distance h. The "grid boundary" Γ_h is formed by the points of intersection of the mesh lines with the boundary Γ and further intersection points of certain "x_3-segments" (cf. fig. 1a) of the stars $S(P_0)$ with Γ.

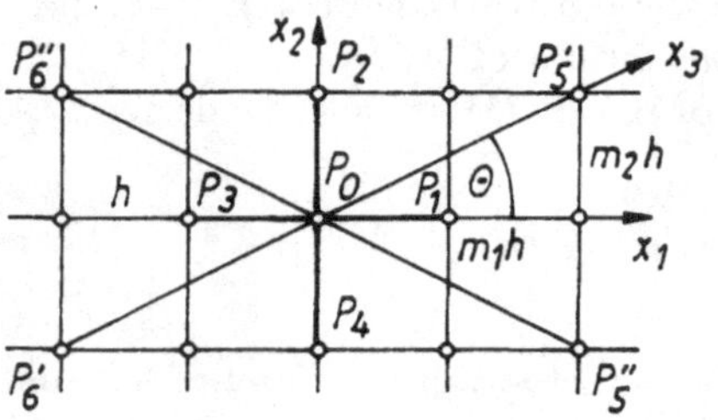

Fig. 1a

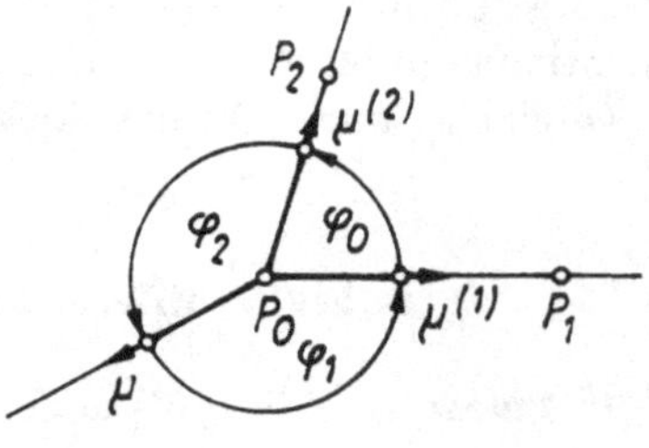

Fig. 1b

The star $S(P_0)$, $P_0 \in \Omega_h$, is defined as the union of P_0 with other points $P_1, P_2, \ldots, P_6$ from $\bar{\Omega}_h$ (cf. fig. 1a):

$$S(P_0) = \{P_0, P_1, \ldots, P_4\} \text{ if } a_{12}(P_0) = 0,\ S(P_0) = \{P_0, P_1, \ldots, P_6\} \text{ if } a_{12}(P_0) \neq 0, \tag{15}$$

$$P_0 = (x_1, x_2),\ P_1 = (x_1 + \lambda_1 h, x_2),\ P_2 = (x_1, x_2 + \lambda_2 h,)\ P_3 = (x_1 - \lambda_3 h, x_2),$$

$$P_4 = (x_1, x_2 - \lambda_4 h),\ P_5 = (x_1 + \lambda_5 m_1 h, x_2 \pm \lambda_5 m_2 h),\ P_6 = (x_1 - \lambda_6 m_1 h, x_2 \mp \lambda_6 m_2 h) \tag{16}$$
$$\text{for } a_{12}(P_0) \gtrless 0,$$

$0 < \lambda_i \leqq 1$ $(i = 1, 2, \ldots, 6)$, $m_i \geqq 1$ $(i = 1, 2)$, m_1 and m_2 are relatively prime integers.

If at $P_0 \in \Omega_h$ the condition $S(P_0) \subset \bar{\Omega}$ with $S(P_0)$ from (15), (16) and $\lambda_i = 1$ for all i in question is fulfilled, then P_0 is a so-called regular point of Ω_h, otherwise an irregular one. In [2] m_1, m_2 are chosen depending on $a_0, a_1, a_{11}, a_{12}, a_{22}$ such that with the abbreviations $d_1 = a_{12}d^{-1}$, $d_2 = a_{12}d$ the following relations are fulfilled

$$\left.\begin{array}{l} d_1 + d_2 \geqq 0,\ a_{ii} - d_i > \bar{a}_0 > 0\ (i = 1, 2) \quad in \quad \Omega,\ |d| = \dfrac{m_2}{m_1}, \\ 1 \leqq m_i < K\ (i = 1, 2), \\ \textit{the } m_i \textit{ are relatively prime integers and } K, \bar{a}_0 \textit{ do not depend on} \\ P \in \Omega,\ d = \tan\theta,\ \theta \textit{ from fig. } 1\text{a}. \end{array}\right\} \tag{17}$$

The choice of m_i with $m_i \geqq 1$ instead of $m_i = 1$ $(i = 1, 2)$ is essential for assuring $\mathrm{M}(A_h, \bar{\Omega}_h)$. In special cases taking $m_1 = m_2 = 1$ may be sufficient, e.g. if $|a_{12}| < a_{ii}$ $(i = 1, 2)$ on $\bar{\Omega}$. For details, particularly for the choice of m_1, m_2, cf. [2].

3⁰ Approximation of R. Now we can express the mixed derivative $\partial^2 u/\partial x_1\,\partial x_2$ at $P_0 \in \Omega_h$ by $\partial^2 u/\partial x_i^2$ $(i = 1, 2, 3)$ and d from (17). The finite-difference operator R_h approximating R is defined at $P_0 \in \Omega_h$ in the following way:

$$(R_h v)\,(P_0) = \left(-\sum_{i=1}^{2}\left\{(a_{ii} - d_i)\,\frac{\delta^2 v}{\delta x_i^2} + d_i\,\frac{\delta^2 v}{\delta x_3^2}\right\} + c\left(v, \frac{\delta v}{\delta x_1}, \frac{\delta v}{\delta x_2}\right)\right)(P_0). \quad (18)$$

All functions and expressions of (18) are to be taken at $P_0 \in \Omega_h$. The difference quotients are defined as usual by

$$\frac{\delta^2 v(P_0)}{\delta x_i^2} = 2\,\frac{h_{i+2} v_i - (h_i + h_{i+2})\,v_0 + h_i v_{i+2}}{h_i h_{i+2}(h_i + h_{i+2})},\quad \frac{\delta v(P_0)}{\delta x_i} = \frac{v_i - v_{i+2}}{h_i + h_{i+2}} \quad (i = 1, 2),$$

$$\frac{\delta^2 v(P)}{\delta x_3^2} = 2\,\frac{k_6 v_5 - (k_5 + k_6)\,v_0 + k_5 v_6}{k_5 k_6 (k_5 + k_6)} \quad (19)$$

with $v_j = v(P_j)$, $h_j = \lambda_j h$, $k_j = \lambda_j k$, $k = h\sqrt{m_1^2 + m_2^2}$, cp. (19) with fig. 1a. For $u \in C^k(\bar{\Omega})$, $k = 3$ or $k = 4$, the local approximation error at P_0 is

$$|R_h u - Ru|_{P_0} = \begin{cases} O(h^{k-2}) \\ O(h) \end{cases} \quad \text{if} \quad P_0 \in \Omega_h \quad \text{is} \quad \begin{cases} \text{regular} \\ \text{irregular} \end{cases}. \quad (20)$$

4⁰ Approximation of r. Note that the inclusion $\Gamma_h \subset \Gamma$ holds. For $P_0 \in \Gamma_h$ the star $S(P_0)$ with $\beta(P_0) \neq 0$ is defined by $S(P_0) = \{P_0, P_1, P_2\}$, $\{P_1, P_2\} \subset \Omega_h$. The points P_1, P_2 are located "near" the boundary Γ, P_1 on the one side and P_2 on the other side of the straight line g_μ spanned by μ from (5b). If we can find "near" $P_0 \in \Gamma_h$ a point $P_1 \in \Omega_h$ with $P_1 \in g_\mu$, then we may take $S(P_0) = \{P_0, P_1\}$. For details and exact formulations cf. [16]. The operator r_h approximating r is given by

$$(r_h v)\,(P_0) = \alpha\big(P_0, v(P_0)\big) + \beta(P_0)\,\frac{\delta v(P_0)}{\delta\mu},\quad P_0 \in \Gamma_h,\quad \text{cf. fig. 1b}, \quad (21)$$

$$\frac{\delta v(P_0)}{\delta\mu} = -\frac{1}{\sin\varphi_0}\left\{\sin\varphi_2\,\frac{\delta v(P_0)}{\delta\mu^{(1)}} + \sin\varphi_1\,\frac{\delta v(P_0)}{\delta\mu^{(2)}}\right\} \quad \text{for } \{P_1, P_2\} \cap g_\mu = \emptyset,$$

$$\frac{\delta v(P_0)}{\delta\mu_{(i)}} = \frac{v_i - v_0}{\tau_i},\ \tau_i\text{: length of } \overline{P_0 P_i}\ (i = 1, 2),\ \frac{\delta v(P_0)}{\delta\mu} = -\frac{\delta v(P_0)}{\delta\mu^{(1)}} \ \text{for } P_1 \in g_\mu, \quad (22)$$

cp. (22) with fig. 1b. Here we obtain the local approximation error

$$|r_h u - ru|_{P_0} = O(h) \quad \text{for} \quad u \in C^k(\bar{\Omega}),\quad k \geqq 3,\quad P_0 \in \Gamma_h. \quad (23)$$

5⁰ Finite-difference scheme (F.D.S.). The discrete analogue to the B.V.P. (1), (2) is defined by

$$\begin{aligned} R_h u_h &= g_h \quad \text{on} \quad \Omega_h \quad R_h \text{ from (18)},\quad g_h(P_0) = g(P_0) \quad \text{at} \quad P_0 \in \Omega_h,\quad g \text{ from (1)}, \\ r_h u_h &= \varphi_h \quad \text{on} \quad \Gamma_h,\quad r_h \text{ from (21)},\quad \varphi_h(P_0) = \varphi(P_0) \quad \text{at} \quad P_0 \in \Gamma_h,\quad \varphi \text{ from (2)}, \end{aligned} \quad (24)$$

i.e. the right-hand sides g, φ are exactly approximated on Ω_h, Γ_h. The brief denotation of the F.D.S. from (24) is

$$A_h u_h = G_h \quad \text{with} \quad A_h = (R_h, r_h)^T,\quad G_h = (g_h, \varphi_h)^T. \quad (25)$$

The "grid boundary" Γ_h is subdivided into $\Gamma_{h1} = \{P_0 \in \Gamma_h : \beta(P_0) = 0\}$, $\Gamma_{h2} = \{P_0 \in \Gamma_h : \beta(P_0) \neq 0\}$. Then r_{h1}, r_{h2} are the corresponding discrete boundary operators, $r_h = (r_{h1}, r_{h2})^T$. For the sake of simplicity, we assume without loss of generality for the following results that r_{h1} is linear. Due to (5a) we can principally solve $\alpha\big(P_0, u_h(P_0)\big) = \varphi_h(P_0)$ for $u_h(P_0)$ at each $P_0 \in \Gamma_{h1}$. Now r_h of (24) is defined by

$$r_h = (r_{h1}, r_{h2})^T, \quad r_{h1} v \equiv v \quad \text{on} \quad \Gamma_{h1}, \quad r_{h2} v \equiv \alpha(v) + \beta \frac{\delta v}{\delta \mu} \quad \text{on} \quad \Gamma_{h2} \text{ (cf. (21))}. \tag{26}$$

From (5a) we also obtain $\Gamma_{h1} \neq \emptyset$.

Now we number the N_2 points $P_0 \in \bar{\Omega}_h = \Omega_h + \Gamma_{h1} + \Gamma_{h2}$ in a one-to-one fashion with integers i, $i = 1, 2, \ldots, N_2$. We obtain $\bar{\Omega}_h = \{P_{i0}\}_{i=1}^{N_2}$, and instead of P_{i0} we may write simply P_i:

$$\Omega_h = \{P_1, P_2, \ldots, P_{N_0}\}, \quad \Gamma_{h1} = \{P_{N_0+1}, \ldots, P_{N_1}\}, \quad \Gamma_{h2} = \{P_{N_1+1}, \ldots, P_{N_2}\}. \tag{27}$$

The "neighbours" of the "center" $P_i = P_{i0}$ of a star $S(P_i)$ are denoted by $P_{i\varkappa}$, $\varkappa = 1, 2, \ldots, n(P_i)$.

3. Properties of A_h

Now we describe properties of A_h used essentially in section 4, 5. The operators A_h from (25) and F_h defined by $F_h u_h \equiv A_h u_h - G_h$, G_h also from (25), act on grid functions u_h defined on $\bar{\Omega}_h$. In connection with (27) we may interpret A_h, F_h, u_h as

$$\left.\begin{aligned} &A_h : R^{N_2} \to R, \quad A_h = \big(a_{P_1}, \ldots, a_{P_{N_2}}\big)^T, \quad a_{P_i} : R^{N_2} \to R^1, \\ &a_{P_i}(u_h) := (A_h u_h)(P_i), \quad A_h \text{ from (25)}, \quad u_h = \big(u_h(P_1), \ldots, u_h(P_{N_2})^T\big), \\ &F_h : R^{N_2} \to R^{N_2}, \quad F_h = \big(f_{P_1}, \ldots, f_{P_{N_2}}\big)^T, \quad f_{P_i}(u_h) = a_{P_i}(u_h) - G_h(P_i). \end{aligned}\right\} \tag{28}$$

Lemma 2. *There exists the Fréchet derivative $A_h'(v_h)$ at each $v_h \in R^{N_2}$ and A_h' is continuous on R^{N_2}, Moreover, the linear finite-difference operator $A_h'(v_h)$ fulfils for all $v_h \in R^{N_2}$ the property* $\mathrm{M}\big(A'_h(v_h), \bar{\Omega}_h\big)$ *from section* 2.

Proof. From $\mathrm{P}(A, \bar{\Omega})$, the existence and continuity of the partial derivatives $\partial a_{P_i}(v_h)/\partial \zeta^j$ for $i, j = 1, 2, \ldots, N_2$ follow, with $\zeta^j\big(= u_h(P_j)\big)$ from $\zeta = (\zeta^1, \ldots, \zeta^{N_2})^T \in R^{N_2}$. Therefore the F-derivative of A_h exists and is also continuous on R^{N_2}, $A_h'(v_h)$ is represented by $A_h'(v_h) = \big(\alpha_{ij}(v_h)\big)_{N_2 \times N_2}$,

$$\alpha_{ij}(v_h) = \frac{\partial a_{P_i}(v_h)}{\partial \zeta^j} \begin{cases} > 0 & \text{for} \quad j = i \\ < 0 & \text{for} \quad j = i\varkappa,\ P_{i\varkappa} \in N(P_i);\ i, j = 1, 2, \ldots, N_2. \\ = 0 & \text{otherwise} \end{cases} \tag{29}$$

We note that $N(P_i) = \emptyset$ for $P_i \in \Gamma_{h1}$, and the inequalities $\partial a_{P_i}(v_h)/\partial \zeta^j < 0$ from (29) for i with $P_i \in \Omega_h$, $P_{i\varkappa} \in N(P_i)$, in general hold for $h \in (0, h_0]$ only, h_0 suitably chosen. Then it is obvious that (12), (13), (14) and (iv) of $\mathrm{M}\big(A_h'(v_h), \bar{\Omega}_h\big)$ are fulfilled uniformly with respect to $v_h \in R^{N_2}$, $h \in (0, h_0]$.

Corollary 1. *There exists the F-derivative F_h' of F_h from* (28) *and obviously $A_h'(v_h) = F_h'(v_h)$ holds for all $v_h \in R^{N_2}$.*

Corollary 2. *The linear operator $A'_h(v_h)\colon R^{N_2} \to R^{N_2}$ is, uniformly with respect to $v_h \in R^{N_2}$, "of monotonic type", i.e. $A'_h(v_h)\, u_h \geqq 0$ on $\bar{\Omega}_h$ implies $u_h \geqq 0$ on $\bar{\Omega}_h$ for all $v_h \in R^{N_2}$, $h \in (0, h_0]$.*

Corollary 3. *There exists $[A'_h(v_h)]^{-1}$ with $[A'_h(v_h)]^{-1} \geqq 0$ for any $v_h \in R^{N_2}$, $h \in (0, h_0]$.*

The assertion of Corollary 2 follows from Lemmata 1, 2 and a consequence of Corollary 2 is that $A'_h(v_h)\, u_h = 0$ implies $u_h = 0$ on $\bar{\Omega}_h$ and all entries of $[A'_h(v_h)]^{-1}$ are nonnegative (cf. Corollary 3).

Theorem 3. *The nonlinear operator A_h from (24), (25), (26) is "of monotonic type" i.e. for all $u_{h1}, u_{h2} \in R^{N_2}$, $h \in (0, h_0]$, the inequality $A_h u_{h1} \geqq A_h u_{h2}$ on $\bar{\Omega}_h$ yields $u_{h1} \geqq u_{h2}$ on $\bar{\Omega}_h$. The same is valid for F_h from (28).*

Proof. We may write $A_h u_{h1} - A_h u_{h2} = B_h(u_{h1}, u_{h2})\,(u_{h1} - u_{h2})$ with

$$B_h = (\beta_{ij})_{N_2 \times N_2},\ \beta_{ij} = \frac{\partial a_{P_i}(z_h^{(i)})}{\partial \zeta^j},\ z_h^{(i)} = u_{h1} + \theta_i(u_{h2} - u_{h1}),\ \theta_i \in [0, 1] \tag{30}$$

$(i, j = 1, 2, \ldots, N_2)$. The linear operator $B_h\colon R^{N_2} \to R^{N_2}$ depending on u_{h1}, u_{h2} fulfils the property $\mathrm{M}(B_h, \bar{\Omega}_h)$ for all $u_{h1}, u_{h2} \in R^{N_2}$. The proof is similar to that of Lemma 2. There exists B_h^{-1} with $B_h^{-1} \geqq 0$ for all $u_{h1}, u_{h2} \in R^{N_2}$, and obviously the inequality $u_{h1} - u_{h2} \geqq 0$ follows from $A_h u_{h1} - A_h u_{h2} \geqq 0$.

Note. *If c from (1) is especially a function of type $c(x, u)$, then the relation $B_h(u_{h1}, u_{h2}) = A'_h(z_h)$ with $z_h \in R^{N_2}$, in general $z_h \notin [u_{h1}, u_{h2}]$, holds.*

Theorem 4. *For $A_h = (R_h, r_h)^T$ from (24), (25), (26) and all u_{h1}, u_{h2} on $\bar{\Omega}_h$, $h \in (0, h_0]$, the inequality*

$$\|u_{h1} - u_{h2}\|_{C(\bar{\Omega}_h)} \leqq N_0\, \|R_h u_{h1} - R_h u_{h2}\|_{C(\Omega_h)} + N_1\, \|r_h u_{h1} - r_h u_{h2}\|_{C(\Gamma_h)} \tag{31}$$

holds and the positive constants N_i $(i = 1, 2)$ do not depend on u_{h1}, u_{h2}, $h \in (0, h_0]$.

Proof. We apply the components $\tilde{L}_h, \tilde{l}_h$ of $B_h = (\tilde{L}_h, \tilde{l}_h)^T$ (cf. (30)) to the functions $w_1 = \max\,(1, \alpha_0^{-1})$ and w_0 from (8), with c_i $(i = 0, 1, 2)$ and e_1, e_2 chosen suitably. This yields

$$\tilde{L}_h w_0 \geqq 1 \quad\text{and}\quad \tilde{L}_h w_1 \geqq 0 \quad\text{on}\quad \Omega_h, \quad \tilde{l}_h w_0 \geqq 0 \quad\text{and}\quad \tilde{l}_h w_1 \geqq 1 \quad\text{on}\quad \Gamma_h \tag{32}$$

uniformly with respect to u_{h1}, u_{h2} and $h \in (0, h_0]$. Because $\mathrm{M}(B_h, \bar{\Omega}_h)$ and (32) are fulfilled we obtain (cf. [17]) $\|v_h\|_{C(\bar{\Omega}_h)} \leqq N_0\, \|\tilde{L}_h v_h\|_{C(\Omega_h)} + N_1\, \|\tilde{l}_h v_h\|_{C(\Gamma_h)}$ with $v_h = u_{h1} - u_{h2}$, $N_i = \|w_i\|_{C(\bar{\Omega})}$.

Now we describe the properties of A_h, F_h (cf. (28)) by means of notions developed in [10].

Note 1. *The derivative $A'_h(v_h)\ \big(= F'_h(v_h)\big)$ is for any $v_h \in R^{N_2}$ an M-matrix in the sense of [10].*

The property $\mathrm{M}\big(A'_h(v_h), \bar{\Omega}_h\big)$ is fulfilled. This is sufficient for $A'_h(v_h)$ to be an M-matrix.

Note 2. *A_h and F_h are inverse isotone on R^{N_2} in the sense of [10].*

The notions of being "of monotonic type" and "inverse isotone" used here are identical.

Note 3. *$A_h : R^{N_2} \to R^{N_2}$, F_h too, is an M-function in the sense of* [10].

The proof consists in showing that F_h is "inverse isotone" and "off-diagonal antitone" (cf. [10]). Indeed, this is true.

4. Existence of solutions, convergence

Theorem 5. *The F.D.S. $A_h u_h = G_h$ of (24), (25), (26) has a unique solution u_h for any $G_h \in R^{N_2}$.*

Proof. The operator $A_h^{L_0} = (R_h^L, r_h^{L_0})^T$ with the linear main part R_h^L of R_h and $r_h^{L_0}$ defined by

$$r_h^{L_0} u_h = \begin{cases} u_h & \text{on } \Gamma_{h1} \\ \alpha_0 u_h + \beta \dfrac{\delta u_h}{\delta \mu} & \text{on } \Gamma_{h2} \end{cases}, \quad \alpha_0 \text{ from (5a)},$$

fulfils $M(A_h^{L_0}, \bar{\Omega}_h)$. Hence, $(A_h^{L_0})^{-1}$ exists and $u_h = T_h u_h$ with continuous $T_h = -(A_h^{L_0})^{-1} \times (A_h^{N_0} u_h - G_h)$, $A_h^{N_0} = A_h - A_h^{L_0}$, is equivalent to $A_h u_h = G_h$ with respect to solutions. We also consider the family of F.D.S. $F_{ht} u_h = A_h^{L_0} u_h + t(A_h^{N_0} u_h - G_h) = 0$, $t \in [0, 1]$. For all u_h, v_h on $\bar{\Omega}_h$ the a priori estimate

$$\|u_h - v_h\|_{C(\bar{\Omega}_h)} \leqq N_0 \|R_{ht} u_h - R_{ht} v_h\|_{C(\Omega_h)} + N_1 \|r_{ht} u_h - r_{ht} v_h\|_{C(\Gamma_h)}, \; t \in [0, 1], \tag{33}$$

with $(R_{ht}, r_{ht})^T = A_{ht}$, $A_{ht} u_h = F_{ht} u_h + t G_h$, uniformly holds in analogy to (31). If there are solutions of $F_{ht} u_h = 0$, then these solutions u_{ht} are located in a ball $\bar{S}(0, r)$ with $r = \sqrt{N_2}\, K$. The constant K bounds u_{ht} by $\|u_{ht}\|_{C(\bar{\Omega}_h)} \leqq K$ for all $t \in [0, 1]$, $h \in (0, h_0]$ and may be obtained from (33) for $v_h \equiv 0$. Therefore, no solution of $F_{ht} u_h = 0$ can lie on the boundary of $S(0, r + 1)$. If there exist solutions of $\lambda u_h = T_h u_h$ for $\lambda^{-1} = t \in (0, 1)$ then they lie in $\bar{S}(0, r)$, besides we have $\lambda u_h \neq T_h u_h$ for $u_h \in \partial S(0, r + 1)$. Applying the Leray-Schauder Theorem from [10, 6.3.3.] we conclude that T_h has a fixed point. The uniqueness may be derived from (31).

The Hadamard Theorem in [10, 5.3.10.] may be applied to prove Theorem 5, too. We only show that A_h is continuously differentiable on R^{N_2} and $\|[A_h'(v_h)]^{-1}\| \leqq \gamma < \infty$ is fulfilled uniformly with respect to $v_h \in R^{N_2}$, $h \in (0, h_0]$. Indeed, this is true and Lemma 2 and the way of proving (31) give the tool to do it.

Theorem 6. *Let u_h be the solution of the F.D.S. $A_h u_h = G_h$ from (24), (25), (26), u the solution of the B.V.P. $Au = G$ from (1), (2), (3). Provided $u \in C^k(\bar{\Omega})$, $k \geqq 3$, the error estimate*

$$\|u_h - u\|_{C(\bar{\Omega}_h)} \leqq Mh, \quad M = O(1), \quad h \in (0, h_0],$$

holds.

Proof. The solutions u_h, u are unique. We use (31) with $v_h = u$ on $\bar{\Omega}_h$, the relation $A_h u_h - A_h u = Au - A_h u$ and (20), (23) and obtain the assertion.

5. Iterative methods

In [10] several iterative methods (I.M.) for solving nonlinear equations $A_h u_h = G_h$ or $F_h u_h = 0$ are presented. Now we mainly use the fashion $F_h u_h = 0$ with the components $f_{P_i}(u_h) = 0$ $(i = 1, 2, \ldots, N_2)$ and the vector (grid function) $u_h = (u_{h1}, \ldots, u_{hN_2})^T$, cf. (28).

1⁰ SOR.
$$\left.\begin{aligned} &\textit{Solve } f_{P_i}(u_{h1}^{k+1}, \ldots, u_{h(i-1)}^{k+1}, u_{hi}, u_{h(i+1)}^k, \ldots, u_{hN_2}^k) = 0 \quad \textit{for} \quad u_{hi}, \\ &\textit{set} \quad u_{hi}^{k+1} = u_{hi}^k + \omega(u_{hi} - u_{hi}^k), \quad i = 1, 2, \ldots, N_2, \quad k = 0, 1, \ldots; \\ &\omega \in (0, 1] \quad \textit{and} \quad u_h^0 \in R^{N_2} \quad \textit{are given.} \end{aligned}\right\} \tag{34}$$

2⁰ Newton. *Solve* $F_h'(u_h^k)\, \Delta u_h^k = -F_h(u_h^k)$ *for* Δu_h^k, *set* $u_h^{k+1} = u_h^k + \Delta u_h^k$, (35)

$k = 0, 1, \ldots;$ u_h^0 *is given.*

3⁰ SOR-Newton (one-step). *Set*

$$u_{hi}^{k+1} = u_{hi}^k - \omega \frac{f_{P_i}(u_{h1}^{k+1}, \ldots, u_{h(i-1)}^{k+1}, u_{hi}^k, \ldots, u_{hN_2}^k)}{\dfrac{\partial f_{P_i}}{\partial u_{hi}}(u_{h1}^{k+1}, \ldots, u_{h(i-1)}^{k+1}, u_{hi}^k, \ldots, u_{hN_2}^k)}, \quad i = 1, 2, \ldots, N_2$$

$k = 0, 1, \ldots;$ *and* $\omega \in (0, 1]$, $u_h^0 \in R^{N_2}$ *are given.*

4⁰ Newton-SOR (m-step). *We only use m SOR-steps for the approximate solution of equations of type* (35), *cf.* [10, p. 216] *for details.*

Lemma 3. *The I.M.* 1⁰, 3⁰, 4⁰ *for solving $F_h u_h = 0$ with A_h (cf.* (28)) *from* (24), (25), (26) *converge locally, i.e. $u_h^k \to u_h^*$, where u_h^* is the solution of $F_h u_h = 0$ and u_h^0 is to be chosen in a sufficiently small neighbourhood of u_h^*.*

Proof. The operator $F_h \colon R^{N_2} \to R^{N_2}$ is continuously differentiable on R^{N_2}, $F_h'(u_h^*)$ is also an M-matrix (cf. note 1). Using assertion 10.3.6. of [10] we obtain at least the local convergence property of the I.M. 1⁰, 3⁰, 4.⁰

Lemma 4. *For any $G_h \in R^{N_2}$, $u_h^0 \in R^{N_2}$, $\omega \in (0, 1]$ the I.M.* 1⁰ *(cf.* (34)) *converges, i.e. the I.M.* 1⁰ *is globally convergent.*

Proof. Since F_h is a continuous M-function (cf. note 3), the assertion can be derived from 13.5.9. in [10].

Lemma 5. *If $c\left(x, u, \dfrac{\partial u}{\partial x_1}, \dfrac{\partial u}{\partial x_2}\right)$, $\alpha(x, u)$ from* (1), (2) *are convex with respect to $\left(u, \dfrac{\partial u}{\partial x_1}, \dfrac{\partial u}{\partial x_2}\right) \in R^3$ and $u \in R^1$, respectively, then the I.M.* 2⁰ *from* (35) *converges globally and monotonically, i.e. for any $u_h^0 \in R^{N_2}$ the iterates $u_h^{k+1} = u_h^k - [F_h'(u_h^k)]^{-1} F_h(u_h^k)$, $k = 0, 1, \ldots$, converge to the solution u_h^* of $F_h u_h = 0$ in the sense of*

$$u_h^k \geqq u_h^{k+1} \geqq \cdots \geqq u_h^* \quad \textit{for} \quad k = 1, 2, \ldots$$

Proof. We may derive that F_h is also convex on R^{N_2} and use the previous results: $[F_h'(v_h)]^{-1} \geqq 0$ holds for all $v_h \in R^{N_2}$, a solution u_h^* of $F_h u_h = 0$ exists, F_h' is continuous on R^{N_2}. Then all assumptions of the global Newton Theorem 13.3.7. in [10] are fulfilled and Lemma 5 is proved.

Note. (i) *Since in general the I.M.* 1^0 *requires an inner procedure the SOR method is implementable only in a limited way.* (ii) *For the solution of equations of type* (35) *with the sparse matrix* $F_h'(u_h^k)$ *we can use iterative as well as adapted direct methods.*

References

[1] Bers, L., On mildly nonlinear partial difference equations of elliptic type. J. Res. Nat. Bur. Standards **51** (1953), 229–236.

[2] Bramble, J. H., and B. E. Hubbard, A theorem on error estimation for finite-difference analogues of the Dirichlet problem for elliptic equations. Contrib. Differential Equations **2** (1963), 319–340.

[3] Ciarlet, Ph. G., Discrete maximum principle for finite-difference operators. Aequationes Math. **4** (1970), 338–352.

[4] Collatz, L., Funktionalanalysis und numerische Mathematik. Springer-Verlag, Berlin–Göttingen–Heidelberg 1964.

[5] Gajewski, H., K. Gröger und K. Zacharias, Nichtlineare Operatorgleichungen und Operatordifferentialgleichungen. Akademie-Verlag, Berlin 1974.

[6] Карчевский, М. М., и А. Д. Ляшко, Разностные схемы для нелинейных задач математической физики. Казань 1976 (Ротапринт).

[7] McAllister, G. T., Quasilinear uniformly elliptic partial differential equations and difference equations. SIAM J. Numer. Anal. **3** (1966), 13–33.

[8] Meyer, A. G., Schranken für die Lösung von Randwertaufgaben mit elliptischer Differentialgleichung. Arch. Rat. Mech. Anal. **6** (1960), 277–298.

[9] Michlin, S. G., und Ch. L. Smolizki, Näherungsmethoden zur Lösung von Differential- und Integralgleichungen. B. G. Teubner Verlagsges., Leipzig 1969.

[10] Ortega, J. M., and W. C. Rheinboldt, Iterative solution of nonlinear equations in several variables. Academic Press, New York–London 1970.

[11] Protter, M., and H. Weinberger, Maximum principles in differential equations. Prentice-Hall Inc. Englewood Cliffs, N. J. 1967.

[12] Redheffer, R. M., Fehlerabschätzung bei nichtlinearen Differentialungleichungen mit Hilfe linearer Differentialungleichungen. Numer. Math. **28** (1977), 393–405.

[13] Sachs, A., Über Differenzenapproximationen des Dirichletproblems nichtlinearer elliptischer Operatoren in Divergenzform. Dissertation, München 1971.

[14] Самарский, А. А., и В. Б. Андреев, Разностные методы для эллиптических уравнений. Наука, Москва 1976.

[15] Stepleman, R. S., Difference analogues of quasi-linear elliptic Dirichlet problems with mixed derivatives. Math. of Comp. **25** (1971), 257–269.

[16] Heinrich, B., Monotone Differenzenapproximationen für lineare elliptische Differentialgleichungen mit gemischten Randbedingungen. Beiträge zur Numer. Math. **7** (1979), 49–64.

[17] Heinrich, B., Zur Fehlerabschätzung beim Differenzenverfahren für elliptische Differentialgleichungen nach Gerschgorin-Batschelet, Beiträge zur Numer. Math. **9** (1981), 73–85.

Address of the author:

Technische Hochschule Karl-Marx-Stadt
Sektion Mathematik
9010 Karl-Marx-Stadt
Reichenhainer Str. 41

Finite element analysis of the unilateral contact between elastic bodies

Ivan Hlaváček, Prague

Abstract. The problem of unilateral frictionless contact between two solid bodies can be formulated by means of variational inequalities. Within the range of linear plane elastostatics, the solvability of the problem is studied in semi-coercive cases (virtual rigid displacements are not excluded). The problem represents a natural generalization of the famous Signorini problem or of the contact between a rigid stamp and an elastic halfplane.

Piecewise linear finite elements are applied to both the primary (potential energy) and the dual (complementary energy) variational formulations and some convergence theorems stated. If the exact solutions are regular enough, some a priori error estimates can also be deduced.

1. Formulations of the continuous problems

Let us consider two elastic bodies Ω' and Ω'' in the plane $\mathbb{R}^2$, where Ω', Ω'' are bounded domains with Lipschitz boundaries. Let $x = (x_1, x_2)$ be Cartesian coordinates,

$$u' = (u_1', u_2') \quad \text{and} \quad u'' = (u_1'', u_2'')$$

the displacement vector in Ω' and Ω'', respectively.

Assume the strain-displacement relations

$$e_{ij}(u) = \frac{1}{2}\left(\frac{\partial u_i}{\partial x_j} + \frac{\partial u_j}{\partial x_i}\right), \quad i, j = 1, 2,$$

the stress-strain relations

$$\tau_{ij} = c_{ijkm} e_{km},$$

where the repeated subscript implies summation over 1, 2,

$$c_{ijkm} \in L^\infty(\Omega' \cup \Omega''),$$

$$c_{ijkm} = c_{jikm} = c_{kmij},$$

$$c_{ijkm} e_{ij} e_{km} \geqq c_0 e_{ij} e_{ij} \qquad \forall\, e_{ij} = e_{ji}$$

holds almost everywhere in $\Omega' \cup \Omega''$, with some positive constant c_0.

Let the equilibrium equations be

$$\frac{\partial \tau_{ij}}{\partial x_j} + F_i = 0, \quad i = 1, 2,$$

where $F_i \in L^2(\Omega^M)$, $M = \mathrm{I}, \mathrm{II}$, are the body forces.

Assume the following boundary conditions:

$$u' = 0 \qquad \text{on} \quad \Gamma_u \subset \partial\Omega',$$

$$\tau_{ij} n_j = P_i \qquad \text{on} \quad \Gamma'_\tau \cup \Gamma''_\tau \equiv \Gamma_\tau, \quad P_i \in L^2(\Gamma^M_\tau),$$

$$u_n = 0, \quad T_t = 0 \qquad \text{on} \quad \Gamma_0 \subset \partial\Omega'',$$

where

$$u_n = u_i n_i, \quad T_t = \tau_{ij} n_j t_i, \quad t = (t_1, t_2) = (-n_2, n_1)$$

and n is the unit outward normal to the boundary.

Note that the conditions on Γ_0 describe a bilateral contact with the support, which may occur on the axis of symmetry, if any exists.

On the remaining part of $\partial\Omega' \cup \partial\Omega''$ a possible unilateral contact may be realized. We distinguish two classes of contact problems: (i) with a bounded contact zone, (ii) with an enlarging contact zone.

(i) Problems with a bounded contact zone are determined by the geometry of the boundaries in a neighbourhood of a possible contact zone. We may define

$$\Gamma_K = \partial\Omega' \cap \partial\Omega''.$$

Consequently,

$$\partial\Omega' = \bar{\Gamma}_u \cup \bar{\Gamma}'_\tau \cup \Gamma_K,$$

$$\partial\Omega'' = \bar{\Gamma}_0 \cup \bar{\Gamma}''_\tau \cup \Gamma_K.$$

We say that a unilateral bounded contact on Γ_K occurs if

$$u'_n + u''_n \leqq 0 \quad \text{on} \quad \Gamma_K \tag{1.1}$$

(condition of non-penetrating). Besides, we have

$$T'_n = T''_n, \quad T'_t = T''_t$$

and the normal contact forces cannot be tensile, i.e.

$$T_n \leqq 0 \quad \text{on} \quad \Gamma_K. \tag{1.2}$$

If the contact is not realized, the contact force vanishes, i.e.

$$u'_n + u''_n < 0 \Rightarrow T_n = 0. \tag{1.3}$$

The conditions (1.1)—(1.3) can be written in the form of a complementary set

$$u'_n + u''_n \leqq 0, \quad T_n \leqq 0, \quad (u'_n + u''_n)\, T_n = 0 \quad \text{on} \quad \Gamma_K. \tag{1.4}$$

Moreover, we assume that the friction is zero:

$$T_t = 0. \tag{1.5}$$

Remark 1.1. Assuming that one of the two bodies becomes rigid, we obtain the problem of Signorini (1933), studied later by Fichera (1964).

The case of a rigid body in contact with an elastic, viscoelastic or elasto-plastic halfplane was analyzed by Duvaut (1976) and C. Johnson (1977).

(ii) Problems with an enlarging contact zone

If $\partial\Omega'$ and $\partial\Omega''$ are smooth in the neighbourhood of $\partial\Omega' \cap \partial\Omega''$, the zone of the contact may enlarge during the deformation process. Let us introduce a local Cartesian system

(ξ, η) such that the η-axis coincides with a common tangent of $\partial\Omega'$ and $\partial\Omega''$ at a "central" point $P \in \partial\Omega' \cap \partial\Omega''$ (before the deformation). We estimate a priori the parts of $\partial\Omega'$, $\partial\Omega''$, which may come into a contact during the deformation:

$$\Gamma_K^M = \{(\xi, \eta) \mid a \leqq \eta \leqq b, \quad \xi = f^M(\eta)\}, \quad M = I, II,$$

where f^M are functions, describing the arcs of $\partial\Omega^M$.

Then the condition of non-penetrating is

$$u_\xi'' - u_\xi' \leqq \varepsilon \quad \forall\, \eta \in [a, b], \tag{1.6}$$

where $\varepsilon(\eta) = f'(\eta) - f''(\eta)$.

Contact force cannot be tensile:

$$-T_\xi'(\cos\alpha')^{-1} = T_\xi''(\cos\alpha'')^{-1} \leqq 0 \quad \forall\, \eta \in [a, b] \tag{1.7}$$

where α^M is the angle between η-axis and the tangent to Γ_K^M.

No contact implies no contact force:

$$T_\xi''(u_\xi'' - u_\xi' - \varepsilon) = 0, \quad \forall\, \eta \in [a, b], \tag{1.8}$$

the friction vanishes:

$$T_\eta' = T_\eta'' = 0.$$

2. The primary variational formulations

We introduce the space of displacements with a finite energy

$$W = \{u \mid u = (u', u'') \in [H^1(\Omega')]^2 \times [H^1(\Omega'')]^2\},$$

the space of virtual displacements

$$V = \{v \in W \mid v' = 0 \quad \text{on} \quad \Gamma_u, v_n'' = 0 \quad \text{on} \quad \Gamma_0\},$$

the potential energy

$$\mathscr{L}(v) = \frac{1}{2}\, a(v, v) - \mathscr{F}(v),$$

$$a(u, v) = \frac{1}{2} \int_\Omega c_{ijkm} e_{ij}(u)\, e_{km}(v)\, dx, \quad \Omega = \Omega' \cup \Omega'',$$

$$\mathscr{F}(v) = \int_\Omega F_i v_i\, dx + \int_{\Gamma_\tau} P_i v_i\, ds, \quad \Gamma_\tau = \Gamma_\tau' \cup \Gamma_\tau''.$$

For the problem with a *bounded contact zone* we define the set of admissible displacements

$$K = \{v \in V \mid v_n' + v_n'' \leqq 0 \quad \text{on} \quad \Gamma_K\}$$

and the problem (minimum of potential energy)

$$\mathscr{L}(u) = \min \text{ over } K. \tag{$\mathscr{P}_1$}$$

For an *enlarging contact zone* we define

$$K_\varepsilon = \{v \in V \mid v_\xi'' - v_\xi' \leqq \varepsilon \quad \forall\, \eta \in [a, b]\},$$
$$\mathscr{L}(u) = \min \text{ over } K_\varepsilon. \qquad (\mathscr{P}_2)$$

2.1. Solvability of the problem ($\mathscr{P}_1$)

We introduce first the subspace of rigid bodies displacements

$$\mathscr{R} = \{z \in W \mid z_1^M = a_1^M - b^M x_2, \quad z_2^M = a_2^M + b^M x_2, \quad M = I, II\}$$

where a_i^M, b^M are arbitrary constants, and the set of "bilateral" admissible rigid displacements

$$\mathscr{R}^* = \{z \in \mathscr{R} \cap K \mid z \in \mathscr{R}^* \Rightarrow -z \in \mathscr{R}^*\}$$
$$= \{z \in \mathscr{R} \cap V \mid z_n' + z_n'' = 0 \quad \text{on} \quad \Gamma_K\}.$$

Theorem 2.1. *Assume that*

$$\mathscr{F}(y) \leqq 0 \quad \forall\, y \in K \cap \mathscr{R}, \tag{2.1}$$
$$\mathscr{F}(y) < 0 \quad \forall\, y \in K \cap \mathscr{R} \dot{-} \mathscr{R}^*. \tag{2.2}$$

Then there exists a solution u of ($\mathscr{P}_1$). Any other solution $\tilde{u}$ has the form $\tilde{u} = u + y$, where $y \in \mathscr{R} \cap V$ is such that $u + y \in K$, $\mathscr{F}(y) = 0$.

Proof is based on an abstract theorem by Fichera [1].

The nonuniqueness, however, is an obstacle for numerical analysis. Moreover, the proof of convergence of finite element approximations requires also the coerciveness of $\mathscr{L}$ over K. Therefore, in the study of the primary problems, we restrict ourselves to cases when $\dim(\mathscr{R} \cap V) \leqq 1$. This enables us to obtain both the uniqueness and the coerciveness.

Theorem 2.2. *Assume that*

$$\mathscr{R}^* = \{0\}, \quad \mathscr{R}_V \equiv \mathscr{R} \cap V \neq \{0\}, \tag{2.3}$$
$$\mathscr{F}(y) \neq 0 \quad \forall\, y \in \mathscr{R}_V \dot{-} \{0\} \tag{2.4}$$

and either $K \cap \mathscr{R} = \{0\}$ or,

$$K \cap \mathscr{R} \neq \{0\}, \quad \mathscr{F}(y) < 0 \quad \forall\, y \in K \cap \mathscr{R} \dot{-} \{0\}. \tag{2.5}$$

Then $\mathscr{L}$ is coercive over K and the problem ($\mathscr{P}_1$) is uniquely solvable.

Remark 2.1. The assumption (2.4) implies that $\dim \mathscr{R}_V \leqq 1$. In fact, if $\Gamma_0 = \emptyset$, then $\dim \mathscr{R}_V = 3$ and

$$0 = \mathscr{F}(y) = a_1 V_1 + a_2 V_2 + bM,$$

where

$$V_i = \int_{\Omega''} F_i\, dx + \int_{\Gamma_\tau''} P_i\, ds, \quad i = 1, 2,$$
$$M = \int_{\Omega''} (x_1 F_2 - x_2 F_1)\, dx + \int_{\Gamma_\tau''} (x_1 P_2 - x_2 P_1)\, ds,$$

holds for any vector (a_1, a_2, b) orthogonal to (V_1, V_2, M) in $\mathbb{R}^3$.

Let Γ_0 be parallel with x_1-axis. Assume that $V_1 > 0$. Then

$$y \in \mathscr{R}_V \dot{-} \{0\} \Rightarrow F(y) = a_1'' V_1 \neq 0 \Rightarrow (2.4),$$
$$y \in K \cap \mathscr{R} \dot{-} \{0\} \Rightarrow a_1'' < 0 \Rightarrow \mathscr{F}(y) < 0 \Rightarrow (2.5).$$

Proof of Theorem 2.2 is based on some abstract results by Nečas [2]. For the details see [3] — I., Theorem 2.4.

2.2 Solvability of the problem $(\mathscr{P}_2)$

We were not able to apply the same abstract theorem as for the bounded contact zone. Nevertheless, we have

Theorem 2.3. *Let Γ_0 be parallel with the x_1-axis, $\cos(\xi, x_1) > 0$ and $V_1 > 0$. Then $\mathscr{L}$ is coercive over K_ε and the problem $(\mathscr{P}_2)$ is uniquely solvable.*

Proof. Introduce the functional

$$p(v) = \int_a^b (v_\xi'' - v_\xi')\, d\eta$$

and the subspace

$$V_p = \{v \in V \mid p(v) = 0\}.$$

Then $\mathscr{R} \cap V_p = \{0\}$ and an inequality of Korn's type holds for $v \in V_p$ — see e.g. [4]. Define the mapping $Q : V \to \mathscr{R}_V$ by

$$(Qv)_1'' = p(v) \left(\int_a^b \cos(\xi, x_1)\, d\eta \right)^{-1} \equiv y_1''$$

and $P \equiv I - Q : V \to V_p$. Then we may write

$$\mathscr{L}(v) = \frac{1}{2} a(Pv, Pv) - \mathscr{F}(Pv) - \mathscr{F}(Qv) \geqq C_1 \|Pv\|_1^2 - C_2 \|Pv\|_1 - y_1'' V_1,$$

$$v \in K_\varepsilon \Rightarrow p(v) \leqq \int_a^b \varepsilon\, d\eta < +\infty,$$

$$\|y\|_1 = \|Qv\|_1 = |p(v)| \text{ const.}$$

If $\|v\|_1 \to +\infty$, at least one of the norms $\|Pv\|_1$, $\|Qv\|_1$ tends to infinity.

Let $\|y\|_1 \to \infty$. Then $p(v) \to -\infty$, $y_1'' \to -\infty$, $\mathscr{L}(v) \to +\infty$. Second, let $\|Pv\|_1 \to +\infty$. Then

$$C_1 \|Pv\|_1^2 - C_2 \|Pv\|_1 \to +\infty,$$

$$-y_1'' V_1 = -V_1 p(v) \left(\int_a^b \cos(\xi, x_1)\, d\eta \right)^{-1} > -\infty \quad \forall\, v \in K_\varepsilon.$$

Hence the coerciveness of $\mathscr{L}$ over K_ε follows. Since K_ε is closed and convex, $\mathscr{L}$ convex and continuous, a solution of $(\mathscr{P}_2)$ exists.

To prove the uniqueness, we first show that any two solutions differ by an element $z \in \mathscr{R}_V$ such that $\mathscr{F}(z) = 0$. On the other hand,

$$\mathscr{F}(z) = cV_1, \quad c \in \mathbb{R}.$$

Consequently, $c = 0$ follows from $V_1 > 0$ and therefore $z = 0$.

3. Approximations to the primary problems

Let us consider problems with a bounded contact zone and with enlarging contact zone separately.

3.1. Problems with a bounded contact zone

Assume first that both Ω' and Ω'' are bounded *polygonal* domains. Consider two regular families of triangulations $\{\mathscr{T}'_h\}$ and $\{\mathscr{T}''_h\}$, $0 < h \leqq h_0$, respectively, having common nodes on Γ_K and such that the boundary points of Γ_u, Γ_0, Γ'_τ, Γ''_τ and Γ_K coincide with some nodes.

We define the spaces V_h^M of piecewise linear and continuous in Ω^M functions over $\mathscr{T}_h^M$, $M = I, II$, and introduce

$$K_h = \{v = (v', v'') \mid v^M \in [V_h^M]^2, \; v' = 0 \text{ on } \Gamma_u, \\ v''_n = 0 \text{ on } \Gamma_0, \; v'_n + v''_n \leqq 0 \text{ on } \Gamma_K\}.$$

It is easy to realize that $K_h \subset K$. Since $(v'_n + v''_n)$ is piecewise linear, it is sufficient to satisfy the boundary conditions at the nodes of Γ_u, Γ_0, Γ_K only.

Second, let $\partial\Omega' \cap \partial\Omega'' = \Gamma_K$ be a *curved arc*. We consider again regular families of triangulations, where the triangles adjacent to Γ_K are replaced by "triangular" elements with one curved side. Again, the spaces V_h^M of continuous in Ω^M and piecewise linear functions over $\mathscr{T}_h^M$ will be employed. We define K_h in the same way as previously, with the only change that the condition (1.1) will be satisfied at the nodes of Γ_K only. Then $K_h \not\subset K$, in general.

An approximation to the primary problem $(\mathscr{P}_1)$ is defined as a solution u_h of the following problem

$$\mathscr{L}(u_h) = \min \quad \text{over } K_h. \tag{3.1}$$

Theorem 3.1. *Let Ω' and Ω'' be polygonal, $\mathscr{L}$ be coercive on K and $(\mathscr{P}_1)$ have a unique solution. Assume that $\Gamma_K \cap \bar{\Gamma}_u = \emptyset$, $\Gamma_K \cap \bar{\Gamma}_0 = \emptyset$ and there exists only a finite number of boundary points $\Gamma_K \cap \bar{\Gamma}_\tau$, $\bar{\Gamma}_u \cap \bar{\Gamma}_\tau$, $\bar{\Gamma}_\tau \cap \bar{\Gamma}_0$.*

Then

$$\|u - u_h\|_{1,\Omega} \to 0, \quad h \to 0$$

holds for any regular family of triangulations.

For the detailed proof see [3] — II. The main point is to prove the density of the set

$$K \cap [C^\infty(\bar{\Omega}')]^2 \times [C^\infty(\bar{\Omega}'')]^2$$

in K.

Theorem 3.2. *Let $\Gamma_K \in C^3$, containing a straight segment I. Assume that the conditions of Theorem 2.2 are satisfied provided K is replaced by*

$$K_I = \{v \in V \mid v'_n + v''_n \leqq 0 \quad \text{on} \quad I\}.$$

Let the solutions u and u_h of the problem $(\mathscr{P}_1)$ and (3.1), respectively, exist, u, $\tau(u)$ and $T_n(u)$ on Γ_K being sufficiently regular. Then

$$\|e(u - u_h)\|_{0,\Omega} \leqq C(u)\, h^{3/4}$$

holds for any strongly regular family of triangulations such that $\beta = 2$ and the maximal sides are on Γ_K.

(A regular family $\mathscr{T}_h$ is called *strongly regular*, if a positive constant β exists, independent of h and such that the ratio of any two sides in $\mathscr{T}_h$ is less than β.)

For the proof see [3] — II.

3.2 Problems with enlarging contact zone

Assume that the functions f^M, $M = I, II$, describing Γ_K^M, belong to $C^1([a, b])$. Curved elements will be used along Γ_K^M and for the construction of the trial functions, we use the technique developed in [5] (with linear polynomials on the reference triangle). The vertices of elements along Γ'_K, Γ''_K are placed on the lines parallel with the ξ-axis. Thus we generate the spaces V_h^M.

We define

$$\begin{aligned} K_{\varepsilon h} = \{v = (v', v'') \mid\ & v^M \in [V_h^M]^2, \quad M = I, II, \\ & v' = 0 \quad \text{on} \quad \Gamma_u,\ v''_n = 0 \quad \text{on} \quad \Gamma_0, \\ & v''_\xi(B_j) - v'_\xi(A_j) \leqq \varepsilon(C_j), \quad j = 1, \ldots, m\}, \end{aligned}$$

where A_j, B_j and C_j lie on the same parallel. Thus we have $K_{\varepsilon h} \not\subset K$, in general.

A finite element approximation is the solution u_h of the problem

$$\mathscr{L}(u_h) = \min \quad \text{over}\ K_{\varepsilon h}.$$

Theorem 3.3. *Let $\Gamma'_K \cap \Gamma''_K$ contain a straight segment; let Γ_0 be parallel with the x_1-axis, $\cos(\xi, x_1) > 0$ and $V_1 > 0$. Assume that $f^M \in C^1([a, b])$, $M = I, II$, $\Gamma_K^M \cap \bar{\Gamma}_u = \emptyset$, $\Gamma_K^M \cap \bar{\Gamma}_0 = \emptyset$ and there exists only a finite number of points $\bar{\Gamma}_u \cap \bar{\Gamma}_\tau$, $\bar{\Gamma}_0 \cap \bar{\Gamma}_\tau$.*

Then

$$\|u - u_h\|_{1,\Omega} \to 0, \quad h \to 0$$

holds for any regular family of triangulations.

Proof — see [3] — II.

4. Dual variational formulation for the problem with a bounded contact zone

There are problems when the stress field is more important than the displacement field. Then the use of a dual formulation — an extension of the classical Castigliano's principle — may be more efficient than the primary variational formulation.

In the following we restrict ourselves to the problems with a bounded contact zone and with polygonal boundaries.

Let us define

$$H = \{\tau = (\tau_{ij}), \quad i, j = 1, 2, \tau_{ij} \in L_2(\Omega), \quad \tau_{ij} = \tau_{ji}\},$$

$$\Lambda^+_{F,P} = \left\{\tau \in H \mid \int_\Omega \tau_{ij} e_{ij}(v)\, dx \geqq \mathscr{F}(v) \quad \forall\, v \in K\right\},$$

$$\mathscr{S}(\tau) = \frac{1}{2} \int_\Omega b_{ijkl} \tau_{ij} \tau_{kl}\, dx,$$

where b_{ijkl} are the coefficients of the "inverse" strain-stress relations, i.e.

$$e_{ij} = b_{ijkl} \tau_{kl}.$$

We say that a $\tau \in \Lambda^+_{F,P}$ is a solution of the dual problem if

$$\mathscr{S}(\tau) = \min \quad \text{over } \Lambda^+_{F,P}. \tag{4.1}$$

Theorem 4.1. *Let* (2.1), (2.2) *hold. Then the dual problem* (4.1) *has a unique solution* $\tau = \tau(u)$, *where* u *is a solution of the primary problem* $(\mathscr{P}_1)$.

Moreover, it holds

$$\mathscr{L}(u) + \mathscr{S}\big(\tau(u)\big) = 0.$$

The proof — via a saddle point — is given in [3] — III.

Remark 4.1. We obtained a uniquely solvable formulation of a rather general class of contact problems, in contradistinction to the primary variational formulation, where only the cases with $\dim(\mathscr{R} \cap V) \leqq 1$ have been considered.

Interpretation of the set $\Lambda^+_{F,P}$

Let $\tau \in \Lambda^+_{F,P}$ be sufficiently smooth. Then it holds

$$\frac{\partial \tau_{ij}}{\partial x_j} + F_i = 0 \qquad \text{in} \quad \Omega' \cap \Omega'', \tag{4.2}$$

$$\tau_{ij} n_j = P_i \qquad \text{on} \quad \Gamma'_\tau \cup \Gamma''_\tau, \tag{4.3}$$

$$T''_t(\tau) = 0 \qquad \text{on} \quad \Gamma_0, \tag{4.4}$$

$$T'_t(\tau) = T''_t(\tau) = 0 \quad \text{on} \quad \Gamma_K, \tag{4.5}$$

$$T'_n(\tau) = T''_n(\tau) \leqq 0 \quad \text{on} \quad \Gamma_K. \tag{4.6}$$

Conversely, if $\tau \in H$ is sufficiently smooth and satisfies (4.2)—(4.6), then $\tau \in \Lambda^+_{F,P}$.

4.1 Approximations to the dual problem

First we introduce a particular solution $\bar{\tau}$ of the non-homogeneous conditions (4.2), (4.3) and then write

$$\tau = \bar{\tau} + \sigma,$$

where σ is a self-equilibriated stress field. It is not essential for $\bar{\tau}$ to belong to $\Lambda^+_{F,P}$. In practice, however, $\bar{\tau}$ should be such that $\bar{\tau}_{ij} n_j$ are constant on Γ_K. Then we can construct internal approximations of $\Lambda^+_{F,P}$, which is advantageous.

For example, let Γ_0 be parallel with the x_1-axis and Γ_K be a straight segment such that $n_1'' > 0$ on Γ_K. We can find $\bar{\tau} \in \Lambda^+_{F,P}$ such that

$$T_n'(\bar{\tau}) = T_n''(\bar{\tau}) = g \quad \text{on} \quad \Gamma_K,$$

where

$$g = -V_1 \Big(\int_{\Gamma_K} dx_2 \Big)^{-1}.$$

Assuming that $V_1 \geqq 0$, we obtain $g \leqq 0$ and (4.6) is satisfied.

Let the stress field $\bar{\tau}$ satisfy

$$\int_\Omega \bar{\tau}_{ij} e_{ij}(v)\, dx = \mathscr{F}(v) + \int_{\Gamma_K} g(v_n' + v_n'')\, ds \quad \forall\, v \in V.$$

Obviously, it holds

$$\tau \in \Lambda^+_{F,P} \Leftrightarrow \tau - \bar{\tau} \equiv \sigma \in \mathscr{U}_0,$$

where

$$\mathscr{U}_0 = \Big\{ \sigma \in H \mid \int_\Omega \sigma_{ij} e_{ij}(v)\, dx \geqq -g \int_{\Gamma_K} (v_n' + v_n'')\, ds \quad \forall\, v \in K \Big\}.$$

Elements of $\mathscr{U}_0$ satisfy the homogeneous equations (4.2), (4.3), further (4.4), (4.5) and

$$T_n'(\sigma) = T_n''(\sigma) \leqq -g \quad \text{on} \quad \Gamma_K.$$

We construct subsets of $\mathscr{U}_0$ using spaces $\mathscr{N}_h$ of piecewise linear self-equilibriated triangular block-elements, proposed by Watwood and Hartz (see [6, 7, 8]). We define

$$\mathscr{U}_{0h} = \mathscr{U}_0 \cap N_h(\Omega),$$

where

$$N_h(\Omega) = \{(\sigma', \sigma'') \mid \sigma^M \in \mathscr{N}_h(\Omega^M), \quad M = I, II\}.$$

We say that $\sigma^h \in \mathscr{U}_{0h}$ is an approximation to the dual problem if

$$\Phi(\sigma^h) = \min \text{ over } \mathscr{U}_{0h}, \tag{4.7}$$

$$\Phi(\sigma) = \frac{1}{2} \int_\Omega b_{ijkl} \sigma_{ij} (\sigma_{kl} + 2\bar{\tau}_{kl})\, dx.$$

Theorem 4.2. *If $V_1 \geqq 0$, Γ_0 is parallel with the x_1-axis and Γ_K a straight segment with $n_1'' > 0$ on Γ_K, the problem (4.7) has a unique solution.*

Proof follows from the fact that $\mathscr{U}_0$ contains the zero element, being closed and convex in H, $N_h(\Omega)$ is convex and closed in H.

Using a particular type of strongly regular families of triangulations, we can prove the error estimate

$$\|\sigma - \sigma^h\|_{0,\Omega} \leqq C(\sigma)\, h^{3/2}$$

provided σ is smooth enough.

The proof can be found in the paper [3] — III, which contains also an algorithm for (4.7) and a construction of the element $\bar{\tau}$.

Remark 4.2. Having approximations to both the primary and the dual formulations, one can obtain some a posteriori error estimates and two-sided bounds for the exact energy. For analogous results in some simpler unilateral boundary value problems, we refer to [9] and the references therein.

References

[1] Fichera, G., Boundary value problems of elasticity with unilateral constraints. Encycl. of Physics (ed. S. Flügge), vol. VIa/2, Springer Verlag, Berlin—Heidelberg—New York 1972.

[2] Nečas, J., On regularity of solutions to nonlinear variational inequalities for 2nd order elliptic systems. Rend. di Matem. 2 (1975) 8., VI., 481—498.

[3] Haslinger, J.; Hlaváček, I., Contact between elastic bodies. I. Continuous problems. II. Finite element analysis. III. Dual finite elements analysis. Apl. Mat. **25** (1980), 324—347.

[4] Hlaváček, I., Nečas, J., On inequalities of Korn's type. Arch. Rat. Mech. Anal. **36** (1970), 305—334.

[5] Zlámal, M., Curved elements in the finite element method. SIAM J. Numer. Anal. **10** (1973), 229—240.

[6] Watwood, V. B., Hartz, B. J., An equilibrium stress field model for finite element solution of two-dimensional elastostatic problems. Int. J. Solids Structures **4** (1968), 857—873.

[7] Hlaváček, I., Convergence of an equilibrium finite element model for plane elastostatics. Apl. Mat. **24** (1979), 427—457.

[8] Johnson, C., Mercier, B., Some equilibrium finite element methods for two-dimensional elasticity problems. Numer. Math. **30** (1978), 103—116.

[9] Hlaváček, I., Dual finite element analysis for some unilateral boundary value problems. *In:* Theory of Nonlinear Operators (Proc. Internat. Summer School Berlin 1977). Abh. Akad. Wiss. DDR, Jahrg. 1978 Nr. 6N, Akademie-Verlag, Berlin 1978, 109—116.

Address of the author:

Mathematical Institute of the Czechoslovak Academy of Sciences
Žitná 25
11567 Praha 1
Czechoslovakia

An inverse problem for "coefficients" in linear equations. Uniqueness and iterative solution

Reinhard Kluge, Berlin

We consider here a parameter determination problem (inverse problem) in a linear operator equation on a real Hilbert space Y which has the form of a minimum problem. In applications such parameters may be e.g. the coefficients (which may depend on the space variables) in the main parts of the partial differential operators of some boundary value problems (cf. the example in part 5).

The functional in our optimality criterion has the form $h(y) + f(v)$, $y \in Y$, $v \in V$, V denotes the parameter space, with $h(y) = \frac{l}{2} \|y - x_0\|_Y^2$, $x_0 \in Y$, and $f(v) = \frac{k}{2} \|v - v_0\|_V^2$, $v_0 \in V$, $k, l > 0$.

For right hand sides of the basic parametric equations from a ball around the zero element of Y with a radius depending on k (cf. (12)) we prove the uniqueness of optimal solutions ("coefficients") using a gradient-like (iteration) method. For this method we obtain results on the strong convergence and on the linear speed of this convergence to the unique "coefficient" (part 3). In part 4 we control the uniqueness of this optimal parameter for $k \to 0$ by the constraints. In part 5 we give an example, an inverse problem for the coefficients in a one-dimensional homogeneous Dirichlet problem for elliptic differential equations of second order.

1. Existence of optimal parameters

Let H be a real Hilbert space with scalar product $(\cdot, \cdot)_H$ and norm $\|\cdot\|_H$. We denote the set of all single-valued mappings A of H into H with domain $D(A) = H$ by $(H \to H)$ and the subset of all linear and bounded mappings of $(H \to H)$ by $[H \to H]$.

Let $U \subset V$ be a nonempty, convex and closed subset, $T(\cdot, v) \in (Y \to Y)$ for $v \in V$ and $g \in Y$ fixed. We consider the parametric operator equation

$$T(y, v) = g, \quad v \in U. \tag{1}$$

For fixed $v \in U$ we denote the set of all solutions of (1) by Sv. S is the so called solution operator of the parametric equation (1). By $G(S)$ we denote the graph of S.

Let be $x_0 \in Y, v_0 \in V, k, l > 0$ fixed, $h(y) = \frac{l}{2} \|y - x_0\|_Y^2$ for $y \in Y$ and $f(v) = \frac{k}{2} \|v - v_0\|_V^2$ for $v \in V$.

We consider the minimum problem (P):

$$\text{(P)} \qquad h(x) + f(u) = \inf_{[y,v]\in G(S)} [h(y) + f(v)]. \tag{2}$$

The component u of each solution $[x, u]$ of (P) is called an *optimal parameter*.

We formulate the following conditions (E_1)—(E_3):

(E_1) *T is bilinear on $Y \times V$.*

(E_2) $\|T(y, v)\|_Y \leqq L \|y\|_Y \|v\|_V$ *for* $y \in Y, v \in V$ *with fixed* $L > 0$.

(E_3) $(T(y, v), y)_Y \geqq c \|y\|_Y^2$ *for* $y \in Y, v \in U$ *with fixed* $c > 0$.

Lemma 1. *Let the conditions* (E_1)—(E_3) *be satisfied. Then S is a single-valued mapping with domain $D(S) = U$, where*

$$Sv - Su = S'(u)(v - u) + Q(u, v - u)$$

with

$$S'(u) = -T(\cdot, u)^{-1} T(Su, \cdot) \in [V \to V],$$

$$Q(u, v - u) = -T(\cdot, u)^{-1} T(Sv - Su, v - u),$$

$$\|Sv\|_Y \leqq \|g\|_Y / c, \tag{3}$$

$$\|Sv - Su\|_Y \leqq \frac{L}{c^2} \|g\|_Y \|v - u\|_V, \tag{4}$$

$$\|Q(u, v - u)\|_Y \leqq \frac{L^2}{c^3} \|g\|_Y \|v - u\|_V^2, \tag{5}$$

$$\|S'(v)\| \leqq \frac{L}{c^2} \|g\|_Y \tag{6}$$

and

$$\|S'(v) - S'(u)\| \leqq \frac{2L^2}{c^3} \|g\|_Y \|v - u\|_V \tag{7}$$

hold for $v, u \in U$.

Proof. Let be $v, u \in U$, $x = Su$ and $y = Sv$. Obviously, $T(\cdot, v)^{-1}$ exists for each $v \in U$. Because of $c \|y\|^2 \leqq (T(y, v), y) \leqq \|g\| \|y\|$ we prove (3)

$$\|Sv\| = \|T(\cdot, v)^{-1} g\| \leqq \frac{\|g\|}{c}$$

and at the same time

$$\|T(\cdot, v)^{-1}\| \leqq \frac{1}{c} \quad \text{for} \quad v \in U. \tag{8}$$

Furthermore $T(y, v) = T(x, u) + T(y - x, u) + T(x, v - u) + T(y - x, v - u)$ and consequently

$$\begin{aligned} Sv - Su &= -T(\cdot, u)^{-1} T(Su, v - u) - T(\cdot, u)^{-1} T(Sv - Su, v - u) \\ &= S'(u)(v - u) + Q(u, v - u) \end{aligned}$$

holds. We have further

$$\begin{aligned} 0 = (T(x, u) - T(y, v), x - y) &= (T(x, u) - T(y, u), x - y) \\ &+ (T(y, u) - T(y, v), x - y) \geqq c \|x - y\|^2 - L \|y\| \|u - v\| \|x - y\| \end{aligned}$$

and because of (3) we find the estimation (4). By (8) and (4) we prove (5):

$$\|Q(u, v - u)\| \leqq \|T(\cdot, u)^{-1}\| \, \|T(Sv - Su, v - u)\| \leqq \frac{L}{c} \|Sv - Su\| \, \|v - u\|$$
$$\leqq \frac{L^2}{c^3} \|g\| \, \|v - u\|^2 .$$

From $\|T(Sv, h)\| \leqq L \|Sv\| \, \|h\|$ for $h \in V$ follows $\|T(Sv, \cdot)\| \leqq L \|Sv\| \leqq \frac{L}{c} \|g\|$ and from

$$\|T(\cdot, u)^{-1} g - T(\cdot, v)^{-1} g\| = \|Su - Sv\| \leqq \frac{L}{c^2} \|g\| \, \|v - u\| \quad \text{for} \quad g \in Y$$

follows (S depends on g)

$$\|T(\cdot, u)^{-1} - T(\cdot, v)^{-1}\| \leqq \frac{L}{c^2} \|v - u\| .$$

So we find the estimations (6)

$$\|S'(v)\| = \|-T(\cdot, v)^{-1} \, T(Sv, \cdot)\| \leqq \frac{L}{c^2} \|g\|$$

and (7)

$$\|S'(v) - S'(u)\| = \|T(\cdot, u)^{-1} \, T(Su, \cdot) - T(\cdot, v)^{-1} \, T(Sv, \cdot)\|$$
$$\leqq \|T(\cdot, u)^{-1} - T(\cdot, v)^{-1}\| \, \|T(Su, \cdot)\| + \|T(\cdot, v)^{-1}\| \, \|T(Su, \cdot) - T(Sv, \cdot)\|$$
$$\leqq \frac{2L^2}{c^3} \|g\| \, \|v - u\|, \quad \text{q.e.d.}$$

We formulate a further condition (E_4):

(E_4) *The operator $T(y, \cdot)$ is compact on U for each $y \in Y$.*

We now prove the existence of solutions of (P).

Theorem 2. *Let the conditions* (E_1)—(E_4) *be satisfied. Then the problem* (P) *has at least one solution.*

Proof. The functional $h(y) + f(v)$ is (weakly, weakly)-lower semicontinuous on $Y \times V$ and $h(y) + f(v) \to +\infty$ if $\|y\| + \|v\| \to \infty$. The set $G(S)$ is (weakly, weakly)-closed in $Y \times V$ (cf. [2—4]). Then (P) has at least one solution by the Weierstraß Theorem (cf. [1, 2]).

Remark. The assertion of Theorem 1 holds too, if in condition (E_3) $c = 0$ holds.

We remark further that under the conditions (E_1)—(E_3) the problem (P) ist equivalent to the minimum problem (P_0):

(P_0) $$F(u) = \inf_{v \in U} F(v) \quad \text{with} \quad F(v) = h(Sv) + f(v), \quad v \in U.$$

So each solution of (P_0) is an optimal parameter of (P).

2. A necessary condition for an optimal parameter

Let be $G \subset H$ convex, $e: G \to R^1$ a proper functional $\big(e(b) > -\infty,\ e \not\equiv +\infty\big)$. Then we set

$$e'_+(a, b-a) = \lim_{\substack{t>0 \\ t\to+0}} \frac{e\big(a + t(b-a)\big) - e(a)}{t}, \quad b, a \in G.$$

Lemma 3. *Let $a_0 \in G$ be a solution of the minimum problem*

$$e(a_0) = \inf_{b \in G} e(b)$$

and let there exist a finite $e'_+(a_0, b - a_0)$ for each $b \in G$. Then

$$e'_+(a_0, b - a_0) \geqq 0 \quad \textit{for} \quad b \in G.$$

holds.

Proof. We suppose the existence of an element $b \in G$ with $e'_+(a_0, b - a_0) < 0$. Then for sufficiently small t we have the contradiction

$$0 \leqq e\big(a_0 + t(b - a_0)\big) - e(a_0) = t\left[e'_+(a_0, b - a_0) + \frac{o(t)}{t}\right] < 0.$$

Lemma 4. *Under the conditions* (E_1)—(E_3) *each solution u from* (P_0) *satisfies the variational inequality*

$$\big(k(u - v_0) + lS'(u)^* (Su - x_0), v - u\big) \geqq 0, \quad v \in U, \tag{9}$$

with the adjoint operator $S'(u)^*$ of $S'(u)$.

Proof. We remark that $f'_+(u, v - u) = k(u - v_0, v - u)$ holds. We set $H(v) = h(Sv)$, $v \in U$. Because of $\|x - z\|^2 - \|y - z\|^2 = \|x - y\|^2 + 2(x - y, y - z)$ we find

$$\begin{aligned}
&\big\|S\big(u + t(v-u)\big) - x_0\big\|^2 - \|Su - x_0\|^2 \\
&= \big\|S\big(u + t(v-u)\big) - Su\big\|^2 + 2\big(S\big(u + t(v-u)\big) - Su,\ \ Su - x_0\big) \\
&= \big\|tS'(u)(v-u) + Q\big(Su, t(v-u)\big)\big\|^2 + 2\big(tS'(u)(v-u) + Q(Su, t(v-u)),\ Su - x_0\big) \\
&= t^2 \big\| - T(\cdot, u)^{-1} \big[T(Su, v-u) + T\big(S(u + t(v-u)) - Su, v-u\big)\big]\big\|^2 \\
&\quad + 2t\big(S'(u)(v-u), Su - x_0\big) + 2t\big(-T(\cdot, u)^{-1}\, T\big(S(u + t(v-u)) - Su, v-u\big), Su - x_0\big)
\end{aligned}$$

and consequently

$$\begin{aligned}
H'_+(u, v-u) &= \lim_{\substack{t>0 \\ t\to+0}} \left[\frac{\frac{l}{2}\big\|S\big(u + t(v-u)\big) - x_0\big\|^2 - \frac{l}{2}\|Su - x_0\|^2}{t}\right] \\
&= l\big(S'(u)^* (Su - x_0), v - u\big).
\end{aligned}$$

Furthermore we have $F'_+(u, v - u) = H'_+(u, v - u) + f'_+(u, v - u)$ and so, by Lemma 3, (9) holds.

3. The iteration method

Under the conditions (E_1)—(E_3) each solution of (P) is a solution of the following fixed point problem

$$u = P(U)\big(u - t(k(u - v_0) + lS'(u)^*(Su - x_0))\big), \quad t > 0, \tag{10}$$

with the projector $P(U)$ onto the subset U.

On the base of (10) we define an iteration method

$$w_{n+1} = P(U)\big(w_n - t_n(k(w_n - v_0) + lS'(w_n)^*(Sw_n - x_0))\big), \quad w_1 \in U, \tag{11}$$

to solve (P_0).

Theorem 5. *Let the conditions* (E_1)—(E_4) *be satisfied,* $0 < s < 1$, $r > 0$,

$$k \geqq \frac{2lL^2r(r + c\,\|x_0\|)}{(1 - s)\,c^4}, \quad g \in K(o, r) = \{y \in Y : \|y\|_Y \leqq r\}, \tag{12}$$

and

$$0 < t_n \leqq t_0 = \frac{skc^8}{6k^2c^8 + 4l^2L^4r^4}. \tag{13}$$

Then the iteration method (11) *converges with the speed*

$$\|w_n - u\| = O\big(\big[\sqrt{1 - t_0sk'}\big]^n\big) \tag{14}$$

to the unique optimal parameter of (P).

Proof. Because of the monotonicity of h' we have

$$0 \leqq \big(h'(Sw_n) - h'(Su),\; Sw_n - Su\big) = \big(h'(Sw_n), Sw_n - Su\big)$$
$$- \big(h'(Su),\; S'(u)(w_n - u) + Q(u, w_n - u)\big),$$

furthermore

$$\big(h'(Su), S'(u)(w_n - u)\big) \leqq \big(h'(Sw_n), Sw_n - Su\big) - \big(h'(Su), Q(u, w_n - u)\big)$$
$$= \big(-h'(Sw_n), S'(w_n)(u - w_n)\big) - \big(h'(Sw_n), Q(w_n)(u - w_n)\big)$$
$$- \big(h'(Su), Q(u, w_n - u)\big)$$

and consequently

$$\big(S'(w_n)^*\,h'(Sw_n) - S'(u)^*\,h'(Su), w_n - u\big)$$
$$\geqq \big(h'(Sw_n), Q(w_n, u - w_n)\big) + \big(h'(Su), Q(u, w_n - u)\big).$$

We now find

$$\|w_{n+1} - u\|^2 \leqq \big\|w_n - t_n(f'(w_n) + S'(w_n)^*\,h'(Sw_n)) - (u - t_n(f'(u) + S'(u)^*\,h'(Su)))\big\|^2$$
$$= \|w_n - u\|^2 - 2t_n\big(f'(w_n) - f'(u), w_n - u\big)$$
$$- 2t_n\big(S'(w_n)^*\,h'(Sw_n) - S'(u)^*\,h'(Su), w_n - u\big) + t_n^2\,\|f'(w_n) - f'(u)$$
$$+ S'(w_n)^*\,h'(Sw_n) - S'(u)^*\,h'(Su)\|^2$$
$$\leqq (1 - 2kt_n)\,\|w_n - u\|^2 - 2t_n\big[\big(h'(Sw_n), Q(w_n, u - w_n)\big) + \big(h'(Su), Q(u, w_n - u)\big)\big]$$
$$+ 2t_n^2\,\|f'(w_n) - f'(u)\|^2 + 2t_n^2\,\|S'(w_n)^*\,h'(Sw_n) - S'(u)^*h'(Su)\|^2.$$

For $v \in U$ we have $\|Sv - x_0\| \leqq \frac{1}{c}(\|g\| + c\,\|x_0\|)$. Then we find

$$\begin{aligned}
&\|S'(w_n)^*\,h'(Sw_n) - S'(u)^*\,h'(Su)\|^2 \\
&\leqq 2l^2\Big\{\big\|\big(S'(w_n)^* - S'(u)^*\big)\,(Sw_n - x_0)\big\|^2 + \|S'(u)^*\,(Sw_n - Su)\|^2\Big\} \\
&\leqq 2l^2\,\{\|S'(w_n) - S'(u)\|^2\,\|Sw_n - x_0\|^2 + \|S'(u)\|^2\,\|Sw_n - Su\|^2\} \\
&\leqq 2l^2\left\{\frac{4L^4}{c^6}\,\|g\|^2\,\|w_n - u\|^2\,\|Sw_n - x_0\|^2 + \frac{L^4}{c^8}\,\|g\|^4\,\|w_n - u\|^2\right\}
\end{aligned}$$

and furthermore

$$\begin{aligned}
\|w_{n+1} - u\|^2 \leqq \Bigg(&1 - 2t_nk + 4t_nl\,\frac{L^2}{c^4}\,\|g\|\,(\|g\| + c\,\|x_0\|) + 2t_n^2k^2 \\
&+ 4t_n^2l^2\left\{\frac{4L^4}{c^8}\,\|g\|^2\,(\|g\| + c\,\|x_0\|)^2 + \frac{L^4}{c^8}\,\|g\|^4\right\}\Bigg)\,\|w_n - u\|^2.
\end{aligned}$$

From (12) follows

$$\|g\|\,(\|g\| + c\,\|x_0\|) \leqq \frac{kc^4 2(1-s)}{4L^2l}$$

and further

$$-t_n2(1-s)\,k + 4t_n\,\frac{L^2}{c^4}\,l(\|g\| + c\,\|x_0\|)\,\|g\| \leqq 0\,;$$

by (13) we find

$$2t_n^2k^2 + 4t_n^2\,\frac{L^4l^2}{c^8}\,\{4\,\|g\|^2\,(\|g\| + c\,\|x_0\|)^2 + \|g\|^4\} \leqq skt_n.$$

Finally we obtain $\|w_{n+1} - u\|^2 \leqq (1 - skt_n)\,\|w_n - u\|^2$ and so (14). The uniqueness of the solution follows immediately, q.e.d.

4. One special case

Let the conditions (E_1)—(E_4) be satisfied. Furthermore, let k satisfy the condition

$$k > \frac{2lL^2r(r + c\,\|x_0\|)}{c^4}. \tag{15}$$

If we set

$$0 < s \leqq \frac{kc^4 - 2lL^2r(r + c\,\|x\|_0)}{kc^4},$$

then (12) is satisfied, too. By Theorem 5 for each $g \in K(o, r)$ there exists a unique optimal parameter of (P).

We discuss one special case of (15). We set $r = c\,\|x_0\|$. Then (15) has the form

$$k > 4l\left(\frac{L\,\|x_0\|}{c}\right)^2. \tag{16}$$

Let be $c = k^\alpha$, $\alpha \in R^1$, and $\beta = 1 + 2\alpha$. Then (16) is equivalent to

$$k^\beta > 4l(L\,\|x_0\|)^2$$

or to

$$\left.\begin{aligned} &k > (4l)^{1/\beta}\,(L\,\|x_0\|)^{2/\beta} && \text{for } \alpha > -\frac{1}{2},\\ &\|x_0\| \leqq 1/2L\sqrt{l} && \text{for } \alpha = -\frac{1}{2},\\ &\text{and}\\ &k < \frac{1}{(4l)^{1/|\beta|}\,(L\,\|x_0\|)^{2/|\beta|}} && \text{for } \alpha < -\frac{1}{2}. \end{aligned}\right\} \tag{17}$$

From (E_2) and (E_3) follows

$$c \leqq L\,\|v\|,\ v \in U. \tag{18}$$

We set

$$|U|_0 = \min_{v\in U} \|v\| \qquad \text{and} \qquad K = \sup_{v\in U} \|v\|.$$

Obviously $|U|_0 \neq 0$ and $c \leqq L\,|U|_0$ hold. With $c = k^\alpha$ we have $k^\alpha \leqq L\,|U|_0$ and in the case $\alpha = -1$ we get a further condition on k:

$$k \geqq 1/L\,|U|_0.$$

So the following corollary is proved:

Corollary 6. *Let the conditions* $(E_1)-(E_4)$ *be satisfied. Furthermore, let k be such that*

$$\|g\| \leqq \frac{1}{k}\,\|x_0\|$$

and

$$\frac{1}{L\,|U|_0} \leqq k < \frac{1}{4lL^2\,\|x_0\|^2}. \tag{19}$$

Then there exists a unique optimal parameter of (P).

In our example (part 5) the left hand side does not give any restriction for k.

Besides of (P) we consider the minimum problem

$$(\mathrm{P})_0 \qquad h(x) = \inf_{[y,v]\in G(S)} h(y) = \inf_{v\in U} h(Sv) = h(Su_{0k}).$$

Here we take into account that the set U may depend on k.

If U is bounded, i.e. $K < +\infty$ holds, then by the Weierstraß Theorem there exists at least one solution u_{0k} of $(\mathrm{P})_0$. We denote the corresponding unique solution of (P) for $k > 0$ by u_k.

We then find the following estimation:

$$|h(Su_k) + f(u_k) - h(Su_{0k})|$$
$$\leqq f(u_k) + [h(Su_k) - h(Su_{0k})| \leqq \frac{k}{2} \|u_k - v_0\|^2 + \frac{l}{2} \left| \|Su_k - x_0\|^2 - \|Su_{0k} - x_0\|^2 \right|$$
$$= \frac{k}{2} \|u_k - v_0\|^2 + \frac{l}{2} \left| \|Su_k - Su_{0k}\|^2 + 2(Su_k - Su_{0k}, Su_{0k} - x_0) \right|$$
$$\leqq \frac{k}{2} \|u_k - v_0\|^2 + \frac{l}{2} |(L^2/c^4) \|g\|^2 \|u_k - u_{0k}\|^2 + (2L \|g\|/c^3) \|u_k - u_{0k}\| (\|g\| + c \|x_0\|)$$
$$\leqq k \left[K^2 + \|v_0\|^2 + \frac{l}{2} \|x_0\|^2 (4kL^2K^2 + 8LK) \right] \leqq k \left[\frac{3}{2} K^2 + \|v_0\|^2 + 4LKl \|x_0\|^2 \right],$$

i.e.

$$|h(Su_k) + f(u_k) - h(Su_{0k})| \leqq k \left[\frac{3}{2} K^2 + \|v_0\|^2 + 4LKl \|x_0\|^2 \right]. \tag{20}$$

Remark. In the case $r = c \|x_0\|$ and $c = 1/k$ we get from (3)—(7) the estimations

$$\|Sv\| \leqq \|x_0\|, \qquad \|S'(v)\| \leqq Lk \|x_0\|$$

and

$$h(Su_{0k}) \leqq 2l \|x_0\|^2$$

and the Lipschitz constants

$$L(S) = Lk \|x_0\| \quad \text{and} \quad L(S') = 2L^2k^2 \|x_0\|, \quad \|g\| \leqq r.$$

5. An example

Let be $a, b \in R^1$, $a < b$, Y the Sobolev space $\mathring{W}_2^1((a, b))$ with norm

$$\|y\| = \left(\int_a^b \left(\frac{\partial y(s)}{\partial s} \right)^2 ds \right)^{1/2}, \quad y \in Y,$$

and V the Sobolev space $W_2^1((a, b))$ with norm

$$\|v\| = \left[\int_a^b (v(s))^2 ds + \int_a^b \left(\frac{\partial v(s)}{\partial s} \right)^2 ds \right]^{1/2}, \quad v \in V.$$

We consider the parametric homogeneous Dirichlet problem

$$-\frac{\partial}{\partial s} v(s) \frac{\partial y(s)}{\partial s} = \tilde{g}(s), \quad y(a) = y(b) = 0, \quad s \in (a, b) \tag{21}$$

with parameters $v \in V$ and $\tilde{g} \in L_2(a, b)$. By (21) we generate the parametric mapping $T(\cdot, v)$ by

$$\left(T(y, v), h\right) = \int_a^b v(s) \frac{\partial y(s)}{\partial s} \frac{\partial h(s)}{\partial s} ds, \quad h \in Y,$$

and the element $g \in Y$ by

$$(g, h) = \int_a^b \bar{g}(s)\, h(s)\, ds, \quad h \in Y.$$

Using the Sobolev imbedding theorem ([6]) we get

$$\left|(T(y, v), h)\right| \leqq \max_{s \in [a,b]} |v(s)|\, \|y\|\, \|h\| \leqq L\, \|v\|\, \|y\|\, \|h\|$$

and consequently

$$\|T(y, v)\| \leqq L\, \|y\|\, \|v\|, \qquad y \in Y, \quad v \in V$$

with the imbedding constant L. The representatives of the elements $y \in Y$ are absolutely continuous. Then via the integral representation for such functions we find

$$L = \sqrt{2 \cdot \max\left\{b - a, \frac{1}{b - a}\right\}}.$$

We define the subset $U \subset V$ by

$$U = \{v \in V : v(s) \geqq c, \quad s \in [a, b]\}$$

with fixed $c \in (0, +\infty)$. Then we get the estimation

$$(T(y, v), y) = \int_a^b v(s) \left(\frac{\partial y(s)}{\partial s}\right)^2 ds \geqq c\, \|y\|^2, \quad y \in Y, v \in U.$$

So the conditions (E_1)—(E_3) are proved. (E_4) follows from the compactness of the imbedding of V into $C([a, b])$ ([6]).

We define the functionals h and f by

$$h(y) = \frac{l}{2} \int_a^b \left(\frac{\partial y(s)}{\partial s} - \frac{\partial x_0(s)}{\partial s}\right)^2 ds$$

with $x_0 \in Y$, x_0 fixed, and

$$f(v) = \frac{k}{2} \left[\int_a^b \left(\frac{\partial v(s)}{\partial s} - \frac{\partial v_0(s)}{\partial s}\right)^2 ds + \int_a^b (v(s) - v_0(s))^2\, ds\right].$$

with $v_0 \in V$, v_0 fixed.

Now we define $|U|_0$. Firstly,

$$\|v\|^2 \geqq \int_a^b (v(s))^2\, ds \geqq c^2(b - a)$$

holds, so $|U_0| \geqq c\sqrt{b - a}$. But for $v = c$ the equality holds, i.e. $|U|_0 = c\sqrt{b - a}$. So we can use the results of the parts 3 and 4.

In the case where $c = 1/k$ and k satisfies $\|g\| \leqq (1/k)\, \|x_0\|$ and

$$k < \frac{1}{8l(b-a)\,\|x_0\|^2} \quad \text{if} \quad b-a \geqq 1, \quad k < \frac{b-a}{8l\,\|x_0\|^2} \quad \text{if} \quad b-a < 1,$$

(P) has a unique optimal parameter. For the estimation (20) we have to accept a further condition

$$k > \frac{1}{K\sqrt{2(b-a)}} \quad \text{if} \quad b-a \geqq 1,$$

$$k > \frac{\sqrt{b-a}}{K\sqrt{2}} \quad \text{if} \quad b-a < 1.$$

References

[1] Kluge, R., Über eine Klasse von Minimumproblemen mit Nebenbedingungen. Math. Nachr. 78 (1977), 37–64.
[2] —, Nichtlineare Variationsungleichungen und Extremalaufgaben. Theorie und Näherungsverfahren. VEB Deutscher Verlag der Wissenschaften, Berlin 1979.
[3] —, On some parameter determination problems in variational and quasi-variational inequalities. I. Math. Nachr. 89 (1979), 127–148.
[4] —, On some inverse problems in variational and quasi-variational inequalities. *In:* Mathematical Research, Vol. 1, Akademie-Verlag, Berlin 1979, 141–149.
[5] Kluge, R., and H. Langmach, On some problems of determination of functional parameters in partial differential equations. *In:* Distributed Parameter Systems: Modelling and Identification (IFIP Working Conference, Rome, Italy, 1976). Lecture Notes in Control and Information Sciences, Vol. 1, Springer-Verlag, Berlin 1978, 298–309.
[6] Sobolew, S. L., Einige Anwendungen der Funktionalanalysis auf Gleichungen der mathematischen Physik. Akademie-Verlag, Berlin 1964.

Remark added in proof

For some essential extensions and for other proofs of the uniqueness see the following papers of the author: (i) Zur Koeffizientenbestimmung in linearen partiellen Differenzialgleichungen. I. Math. Nachr. (to appear); (ii) Zur starken Konvexität eines Funktionals bei der „Koeffizienten"-bestimmung in linearen Operatorgleichungen. Herbstschule über Variationsrechnung und optimale Prozesse, Vitte/Hiddensee 1980. Preprint-Reihe Mathematik, Ernst-Moritz-Arndt-Universität Greifswald (to appear).

Address of the author:

Akademie der Wissenschaften der DDR
Institut für Mathematik
1080 Berlin
Mohrenstr. 39
DDR

Nonlinear operators and solvability of operator equations

Josef Kolomý, Prague

This paper concerns with surjectivity results, interior mapping principle and solvability of nonlinear equations in abstract spaces.

In the last period a series of surjectivity results and mapping theorems have been obtained on the base of (*i*) monotonicity methods, (*ii*) Fredholm alternative for nonlinear opterators, (*iii*) normal solvability and fixed point theory (see for instance [1, 12, 19]).

We shall use some known definitions and results from the geometry of Banach spaces. We shall say that:

(i) X is *uniformly rotund* (UR), if for each $\varepsilon > 0$, $0 < \varepsilon \leqq 2$ there is $\delta(\varepsilon) > 0$ such that $\|u\| \leqq 1$, $\|v\| \leqq 1$ and $\|u - v\| \leqq \varepsilon$ implies $\left\|\frac{1}{2}(u + v)\right\| \leqq 1 - \delta(\varepsilon)$;

(ii) X is an $(2R)$-*space*, if for each sequence $(u_n) \in X$ such that $\|u_n + u_m\| \to 2$ as $n, m \to \infty$, then (u_n) is a Cauchy sequence;

(iii) X is an (H)-*space*, if X is strictly convex and the following condition is satisfied: if (u_n) converges weakly to u and $(\|u_n\|)$ converges to $\|u\|$, then (u_n) converges to u in norm;

(iv) X is *uniformly smooth* (US) if for a given $\varepsilon > 0$ there is $\varrho(\varepsilon) > 0$ such that $\|u\| \leqq 1$, $\|v\| \leqq \varrho \Rightarrow \|u + v\| + \|u - v\| \leqq 2 + \varepsilon \|v\|$;

(v) X *admits nearest points* if for each closed subset $E \subset X$ the set $\{u \in X$: there is a point $v \in E$ such that $\|u - v\| = \inf \|u - z\| : z \in E\}$ is dense in X.

It is well-known that $(UR) \Rightarrow (2R) \Rightarrow (H)$. Neither reverse implication is true. Each of (UR), $(2R)$ and (US) imply reflexivity of X. One can construct a $(2R)$-space which is not (UR). Moreover, X is a $(2R)$-space $\Leftrightarrow$ X is an (H)-reflexive space ([11]). X is $(US) \Leftrightarrow \|\cdot\|$ of X is uniformly Fréchet differentiable on each bounded subset of X away the origin. A space X is (US) (or (UR)) $\Leftrightarrow$ X^* is (UR) (or (US)). Wulbert [21] has proved that a Banach space X admits nearest points if either a) X is $(2R)$-space (in particular (UR)-space), or b) X is (US) and (H)-space.

A mapping $F: X \to Y$, where X, Y are normed linear spaces, is said to be:

(i) *demicontinuous* at $u_0 \in X$ if $u_n \to u_0$ implies $F(u_n) \to F(u_0)$ weakly in Y;

(ii) sequentially weakly continuous at $u_0 \in X$ if $u_n \to u_0$ weakly implies $F(u_n) \to F(u_0)$ weakly.

A mapping F of X into its dual space X^* satisfies the condition (P) ([9]), if for each $(u_n) \in X$, $u_n \to u \in X$ weakly, then $\limsup_{n\to\infty} \langle F(u_n), u_n - u\rangle \geqq 0$.

We shall say that a mapping F from an algebraically open subset $U \subset X$ into Y is *one-sided Gâteaux differentiable on* U if there exists the limit $\lim_{t\to 0_+} \frac{1}{t} [F(u + th) - F(u)] = V_+F(u, h)$ for each $u \in U$ and $h \in X$. A functional $f: M \to R$ is said to be *quasi-convex on a convex set* $M \subset X$ if $u, v \in M$, $\lambda \in [0, 1]$ implies that $f(\lambda u + (1 - \lambda) v) \leqq \max [f(u), f(v)]$. By $B_r(u)$ we denote an open ball centered about the point $u \in x$ and with the radius $r > 0$. We introduce the following

Definition 1. Let X, Y be normed linear spaces (NLS). We shall say that a mapping G_u (depending on $u \in X$) of X onto Y *has the property* (Q), if there exists a constant $\alpha_u > 0$ with the following property: for each $v \in Y$ there exists $u_0 \in X$ such that $G_u(u_0) = v$ and

$$\alpha_u \|u_0\| \leqq \|v\|. \tag{1}$$

Definition 2. Let X, Y be NLS, M an algebraically open subset of X, $F: M \to Y$. We shall say that F *has an approximation property* (AP) *on* M, if for each fixed $u \in M$ there exists a positively homogeneous mapping G_u of X onto Y having the property (Q) that for a given $h \in X$ there is a constant $\delta = \delta(u, h) > 0$ such that $0 < t < \delta \Rightarrow \|F(u + th) - F(u) - G_u(th)\| \leqq \alpha_u' \|th\|$, where $0 \leqq \alpha_u' < \alpha_u$ and α_u is a constant from (1).

Definition 3. Let X, Y be NLS. We shall say that a mapping $F: X \to Y$ *has a uniform approximation property* (UAP) *on an algebraically open subset* $M \subset X$ if there exists a positively homogeneous mapping G of X onto Y with the property (Q) such that for a given $h \in X$ there exists a constant $\delta = \delta(h) > 0$ such that

$$\|F(u + th) - F(u) - G(th)\| \leqq \alpha' \|th\|$$

holds for each $0 < t < \delta(h)$ and $u \in M$, where $0 \leqq \alpha' < \alpha$ and α is a constant from Definition 1 (here $\alpha = \alpha_u$).

Remark 1. Let X, Y be NLS. Then $G_{u_0}: X \to Y$ has the property (Q) when one of the following two conditions is satisfied:

(i) G_u is positively homogeneous and $B_{r_u}^Y(0) \subset G_u(B_1^X(0))$ for some $r_u > 0$, where $B_{r_u}^Y(0) = \{v \in Y: \|v\| < r_u\}$, $B_1^X(0) = \{u \in X: \|u\| < 1\}$;

(ii) X is complete, G_u is a linear continuous mapping and the range $G_u(X)$ is of the second category in Y.

The assertion (ii) is a well-known result of Banach concerning the open mappings. Let X, Y be (NLS), $M \subset X$ an algebraically open subset of X, $F: M \to Y$ a mapping having a one-sided Gâteaux differential $V_+F(u, \cdot)$ for each $u \in M$. If $V_+F(u, \cdot)$ has the property (Q) for each (fixed) $u \in M$, then F possesses the approximation property on M. In particular, F has this property, when X and Y are both complete, F has the Gâteaux derivative $F'(u)$ on M and $F'(u)$ is onto for each (fixed) $u \in M$. Similarly, a mapping F has a uniform approximation property on $B_r(0) \subset X$, if X, Y are both Banach spaces, $F: B_r(0) \to Y$ has a continuous Gâteaux derivative $F'(u)$ on $B_r(0)$, $F'(0)$ is onto and r is sufficiently small.

Let $M \subset X$ be an open set, X, Y normed linear spaces. A mapping $F: M \to Y$ is said to *have a local uniform approximation property* (*LUAP*) *on* M if for each $u \in M$ there exists a constant $\eta_u > 0$ such that F has (*UAP*) on $B\eta_u(u) \cap M$. Obviously (UAP) $\Rightarrow$ (LUAP) $\Rightarrow$ (AP).

Theorem 1. *Let* X, Y *be Banach spaces,* $F: X \to Y$ *a mapping having an approximation property on* X. *Moreover, assume that one of the following three conditions is fulfilled:*

(i) *Y has nearest points and $F(X)$ is closed,*
(ii) *Y is reflexive and $F(X)$ is sequentially weakly closed;*
(iii) *F is sequentially weakly continuous, $F(0) = 0$ and $E(a) = \{u \in X: \|F(u)\| \leqq a\}$ is relatively sequentially weakly compact for each $a \geqq 0$.*

Then $F(X) = Y$.

Let us remark that the result of Theorem 1 is related to that of Pochoshajev [14, 15], Sabreiko and Krasnosjelski [18], but in comparison with them the concept of differentiability of maps is not used. Compare also the results of Browder [2, 3, 4].

Assume that X, Y are Banach spaces, F is a mapping of X into Y. Then:

(i) $F(X)$ is closed, if F is continuous and $\|F(u) - F(v)\| \geqq \varphi(\|u - v\|)$ for each $u, v \in X$, where $\varphi: R^+ \to R^+$ is such that $\varphi(\varrho) \to 0_+$ as $\varrho \to 0_+$;
(ii) $F(X)$ is sequentially weakly closed, if X is reflexive, F is sequentially weakly continuous and $\|F(u)\| \to +\infty$ as $\|u\| \to \infty$.
(iii) $E(a)$ is relatively sequentially weakly compact for each $a \geqq 0$ when X is reflexive, F is p-positively homogeneous on X and $\|F(u)\| \geqq m > 0$ for each $u \in X$, $\|u\| = r$, where r is some positive number.

For further conditions under which $F(X)$ is closed see Lemma 3.1 [6]. Note that under the assumptions (iii) of Theorem 1 the sets $E(a)$ are already sequentially weakly compact for each $a \geqq 0$. Under the assumption (iii) of Theorem 1 assume that X, Y are Banach spaces and that F has the Gâteaux derivative $F'(u)$ on X. If $\ker [F'(u)]^* = (0)$ for each $u \in X$, then $F(X) = Y$. More generally, we have

Proposition 1. *Let* X, Y *be Banach spaces,* F *a mapping of* X *into* Y *having the one-sided Gâteaux differential* $V_+F(u, \cdot)$ *for each* $u \in X$. *Suppose that the image under the mapping* $V_+F(u, \cdot)$ *of the whole space* X *is dense in* Y *for each (fixed)* $u \in X$. *If one of the conditions* (i), (ii), (iii) *of Theorem* 1 *is satisfied, then* $F(X) = Y$.

This Proposition extends the results of [14], [15].

Theorem 2. *Let* X *be a real Hilbert space,* $D \subset X$ *an open convex set,* $N_\varepsilon(\overline{D})$ *an* ε-*neighbourhood of* $\overline{D}$, $F: N_\varepsilon(\overline{D}) \to X$ *a mapping having a Gâteaux derivative at the points of the boundary* ∂D *of* D. *Assume that* F *admits an approximation property on* D *and that* f, *defined by* $f(u) = \|F(u)\|$, $u \in \overline{D}$, *is sequentially weakly lower semicontinuous on* $\overline{D}$. *If* $\langle T(u), u\rangle > 0$ *for each* $u \in \partial D$, *where* $T(u) = [F'(u)]^* F(u)$, $u \in D$, *then there exists* $u_0 \in D$ *such that* $F(u_0) = 0$.

Remark 2. The assumption that f is sequentially weakly lower semicontinuous on $\overline{D}$ is satisfied when one of the following three conditions is valid:

(i) F is sequentially weakly continuous on $\overline{D}$.
(ii) F is semicontinuous and f is quasi-convex on $\overline{D}$.
(iii) F is continuous and has the Gâteaux derivative $F'(u)$ on $\overline{D}$ and $T(\cdot) = [F'(\cdot)]^*F(\cdot)$ satisfies condition (P) on $\overline{D}$ ([13]).

Theorem 3. *Let X, Y be Banach spaces, F a mapping from a domain $D(F) \subset X$ into Y. Assume that an ε-neighbourhood of the ball $\overline{B_r(u_0)}$ is contained in $D(F)$ and that F has a uniform approximation property on $\overline{B_r(u_0)}$. If F is continuous on $\overline{B_r(u_0)}$ and $\|F(u) - F(v)\| \geqq \beta \|u - v\|$ for each $u, v \in \overline{B_r(u_0)}$, then $F(u_0) \in \operatorname{Int} F(\overline{B_r(u_0)})$, where $\beta + \alpha' \geqq \alpha$ and α, α' are constants from Definition* 3, $\beta > 0$.

Recall that the interior mapping principles have been derived for various classes of mappings for instance by Clarke [5], Fabian [6], Graves [7] and Rheinboldt [17].

References

[1] F. E. Browder, Nonlinear operators and nonlinear equations of evolution in Banach spaces. *In:* Proc. Symp. Pure Math. Vol. 18, part 2, American Math. Soc., Providence, Rhode Island 1976.

[2] F. E. Browder, Normal solvability for nonlinear mappings in Banach spaces. Bull. Amer. Math. Soc. **77** (1971), 73—77.

[3] F. E. Browder, Normal solvability and the Fredholm alternative for mappings into infinite dimensional manifolds. J. Funct. Anal. **8** (1971), 250—274.

[4] F. E. Browder, Normal solvability and Φ-accretive mappings of Banach spaces. Bull. Amer. Math. Soc. **78** (1972), 186—192.

[5] F. H. Clarke, On the inverse function theorem. Pacific J. Math. **64** (1976), 97—102.

[6] J. Daneš, J. Kolomý, Fixed point, surjectivity and invariance of domain theorems for weakly continuous mappings. Boll. Un. Mat. Ital. **13-B** (1976), 369—394.

[7] M. Fabian, Concerning interior mapping theorem. Comment. Math. Univ. Carolinae **20** (1979), 345—356.

[8] L. M. Graves, Some mapping theorems. Duke Math. J. **17** (1950), 111—114.

[9] P. Hess, A variational approach to a class of nonlinear eigenvalue problems. Proc. Amer. Math. Soc. **29** (1971), 272—276.

[10] J. Kolomý, Normal solvability, solvability and fixed point theorems. Colloq. Math. **29** (1974), 253—266.

[11] Ky Fan, I. Glicksberg, Some geometric properties of the spheres in normed linear space. Duke Math. J. **25** (1958), 553—568.

[12] D. Pascali, S. Sburlan, Nonlinear mappings of monotone type. Editura Academici, Bucuresti 1978.

[13] J. P. Penot, A remark on the direct method of the calculus of variations. Proc. Amer. Math. Soc. **67** (1977), 135—141.

[14] С. И. Похожаев, О нелинейных операторах, имеющих слабо замкнутую область значений и квазилинейных эллиптических уравнениях. Матем. сб. **78** (1969), 237—259.

[15] С. И. Похожаев, Нормальная разрешимость нелинейных уравнений в равномерно выпулкых Банаховых пространствах. Функц. анализ и его прилож. **3** (1969), 80—84.

[16] B. H. Pourciau, Analysis and optimization of Lipschitz continuous mappings, J. Opt. Theory Appl. **22** (1977), 311—351.
[17] W. C. Rheinboldt, Local mapping relation and global implicit function theorems. Trans. Amer. Math. Soc. **138** (1969), 183—198.
[18] П. П. Забрейко, М. А. Красносельский, О разрешимости нелинейных операторных уравнений. Функц. анализ и его прилож. **5** (1971), 42—44.
[19] S. Swaminathan (ed.), Fixed point, theory and its aplications. Academic Press, New York, 1976.
[20] S. Swaminathan, On the closed range theorem for nonlinear operators. *In:* Proc. of the Third Prague Topol. Sympos. 1971, Academia, Prague 1972, 417—418.
[21] D. E. Wulbert, Uniqueness and differential characterization from manifolds of functions. Amer. J. Math. **93** (1971), 350—366.
[22] S. Yamamuro, Differential calculus in topological linear spaces. Lecture Notes in Mathematics Vol. 374, Springer Verlag, Berlin—Heidelberg—New York, 1974.

Address of the author:

Mathematical Institute of Charles University
Sokolovská 83
18600 Praha 8-Karlín
Czechoslovakia

On convergence properties of families of maximal monotone operators

Eckehard Krauß, Berlin

1. Introduction

In the last years several notions of convergence for families of maximal monotone operators have received some attention in the literature (cf. [1, 2, 3, 5, 6, 12, 13, 17]). Especially the convergence of graphs, resolvents and Yoshida-approximations was extensively investigated. These considerations are useful both in the nonlinear semigroup theory ([1, 5, 6]) and in examinations on stability. They apply also to projection-iteration methods for the solution of problems involving maximal monotone operators (see [14]).

In this paper we extend the notion of a generalized pseudomonotone operator (cf. [10]) to a family of operators. Moreover, we study the relations between this concept and the types of convergence mentioned above. Our investigations were motivated by the treatment of existence problems for multivalued operator equations, among them variational and quasivariational inequalities (see [16]).

2. Notations and preliminaries

Let be given a *real* reflexive Banach space E with the topologically dual space E'. Without loss of generality we will always assume $(E, \|\cdot\|)$ and $(E', \|\cdot\|)$ to be *locally uniformly convex* (cf. [18]). In particular this implies that the *duality mapping* $J: E \to 2^{E'}$, $Jx \underset{\text{df.}}{=} \{f \in E';\ (f, x) = \|f\|\,\|x\| \text{ and } \|f\| = \|x\|\}$ is a homeomorphism between $(E, \|\cdot\|)$ and $(E', \|\cdot\|)$ (compare e.g. [11]). The strong and the weak convergence in each of the spaces E and E' are denoted by "$\to$" and "$\rightharpoonup$", respectively.

Frequently we will identify a multivalued mapping $A: E \supseteq D(A) \to 2^{E'}$ with its graph in $E \times E'$, i.e. $A = \{[x, f] \in E \times E';\ f \in Ax\}$. Let now $A \subseteq E \times E'$ be *maximal monotone*. Then we can attach to A the single-valued selection $A^0: D(A) \to E'$, $A^0x \underset{\text{df.}}{=} P(Ax)\,0$.[1] The *resolvent* $J_\lambda: E \to E$ $(\lambda > 0)$ for A is defined as the unique solution $J_\lambda x$ of the equation $J(J_\lambda x - x) + \lambda A J_\lambda x \ni 0$. The *Yoshida approximation* $A_\lambda: E \to E'$ is given by $A_\lambda x \underset{\text{df.}}{=} \dfrac{J(x - J_\lambda x)}{\lambda}$ (see [4], [7]).

The following lemma summarizes some properties of the mappings A^0, J_λ and A_λ which will be needed in the sequel.

---

[1] The metric projection $P(Ax)$ is well-defined due to the convexity and closedness of the sets $\{Ax\}$.

Lemma 1. *For a maximal monotone mapping $A \subseteqq E \times E'$ in the reflexive Banach space E the following statements are true:*

1) *If $x \in \overline{D(A)}$ and $(f - A^0y, x - y) \geqq 0$ for any $y \in D(A)$ then $[x, f] \in A$.*
2) *$A_\lambda x \in AJ_\lambda x$ for any $x \in E, \lambda > 0$.*
3) *For any $\lambda > 0$ and $[x, f] \in A$ there is a $y \in E$ with $J_\lambda y = x$ and $A_\lambda y = f$.*
4) *$A_{\lambda+\mu} = (A_\lambda)_\mu = (A_\mu)_\lambda$ for arbitrary $\lambda, \mu > 0$.*
5) $\lim\limits_{0<\lambda\to 0} A_\lambda x = A^0 x, x \in D(A)$.
6) *$\|J_\lambda x - x\| \leqq \lambda \|A^0 v\| + 2\|v - x\|$ for any $\lambda > 0, x \in E$ and $v \in D(A)$.*

Proof. Ad 1) cf. [8, 9]; ad 2), 5) cf. [4, 7]; ad 3), 4), 6) cf. [16]. q.e.d.

We conclude this section with some definitions concerning generalized sequences. Using the terminology of [15] we call a subset I' of a *directed set I confinal* if for any $\alpha \in I$ there is a $\beta \geqq \alpha$ with $\beta \in I'$. Thus I' becomes a directed set itself. If K is another directed set we call $\tau: I \to K$ a *confinal mapping* if it maps confinal subsets of I onto confinal subsets of K. In the special case $K \underset{\text{df.}}{=} \{1, 2, \ldots\}$ these mappings are characterized by $\lim\limits_{\alpha\in I} \tau(\alpha) = +\infty$.

Let now be given a *generalized sequence* $\{x_\alpha\}_{\alpha\in I}$ in the locally convex space E, i.e. a mapping $I \ni \alpha \mapsto x_\alpha \in E$. We say that $\{x_\alpha\}_{\alpha\in I}$ is *almost bounded* if $\{x_\alpha\}_{\alpha\geqq\alpha_0}$ is bounded for some $\alpha_0 \in I$. In the case $I \underset{\text{df.}}{=} \{1, 2, \ldots\}$ this condition coincides with the usua boundedness definition of a sequence. Finally, $x \in E$ is called an *adherent point* ol $\{x_\alpha\}_{\alpha\in I}$ if any neighbourhood $U(x)$ of x contains a confinal subsequence of $\{x_\alpha\}_{\alpha\in I}$.

3. Generalized pseudomonotone operator families

In this section we give a natural extension of the concept of generalized pseudomonotone operator (cf. [10]) to a family of operators. It will be shown later (compare section 4) that this notion may be a useful tool for the derivation of different existence results.

First we recall the above mentioned definition of Browder and Hess together with some modifications.

Definition 1. An operator $A \subseteqq E \times E'$ in the reflexive Banach space E is called

a) *generalized pseudomonotone* if $x_\beta \rightharpoonup x$, $Ax_\beta \ni f_\beta \rightharpoonup f$, $\overline{\lim\limits_\beta}\,(f_\beta, x_\beta) \leqq (f, x)$ implies $[x, f] \in A$ and $\lim\limits_\beta (f_\beta, x_\beta) = (f, x)$,

b) *generalized pseudomonotone on bounded sets* if $x_\beta \rightharpoonup x$, $Ax_\beta \ni f_\beta \rightharpoonup f$, $\overline{\lim\limits_\beta}\,(f_\beta, x_\beta) \leqq (f, x)$, with $\{[x_\beta, f_\beta]\}_\beta$ bounded, implies $[x, f] \in A$ and $\lim\limits_\beta (f_\beta, x_\beta) = (f, x)$,

c) *sequentially generalized pseudomonotone* if $x_n \rightharpoonup x$, $Ax_n \ni f_n \rightharpoonup f$, $\overline{\lim\limits_{n\to\infty}}\,(f_n, x_n) \leqq (f, x)$ implies $[x, f] \in A$ and $\lim\limits_{n\to\infty} (f_n, x_n) = (f, x)$.

The original definition of a generalized pseudomonotone mapping coincides with the condition c). The following lemma establishes some connections between these different notions.

Lemma 2. *For the conditions* a)—c) *the implications* a) $\to$ b) $\leftrightarrow$ c) *hold.*

Proof. The implications a) $\to$ b) $\to$ c) are obvious. We show c) $\to$ b). For this purpose we assume $x_i \rightharpoonup x$, $Ax_\beta \ni f_\beta \rightharpoonup f$ and $\overline{\lim\limits_\beta}\,(f_\beta, x_\beta) \leqq (f, x)$, with $\{[x_\beta, f_\beta]\}_\beta$ bounded. We set $c \underset{\text{df.}}{=} \underline{\lim\limits_\beta}\,(f_\beta, x_\beta)$. Then $[x, f, c]$ is an adherent point of the relatively compact set $M \underset{\text{df.}}{=} \{[x_\beta, f_\beta, (f_\beta, x_\beta)]\}_\beta \subseteqq E_w \times E'_w \times R$.[1] By a result of Smulian-Kaplansky (cf. [15], p. 313) then there exists a sequence $\{[x_n, f_n, (f_n, x_n)]\}$ in $M \subseteqq E \times E' \times R$ with $[x_n, f_n, (f_n, x_n)] \rightharpoonup [x, f, c]$. This means $x_n \rightharpoonup x$, $Ax_n \ni f_n \rightharpoonup f$ and $\lim\limits_{n\to\infty} (f_n, x_n) = c = \underline{\lim\limits_\beta}\,(f_\beta, x_\beta) \leqq (f, x)$. The condition c) implies now $[x, f] \in A$ and $\lim\limits_{n\to\infty} (f_n, x_n) = \underline{\lim\limits_\beta}\,(f_\beta, x_\beta) = (f, x)$, i.e. A is generalized pseudomonotone on bounded sets. q.e.d.

As already announced, the definitions a) and b) admit the following natural extensions to families of operators:

Definition 2. Let $A \subseteqq E \times E'$ be an operator in the reflexive Banach space E. Moreover, for each element α of the directed set I let be given an operator $A^\alpha \subseteqq E \times E'$. Then the family $(A^\alpha, A)_{\alpha \in I}$ is called

a) *generalized pseudomonotone* if for any directed set K and for any confinal mapping $\tau: K \to I$ the relations $x_\beta \underset{K}{\rightharpoonup} x$, $A^{\tau(\beta)}x_\beta \ni f_\beta \underset{K}{\rightharpoonup} f$, $\overline{\lim\limits_\beta}\,(f_\beta, x_\beta) \leqq (f, x)$ imply $[x, f] \in A$ and $\lim (f_\beta, x_\beta) = (f, x)$,

b) *generalized pseudomonotone on bounded sets* if for any directed set K and for any confinal mapping $\tau: K \to I$ the relations $x_\beta \underset{K}{\rightharpoonup} x$, $A^{\tau(\beta)}x_\beta \ni f_\beta \underset{K}{\rightharpoonup} f$, $\overline{\lim\limits_\beta}\,(f_\beta, x_\beta) \leqq (f, x)$, with $\{[x_\beta, f_\beta]\}_\beta$ bounded, imply $[x, f] \in A$ and $\lim (f_\beta, x_\beta) = (f, x)$.

Remarks. The introduction of the different definitions a) and b) is useful for many reasons. It is suggested, for instance, by the fact that for a sum $(A^\alpha + B^\alpha, A + B)_{\alpha \in I}$ of families the property b) is easier to check (cf. [16]). Let us mention that a family $(A^n, A)_{n=1,2,\ldots}$ is generalized pseudomonotone iff $x_\beta \rightharpoonup x$, $A^{n_\beta}x_\beta \ni f_\beta \to f$, $\overline{\lim\limits_\beta}\,(f_\beta, x_\beta) \leqq (f, x)$, $\lim n_\beta = +\infty$ implies $[x, f] \in A$ and $\lim\limits_\beta (f_\beta, x_\beta) = (f, x)$. The Definition 2, b) then simplifies in an analogous manner. Note that A is generalized pseudomonotone (on bounded sets) if and only if the family $(A^\alpha, A)_{\alpha \in I}$, with $A^\alpha = A$, possesses this property.

4. An existence theorem

In this section we illustrate how the concept of generalized pseudomonotone families may be used to obtain existence results for the multi-valued equation $Ax \ni f$. Here we pursue the idea to approximate an operator A, which is *not continuous* even in the mildest sense, by "continuous" operators $A^\alpha (\alpha \in I)$. For these approximating operators A^α the customary topological methods for finding solutions are applicable. The point is now that we describe the approximation by a generalized pseudomonotone family $(A^\alpha, A)_{\alpha \in I}$. As examples may serve the approximations $(A_{\lambda_n}, A)_{0<\lambda_n\to 0}$ of a *maximal*

[1] In the sequel the notations E_w and E'_w indicate that E resp. E' is equipped with the weak topology.

monotone operator A and $(A_{\lambda_n}(x, x), A(x, x))_{0<\lambda_n\to 0}$ of a *semimonotone operator* $A(x, x)$ (here we call $B \subseteq E \times E'$, $Bx \underset{\mathrm{df.}}{=} A(x, x)$, *semimonotone* if $A(\cdot, x) \subseteq E \times E'$ is maximal monotone for any $x \in M \subseteq E$; by $A_\lambda(x, x) \underset{\mathrm{df.}}{=} A_\lambda(\cdot, x)\,(x)$ we denote the *pointwise Yoshida approximation* of $A(x, x)$.

Theorem 1. *Let be E a reflexive Banach space and I a directed set. Suppose that the family $(A^\alpha, A)_{\alpha\in I}$ of operators $A \subseteq E \times E'$, $A^\alpha \subseteq E \times E'$ $(\alpha \in I)$ satisfies the following conditions:*

1) *$(A^\alpha, A)_{\alpha\in I}$ is generalized pseudomonotone.*
2) *For any $\alpha \in I$ and $x \in E$ the image set $\{Ax\}$ is nonempty, convex, closed and bounded. Moreover, for any finite-dimensional subspace $E_0 \subseteq E$ the mappings $A^\alpha|_{E_0} : E_0 \to 2^{E'_w}$ are upper semicontinuous.*
3) *(Coercivity) There is an element $v_0 \in E$ with*

$$\lim_{\substack{\alpha\in I\\ \|x\|\to\infty}} \inf_{f\in A^\alpha x} \frac{(f, x - v_0)}{\|x\|} = +\infty.$$

Then the equation $Ax \ni f$ is solvable for any $f \in E'$.

Remark. The assumption 1) can be replaced by a slightly weaker one. The condition 3) can be weakened considerably such that this existence theorem allows also the solution of quasivariational inequalities (cf. [16]).

5. Generalized pseudomonotonicity and convergence properties of families of maximal monotone operators

It is of special interest to give suitable criteria for the generalized pseudomonotonicity of a family of *maximal monotone* operators. Let us mention that also the consideration of families of semimonotone operators reduces to this case. The following theorem contains the main result of this paper. It shows that the generalized pseudomonotonicity of such a family is essentially equivalent both to the convergence of the graphs and to the convergence of the resolvents of the operators.

Theorem 2. *Let be E a reflexive Banach space and I a directed set. Suppose that there are given maximal monotone operators $A \subseteq E \times E'$ and $A^\alpha \subseteq E \times E'(\alpha \in I)$. Then the conditions* (i)—(viii) *are equivalent:*

(i) *The family $(A^\alpha, A)_{\alpha\in I}$ is generalized pseudomonotone. Moreover, there exists an almost bounded sequence $\{[v_\alpha, g_\alpha] \in A^\alpha\}_{\alpha\in I}$.*

(ii) *The family $(A^\alpha, A)_{\alpha\in I}$ is generalized pseudomonotone on bounded sets. Moreover, there is an almost bounded sequence $\{[v_\alpha, g_\alpha] \in A^\alpha\}_{\alpha\in I}$.*

(iii) *(Approximation of the graph of A)*
For all $[x, f] \in A$ there exists a sequence $A^\alpha \ni [x_\alpha, f_\alpha] \to [x, f]$.

(iv) *(Approximation of the graph of A^0)*
For all $x \in D(A)$ there exists a sequence $A^\alpha \ni [x_\alpha, f_\alpha] \to [x, A^0x]$. Moreover, for any directed set K and confinal mapping $\tau : K \to I$ the relations $x_\beta \underset{K}{\rightharpoonup} x$, $f_\beta \underset{K}{\rightharpoonup} f$ and $\overline{\lim_K}\,(f_\beta, x_\beta) \leq (f, x)$, with $\{[x_\beta, f_\beta] \in A^{\tau(\beta)}\}_{\beta\in K}$ bounded, imply $x \in \overline{D(A)}$.

(v) (*Convergence of the resolvents and Yoshida approximations for all* $\lambda > 0$)
For all $x \in E$, $\lambda > 0$ *we have* $J_\lambda^\alpha x \to J_\lambda x$ *and* $A_\lambda^\alpha x \to A_\lambda x$.

(vi) (*Convergence of the resolvents or Yoshida approximations for a fixed* $\lambda > 0$)
There exists a $\lambda > 0$ *with* $J_\lambda^\alpha x \to J_\lambda x$ *or* $A_\lambda^\alpha x \to A_\lambda x$ *for all* $x \in E$.

(vii) *For all* $\lambda > 0$ *the family* $(A_\lambda^\alpha, A_\lambda)_{\alpha \in I}$ *of Yoshida approximations is generalized pseudomonotone. Moreover, there exists an almost bounded sequence* $\{[v_\alpha, g_\alpha] \in A^\alpha\}_{\alpha \in I}$.

(viii) *There exists a real number* $\lambda > 0$ *such that* $(A_\lambda^\alpha, A_\lambda)_{\alpha \in I}$ *is generalized pseudomonotone on bounded sets. Moreover, there is an almost bounded sequence* $\{[v_\alpha, g_\alpha] \in A^\alpha\}_{\alpha \in I}$.

Remarks. The main result of this theorem consists in the proof that the convergence properties (iii), (iv) and (v) are *necessary* for the generalized pseudomonotonicity of the family $(A^\alpha, A)_{\alpha \in I}$. Conversely, the sufficiency of these conditions is easy to check. The condition (iv) may be used if we are concerned with existence problems for variational or quasivariational inequalities (note the identity $(\partial I_C)^0 \equiv 0$, where ∂I_C is the subdifferential of the indicatrix $I_C: E \to \bar{R}$, $I_C(x) \underset{\text{df.}}{=} \begin{cases} 0 \text{ for } x \in C \\ +\infty \text{ for } x \notin C \end{cases}$. The assertions of the theorem keep true if we replace the operators $(A^\alpha, A)_{\alpha \in I}$ in one of the conditions (i) to (viii) by $\left((A^\alpha)^{-1}, A^{-1}\right)_{\alpha \in I}$ or $(A_\lambda^\alpha, A_\lambda)_{\alpha \in I}$ $(\lambda > 0)$, respectively. Indeed, the condition (iii) is not changed by the transformation $A \mapsto A^{-1}$, $A^\alpha \mapsto (A^\alpha)^{-1}$ and the possibility of the transformation $A \mapsto A_\lambda$, $A^\alpha \mapsto A_\lambda^\alpha$ is guaranteed by the equivalence (i) $\leftrightarrow$ (vii). For the case $I \underset{\text{df.}}{=} \{1, 2, \ldots\}$ the equivalence (iii) $\leftrightarrow$ (v) is already contained in [5]. A result comparable with the implication (iii) $\to$ (i) can be found in [11, 12, 13]. If E is a Hilbert space and $A^n = \partial\varphi_n$, $A = \partial\varphi$ are subdifferentials of convex lower semicontinuous functions then the condition (v) is closely related to the *Mosco convergence* of the corresponding functionals (cf. [1, 2, 3, 17]). A more comprehensive investigation of families of maximal monotone operators was given in [16].

Before proving the theorem we give a simple consequence which illustrates how the obtained results may be applied.

Corollary. *Let* $A \subseteq E \times E'$ *be maximal monotone in the reflexive Banach space* E. *Then each family* $(A_{\lambda_\alpha}, A)_{0 < \lambda_\alpha \to 0}$ *is generalized pseudomonotone.*

Remark. Under suitable assumptions one can derive a similar result for a semimonotone operator A (cf. [16]).

Proof of the Corollary. For any $\lambda > 0$, $x \in E$ we have (compare Lemma 1; 4) and 5)) $(A_{\lambda_\alpha})_\lambda x = (A_\lambda)_{\lambda_\alpha} x \underset{\alpha}{\to} (A_\lambda)^0 x = A_\lambda x$, i.e., the family $(A_{\lambda_\alpha}, A)_{0 < \lambda_\alpha \to 0}$ satisfies the condition (vi) of Theorem 2. The assertion results now immediately from the implication (vi) $\to$ (i).

Proof of Theorem 2. The proof is based on the following scheme:

$$\text{(vi)} \to \text{(iii)} \to \{\text{(ii)} \wedge \text{(iii)}\} \to \text{(iv)} \to \text{(ii)} \to \text{(v)} \to \text{(vii)} \to \text{(viii)} \to \text{(vi)}.$$

$$\text{(iii)} \to \text{(i)} \to \text{(ii)}$$

ad (vi) → (iii). Under our assumptions on the norms in E and E' the convergence $J_\lambda^\alpha x \underset{I}{\to} J_\lambda x$ is equivalent to $A_\lambda^\alpha x \underset{I}{\to} A_\lambda x$. Therefore (vi) implies the existence of a real number $\lambda_0 > 0$ with $J_{\lambda_0}^\alpha x \to J_{\lambda_0} x$ and $A_{\lambda_0}^\alpha x \to A_{\lambda_0} x$, $x \in E$. Let now be given an arbitrary element $[x, f] \in A$. Because of Lemma 1; 3) there is an element $y \in E$ with $[x, f] = [J_{\lambda_0} y, A_{\lambda_0} y]$. Thus we obtain $A^\alpha \ni [J_{\lambda_0}^\alpha y, A_{\lambda_0}^\alpha y] \to [J_{\lambda_0} y, A_{\lambda_0} y] = [x, f]$, as desired.

ad (iii) → (i). By the assumption (iii) for a fixed element $[y_0, g_0] \in A$ there exists a sequence $A^\alpha \ni [y_\alpha, g_\alpha] \to [y_0, g_0]$. It is easy to see that $\{[y_\alpha, g_\alpha] \in A^\alpha\}_{\alpha \in I}$ is almost bounded. Thus is remains to show that $(A^\alpha, A)_{\alpha \in I}$ is generalized pseudomonotone. For this purpose we suppose for a directed set K and for a confinal mapping $\tau: K \to I$

$$x_\beta \underset{K}{\rightharpoonup} x, \ f_\beta \underset{K}{\rightharpoonup} f \ \text{ and } \ \overline{\lim_K} (f_\beta, x_\beta) \leqq (f, x), \ \text{ with } \ [x_\beta, f_\beta] \in A^{\tau(\beta)}. \tag{1}$$

For any $[y, g] \in A$ wechoose a sequence $A^\alpha \ni [y_\alpha, g_\alpha] \underset{I}{\to} [y, g]$. Then we have also $A^{\tau(\beta)} \ni [y_{\tau(\beta)}, g_{\tau(\beta)}] \underset{K}{\to} [y, g]$. The monotonicity of $A^{\tau(\beta)}$ now implies

$$0 \leqq \underline{\lim_K} (f_\beta - g_{\tau(\beta)}, x_\beta - y_{\tau(\beta)}) \leqq (f - g, x - y) \text{ for any } [y, g] \in A. \tag{2}$$

As a consequence of the maximal monotonicity of A we can conclude $[x, f] \in A$. Especially we can choose in (2) $[y_{\tau(\beta)}, g_{\tau(\beta)}] \underset{K}{\to} [y, g]$ and obtain $\underline{\lim_K} (f_\beta, x_\beta) \geqq (f, x)$. Hence the family $(A^\alpha, A)_{\alpha \in I}$ is generalized pseudomonotone.

ad (i) → (ii) and $\{$(ii) $\wedge$ (iii)$\}$ → (iv). These implications are obvious.

ad (iii) → $\{$(ii) $\wedge$ (iii)$\}$. This implication follows immediately from the just proved relation (iii) → (i) → (ii).

ad (iv) → (ii). First we observe that any sequence $A^\alpha \ni [y_\alpha, g_\alpha] \underset{I}{\to} [y_0, A^0 y_0]$, $y_0 \in D(A)$ fixed, is almost bounded. In order to check the generalized pseudomonotonicity on bounded sets of $(A^\alpha, A)_{\alpha \in I}$ we assume for a directed set K and for a confinal mapping $\tau: K \to I$

$$x_\beta \underset{K}{\rightharpoonup} x, \ f_\beta \underset{K}{\rightharpoonup} f \ \text{ and } \ \overline{\lim_\beta} (f_\beta, x_\beta) \leqq (f, x), \ \text{ with } \ \{[x_\beta, f_\beta] \in A^{\tau(\beta)}\} \text{ bounded.} \tag{3}$$

First of all, by the assumption (vi) we obtain $x \in \overline{D(A)}$. We choose now for any $y \in D(A)$ a sequence $A^\alpha \ni [y_\alpha, g_\alpha] \underset{I}{\to} [y, A^0 y]$. Then we have also $A^{\tau(\beta)} \ni [y_{\tau(\beta)}, g_{\tau(\beta)}]) \underset{K}{\to} [y, A^0 y]$ and the monotonicity of $A^{(\tau\beta)}$ implies

$$x \in \overline{D(A)} \ \text{ and } \ 0 \leqq \underline{\lim_K} \left(f_\beta - g_{\tau(\beta)}, x_\beta - y_{\tau(\beta)}\right) \leqq (f - A^0 y, x - y) \ \text{ for all } \ y \in D(A). \tag{4}$$

From (4) we conclude $[x, f] \in A$ (compare Lemma 1; 1)). Especially we can choose in (4) $A^{\tau(\beta)} \ni [y_{\tau(\beta)}, g_{\tau(\beta)}] \underset{K}{\to} [x, A^0 x]$ and obtain $\underline{\lim_K} (f_\beta, x_\beta) \geqq (f, x)$. Hence the family $(A^\alpha, A)_{\alpha \in I}$ is generalized pseudomonotone on bounded sets.

ad (ii) → (v). Let be $x \in E$ and $\lambda > 0$ given arbitrarily. We have to check $u_\alpha \underset{\text{df.}}{=} J^\alpha_\lambda x \underset{I}{\to} J_\lambda x$ and $g_\lambda \underset{\text{df.}}{=} A^\alpha_\lambda x \underset{I}{\to} A_\lambda x$. By the definition of J^α_λ and A^α_λ we have

$$J(u_\alpha - x) + \lambda g_\alpha = 0 \text{ and } g_\alpha \in A^\alpha u_\alpha \text{ for any } \alpha \in I. \tag{5}$$

Let be $\{[v_\alpha, f_\alpha] \in A^\alpha\}_{\alpha \in I}$ an almost bounded sequence which exists by the assumption (ii). From (5) and the estimation $\|u_\alpha - x\| \leqq \lambda \|f_\alpha\| + 2 \|v_\alpha - x\|$ (compare Lemma 1; 6)) we can conclude that the sequences $\{u_\alpha\}_{\alpha \geqq \alpha_0}$, $\{J(u_\alpha - x)\}_{\alpha \geqq \lambda_0}$ and $\{g_\alpha\}_{\alpha \geqq \alpha_0}$ are bounded (and hence relatively weakly compact) for some $\alpha_0 \in I$.

Let now be $[u, g]$ an arbitrary adherent point of the sequence $\{[u_\alpha, g_\alpha]\}_{\alpha \geqq \alpha_0} \subseteqq E_w \times E'_w$. Let further be $\mathfrak{W}$ the family of the weak neighbourhoods of zero in $E \times E'$. Then we introduce the directed set $K \underset{\text{df.}}{=} \{\alpha \in I;\ \alpha \geqq \alpha_0\} \times \mathfrak{W}$ by agreeing $[\alpha_1, W_1] \leqq [\alpha_2, W_2] \underset{\text{df.}}{=} \alpha_1 \leqq \alpha_2 \wedge W_1 \supseteqq W_2$. According to the definition of an adherent point one finds for any $[\alpha, W] \in K$ an element $\tau(\alpha, W) \in I$ with $\tau(\alpha, W) \geqq \alpha$ and $[u_{(\alpha,W)}, g_{(\alpha,W)}] \underset{\text{df.}}{=} [u_{\tau(\alpha,W)}, g_{\tau(\alpha,W)}] \in W + [u, g]$. These relations yield immediately

$$\left.\begin{array}{l} u_{(\alpha,W)} \underset{K}{\rightharpoonup} u,\ g_{(\alpha,W)} \underset{K}{\rightharpoonup} g, \text{ with } \tau: K \to I \text{ confinal and} \\ \{[u_{(\alpha,W)}, g_{(\alpha,W)}] \in A^{\tau(\alpha,W)}\} \text{ bounded.} \end{array}\right\} \tag{6}$$

Obviously (5) and (6) imply $J(u_{(\alpha,W)} - x) \underset{K}{\rightharpoonup} -\lambda g$. From the monotonicity of J we get

$$\big(J(u_{(\alpha,W)} - x), u_{(\alpha,W)}\big) \leqq \big(J(u_{(\alpha,W)} - x), u\big) + \big(J(u - x), u_{(\alpha,W)} - u\big)$$

and hence

$$\underline{\lim}_{K} \big(J(u_{(\alpha,W)} - x), u_{(\alpha,W)}\big) \geqq -\lambda(g, u).$$

With regard to the equation (5) we can conclude

$$\overline{\lim}_{K} (g_{(\alpha,W)}, u_{(\alpha,W)}) \leqq (g, u). \tag{7}$$

Because of the generalized pseudomonotonicity on bounded sets of $(A^\alpha, A)_{\alpha \in I}$ the relations (6) and (7) imply

$$[u, g] \in A \quad \text{and} \quad \lim_{K} (g_{(\alpha,W)}, u_{(\alpha,W)}) = (g, u). \tag{8}$$

As a consequence of (5) and (8) we get

$$\lim_{K} \big(J(u_{(\alpha,W)} - x), u_{(\alpha,W)}\big) = (-\lambda g, u)$$

and hence

$$\big(-\lambda g - Jv, (u - x) - v\big) = \lim_{K} \big(J(u_{(\alpha,W)} - x) - Jv, (u_{(\alpha,W)} - x) - v\big) \geqq 0$$

for any $v \in E$. With respect to the maximal monotonicity of the operator J we find $-\lambda g = J(u - x)$. This implies (compare (8)) $J(u - x) + \lambda A u \ni 0$ and, according to the definition of J_λ and A_λ, $u = J_\lambda x$ and $g = A_\lambda x$.

Summarizing what we have shown up to now we can say that the sequence $\{[u_\alpha, g_\alpha]\}_{\alpha \geqq \alpha_0}$ is relatively weakly compact and has exactly one weak adherent point $[u, g] = [J_\lambda x, A_\lambda x]$. This yields immediately

$$u_\alpha \underset{I}{\rightharpoonup} J_\lambda x \quad \text{and} \quad g_\alpha \underset{I}{\rightharpoonup} A_\lambda x. \tag{9}$$

Moreover, we have

$$\underline{\lim}_{I} \, (g_\alpha, u_\alpha) \leqq (A_\lambda x, J_\lambda x). \tag{10}$$

Indeed, the assumption

$$\underline{\lim}_{I} \, (g_\alpha, u_\alpha) < (A_\lambda x, J_\lambda x)$$

implies the existence of a confinal subset $I' \subseteqq \{\alpha \in I; \alpha \leqq \alpha_0\}$ with

$$u_{\alpha'} \underset{I'}{\rightharpoonup} J_\lambda x, \, g_{\alpha'} \underset{I'}{\rightharpoonup} A_\lambda x \quad \text{and} \quad \overline{\lim}_{I} \, (g_{\alpha'}, u_{\alpha'}) < (A_\lambda x, J_\lambda x). \tag{11}$$

But this contradicts the generalized pseudomonotonicity on bounded sets of the family $(A^\alpha, A)_{\alpha \in I}$.

With regard to the definitions $u_\alpha \underset{\text{df.}}{=} J_\lambda^\alpha x$ and $g_\alpha \underset{\text{df.}}{=} A_\lambda^\alpha x$ we get from (9) and (10) both $x - J_\lambda^\alpha x \underset{I}{\rightharpoonup} x - J_\lambda x$ and

$$\overline{\lim}_{I} \, \|x - J_\lambda^\alpha x\|^2 = \overline{\lim}_{I} \, \lambda(A_\lambda^\alpha x, x - J_\lambda^\alpha x) = \overline{\lim}_{I} \, \lambda(g_\alpha, x - u_\alpha) \leqq \lambda(A_\lambda x, x - J_\lambda x) = \|x - J_\alpha x\|^2.$$

The assumptions made on the norm in E now imply the strong convergence $x - J_\lambda^\alpha x \underset{I}{\to} x - J_\lambda x$. This leads immediately to $J_\lambda^\alpha x \underset{I}{\to} J_\lambda x$ and

$$A_\lambda^\alpha x \underset{\text{df.}}{=} \frac{J(x - J_\lambda^\alpha x)}{\lambda} \underset{I}{\to} \frac{J(x - J_\lambda x)}{\lambda} \underset{\text{df.}}{=} A_\lambda x,$$

as desired.

ad (v) → (vii). Let (v) be satisfied. Then for fixed $\lambda > 0$ and $x \in E$ the sequence $A^\alpha \ni [J_\lambda^\alpha x, A_\lambda^\alpha x] \underset{I}{\to} [J_\lambda x, A_\lambda x]$ is almost bounded. Moreover, the convergence $A_\lambda^\alpha x \underset{I}{\to} A_\lambda x$ $(x \in E, \lambda > 0)$ implies that any family $(B^\alpha, B)_{\alpha \in I} \underset{\text{df.}}{=} (A_\lambda^\alpha, A_\alpha)_{\alpha \in I}$ fulfils the condition (iii) of the Theorem. The generalized pseudomonotonicity of the families $(B^\alpha, B)_{\alpha \in I} \underset{\text{df.}}{=} (A_\lambda^\alpha, A_\lambda)_{\alpha \in I}$ $(\lambda > 0)$ results now from the implication (iii) → (i) we already have proved.

ad (vii) → (viii). This implication is obvious.

ad (viii) → (vi). Let the family $(A_{\lambda_0}^\alpha, A_{\lambda_0})_{\alpha \in I}$ be generalized pseudomonotone on bounded sets and let $\{[v_\alpha, g_\alpha] \in A^\alpha\}_{\alpha \in I}$ be almost bounded. By Lemma 1; 3) we can choose a sequence $\{y_\alpha\} \subseteqq E$ with $[v_\alpha, g_\alpha] = [J_{\lambda_0}^\alpha y_\alpha, A_{\lambda_0}^\alpha y_\alpha]$. Taking account of

$$A_{\lambda_0}^\alpha y_\alpha = \frac{J(y_\alpha - J_{\lambda_0}^\alpha y_\alpha)}{\lambda_0} = \frac{J(y_\alpha - v_\alpha)}{\lambda_0},$$

we see that $\{[y_\alpha, A^\alpha_{\lambda_0} y_\alpha]\}_{\alpha\in I}$ is almost bounded, too. Thus the family $(B^\alpha, B)_{\alpha\in I} \underset{\text{df.}}{=} (A^\alpha_{\lambda_0}, A_{\lambda_0})_{\alpha\in I}$ satisfies the condition (ii) of the Theorem. By the implication (ii) $\to$ (v) which we already have proved, $(B^\alpha, B)_{\alpha\in I}$ fulfils also the condition (v). This implies $A^\alpha_{\lambda+\lambda_0} x = B^\alpha_\lambda x \underset{I}{\to} B_\lambda x = A_{\lambda+\lambda_0} x$ for any $\lambda > 0$, $x \in E$. In particular the family $(A^\alpha, A)_{\alpha\in I}$ satisfies the condition (vi). q.e.d.

References

[1] Attouch, H., Convergence de fonctions convexes, des sous-différentiels et semi-groups associés, C. R. Acad. Sci. Paris, Sér. A, **284** (1977), 539–542.

[2] Attouch, H., Suites convergentes d'operateurs maximaux monotones et inéquations variationelles, Seminaires Lions-Brézis, 1975–1976.

[3] Attouch, H., Konishi, Y., Convergence d'operatuers maximaux monotones et inéquations variationelles, C. R. Acad. Sci. Paris, Sér. A, **282** (1976), 467–469.

[4] Barbu, V., Nonlinear semigroups and differential equations in Banach spaces, Editure Acad., Bucuresti and Noordhoff Int. Comp., Leyden 1976.

[5] Brezis, H., Cours de 3e cycle, Paris VI, 1974–1975.

[6] Brezis, H., Operateurs maximaux monotones et semi-groupes de contractions dans les espaces de Hilbert, North-Holland Publ. Comp., Amsterdam–London 1973.

[7] Brezis, H., Crandall, M., Pazy, A., Perturbations of nonlinear maximal monotone sets, Comm. Pure Appl. Math. **23** (1970), 123–144.

[8] Brezis, H., Pazy, A., Semi groups of nonlinear contractions on convex sets, J. Funct. Anal. **6** (1970), 237–281.

[9] Browder, F., Nonlinear monotone and accretive operators in Banach spaces, Proc. Nat. Acad. Sci. U.S.A. **61** (1968), 388–393.

[10] Browder, F., Hess. P., Nonlinear mappings of monotone type in Banach spaces, J. Funct. Anal. **11** (1974), 251–294.

[11] Kluge, R., Nichtlineare Variationsungleichungen und Extremalaufgaben, Deutscher Verlag der Wissenschaften, Berlin 1979.

[12] Kluge, R., On some parameter determination problems and quasi-variational inequalities. *In:* Theory of Nonlinear Operators (Proc. Fifth Internat. Summer School Berlin 1977), Abh. Akad. Wiss. DDR, Jahrg. 1978 Nr. 6N, Akademie-Verlag, Berlin 1978, 129–139.

[13] Kluge, R., On some parameter determination problems in variational and quasi-variational inequalities. I. Math. Nachr. **89** (1979), 127–148.

[14] Kluge, R., Telschow, G., On the convergence and speed of convergence of some iteration methods for variational inequalities, II, Math. Nachr. **75** (1976), 73–89.

[15] Köthe, G., Topological vector spaces I, Grundlehren d. math. Wiss., Band 159, Springer-Verlag, Berlin–Heidelberg–New York 1969.

[16] Krauß, E., Dissertation A, Berlin, 1979.

[17] Mosco, U., Convergence of convex sets and solution of variational inequalities, Advances in Math. **3** (1969), 510–585.

[18] Troyanski, S., On locally uniformly convex and differentiable norms in certain non-separable Banach spaces, Studia Math. **37** (1971), 173–180.

Address of the author:

Akademie der Wissenschaften der DDR
Institut für Mathematik
1080 Berlin
Mohrenstr. 39
DDR

On perturbation of branching problems and secondary branching in the theory of plates

Arno Langenbach and Lutz Recke, Berlin

1. Introduction

We consider the domain $\Omega_s := \{(x_1, x_2) \in \Re^2;\ 0 < x_1 < s,\ 0 < x_2 < 1\}$, the middle surface of a thin elastic plate. This plate is simply supported and subject to partly compressive forces acting on the edges in the plane of the middle surface normally to the edges. The thrust in direction x_i ($i = 1, 2$) has the magnitude λ_i.

We describe equilibrium states of the plate by Airy's stress function $f: \Omega_s \to \Re$ and the deflection $w: \Omega_s \to \Re$ such that v. Kármán's equations

$$\begin{aligned} \Delta^2 w &= 2\frac{\partial^2 w}{\partial x_1\,\partial x_2}\frac{\partial^2 f}{\partial x_1\,\partial x_2} - \frac{\partial^2 w}{\partial x_1^2}\frac{\partial^2 f}{\partial x_2^2} - \frac{\partial^2 w}{\partial x_2^2}\frac{\partial^2 f}{\partial x_1^2} - \lambda_1\frac{\partial^2 w}{\partial x_1^2} - \lambda_2\frac{\partial^2 w}{\partial x_2^2}, \\ \Delta^2 f &= \frac{\partial^2 w}{\partial x_1^2}\frac{\partial^2 w}{\partial x_2^2} - \left[\frac{\partial^2 w}{\partial x_1\,\partial x_2}\right]^2, \qquad (x_1, x_2) \in \Omega_s \end{aligned} \tag{1.1}$$

under boundary conditions

$$w = \frac{\partial^2 w}{\partial x_1^2} = \frac{\partial^2 w}{\partial x_2^2} = f = \frac{\partial^2 f}{\partial x_1^2} = \frac{\partial^2 f}{\partial x_2^2} = 0, \quad (x_1, x_2) \in \partial\Omega_s \tag{1.2}$$

are satisfied. Besides the geometrically nonlinear and physically linear model (1.1), (1.2) we consider a geometrically linear and physically nonlinear model described only by the deflection w satisfying the variational equation

$$\int_{\Omega_s}\left\{\Delta w \Delta\varphi + \sum_{k=1}^{n} a_k (H[w, w])^k\, H[w, \varphi]\right\} dx = \int_{\Omega_s}\left\{\lambda_1 \frac{\partial w}{\partial x_1}\frac{\partial \varphi}{\partial x_1} + \lambda_2 \frac{\partial w}{\partial x_2}\frac{\partial \varphi}{\partial x_2}\right\} dx = 0 \tag{1.3}$$

for all functions $\varphi \in \mathring{C}^{(4)}(\bar{\Omega}_s)$. Here we set

$$H[u, v] := \frac{\partial^2 u}{\partial x_1^2}\frac{\partial^2 v}{\partial x_2^2} + \alpha\left(\frac{\partial^2 u}{\partial x_1^2}\frac{\partial^2 v}{\partial x_2^2} + \frac{\partial^2 u}{\partial x_2^2}\frac{\partial^2 v}{\partial x_1^2}\right) + \beta^2 \frac{\partial^2 u}{\partial x_1\,\partial x_2}\frac{\partial^2 v}{\partial x_1\,\partial x_2}. \tag{1.4}$$

The physical constants α and β of the material are connected by the condition

$$2\alpha + \beta^2 = 2. \tag{1.5}$$

The problems (1.1), (1.2) and (1.3) have a common linearisation, namely the classical equation

$$\Delta^2 w + \lambda_1 \frac{\partial^2 w}{\partial x_1^2} + \lambda_2 \frac{\partial^2 w}{\partial x_2^2} = 0. \tag{1.6}$$

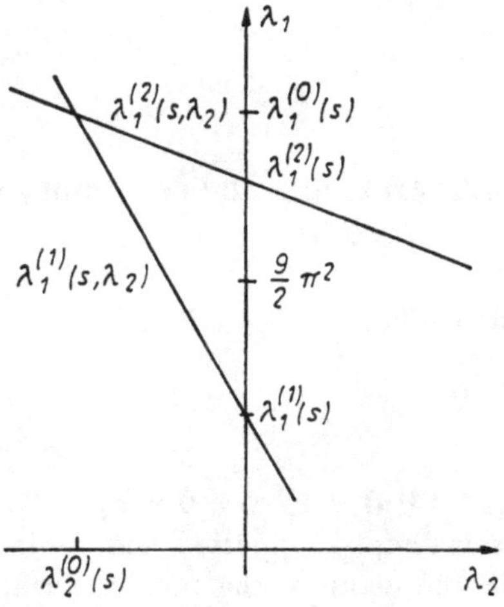

Fig. 1. $s < \sqrt{2}$

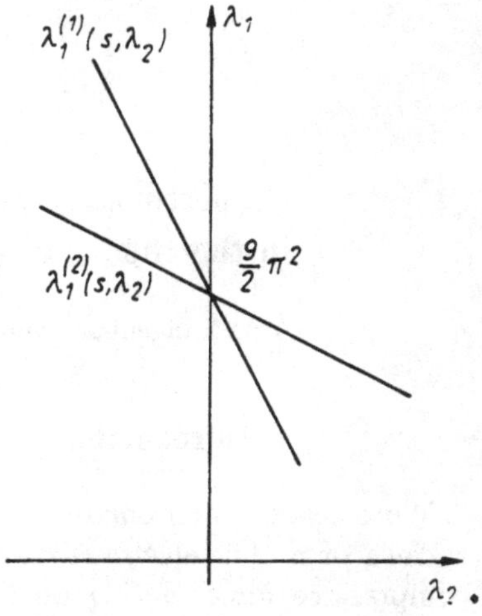

Fig. 2. $s = \sqrt{2}$

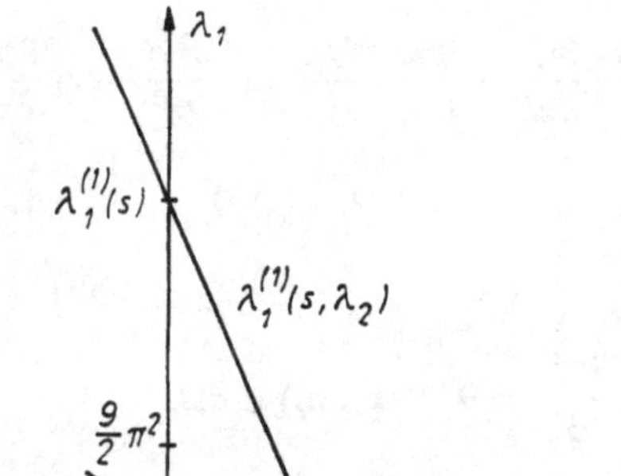

Fig. 3. $s > \sqrt{2}$

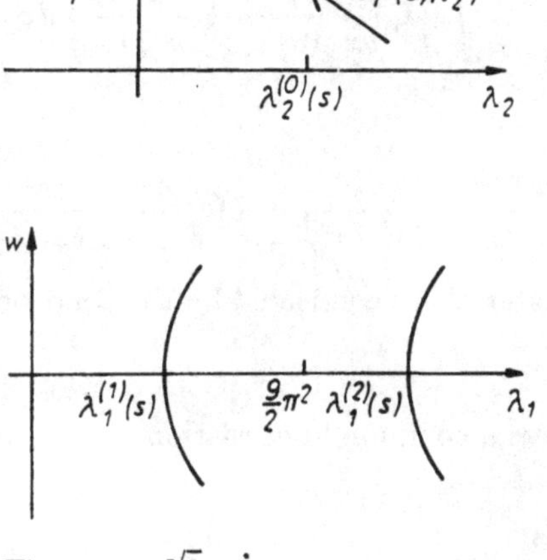

Fig. 4. $s < \sqrt{2}$

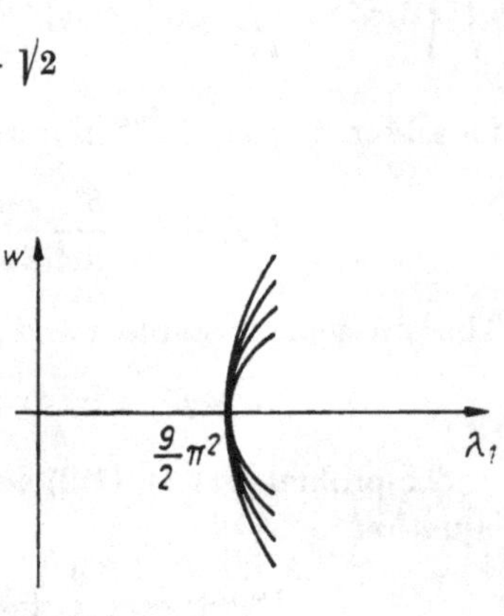

Fig. 5

For derivations of the different equations see [6]. The eigenvalues of the problem (1.6), (1.2) are certainly known, see [3]. They lie on a family of straight lines in the (λ_1, λ_2)-plane,

$$\left(\frac{m}{s}\right)^2 \lambda_1 + n^2\lambda_2 = \pi^2 \left(\left(\frac{m}{s}\right)^2 + n^2\right)^2, \qquad m, n = 1, 2, \ldots \tag{1.7}$$

The eigenfunctions are $\sin \frac{mx_1}{s} \pi \sin nx_2\pi$ accordingly.

Of special interest for us are the eigenfunctions corresponding to numbers $m = n = 1$ and $m = 2, n = 1$. They belong to eigenvalues

$$\begin{aligned} \lambda_1 &= \lambda_1^{(1)}(s, \lambda_2) := \pi^2 \left(\frac{1}{s} + s\right)^2 - s^2\lambda_2, \\ \lambda_1 &= \lambda_1^{(2)}(s, \lambda_2) := \pi^2 \left(\frac{2}{s} + \frac{s}{2}\right)^2 - \frac{s^2}{4} \lambda_2. \end{aligned} \tag{1.8}$$

For $s \approx \sqrt{2}$ and $\lambda_2 = 0$ the smallest two eigenvalues of (1.6), (1.2) are

$$\lambda_1^{(1)}(s) := \frac{(1 + s^2)^2}{s^2} \pi^2, \quad \lambda_1^{(2)}(s) := \frac{(4 + s^2)^2}{4s^2} \pi^2,$$

they coincide for $s = \sqrt{2}$ in $\frac{9}{2} \pi^2$, (see fig. 1—3).

Problem (1.1), (1.2) was studied by different authors under different restrictions:

Kluge [3] proved that every point on the family (1.7) is a bifurcation point for (1.1), (1.2).

Knightly and Sather [4] investigated (1.1), (1.2) with $\lambda_2 = 0$. They show branching in every simple eigenvalue of the linearized equation, where exactly two nontrivial branches start from the trivial solution, (fig. 4).

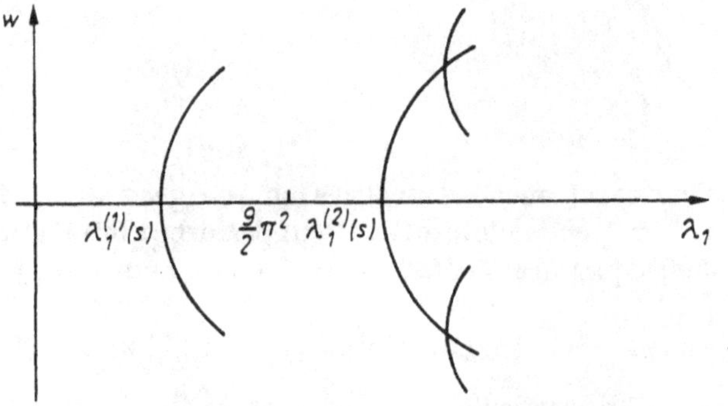

Fig. 6. $s < \sqrt{2}$

Matkowsky and Putnik [9], Knightly and Sather [5] considered (1.1), (1.2) with $\lambda_2 = 0$, $s = \sqrt{2}$. They show branching of exactly eight nontrivial branches in the eigenvalue $\frac{9}{2} \pi^2$ of multiplicity two, (fig. 5).

Bauer, Keller and Reiss [1] studied the generation of four "new" branches in the limit case $s \to \sqrt{2}$ shown in fig. 5, (compare with fig. 4). They formulated the hypothesis that the new primary branches in $s = \sqrt{2}$ are secondary branches in $s \neq \sqrt{2}$, (see fig. 6). They proved such a situation in a special finite dimensional model.

The correctness of this hypothesis was proved independently by List [8] and Recke [11] in 1978. These authors use methods of the analytical theory of bifurcation developed by Chow, Hale and Mallet-Paret [2] for multidimensional parameters of bifurcation. They further use certain symmetries which leave v. Kármán's equations for a rectangular

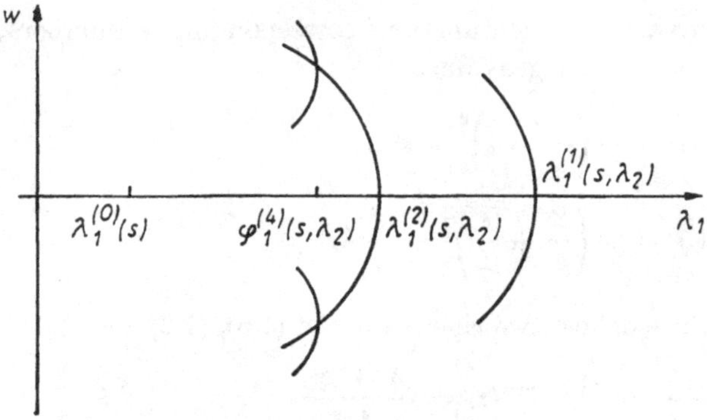

Fig. 7. $\lambda_2 < \lambda_2^{(0)}(s)$, $s > \sqrt{2}$

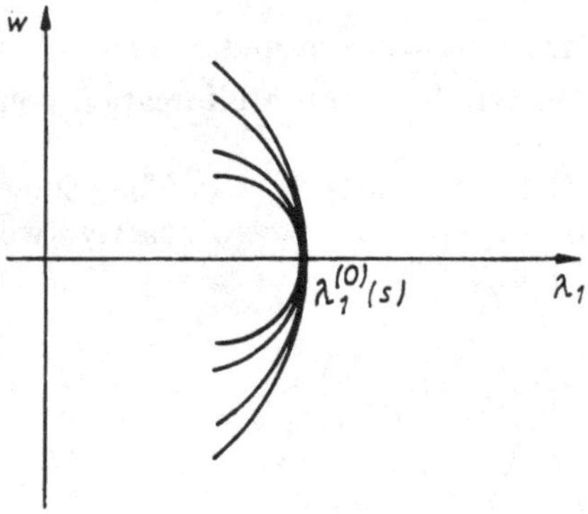

Fig. 8. $\lambda_2 = \lambda_2^{(0)}(s)$

plate invariant. The influence of such symmetries on the generation of secondary bifurcation ("non generic" or "generic bifurcation") in perturbed bifurcation problems was studied for v. Kármán's equations by Mallet-Paret in [10] and later by Vanderbauwhede [11].

In the present paper we describe all small solutions of (1.1), (1.2) and (1.3) for small $|s - \sqrt{2}|$, $\left|\lambda_1 - \frac{9}{2}\pi^2\right|$ and $|\lambda_2|$. In particular, we show that for fixed $s \neq \sqrt{2}$ and for

$$\lambda_2^{(0)}(s) := \pi^2 \left(1 - \frac{4}{s^2}\right) \tag{1.9}$$

(see fig. 1 and 3)

$$\lambda_1 = \lambda_1^{(0)}(s) := \pi^2 \left(\frac{5}{s^4} + 2\right)$$

is an eigenvalue of multiplicity two to (1.6), (1.2) with exactly eight branches of nontrivial solutions of (1.1), (1.2) branching in this eigenvalue from the trivial solution.

Comparing with the situation in fig. 5 we notice that the perturbation $s \neq \sqrt{2}$ is recompensed by the perturbation $\lambda_2 = \lambda_2^{(0)}(s) \neq 0$ with the effect that the bifurcation diagram remains qualitatively the same as in fig. 5.

For fixed $s \neq \sqrt{2}$ and fixed $\lambda_2 \neq \lambda_2^{(0)}(s)$ the eight branches divide into four primary and four secondary branches. As λ_2 tends to zero we arrive at the situation shown in fig 6. Equation (1.3) shows for $a_1 > 0$ similar bifurcation diagrams. In the case $a_1 < 0$ all branches of nontrivial solutions go into the opposite direction, (fig. 7 and 8).

2. Operator equations in Hilbert spaces for the problems (1.1), (1.2) and (1.3)

We set $\Omega := \Omega_{\sqrt{2}}$. By means of the transformation

$$y_1 := x_1 \frac{\sqrt{2}}{s}, \quad y_2 := x_2$$

we change in (1.1), (1.2) and (1.3) to new variables $(y_1, y_2) \in \Omega$. Instead of (1.1), (1.2) we get

$$\left.\begin{aligned}
&\left(\frac{\sqrt{2}}{s}\right)^4 \frac{\partial^4 w}{\partial y_1^4} + 2\left(\frac{\sqrt{2}}{s}\right)^2 \frac{\partial^4 w}{\partial y_1^2\,\partial y_2^2} + \frac{\partial^4 w}{\partial y_2^4}\\
&= \left(\frac{\sqrt{2}}{s}\right)^2 \left[2\frac{\partial^2 w}{\partial y_1\,\partial y_2}\frac{\partial^2 f}{\partial y_1\,\partial y_2} - \frac{\partial^2 w}{\partial y_1^2}\frac{\partial^2 f}{\partial y_2^2} - \frac{\partial^2 w}{\partial y_2^2}\frac{\partial^2 f}{\partial y_1^2}\right] - \left(\frac{\sqrt{2}}{s}\right)^2 \lambda_1 \frac{\partial^2 w}{\partial y_1^2} - \lambda_2 \frac{\partial^2 w}{\partial y_2^2},\\
&\left(\frac{\sqrt{2}}{s}\right)^4 \frac{\partial^4 f}{\partial y_1^4} + 2\left(\frac{\sqrt{2}}{s}\right)^2 \frac{\partial^4 f}{\partial y_1^2\,\partial y_2^2} + \frac{\partial^4 f}{\partial y_2^4} = \left(\frac{\sqrt{2}}{s}\right)^2 \left[\frac{\partial^2 w}{\partial y_1^2}\frac{\partial^2 w}{\partial y_2^2} - \left(\frac{\partial^2 w}{\partial y_1\,\partial y_2}\right)^2\right],
\end{aligned}\right\} \tag{2.1}$$

$(y_1, y_2) \in \Omega$,

$$w = \frac{\partial^2 w}{\partial y_1^2} = \frac{\partial^2 w}{\partial y_2^2} = f = \frac{\partial^2 f}{\partial y_1^2} = \frac{\partial^2 f}{\partial y_2^2} = 0, \quad (y_1, y_2) \in \partial\Omega. \tag{2.2}$$

Instead of (1.3) we get

$$\begin{aligned}
&\int_\Omega \left\{\left(\frac{\sqrt{2}}{s}\right)^4 \frac{\partial^2 w}{\partial y_1^2}\frac{\partial^2 \varphi}{\partial y_1^2} + 2\left(\frac{\sqrt{2}}{s}\right)^2 \frac{\partial^2 w}{\partial y_1\,\partial y_2}\frac{\partial^2 \varphi}{\partial y_1\,\partial y_2} + \frac{\partial^2 w}{\partial y_2^2}\frac{\partial^2 \varphi}{\partial y_2^2}\right.\\
&\qquad \left. - \left(\frac{\sqrt{2}}{s}\right)^2 \lambda_1 \frac{\partial w}{\partial y_1}\frac{\partial \varphi}{\partial y_1} - \lambda_2 \frac{\partial w}{\partial y_2}\frac{\partial \varphi}{\partial y_2}\right\} dy\\
&= \int_\Omega \left\{\sum_{k=1}^n a_k \left[\left(\frac{\sqrt{2}}{s}\right)^4 \left(\frac{\partial^2 w}{\partial y_1^2}\right)^2 + \left(\frac{\sqrt{2}}{s}\right)^2 \left(2\alpha \frac{\partial^2 w}{\partial y_1^2}\frac{\partial^2 w}{\partial y_2^2} + \beta^2 \left(\frac{\partial^2 w}{\partial y_1\,\partial y_2}\right)^2\right)\right.\right.\\
&\qquad \left. + \left(\frac{\partial^2 w}{\partial y_2^2}\right)^2\right]^k \left[\left(\frac{\sqrt{2}}{s}\right)^4 \frac{\partial^2 w}{\partial y_1^2}\frac{\partial^2 \varphi}{\partial y_1^2} + \left(\frac{\sqrt{2}}{s}\right)^2 \left(\alpha\left(\frac{\partial^2 w}{\partial y_1^2}\frac{\partial^2 \varphi}{\partial y_2^2} + \frac{\partial^2 w}{\partial y_1^2}\frac{\partial^2 \varphi}{\partial y_2^2}\right)\right.\right.\\
&\qquad \left.\left.\left. + \beta^2 \frac{\partial^2 w}{\partial y_1\,\partial y_2}\frac{\partial^2 \varphi}{\partial y_1\,\partial y_2}\right) + \frac{\partial^2 w}{\partial y_2^2}\frac{\partial^2 \varphi}{\partial y_2^2}\right]\right\} dy \quad \text{for all} \quad \varphi \in \mathring{C}^{(4)}(\bar{\Omega}).
\end{aligned} \tag{2.3}$$

By H_0 we denote the space $L_2(\Omega)$, by $(u, v)_0 := \int_\Omega uv\,dy$ and $\|\cdot\|_0$ its scalar product and norm.

The space H_2 is the closure of $\mathring{C}^{(4)}(\bar{\Omega})$ in the Sobolev space $W_2^2(\Omega)$ with the scalar product $(u, v)_2 := \int_\Omega \Delta u \Delta v\,dy$, $\|u\|_2 = (u, u)_2^{1/2}$ is the norm of $u \in H_2$,

Last we denote by H_4 the closure of $\mathring{C}^{(4)}(\bar{\Omega})$ in $W_2^4(\Omega)$.

For every $f \in H_0$ the equation $\Delta^2 w = f$ with boundary conditions (1.2) has a uniquely determined weak solution $w_f \in H_2$. In H_0 we use the orthonormal basis

$$u_{mn}(y) := 2^{3/4} \sin \frac{y_1}{\sqrt{2}} m\pi \sin y_2 n\pi, \quad m, n = 1, 2, \ldots$$

In H_2 we may use the orthonormal basis w_{mn},

$$w_{mn} := \pi^2 \left(\frac{m^2}{2} + n^2\right)^{-1} u_{mn}.$$

Now we easily calculate the Fourier expansion

$$w_f = \sum_{m,n=1}^{\infty} (f, w_{mn})_0\, w_{mn}.$$

Calculating from this expansion the partial derivatives of w_f we get $w_f \in H_4$ with $\|w_f\|_{W_2^4}$ bounded by $\|f\|_0$. Using this result we choose on H_4 the scalar product

$$(u, v)_4 := \int_\Omega \Delta^2 u \Delta^2 v\,dy, \quad \|u\|_4 = (u, u)_4^{1/2}.$$

Next we consider the linear bounded selfadjoint positive operators $K_{ij}^{(0)} : H_4 \to H_0$, $K_{ij}^{(2)} : H_2 \to H_2$, $i, j = 1, 2$, $L_i^{(0)} : H_4 \to H_0$, $L_i^{(2)} : H_2 \to H_2$, $i = 1, 2$ defined as follows:

$$\left.\begin{aligned} (K_{ij}^{(\nu)}u, v)_\nu &= \int_\Omega \frac{\partial^2 u}{\partial y_i\, \partial y_j} \frac{\partial^2 v}{\partial y_i\, \partial y_j}\, dy \\ (L_{ij}^{(\nu)}u, v)_\nu &= \int_\Omega \frac{\partial u}{\partial y_i} \frac{\partial v}{\partial y_i}\, dy \end{aligned}\right\} v \in \mathring{C}^{(4)}(\bar{\Omega}), \quad \nu = 0, 2.$$

As a consequence of the compact embeddings $H_4 \subseteq H_2$, $W_2^1 \subseteq H_0$ the operators $L_i^{(0)}$ and $L_i^{(2)}$ are completely continuous. Further we consider the bilinear bounded operator $B : H_2 \times H_2 \to H_2$ defined by

$$\begin{aligned} \big(B(u, v), w\big)_2 := \int_\Omega \Bigg\{ &\left[\frac{\partial^2 u}{\partial y_1\, \partial y_2} \frac{\partial v}{\partial y_2} - \frac{\partial^2 u}{\partial y_2^2} \frac{\partial v}{\partial y_1}\right] \frac{\partial w}{\partial y_1} \\ &+ \left[\frac{\partial^2 u}{\partial y_1\, \partial y_2} \frac{\partial v}{\partial y_1} - \frac{\partial^2 u}{\partial y_1^2} \frac{\partial v}{\partial y_2}\right] \frac{\partial w}{\partial y_2} \Bigg\}\, dy. \end{aligned}$$

As shown in [7] the relations

$$\begin{aligned}\big(A_k(s)\,(u), v\big)_0 := a_k \int_\Omega \Bigg\{\left(\frac{\sqrt{2}}{s}\right)^4 \left(\frac{\partial^2 u}{\partial y_1^2}\right)^2 + \left(\frac{\sqrt{2}}{s}\right)^2 \Bigg[2\alpha\, \frac{\partial^2 u}{\partial y_1^2}\,\frac{\partial^2 u}{\partial y_2^2} \\ + \beta^2 \left(\frac{\partial^2 u}{\partial y_1\,\partial y_2}\right)^2\Bigg]\Bigg\}^k \Bigg\{\frac{\partial^2 u}{\partial y_1^2}\,\frac{\partial^2 v}{\partial y_1^2} + \left(\frac{s}{\sqrt{2}}\right)^2 \Bigg[\alpha\left(\frac{\partial^2 u}{\partial y_1^2}\,\frac{\partial^2 v}{\partial y_2^2} + \frac{\partial^2 u}{\partial y_2^2}\,\frac{\partial^2 v}{\partial y_1^2}\right) \\ + \beta^2 \frac{\partial^2 u}{\partial y_1\,\partial y_2}\,\frac{\partial^2 v}{\partial y_1\,\partial y_2}\Bigg] + \left(\frac{s}{\sqrt{2}}\right)^4 \frac{\partial^2 u}{\partial y_2^2}\,\frac{\partial^2 v}{\partial y_2^2}\Bigg\}\,dy, \quad v \in \mathring{C}^{(4)}(\bar{\Omega})\end{aligned}$$

define for $|s - \sqrt{2}| < \sqrt{2}$ and $k = 1, \ldots, n$ analytic families of operators $A_k(s)\colon H_4 \to H_0$ with powers $2k + 1$.

We are searching to find weak solutions $(w, f) \in H_2 \times H_2$ of (2.1), (2.2) as solutions of the following operator equations

$$\begin{aligned}\left(K_{11}^{(2)} + 2\left(\frac{s}{\sqrt{2}}\right)^2 K_{12}^{(2)} + \left(\frac{s}{\sqrt{2}}\right)^4 K_{22}^{(2)}\right) w + \left(\frac{s}{\sqrt{2}}\right)^2 [B(f, w) - \lambda_1 L_1^{(2)} w] \\ - \lambda_2 \left(\frac{s}{\sqrt{2}}\right)^4 L_2^{(2)} w = 0,\end{aligned} \tag{2.4}$$

$$\left(K_{11}^{(2)} + 2\left(\frac{s}{\sqrt{2}}\right)^2 K_{12}^{(2)} + \left(\frac{s}{\sqrt{2}}\right)^4 K_{22}^{(2)}\right) f - \frac{1}{2}\left(\frac{s}{\sqrt{2}}\right)^2 B(w, w) = 0. \tag{2.5}$$

The variational equation (2.3) is equivalent in H_4 to the operator equation

$$\begin{aligned}\left(K_{11}^{(0)} + 2\left(\frac{s}{\sqrt{2}}\right)^2 K_{12}^{(0)} + \left(\frac{s}{\sqrt{2}}\right)^4 K_{22}^{(0)}\right) w - \left(\frac{s}{\sqrt{2}}\right)^2 \lambda_1 L_1^{(0)} w \\ - \left(\frac{s}{\sqrt{2}}\right)^4 \lambda_2 L_2^{(0)} w + \sum_{k=1}^{n} A_k(s)\,(w) = 0.\end{aligned} \tag{2.6}$$

The problems (2.4), (2.5) and (2.6) both allow a uniform treatment.

The Fréchet derivative of the left hand operator in (2.5) with respect to f in $w = f = 0$, $s = \sqrt{2}$ is the identity of H_2. According to the implicit function theorem[1] (2.5) has for small $\|w\|_2 + |s - \sqrt{2}|$ a uniquely determined analytic solution

$$f = \frac{1}{2}\, B(w, w) + O\big(\|w\|_2^2\, |s - \sqrt{2}|\big). \tag{2.7}$$

Here and in the following we denote by $O(\eta)$ for $\eta \in \mathfrak{R}$ an abstract function for which $\eta^{-1} O(\eta)$ remains bounded when η tends to zero.

Now we substitute the solution (2.7) of (2.5) into (2.4). With the notations

$$\begin{aligned}& C^{(2)}(w) := \frac{1}{2}\, B\big(w, B(w, w)\big), \quad C^{(0)}(w) := A_1\big(\sqrt{2}\big)\,(w), \\ & L_0^{(\nu)} := K_{11}^{(\nu)} + 2K_{12}^{(\nu)} + K_{22}^{(\nu)} \quad \text{for} \quad \nu = 0, 2, \\ & \varepsilon := \frac{s}{\sqrt{2}} - 1\end{aligned} \tag{2.8}$$

---

[1] See, for example [13].

we derive from (2.4), (2.5) and (2.6) similar equations

$$F^{(\nu)}(\varepsilon, \lambda_1, \lambda_2, w) := (L_0^{(\nu)} - \lambda_1 L_1^{(\nu)} - \lambda_2 L_2^{(\nu)})\, w + \varepsilon L^{(\nu)}(\varepsilon, \lambda_1, \lambda_2)\, w + C^{(\nu)}(w) + O(|\varepsilon|\, \|w\|_\nu^3) = 0 \tag{2.9}$$

for $\nu = 0, 2$. In these equations we have set

$$L^{(\nu)}(\varepsilon, \lambda_1, \lambda_2) := (2 + \varepsilon)\, [2K_{12}^{(\nu)} - \lambda_1 L_1^{(\nu)}] + (4 + 6\varepsilon + 4\varepsilon^2 + \varepsilon^3)\, [K_{22}^{(\nu)} - \lambda_2 L_2^{(\nu)}] \quad \text{for} \quad \nu = 0, 2.$$

3. Liapunov-Schmidt equations of bifurcation

The following Lemma is easily proved by direct computation.

Lemma 1. *The smallest eigenvalue of the problems* $(L_0^{(\nu)} - \lambda_1 L_2^{(\nu)})\, w = 0$, $\nu = 0, 2$, *is* $\frac{9}{2}\pi^2$. *It has multiplicity two. As basis in the corresponding eigenspace we may choose the eigenfunctions*

$$u_1(y_1, y_2) := \frac{2^{7/4}}{3\pi^2} \sin \frac{y_1}{2}\pi \sin y_2\pi, \quad u_2(y_1, y_2) := \frac{2^{7/4}}{6\pi^2} \sin \frac{2y_1}{2}\pi \sin y_2\pi$$

which are orthonormal in H_2. *We have further*

$$\left.\begin{aligned}
&(L_1^{(\nu)}u_i, u_i)_\nu = \frac{2}{9\pi^2} \quad \textit{for} \quad i = 1, 2,\\
&(L_2^{(\nu)}u_1, u_1)_\nu = \frac{4}{9\pi^2}, \qquad (L_2^{(\nu)}u_2, u_2)_\nu = \frac{1}{9\pi^2},\\
&\big(L^{(\nu)}(\varepsilon, \lambda_1, \lambda_2)\, u_1, u_1\big)_\nu = (2 + \varepsilon) \left(\frac{4}{9} - \lambda_1 \frac{2}{9\pi^2}\right)\\
&\qquad + (4 + 6\varepsilon + 4\varepsilon^2 + \varepsilon^3) \left(\frac{4}{9} - \lambda_2 \frac{4}{9\pi^2}\right),\\
&\big(L^{(\nu)}(\varepsilon, \lambda_1, \lambda_2)\, u_2, u_2\big)_\nu = (2 + \varepsilon) \left(\frac{4}{9} - \lambda_1 \frac{2}{9\pi^2}\right)\\
&\qquad + (4 + 6\varepsilon + 4\varepsilon^2 + \varepsilon^3) \left(\frac{1}{9} - \lambda_2 \frac{1}{9\pi^2}\right), \quad \textit{for} \quad \nu = 0, 2.
\end{aligned}\right\} \tag{3.1}$$

The implications

$$(u_i, w)_\nu = 0 \Rightarrow (L_j^{(\nu)}u_i, w)_\nu = (K_{jk}^{(\nu)}u_i, w)_\nu = 0 \tag{3.2}$$

hold for every $w \in H_\nu$; $\nu = 0, 2$; $i, j, k = 1, 2$.

Remark. As a consequence of our substitution (2.8) we derive from (3.1) with the notations (1.8)

$$\big((L_0^{(\nu)} - \lambda_1 L_1^{(\nu)} - \lambda_2 L_2^{(\nu)} + \varepsilon L^{(\nu)}(\varepsilon, \lambda_1, \lambda_2))\, u_i, u_i\big)_\nu = \frac{s^2}{9\pi^2} \big(\lambda_1^{(i)}(s, \lambda_2) - \lambda_1\big) \quad \text{for} \quad \nu = 0, 2 \quad \text{and} \quad i = 1, 2. \tag{3.3}$$

Let $\tilde{H}_\nu = \{w \in H_\nu; (w, u_1)_\nu = (w, u_2)_\nu = 0\}$ for $\nu = 0, 2$ be subspaces and $Q^{(\nu)} : H_\nu \to \tilde{H}_\nu$ the corresponding orthogonal projections. We substitute in (2.9) $w = \beta_1 u_1 + \beta_2 u_2 + \tilde{w}$, $\tilde{w} \in \tilde{H}_\nu$ and get instead of (2.9) the equivalent equations

$$\big(u_i, F^{(\nu)}(\varepsilon, \lambda_1, \lambda_2, \beta_1 u_1 + \beta_2 u_2 + \tilde{w})\big)_\nu = 0 \quad (i = 1, 2), \tag{3.4}$$

$$\big(L_0^{(\nu)} - \lambda_1 L_1^{(\nu)} - \lambda_2 L_2^{(\nu)} + \varepsilon L^{(\nu)}(\varepsilon, \lambda_1, \lambda_2)\big)\, \tilde{w} + Q^{(\nu)} C^{(\nu)}(\beta_1 u_1 + \beta_2 u_2 + \tilde{w}) + O\big(|\varepsilon|(|\beta_1|^3 + |\beta_2|^3 + \|\tilde{w}\|_\nu^3)\big) = 0 \tag{3.5}$$

for $\nu = 0, 2$.

In (3.5) we have made use of the implications (3.2).

Lemma 2. *For small* $\left(\varepsilon, \lambda_1 - \dfrac{9}{2}\pi^2, \lambda_2, \beta_1, \beta_2, \tilde{w}\right) \in \Re^5 \times \tilde{H}_\nu$ *equations* (3.5) *admit uniquely determined analytic solutions* $\tilde{w}_\nu(\varepsilon, \lambda_1, \lambda_2, \beta_1, \beta_2)$; *for these the inequality*

$$\|\tilde{w}_\nu(\varepsilon, \lambda_1, \lambda_2, \beta_1, \beta_2)\|_\nu^2 \leqq \text{const}\, (\beta_1^2 + \beta_2^2)^3 \tag{3.6}$$

holds.

Proof. For the left hand side in (3.5) we calculate the partial Fréchet derivative with respect to $\tilde{w}$ in $\varepsilon = \lambda_2 = \beta_1 = \beta_2 = 0$, $\lambda_1 = \dfrac{9}{2}\pi^2$, $\tilde{w} = 0$. It is $L_0^{(\nu)} - \dfrac{9}{2}\pi^2 L_1^{(\nu)}$.

For $u \in H_4$, $v \in H_0$ we have $(L_0^{(0)} u, v)_0 = (\Delta^2 u, v)_0$, therefore $\|L_0^{(0)} u\|_0 = \|u\|_4$. Consequently $L_0^{(0)}$ has a bounded linear inverse.

The operator $L_0^{(2)}$ is the identity on H_2. It evidently has a bounded linear inverse, too. For these reasons the operators $L_0^{(\nu)} - \dfrac{9}{2}\pi^2 L_1^{(\nu)}$ $(\nu = 0, 2)$ are Fredholm operators. They are homeomorphisms on $\tilde{H}_\nu$. The implicit function theorem (see [13]) provides us with the wanted solutions $\tilde{w}_\nu(\varepsilon, \lambda_1, \lambda_2, \beta_1, \beta_2)$. Further we find for every solution $(\varepsilon, \lambda_1, \lambda_2, \beta_1, \beta_2, \tilde{w})$ of (3.5), $\|\tilde{w}\|_\nu < 1$,

$$\begin{aligned}\|\tilde{w}\|_\nu &\leqq \text{const} \left\| \left(L_0^{(\nu)} - \frac{9}{2}\pi^2 L_1^{(\nu)}\right) \tilde{w} \right\|_\nu \\ &= \text{const} \left\| \left[\left(\frac{9}{2}\pi^2 - \lambda_1\right) L_1^{(\nu)} - \lambda_2 L_2^{(\nu)} + \varepsilon L^{(\nu)}(\varepsilon, \lambda_1, \lambda_2)\right] \tilde{w} \right. \\ &\quad \left. + Q^{(\nu)} C^{(\nu)}(\beta_1 u_1 + \beta_2 u_2 + \tilde{w}) + O(|\varepsilon|\, (\beta_1^3 + \beta_2^3 + \|\tilde{w}\|_\nu^3)) \right\| \\ &\leqq \text{const} \left[|\beta_1|^3 + |\beta_2|^3 + \|\tilde{w}\|_\nu \left(\left|\frac{9}{2}\pi^2 - \lambda_1\right| + |\lambda_2| + |\varepsilon| + \beta_1^2 + \beta_2^2 + \|\tilde{w}\|_\nu \right) \right].\end{aligned}$$

From this inequality we derive (3.6) for small $\left|\dfrac{9}{2}\pi^2 - \lambda_1\right| + |\lambda_2| + |\varepsilon| + \beta_1^2 + \beta_2^2 + \|\tilde{w}\|_\nu$, q.e.d.

In the following we suppose $a_1 \neq 0$.

Lemma 3. *For all* $(\beta_1, \beta_2) \in \Re^2$; $\nu = 0, 2$ *the identities*

$$\big(C^{(\nu)}(\beta_1 u_1 + \beta_2 u_2), \beta_1 u_1 + \beta_2 u_2\big)_\nu = a^{(\nu)} \beta_1^4 + 2b^{(\nu)} \beta_1^2 \beta_2^2 + c^{(\nu)} \beta_2^4 \tag{3.7}$$

hold with

$$b^{(2)} > a^{(2)} > c^{(2)} > 0 \tag{3.8}$$

and

$$a_1 b^{(0)} > a_1 a^{(0)} = a_1 c^{(0)} > 0. \tag{3.9}$$

Proof. The identity (3.7) for $\nu = 2$ with (3.8) was proved by Knightly and Sather [5]. Chow, Hale and Mallet-Paret [2] calculated the numbers in (3.8) approximately:

$$a^{(2)} \approx 3{,}945001 \cdot 10^{-4},\ b^{(2)} \approx 5{,}007428 \cdot 10^{-4},\ c^{(2)} \approx 1{,}623543 \cdot 10^{-4}.$$

In the case $\nu = 0$ we have from (1.4)

$$H[u_1, u_1] = \left(\frac{2^{7/4}}{3\pi^2}\right)^2 \pi^4 \left\{\left(\frac{5}{4} + \alpha\right) \sin^2 \frac{y_1}{\sqrt{2}} \pi \sin^2 y_2\pi + \frac{\beta^2}{2} \cos^2 \frac{y_1}{\sqrt{2}} \pi \cos^2 y_2\pi\right\},$$

$$H[u_2, u_2] = \left(\frac{2^{7/4}}{6}\right)^2 \left\{(5 + 4\alpha) \sin^2 \frac{2y_1}{\sqrt{2}} \pi \sin^2 y_2\pi + 2\beta^2 \cos^2 \frac{2y_1}{\sqrt{2}} \pi \cos^2 y_2\pi\right\},$$

$$H[u_1, u_2] = \frac{2^{7/4}}{3} \frac{2^{7/4}}{6} \left\{\left(2 + \frac{3}{2}\alpha\right) \sin \frac{y_1}{\sqrt{2}} \pi \sin \frac{2y_1}{\sqrt{2}} \pi \sin^2 y_2\pi + \beta^2 \cos \frac{y_1}{\sqrt{2}} \pi \cos \frac{2y_1}{\sqrt{2}} \pi \cos^2 y_2\pi\right\}.$$

Taking into account (1.5) we find

$$a^{(0)} = a_1 \int_\Omega \{H[u_1, u_1]\}^2 \, dy = a_1 \frac{2^7}{3^4} \sqrt{2} \left\{\left(\frac{5}{4} + \alpha\right)^2 \left(\int_0^1 \sin^4 x\pi \, dx\right)^2 + \left(\frac{5}{4} + \alpha\right) \beta^2 \left(\int_0^1 \sin^2 x\pi \cos^2 x\pi \, dx\right)^2 + \frac{\beta^4}{4} \left(\int_0^1 \cos^4 x\pi \, dx\right)^2\right\}$$

$$= a_1 \frac{2^7}{3^4} \sqrt{2} \left\{\left(\frac{9}{4} - \frac{\beta^2}{2}\right)^2 \left(\frac{3}{8}\right)^2 + \left(\frac{9}{4} - \frac{\beta^2}{2}\right) \beta^2 \left(\frac{1}{8}\right)^2 + \frac{\beta^4}{4} \left(\frac{3}{8}\right)^2\right\}$$

$$= a_1\, 2^{15/2} 3^{-4} \left\{\frac{729}{1024} - \beta^2 \frac{9}{32} + \beta^4 \frac{1}{16}\right\} = a_1 2^{-5/2} 3^{-4} (729 - 288\beta^2 + 64\beta^4),$$

$$c^{(0)} = a_1 \int_\Omega \{H[u_2, u_2]\}^2 \, dy = a_1 \frac{2^7}{6^4} \frac{\sqrt{2}}{2} \int_0^2 \int_0^1 \{(5 + 4\alpha) \sin^2 y_1\pi \sin^2 y_2\pi + 2\beta^2 \cos^2 y_1\pi \cos^2 y_2\pi\} \, dy_1 \, dy_2 = a^{(0)}.$$

Further we have

$$\frac{b^{(0)}}{a_1} = 2 \int_\Omega \{H[u_1, u_2]\}^2 \, dy + \int_\Omega H[u_1, u_1] \, H[u_2, u_2] \, dy$$

$$= \frac{2^5}{3^4} \sqrt{2} \left\{2 \left[\left(2 + \frac{3}{2}\alpha\right)^2 \frac{3}{32} + \beta^4 \frac{3}{32}\right] + (5 + 4\alpha) \left(\frac{5}{4} + \alpha\right) \frac{3}{32} + \beta^2 \left(\frac{5}{2} + 2\alpha\right) \left(\frac{1}{16} + \frac{1}{16}\right) + \beta^4 \frac{3}{32}\right\} = 2^{15/2} 3^{-4} \left\{\frac{3}{16} \left(\frac{7}{4} - \frac{3}{8}\beta^2\right)^2 + \beta^4 \frac{3}{64} + \left(\frac{9}{4} - \beta^2/2\right)^2 \frac{3}{32} + \left(\frac{9}{4} - \beta^2/2\right) \beta^2 \frac{1}{16} + \beta^4 \frac{3}{32}\right\}$$

$$= 2^{-5/2} 3^{-4} (1074 - 243\beta^2 + 280\beta^4) > a^{(0)}/a_1,$$ q.e.d.

In the next step we substitute the solutions of (3.5) into (3.4). With the new scaling

$$\delta_1 := \frac{2}{9\pi^2}\lambda_1 - 1, \quad \delta_2 := \frac{1}{9\pi^2}\lambda_2, \tag{3.10}$$

$$\begin{aligned}\varkappa_1(\varepsilon, \delta_1, \delta_2) := \delta_1 + 4\delta_2 - \varepsilon\Bigg[(2+\varepsilon)\left(-\frac{5}{9} - \delta_1\right)\\ + (4 + 6\varepsilon + 4\varepsilon^2 + \varepsilon^3)\left(\frac{4}{9} - 4\delta_2\right)\Bigg],\\ \varkappa_2(\varepsilon, \delta_1, \delta_2) := \delta_1 + \delta_2 - \varepsilon\Bigg[(2+\varepsilon)\left(-\frac{5}{9} - \delta_1\right)\\ + (4 + 6\varepsilon + 4\varepsilon^2 + \varepsilon^3)\left(\frac{1}{9} - \delta_2\right)\Bigg]\end{aligned} \tag{3.11}$$

we get the Liapunov-Schmidt equations for $\nu = 0, 2$:

$$\left.\begin{aligned} G_1^{(\nu)}(\varepsilon, \delta_1, \delta_2, \beta_1, \beta_2) := -\varkappa_1(\varepsilon, \delta_1, \delta_2)\,\beta_1 + a^{(\nu)}\beta_1^3 + b^{(\nu)}\beta_1\beta_2^2\\ + O\big((|\beta_1|^3 + |\beta_2|^3)\,(|\varepsilon| + |\beta_1| + |\beta_2|)\big) = 0,\\ G_2^{(\nu)}(\varepsilon, \delta_1, \delta_2, \beta_1, \beta_2) := -\varkappa_2(\varepsilon, \delta_1, \delta_2)\,\beta_2 + c^{(\nu)}\beta_2^3 + b^{(\nu)}\beta_1^2\beta_2\\ + O\big((|\beta_1|^3 + |\beta_2|^3)\,(|\varepsilon| + |\beta_1| + |\beta_2|)\big) = 0.\end{aligned}\right\} \tag{3.12}$$

4. On the solutions of (1.1), (1.2) and (1.3)

Lemma 4. *The functions $G_i^{(\nu)}$ ($i = 1, 2$; $\nu = 0, 2$) are real analytic for small $(\varepsilon, \delta_1, \delta_2, \beta_1, \beta_2) \in \Re^5$. The functions $G_i^{(\nu)}$ are odd with respect to β_i and even with respect to β_j for $i \neq j$.*

Proof. The left hand side of equation (2.6) is odd with respect to w. If we substitute (2.7) into (2.4) we get an odd equation with respect to w, too. We have then $F^{(\nu)}(\varepsilon, \lambda_1, \lambda_2, w) = -F^{(\nu)}(\varepsilon, \lambda_1, \lambda_2, -w)$. From the uniqueness property of equation (3.5) we conclude $\tilde{w}(\varepsilon, \lambda_1, \lambda_2, \beta_1, \beta_2) = -w(\varepsilon, \lambda_1, \lambda_2, -\beta_1, -\beta_2)$. Let us consider the operators $T^{(0)}: H_4 \to H_0$,

$$(T^{(0)}u, v)_0 := \int_{\dot{\Omega}} u\big(\sqrt{2} - y_1, y_2\big)\, v(y_1, y_2)\, dy$$

and $T^{(2)}: H_2 \to H_2$,

$$(T^{(2)}u)\,(y_1, y_2) := u\big(\sqrt{2} - y_1, y_2\big).$$

Then we have for $\nu = 0, 2$

$$\begin{aligned} &T^{(\nu)}L_i^{(\nu)} = L_i^{(\nu)}T^{(\nu)*}, \quad T^{(\nu)}K_{ij}^{(\nu)} = K_{ij}^{(\nu)}T^{(\nu)*} \quad (i = 1, 2),\\ &T^{(2)}B(\cdot,\cdot) = B(T^{(2)*}\cdot, T^{(2)*}\cdot),\ T^{(0)}A_k(s) = A_k(s)\,T^{(0)*} \quad \text{for} \quad k = 1, \ldots, n.\end{aligned}$$

It follows

$$T^{(\nu)}F^{(\nu)}(\varepsilon, \lambda_1, \lambda_2, w) = F^{(\nu)}(\varepsilon, \lambda_1, \lambda_2, T^{(\nu)*}w).$$

Now we have

$$T^{(\nu)*}\tilde{w}(\varepsilon, \lambda_1, \lambda_2, \beta_1, \beta_2) = \tilde{w}(\varepsilon, \lambda_1, \lambda_2, \beta_1, -\beta_2).$$

Our assertions follow then from the definitions

$$G_i^{(\nu)}(\varepsilon, \delta_1, \delta_2, \beta_1, \beta_2) = \left(u_i, F^{(\nu)}\left(\varepsilon, \frac{2}{9\pi^2}\lambda_1 - 1, \frac{\lambda_2}{9\pi^2}, \beta_1 u_1 + \beta_2 u_2 + \tilde{w}\left(\varepsilon, \frac{2}{9\pi^2}\lambda_1 - 1, \frac{\lambda_2}{9\pi^2}, \beta_1, \beta_2\right)\right)\right)_\nu, \quad \text{q.e.d.}$$

Next we make use of the Weierstraß preparation theorem, see [13]. With the help of lemma 4 we get instead of (3.12)

$$G_i^{(\nu)}(\varepsilon, \delta_1, \delta_2, \beta_1, \beta_2) = \beta_i[\beta_i^2 - f_i^{(\nu)}(\varepsilon, \delta_1, \delta_2, \beta_j^2)]\, g_i^{(\nu)}(\varepsilon, \delta_1, \delta_2, \beta_1, \beta_2) = 0$$
$$\text{for} \quad i \neq j; \;\; i, j = 1, 2, \quad \nu = 0, 2. \tag{4.1}$$

In (4.1) the functions $f_i^{(\nu)}$, $g_j^{(\nu)}$ are real analytic in some neighbourhood of zero,

$$g_1^{(\nu)}(0, 0, 0, 0, 0) = a^{(\nu)}, \quad g_2^{(\nu)}(0, 0, 0, 0, 0) = c^{(\nu)},$$
$$f_i^{(\nu)}(\varepsilon, \delta_1, \delta_2, \beta_j^2) = [\varkappa_i(\varepsilon, \delta_1, \delta_2) - b^{(\nu)}\beta_j^2 + O(\beta_j^3)]\,(g_i^{(\nu)})^{-1}|_{\beta_i=0} \tag{4.2}$$

for $i \neq j$; $i, j = 1, 2$, $\nu = 0, 2$.

The equations (4.1) have exactly the following solutions with $\beta_1 = 0$:

(i) $\beta_1 = \beta_2 = 0$,

(ii) $\beta_1 = 0, \quad \beta_2 = \pm[f_2^{(\nu)}(\varepsilon, \delta_1, \delta_2, 0)^{1/2}]$.

We have yet to ensure solution (ii) to be real. From (4.2), (3.11), (3.3) we easily prove that this is exactly the case for $\nu = 2$ or $\nu = 0$ and $a_1 > 0$ if $\lambda_1 \geqq \lambda_1^{(2)}(s, \lambda_2)$ holds. For $\nu = 0$ and $a_1 < 0$ we have $f^{(0)}(\varepsilon, \delta_1, \delta_2, 0) \geqq 0$ just if the inequality $\lambda_1 \leqq \lambda_1^{(2)}(s, \lambda_2)$ is true.

Except (i) equation (4.1) has exactly the following solutions with $\beta_2 = 0$:

(iii) $\beta_1 = \pm[f_1^{(\nu)}(\varepsilon, \delta_1, \delta_2, 0)^{1/2}], \quad \beta_2 = 0$.

These solutions are real just if $\lambda_1 \geqq \lambda_1^{(1)}(s, \lambda_2)$ holds, ($\nu = 2$ and $\nu = 0$, $a_1 \geqq 0$), and for $\lambda_1 \leqq \lambda_1^{(1)}(s, \lambda_2)$ in the case $\nu = 0$, $a_1 < 0$. Besides that, we have to look for real solutions of the systems

$$\beta_i^2 - f_i^{(\nu)}(\varepsilon, \delta_1, \delta_2, \beta_j^2) = 0 \quad (i \neq j; \;\; i, j = 1, 2, \quad \nu = 0, 2). \tag{4.3}$$

We substitute and once more use the Weierstraß preparation theorem. Then (4.3) turns to

$$\left(\beta_i^2 - \tilde{f}_i^{(\nu)}(\varepsilon, \delta_1, \delta_2)\right) \tilde{g}_i^{(\nu)}(\varepsilon, \delta_1, \delta_2, \beta_i) = 0 \quad (i = 1, 2).$$

Again the functions $\tilde{f}_i^{(\nu)}$, $\tilde{g}_i^{(\nu)}$ are real analytic in some neighbourhood of zero. We have

$$\tilde{g}_i^{(\nu)}(0, 0, 0, 0) = 1 - \frac{(b^{(\nu)})^2}{a^{(\nu)}c^{(\nu)}} < 0$$

and

$$\tilde{f}_1^{(\nu)}(\varepsilon, \delta_1, \delta_2) = \frac{\delta_1(b^{(\nu)} - c^{(\nu)}) + \delta_2(b^{(\nu)} - 4c^{(\nu)}) + \frac{2}{3}\,\varepsilon(b^{(\nu)} + c^{(\nu)})}{(b^{(\nu)})^2 - a^{(\nu)}c^{(\nu)}} + O(\varepsilon^2 + \delta_1^2 + \delta_2^2),$$

$$\tilde{f}_2^{(\nu)}(\varepsilon, \delta_1, \delta_2) = \frac{\delta_1(b^{(\nu)} - a^{(\nu)}) + \delta_2(4b^{(\nu)} - a^{(\nu)}) - \frac{2}{3}\,\varepsilon(b^{(\nu)} + a^{(\nu)})}{(b^{(\nu)})^2 - a^{(\nu)}c^{(\nu)}} + O(\varepsilon^2 + \beta_1^2 + \beta_2^2).$$

The inequality $\tilde{f}_1^{(\nu)}(\varepsilon, \delta_1, \delta_2) \geqq 0$ requires

$$\delta_1 \geqq \tilde{\tilde{f}}_1^{(\nu)}(\varepsilon, \delta_2) := -\delta_2\,\frac{b^{(\nu)} - 4c^{(\nu)}}{b^{(\nu)} - c^{(\nu)}} - \frac{2}{3}\,\varepsilon\,\frac{b^{(\nu)} + c^{(\nu)}}{b^{(\nu)} - c^{(\nu)}} + O(\varepsilon^2 + \delta_2^2) \tag{4.4}$$

for $\nu = 2$ or $\nu = 0$, $a_1 > 0$ and

$$\delta_1 \leqq \tilde{\tilde{f}}_1^{(0)}(\varepsilon, \delta_2) \quad \text{in the case} \quad a_1 < 0. \tag{4.5}$$

Finally, $\tilde{f}_2^{(\nu)}(\varepsilon, \delta_1, \delta_2) \geqq 0$ is equivalent to

$$\delta_1 \geqq \tilde{\tilde{f}}_2^{(\nu)}(\varepsilon, \delta_2) := -\delta_2\,\frac{4b^{(\nu)} - a^{(\nu)}}{b^{(\nu)} - a^{(\nu)}} + \frac{2}{3}\,\varepsilon\,\frac{b^{(\nu)} + a^{(\nu)}}{b^{(\nu)} - a^{(\nu)}} + O(\varepsilon^2 + \delta_2^2) \tag{4.6}$$

for $\nu = 2$ or $\nu = 0$, $a_1 > 0$ and

$$\delta_1 \leqq f_2^{(0)}(\varepsilon, \delta_2) \quad \text{in the case} \quad a_1 < 0. \tag{4.7}$$

Remembering (2.8), (3.10), we introduce the notations

$$\varphi_i^{(\nu)}(s, \lambda_2) := \frac{9}{2}\,\pi^2 \left(\tilde{\tilde{f}}_i^{(\nu)}\left(\frac{s}{\sqrt{2}} - 1, \frac{1}{9\pi^2}\,\lambda_2\right) + 1\right)$$

for $i = 1, 2$, $\nu = 0, 2$. Taking into account (3.8) and (3.9) we prove that the inequalities

$$-\delta_2\,\frac{b^{(\nu)} - 4c^{(\nu)}}{b^{(\nu)} - c^{(\nu)}} - \frac{2}{3}\,\varepsilon\,\frac{b^{(\nu)} + c^{(\nu)}}{b^{(\nu)} - c^{(\nu)}} < -\delta_2 - \frac{2}{3}\,\varepsilon,$$

$$-4\delta_2 + \frac{2}{3}\,\varepsilon < -\delta_2\,\frac{4b^{(\nu)} - a^{(\nu)}}{b^{(\nu)} - a^{(\nu)}} + \frac{2}{3}\,\varepsilon\,\frac{b^{(\nu)} - a^{(\nu)}}{b^{(\nu)} + a^{(\nu)}}$$

hold just for $\delta_2 < \frac{4}{9}\,\varepsilon$. It follows (see (1.8), (1.9) and fig. 9—11)

$$\varphi_1^{(\nu)}(s, \lambda_2) < \lambda_1^{(2)}(s, \lambda_2) < \lambda_1^{(1)}(s, \lambda_2) < \varphi_2^{(\nu)}(s, \lambda_2)$$

exactly for $\lambda_2 < \lambda_2^{(0)}(s)$.

The systems (4.3) have exactly the following four solutions:

(iv) $\qquad \beta_1 = \pm[\tilde{f}_1^{(\nu)}(\varepsilon, \delta_1, \delta_2)^{1/2}], \quad \beta_2 = \pm[f_2^{(\nu)}(\varepsilon, \delta_1, \delta_2)^{1/2}].$

These solutions are real in the case $\lambda_2 \geqq \lambda_2^{(0)}(s)$ exactly under (4.4) or (4.7) and in the case $\lambda_2 \leqq \lambda_2^{(0)}(s)$ under (3.6) or (4.5).

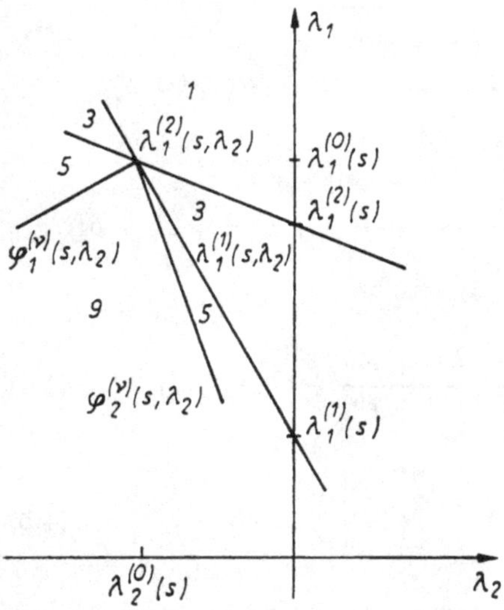

Fig. 9. $s < \sqrt{2}$

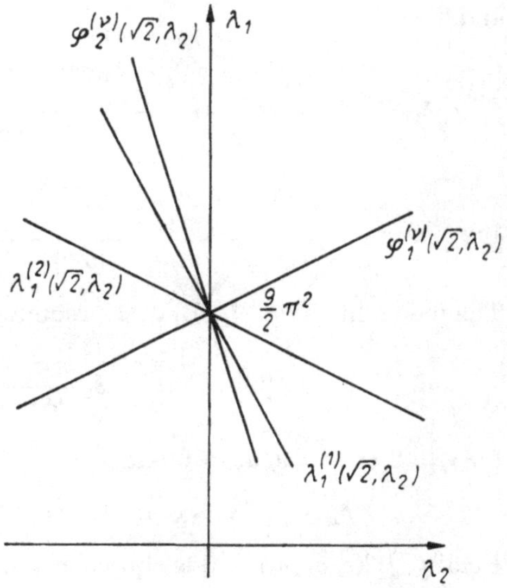

Fig. 10. $s = \sqrt{2}$

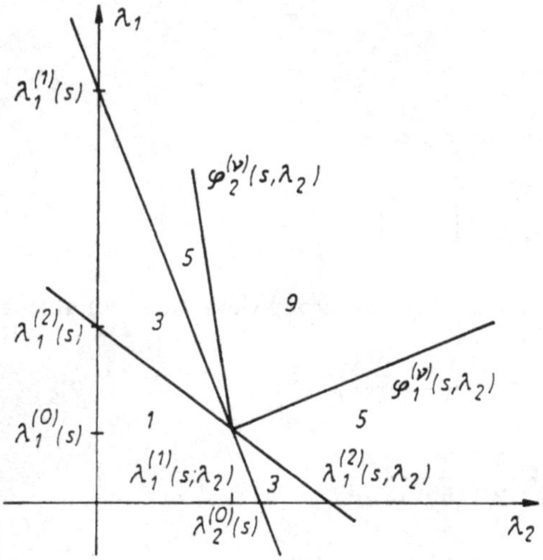

Fig. 11. $s > \sqrt{2}$

Fig. 9 shows us the multiplicities of the solutions of (1.3) in the case $a_1 < 0$, in fig. 11 we see the multiplicities of the solutions to (1.1), (1.2) or (1.3) in the case $a_1 > 0$. Consider the case $a_1 > 0$ with the thrust λ_1 growing. If in this case λ_1 passes through the values $\varphi_2^{(0)}(s, \lambda_2)$ (for $\lambda_2 \leqq \lambda_2^{(0)}(s)$) or $\varphi_1^{(0)}(s, \lambda_2)$ (for $\lambda_2 \geqq \lambda_2^{(0)}(s)$) a branch of secondary solutions

$$\pm \tilde{f}_1^{(0)}(\varepsilon, \delta_1, \delta_2)^{1/2} u_1 \pm \tilde{f}_2^{(0)}(\varepsilon, \delta_1, \delta_2)^{1/2} u_2 + \tilde{w}_0\left(\varepsilon, \delta_1, \delta_2, \pm \tilde{f}_1^{(0)}(\varepsilon, \delta_1, \delta_2)^{1/2}, \pm \tilde{f}_2^{(0)}(\varepsilon, \delta_1, \delta_2)^{1/2}\right) \quad (4.8)$$

of (1.3) corresponding to (iv) starts from the primary solutions of (1.3) corresponding to (iii) or (ii):

$$\pm w_1(\varepsilon, \delta_1, \delta_2) := \pm f_1^{(0)}(\varepsilon, \delta_1, \delta_2, 0)^{1/2} u_1 + \tilde{w}_0\big(\varepsilon, \delta_1, \delta_2, \pm f_1^{(0)}(\varepsilon, \delta_1, \delta_2, 0)^{1/2}, 0\big),$$

$$\pm w_2(\varepsilon, \delta_1, \delta_2) := \pm f_2^{(0)}(\varepsilon, \delta_1, \delta_2, 0)^{1/2} u_2 + \tilde{w}_0\big(\varepsilon, \delta_1, \delta_2, 0, \pm f_2^{(0)}(\varepsilon, \delta_1, \delta_2, 0)^{1/2}\big).$$

A similar behaviour show the solutions of (1.1), (1.2) (see fig. 6). Now consider the case $a_1 < 0$ with the thrust λ_1 falling. If λ_1 passes through the values $\varphi_1^{(0)}(s, \lambda_2)$ (for $\lambda_2 \leqq \lambda_2^{(0)}(s)$), again branches of secondary solutions corresponding to (iv) start from the primary solutions corresponding to (ii) or (iii) (see fig. 7).

Finally we recall the proof of lemma 4. We find

$$T^{(0)} w_1(\varepsilon, \delta_1, \delta_2) = w_1(\varepsilon, \delta_1, \delta_2), \quad T^{(0)} w_2(\varepsilon, \delta_1, \delta_2) = -w_2(\varepsilon, \delta_1, \delta_2).$$

The primary buckling forms w_1 and w_2 are symmetric and antisymmetric respectively with respect to the axis of symmetry $x_1 = \frac{s}{2}$. The secondary buckling forms are neither symmetric nor antisymmetric.

Bibliography

[1] Bauer, L., Keller, H. B., Reiss, E. L., Multiple eigenvalues lead to secondary bifurcation, SIAM Rev. **17** (1975), 101–122.

[2] Chow, S. N., Hale, J. U., Mallet-Paret, J., Applications of generic bifurcation II, Arch. Rat. Mech. Anal. **62** (1976), 209–235.

[3] Kluge, R., Zur Lösung eines Bifurkationsproblems für die Kármánschen Gleichungen im Fall der rechteckigen Platte, Math. Nachr. **44** (1970), 29–54.

[4] Knightly, G. H., Sather, D., On uniqueness of solutions of the von Kármán equations, Arch. Rat. Mech. Anal. **36** (1970), 65–78.

[5] —, Nonlinear buckled states of rectangular plates, Arch. Rat. Mech. Anal. **54** (1974), 356–372.

[6] Langenbach, A. Monotone Potentialoperatoren, Berlin 1976.

[7] —, Über eine Beulgleichung für mäßig nichtlineare Platten, Math. Nachr. **95** (1980), 305–321.

[8] List, S. E., Generic bifurcation with applications to the von Kármán equations, J. Diff. Equations **30** (1978), 89–118.

[9] Matkowsky, B. J., Putnik, L. J., Multiple buckled states of rectangular plates, Int. J. Nonlin. Mech. **9** (1974), 89–103.

[10] Mallet-Paret, J., Buckling of cylindrical shells with small curvature, Quart. Appl. Math. **35** (1977), 383–400.

[11] Recke, L., Anwendungen der Verzweigungstheorie auf geometrisch nichtlineare Schalengleichungen, Dissertation A, Berlin 1978.

[12] Vanderbauwhede, A., Generic and nongeneric bifurcation for the von Kármán equations, J. Math. Anal. Appl. **66** (1978), 550–573.

[13] Weinberg, M. M., Trenogin, W. A., Theorie der Lösungsverzweigung bei nichtlinearen Gleichungen, Berlin 1973.

Address of the authors:

Humboldt-Universität zu Berlin
Sektion Mathematik
1080 Berlin
Unter den Linden 6
DDR

On a program for the computation of matter transport in river networks

Wolfdietrich Müller, Berlin

Abstract. For applications in water quality studies, a computer program has been developed that computes stationary and time-varying distributions of soluted or suspensed matter (e.g. pollutants, planktonic microorganisms) in large river networks. In this paper a short description of the underlying mathematical model is given and some features of the program are outlined.

1. Introduction

Permanently growing industrial, agricultural and communal water demand and decrease of water quality call for decisions on increasing investments, e.g. on sizing, allocation and control of waste water treatment plants or river aeration equipments. Such problems cannot be solved without extensive case studies based on general computer models which are able to simulate the distribution of pollutants and ecological processes in the water.

The program GRAPH[1] corresponds to this demand. It allows the computation of stationary or time-varying distributions of different, possibly nonlinearly interacting water quality components such as, for instance,

- soluted minerals or organic substances,
- suspensed matter,
- algae, bacteria, fungi, zooplankton,
- temperature,
- toxic substances
- radioactivity

in large multiply connected river networks. In this paper a short description of the underlying mathematical model is given (Section 2), some features of the program are outlined (Sections 3—6), and a computational example is sketched (Section 7). For a more detailed description of aspects of modelling we refer to [1, 2].

---

[1] GRAPH has been developed in the framework of an inter-institutional research agreement by H. Baumert and P. Braun from the Institute of Water Management of the G.D.R. and G. Stoyan and the author from the Central Institute of Mathematics and Mechanics of the Academy of Sciences of the G.D.R.

2. The mathematical model

The river system under consideration (which may consist of different reaches or segments of rivers, channels, or lakes) is represented by an oriented graph $\mathcal{G}$ (see fig. 1): the river reaches correspond to the *lines* of the graph (which should be oriented in the predominant direction of water streaming), the junctions to the *nodes*. We have to distinguish between *internal nodes* and *boundary nodes*.

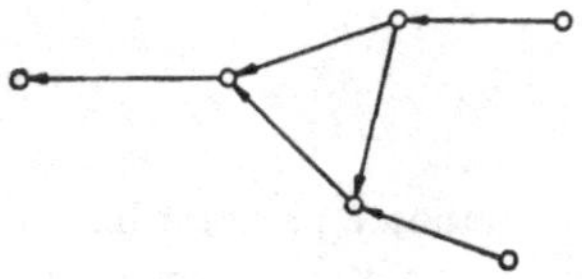

Fig. 1

Let N be the total number of lines of $\mathcal{G}$ and n be the current line number, $n = 1, \ldots, N$ By

$$c_k = c_k(n, x, t), \quad k = 1, \ldots, K,$$

we denote the concentration of the k-th water quality component, K is the total number of components under consideration. The value of c_k is assumed to depend on the line number n, on a one-dimensional spatial coordinate x (for the deduction of the one-dimensional model from the three-dimensional river system and for further details of modelling see Baumert [1]) and on time t. The functions c_k ($k = 1, \ldots, K$) are assumed to satisfy the system of partial differential equations [1]

$$\omega a \frac{\partial c_k}{\partial t} - \frac{\partial}{\partial x}\left(ad \frac{\partial c_k}{\partial x}\right) - av \frac{\partial c_k}{\partial x} + qc_k = e_k + f_k$$
$$(0 \leqq t \leqq T, \quad x \in \mathcal{G}, \quad k = 1, \ldots, K), \tag{1}$$

where

$a = a(n, x, t)$ — cross-sectional area of reach n (varying with x and t),

$v = v(n, x, t)$ — convection coefficient (average over the cross-section of the velocity of the water stream),

$d = d(n, v)$ — dispersion coefficient (modelling the longitudinal mixing that results from a non-uniform velocity distribution over the cross-section),

$q = q(n, x, t)$ — rate of lateral water inflow (modelling a dilution by addition of freshwater from tributaires or from groundwater),

$e_k = \sum_l \delta(x - x_l)\, p_{kl}(t)$ — concentrated source term (modelling a matter import by point sources of intensity $p_{kl}(t)$ at points $x = x_l$),

$f_k = f_k(n, x, t, c_1, \ldots, c_K)$ — distributed sources and nonlinear interactions (modelling matter import by spatially distributed sources, and internal ecological interactions in the water),

$\omega = 1$ for non-stationary problems,

$\omega = 0$ for stationary problems (in this case all coefficients are supposed to be independent of t).

In addition to the equation (1), boundary conditions are specified in all boundary nodes of the graph $\mathcal{G}$ (as a rule, boundary values of concentrations

$$c_k(x_b, t) = c_k^b(t) \tag{2}$$

are prescribed in all inflow nodes, and diffusion streams are supposed to be equal to zero,

$$ad \left. \frac{\partial c_k}{\partial x} \right|_{x=x_b} = 0, \tag{3}$$

in all outflow nodes), and coupling conditions

$$\sum_{+} \left(ad \frac{\partial c_k}{\partial x} + avc_k \right)_{x=x_b} = \sum_{-} \left(ad \frac{\partial c_k}{\partial x} + avc_k \right)_{x=x_b} \tag{4}$$

are assumed to be satisfied in all internal nodes (in formula (4), $\sum_{+}$ denotes summation with respect to all lines which end in the node under consideration, and $\sum_{-}$ denotes summation with respect to all lines which start in this point.) Initial conditions

$$c_k(x, 0) = c_k^0(x), \quad x \in \mathcal{G}, \tag{5}$$

are assumed to be prescribed in the case $\omega = 1$ (non-stationary problem).

3. Structure of the program GRAPH

The program GRAPH was developed to solve the boundary value problem (1)—(5) on a given graph $\mathcal{G}$. Its core is a finite difference scheme of the two-level, three-point type for the integration of equation (1), which was developed by G. Stoyan (see Section 4 for some details).

Assume the problem under investigation to be non-stationary (i.e. $\omega = 1$). In the case where the source terms f_k in (1) do not depend on the concentrations $c_1, \ldots, c_K$ (the equations of the system (1) are then linear and decoupled), GRAPH proceeds in the following steps:

Step 1. On the time level $T_0 = 0$ a time step length τ is chosen.

Step 2. A spatial discretization of the graph $\mathcal{G}$ is fixed in such a way that the set of discretization points contains all nodes of $\mathcal{G}$ and is equidistant along each line (but, as a rule, different along different lines), i.e., a set of spatial step lengths $h(n)$ ($n = 1, \ldots, N$) is chosen. (Denote by NS the total number of spatial discretization points.)

Step 3. The coefficients of the difference scheme are calculated (for a fixed value of k). They form the coefficient matrix and right hand side of a large linear system (of dimension NS) of equations for the calculation of the concentration values c_k in all discretization points.

Step 4. Utilizing an algorithm given by Frjazinov [4], from this large linear system (whose coefficient matrix is sparse and predominantly tridiagonal) a compact linear system of significantly lower dimension is derived, the unknowns of this system being the concentration values c_k in all internal nodes of $\mathcal{G}$ and in all those boundary nodes where no boundary conditions of first kind had been prescribed.

Step 5. The low-dimensional system obtained in step 4 is solved using the procedures DECOMP (Gauss decomposition) and LINSOL (backsubstitution) by Forsythe and Moler [3].

After this, the concentration values c_k in all nodes of the graph $\mathcal{G}$ are known, i.e., the global problem (1)—(5) is totally decomposed into a set of N two-point boundary value problems (1)—(2) on the lines of $\mathcal{G}$. In

Step 6 the corresponding linear systems (with purely tridiagonal coefficient matrices) are solved by the usual shortened Gauss elimination (see, e.g., [6], p. 40).

In this way all (decoupled) equations ($k = 1, \ldots, K$) of system (1) are successively treated on the same time level T_0. (Note that only the coefficients of the right hand sides have to be re-calculated since the values of a, v, d, and q do not change with k.)

In the case where the source terms f_k do depend on the concentrations $c_1, \ldots, c_K$ (such that the equations of system (1) are possibly nonlinearly coupled), steps 3 to 6 of the above described procedure are repeated until the differences between concentration values in successive iterations are sufficiently small or a prescribed maximal iteration number is reached. (Again, only the coefficients of the right hand sides have to be repeatedly calculated in the finite difference scheme.)

Then T_0 is replaced by $T_0 + \tau$, and steps 3 to 6 are repeated on the following time level. (In cases where the values of a, v, d, and q do not depend on t, new values have again to be computed only for the coefficients of the right hand sides.)

4. The finite difference scheme. Reduction of numerical diffusion

The finite difference scheme included in GRAPH was developed by G. Stoyan. It is closely related to the schemes described in [7] and [8]. We refer to [5] for a complete description, but we mention that this scheme

- is maximum norm stable for any combination of time step length τ and spatial step lengths $h(n)$,
- preserves positiveness,
- represents a conservative approximation of the differential equation,
- accepts first and second kind boundary conditions, and
- works for any Peclét number, i.e for any ratio of dispersion d to convection v, including degenerate cases.

Particularly, if d goes to zero, the scheme approaches a maximum norm stable scheme of gas dynamics, and even a simultaneous vanishing of d and v is admitted by the program.

In order to secure all these desirable properties, a certain amount of numerical diffusion had to be accepted in the scheme (cf. [2]). But at the same time a procedure was included into the program GRAPH that made it possible to suppress numerical diffusion if necessary.

Assume v and d to be piecewise constant, $v(n, x, t) \equiv v_n$, $d(n, v) \equiv d_n$. Let

$$q_n = \frac{v_n h(n)}{2d_n}$$

denote the cell Reynolds number of line n and let be

$$\alpha_n = \frac{\sqrt{1 + q_n^2} - 1}{2q_n}.$$

Then it turns out that the finite difference scheme is free of numerical diffusion on each line n of the graph $\mathcal{G}$ where the equality

$$\frac{\tau |v_n|}{h(n)} = 2\,|\alpha_n| \tag{6}$$

is satisfied (the term on the left hand side of this equality being the Courant number of the finite difference scheme). This fact was made use of in the following way:

1. In situations where suppression of numerical diffusion is desirable (i.e. where the propagation of steep concentration fronts or of sharply contoured tracer impulses through the network is to be simulated by the program) all lines of the graph where the front passes are marked in advance, and in step 2 of the algorithm (as described in Section 3) for all marked lines n the number $h(n)$ is chosen (automatically) in such a way that equation (6) is approximately satisfied. (In general it is impossible to satisfy (6) exactly for all n since the length of line n has to be an integer multiple of $h(n)$.)

2. In the particular case of pursuit of a tracer impulse, an additional simply connected path ("trace") through the network can be marked. The maximum of the impulse wave will then be tracked by the program along this trace, the number n_0 of the line where the maximum has been detected will be stored, and between steps 2 and 3 of the algorithm the time step length τ will be fitted exactly to the spatial discretization on line n_0.

In the case of constant coefficients this "time step fitting" has to be repeated whenever the impulse passes a node of the graph, in the case of time-varying coefficients it has to be performed in each time step since v_n, d_n, α_n change their values from step to step.

For a test example treated in this way see Section 7.

5. Working regimes of GRAPH

The run of program as described in Section 3 refers to non-stationary problems, i.e. to situations where $\omega = 1$ in (1) and a system of parabolic equations has to be treated.

But the finite difference scheme remains valid also for $\omega = 0$, and so stationary problems, i.e. systems of elliptic equations (1), are handled by GRAPH, too (in one "time step" and with some obvious simplifications in the computation of the coefficients; in nonlinear problems the "initial conditions" (5) are interpreted as initial approximation to the solution).

Besides, some additional working regimes are realized in GRAPH which combine the solution of stationary and non-stationary problems.

On the one hand, since in the treatment of real non-stationary multi-component water quality studies it may be difficult or even impossible to find initial values of concentrations over the whole network, the possibility is included to generate an initial distribution by solving a stationary problem (by choice nonlinear or linearized) in a first step and to start from this distribution the full transient problem in a second step.

On the other hand, a technique is included to overcome convergence problems in the treatment of strongly nonlinear stationary problems by imbedding them into non-stationary problems where the nonlinearities grow up during time from zero to their original magnitude. This non-stationary problem is then solved in a first step, and its final state is used as the initial approximation of the stationary problem in the second step.

Another interesting feature of the program GRAPH should shortly be mentioned: All data characterizing the whole river network (i.e. the structure of the graph represented by its incidence matrix, the lengths and cross-sections of the different reaches, etc.) are part of the input file. But afterwards calculations can be restricted to any (connected) subgraph by simply marking the lines belonging to this subgraph. Water balance and mass balances will not be destroyed by this procedure. (Of course, all lines where essentially non-zero concentrations are to be expected in a special situation should be included into the actual subgraph.)

6. Computer requirements

The program GRAPH has been implemented on the computers BESM-6 and EC-1040. From a series of test runs the following estimate of CPU time (in seconds) was obtained:

$$T_{\mathrm{CPU}} = \beta \cdot NS \cdot NP \cdot K \cdot ITER,$$

where

NS — number of spatial discretization points,
NP — number of time steps,
K — number of components,
$ITER$ — number of iterations per time step,

$$\beta = \begin{cases} 0.0012 \cdots 0.0035 & \text{for BESM-6,} \\ 0.010 \cdots 0.028 & \text{for EC-1040.} \end{cases}$$

For both computers, the largest values of β correspond to the situation where the coefficients a, v, d, q depend on time, and $K = 1$, $ITER = 1$ (i.e. where all coefficients of the finite difference scheme have to be re-calculated in each step, cf. Section 3). Storage requirements (in machine words) can be roughly estimated by

$$R_{\mathrm{stor}} = NS \begin{cases} (2K + 4) & \text{if } a, d, v, q \text{ depend on } t,\ ITER = 1, \\ (2K + 8) & \text{otherwise.} \end{cases}$$

7. A numerical example

A tracer impulse simulation in a one-component system was performed along a graph (with 38 lines, 25 internal nodes, 8 inflow nodes, 5 outflow nodes); the coefficients a, v were assumed to be constant along each line, $q \equiv 0$, $d \equiv 1$ over the whole graph (see [5] for details). A tracer wave resulting from two unity impulses in the first two time steps at one of the inflow nodes was tracked along a "trace" through the graph, and its values in all nodes along the trace were listed up. Calculations were performed over a time interval of 500000 s (i.e. of about six days) starting with an initial time step length of 2175 s and making use of the time step fitting algorithm as described at the end of Section 4.

For the nodes along the trace the time instants

$t^{(i)}_{\max}$ — time where in the node i the amplitude of the tracer wave has its maximal value $h^{(i)}_{\max}$,

$t^{(i)}_{-1/2}$ — time where the wave reaches 50% of the value $h^{(i)}_{\max}$,

$t^{(i)}_{+1/2}$ — time where the wave goes down again under 50% of the value $h^{(i)}_{\max}$

were determined and the values

$$\tau^{(i)} = t^{(i)}_{+1/2} - t^{(i)}_{-1/2}$$

were calculated. If along the line between the nodes i and j the convection coefficient v has a constant value v_{ij}, then the actual dispersion value d of the tracer wave can be estimated from the computed data by means of the formula ([1])

$$d = \frac{v_{ij}^2}{16 \ln 2} \frac{(\tau^{(j)})^2 - (\tau^{(i)})^2}{t^{(j)}_{\max} - t^{(i)}_{\max}}. \tag{7}$$

In table 1, for the nodes along the trace the node number i, the numbers $t^{(i)}_{\max}$ and $\tau^{(i)}$ (in 1000 s) and the d-values as computed by (7) are listed up.

Table 1

i	$t^{(i)}_{\max}$	$\tau^{(i)}$	d
1	7	3,5	0,984
2	26	10,5	1,002
3	148	27,2	1,015
4	220	42,3	0,998
5	371	75,7	0,916
6	411	82,8	

8. Concluding remarks

A glance at the values in the last column of table 1 (the "theoretical" dispersion value being $d = 1$) shows that the algorithm for suppression of numerical diffusion has worked extremely successfully in this particular example. As a rule, results will be less satisfactory in problems with non-constant values of cross-sections, convection and dispersion. Nevertheless the program GRAPH seems to be a reliable tool for handling a broad range of water quality problems, e.g. short-term and long-term forecast of pollutant concentration distributions, simulation of accident situations, and ecological studies.

References

[1] H. Baumert, Grundlagen der Stofftransport- und Vermischungsprozesse in Fließgewässern. Research Report, Institute of Water Management, Berlin 1977.

[2] H. Baumert, P. Braun, E. Glos. W.-D. Müller, G. Stoyan, Modelling and computation of water quality problems in river networks. *In*: Optimization Techniques. Proceedings of the 9th IFIP Conference on Optimization Techniques, Warsaw 1979. Lecture Notes in Control and Information Sciences Vol. 23, Springer-Verlag, Berlin 1980, 482–491.

[3] G. Forsythe, C. Moler, Computer solutions of linear algebraic systems. Prentice-Hall, New York 1967.

[4] И. В. Фрязинов, Алгоритм решения разностных задач на графах. Ж. вычисл. матем. и матем. физ. **10** (1970), 474–477.

[5] W.-D. Müller, G. Stoyan, Dokumentation zum Algol-Programm GRAPH-1 (Simulation von Stofftransportprozessen in Gewässersystemen). Research Report, Central Institute of Mathematics and Mechanics of the Academy of Sciences of the GDR, Berlin 1979.

[6] А. А. Самарский, Теория разностных схем. Наука, Москва 1977.

[7] G. Stoyan, On a maximum norm stable, monotone and conservative difference approximation of the one-dimensional diffusion-convection equation. *In*: Proceedings of the Conference „Simulation gekoppelter Transport-, Austausch- und Umwandlungsprozesse im Boden- und Grundwasser", Dresden 1979, 139–160.

[8] G. Stoyan, Towards a general-purpose difference scheme for the one-dimensional linear parabolic equation. This volume, 297–314.

Address of the author:

Akademie der Wissenschaften der DDR
Institut für Mathematik
1080 Berlin
Mohrenstr. 39
DDR

Existence and regularity theorems for weak solutions to the equations of motion of visco-plastic media

Joachim Naumann, Berlin

Summary. The first part of this paper is concerned with the constitutive law of a class of visco-plastic media and the formulation of the differential equations and boundary conditions which govern the viscous flow of these media. In the second part we present some recently obtained existence and regularity results for solutions to the resulting variational inequalities.

A characteristic feature of the flow of a visco-plastic medium is the simultaneous appearance of two regions where the medium is in different states: the region of elastic state and the region of viscous state (here the medium behaves as a purely viscous fluid whose constitutive law is assumed to be known). These two regions are separated by the a-priori unknown yield surface. The determination of velocity and pressure of the motion of a homogeneous, incompressible visco-plastic medium within its region of viscous flow thus leads to solving a system of partial differential equations in a domain with partly free boundary. In [4] this problem is reduced to a variational inequality to which the existence of a solution is then proven (Bingham media).

The first part of this lecture (Section1) is devoted to a brief discussion of the constitutive law for a class of visco-plastic media and to the formulation of the differential equations and boundary conditions which govern the motion of these media in the state of viscous flow. In the second part (Sections 2 and 3) I present some recently obtained results on the existence and regularity of solutions to the resulting variational inequalities.

1. Constitutive law and differential equations

1.1. The constitutive law

Let $u = \{u_1, u_2, u_3\}$ (resp. $u = \{u_1, u_2\}$) denote the vector of velocity of a fluid moving in a bounded region of $\mathbb{R}^3$ (resp. $\mathbb{R}^2$), let $D = \{D_{ij}\}$ $(D_{ij} = D_{ij}(u) = \frac{1}{2}(u_{i,j} + u_{j,i})$[1], $i, j = 1, 2, 3$ (resp. $i, j = 1, 2$)) denote the rate of strain tensor and let D_{II}

[1] Throughout we use the notation $v_{,i} = \frac{\partial v}{\partial x_i}$, $v_{,ij} = \frac{\partial^2 v}{\partial x_i \partial x_j}$ (differentiation with respect to a fixed Cartesian frame, $i, j = 1, 2, 3$ (resp. $i, j = 1, 2$)). Further, a repeated subscript means summation over 1, 2, 3 (resp. 1, 2).

$= \frac{1}{2} D_{ij}D_{ij}$. Further, by $S = \{S_{ij}\}$ and $S_{II} = \frac{1}{2} S_{ij}S_{ij}$ we denote the tensor of deviatoric stress and its second invariant, respectively.

Let $a(\cdot)$ be a real continuous function on $(0, \infty)$, $a(\xi) > 0$ for all $\xi \in (0, \infty)$ and $\lim_{\xi \to 0} \times a(\xi^2)\, \xi = 0$, and let $\beta = \text{const} > 0$ (fixed). A *visco-plastic medium (visco-plastic fluid) with yield stress* β is then characterized by the following constitutive law:

$$D_{ij} = 0 \Leftrightarrow 0 \leqq (S_{II})^{1/2} \leqq \beta, \tag{1.1}$$

$$S_{ij} = [a(D_{II}) + \beta(D_{II})^{-1/2}]\, D_{ij} \Leftrightarrow (S_{II})^{1/2} > \beta. \tag{1.2}$$

We complete (1.1) by the assumption that within the region where $0 \leqq (S_{II})^{1/2} \leqq \beta$ the medium under consideration is in an elastic state (the material properties being unspecified there; cf. e.g. [4], [7]).

By (1.2), $(S_{II})^{1/2} - \beta = a(D_{II})\, (D_{II})^{1/2}$. Hence, $(S_{II})^{1/2} \searrow \beta$ if $(D_{II})^{1/2} \to 0$. Further, if the function $\xi \mapsto a(\xi^2)\, \xi \; (\xi \in (0, \infty))$ is uniquely invertible then there is a one-to-one correspondence between $(D_{II})^{1/2} > 0$ and $(S_{II})^{1/2} > \beta$.

Within the region where $(S_{II})^{1/2} > \beta$ the medium is in the state of viscous flow, and it behaves there as a purely viscous non-Newtonian fluid whoserheological behavior is modelled by the materialfunction $a(\cdot)$. If $a(\xi) \equiv \mu = \text{const} > 0$ the medium characterized by (1.1), (1.2) is called a Bingham plastic (Bingham fluid). A more detailed discussion of (1.1), (1.2) (together with further references) will be given in a forthcoming paper.

1.2. Differential equations and boundary conditions (arbitrary profiles)

We consider the motion of a homogeneous (density $\equiv 1$), incompressible visco-plastic medium which is characterized by the constitutive law (1.1), (1.2) through a bounded N-dimensional region Ω ($N = 2, 3$) with (fixed) boundary Γ.

1⁰ *The stationary case*. Let us define

$$\Omega_v = \{x \in \Omega : D_{II}(v) > 0\}$$

($v = \{v_1, \ldots, v_N\}$ any smooth vector field in Ω).

We use the Eulerian description for the motion of the considered visco-plastic medium within its region of viscous flow. The laws of conservation of impulse and mass will be then expressed in terms of velocity u and rate of strain $D = D(u)$. Thus:

$$\left.\begin{aligned} &-\left[\left(a(D_{II}(u)) + \beta(D_{II}(u))^{-1/2}\right) D_{ij}(u)\right]_{,j} + u_j u_{i,j} \\ &= f_i - p_{,i} \quad in \quad \Omega_u \; (i = 1, \ldots, N), \end{aligned}\right\} \tag{1.3}$$

$$u_{i,i} = 0 \quad in \quad \Omega_u \tag{1.4}$$

(p = the constitutively undetermined pressure, $f = \{f_1, \ldots, f_N\}$ = the external force density).

We *assume* that Ω_u contains a boundary strip, in other words, we assume that Γ is the fixed part of the boundary of Ω_u. Then the other part of the boundary of Ω_u, say Γ_u, is a free surface (curve if $N = 2$) and is called the *yield surface*. The following conditions must be satisfied along Γ_u:

$$D_{ij}(u) = 0, \quad (S_{II})^{1/2} = \beta \quad (v.\ \textit{Mises yield criterion}). \tag{1.5}$$

We complete (1.3)—(1.5) by the boundary condition

$$u = 0 \quad on \quad \Gamma. \tag{1.6}$$

2⁰ *The nonstationary case.* Let $0 < T < \infty$ be a fixed instant. We then define

$$\Omega_{v(t)} = \{x \in \Omega : D_{II}(v(t)) > 0\} \ (t \in (0, T))$$

$(v = v(t) = v(x, t))$.

The differential equations governing the viscous flow of the considered visco-plastic medium are then as follows:

$$\left.\begin{aligned} &u_i' - \left[\left(a(D_{II}(u)) + \beta(D_{II}(u))^{-1/2}\right) D_{ij}(u)\right]_{,j} + u_j u_{i,j} \\ &\quad = f_i - p_{,i}{}^1 \quad in \quad \Omega_{u(t)}, t \in (0, T) \ (i = 1, \ldots, N), \end{aligned}\right\} \tag{1.7}$$

$$u_{i,i} = 0 \quad in \quad \Omega_{u(t)}, \quad t \in (0, T) \tag{1.8}$$

(here $p = p(x, t), f = f(x, t)$).

As above, we assume that $\Omega_{u(t)}$ contains a boundary strip for each $t \in [0, T)$. The conditions (1.5) along the yield surface $\Gamma_{u(t)}$ will be maintained, while (1.7), (1.8) will be completed by

$$u = 0 \quad \text{on} \quad \Gamma \times (0, T), \tag{1.9}$$

$$u(0) = u_0 \quad \text{in} \quad \Omega_{u(0)}. \tag{1.10}$$

1.3. Differential equations and boundary conditions (laminar motion through a cylindrical tube)

We now consider the laminar motion of a homogeneous, incompressible visco-plastic medium characterized by (1.1), (1.2) through a cylindrical tube with cross-section $\Omega_0 (\subset \mathbb{R}^2)$ and generators parallel to the x_3-axis. We assume that the external forces can be neglected and that the motion is due to different pressure gradients at the inlet and the outlet of the tube (the pressure being specified there). Then the velocity profile $u = \{0, 0, u_3\}$, $u_3 = u_3(x_1, x_2)$, $\{x_1, x_2\} \in \Omega_0$ (resp. $u_3 = u_3(x_1, x_2, t)$) obeys (1.4) (resp. (1.8)), and p can be calculated explicitly by the aid of (1.3) (resp. (1.7)) and its prescribed values at the inlet and the outlet of the tube.

The equations (1.3) get thus simplified to

$$-\left[\left(\tilde{a}(|\nabla u_3|^2) + \beta\, |\nabla u_3|^{-1}\right) u_{3,j}\right]_{,j} = g_0 \quad in \quad \Omega_{0v_3} \tag{1.11}$$

where $\tilde{a}(\xi) = \frac{1}{2}\, a\left(\frac{1}{4}\,\xi\right) \ (\xi \in (0, \infty))$, $|\nabla u_3| = (u_{3,1}^2 + u_{3,2}^2)^{1/2}$, $\Omega_{0u_3} = \{x \in \Omega_0 : |\nabla u_3| > 0\}$ and $g_0 =$ pressure difference divided by the length of the tube ($g_0 = \text{const} > 0$). The conditions (1.5), and (1.6) are straightforwardly specialized.

---

[1] $v' =$ partial derivative with respect to t.

Analogously, (1.7) takes the form

$$\left.\begin{aligned} &u_3' - \left[\left(\bar{a}(|\nabla u_3|^2) + \beta|\nabla u_3|^{-1}\right) u_{3,j}\right]_{,j} = g \\ &in \quad \Omega_{0u_3(t)},\ t \in (0, T) \end{aligned}\right\} \tag{1.12}$$

where $g = g(t) > 0$ for all $t \in [0, T]$.

2. Existence and regularity of weak solutions (arbitrary profiles)

2.1. Notations. Preliminaries

Let Ω be a bounded domain in $\mathbb{R}^N$ ($N = 2, 3$). Unless otherwise stated, we suppose that the boundary Γ of Ω is Lipschitzian.

Let $W^{m,p}(\Omega)$ ($m \geqq 1$ integer, $1 \leqq p < \infty$) denote the usual Sobolev space of all functions in $L^p(\Omega)$ having all their generalized partial derivatives up to the order m (including) in $L^p(\Omega)$. Further, let denote $H^m(\Omega) = W^{m,2}(\Omega)$, $H_0^m(\Omega)$ = closure of $\mathscr{D}(\Omega)$ in $H^m(\Omega)$ and $H^{-m}(\Omega)$ = dual of $H_0^m(\Omega)$.

Define

$$\mathscr{V} = \{u \in [\mathscr{D}(\Omega)]^N : u_{i,i} = 0 \quad in \quad \Omega\},$$

$$V = closure\ of\ \mathscr{V}\ in\ [H^1(\Omega)]^N,$$

$$H = closure\ of\ \mathscr{V}\ in\ [L^2(\Omega)]^N.$$

V and H are Hilbert spaces with respect to the scalar products

$$((u, v)) = \int_\Omega D(u, v)\, \mathrm{d}x,\ u, v \in V\ \left(D(u, v) = D_{ij}(u)\, D_{ij}(v)\right),$$

$$(u, v) = \int_\Omega u_i v_i\, \mathrm{d}x,\ u, v \in H,$$

respectively. By (u^*, u) we denote the dual pairing between $u^* \in V^*$ and $u \in V$ (V^* = dual of V, $\|\cdot\|_*$ = dual norm on V^*).

2.2. The stationary case

We impose on the function $a(\cdot)$ the following conditions:

$$\left.\begin{aligned} &a \in C([0, \infty)), \\ &a_1 \leqq a(\xi) \leqq a_2 \quad \forall\ \xi \in [0, \infty),\ a_i = \text{const} > 0\ (i = 1, 2); \end{aligned}\right\} \tag{2.1}$$

$$a(\xi^2)\,\xi \geqq a(\eta^2)\,\eta \quad \forall\ \xi \leqq \eta \leqq 0; \tag{2.2}$$

$$\left.\begin{aligned} &[a(\xi^2)\,\xi - a(\eta^2)\,\eta]\,(\xi - \eta) \geqq a_1(\xi - \eta)^2 \quad \forall\ \xi, \eta \in [0, \infty), \\ &a_1 = \text{const} > 0. \end{aligned}\right\} \tag{2.2'}$$

Let us then introduce mappings $A, B: V \to V^*$ and a functional $\Phi: V \to \mathbb{R}^1$ by

$$(Au, v) = \int_\Omega a(D_{II}(u))\, D(u, v)\, \mathrm{d}x, \qquad (Bu, v) = \int_\Omega u_i u_{j,i} v_i\, \mathrm{d}x,$$

$$\Phi(u) = 2\beta \int_\Omega [D_{II}(u)]^{1/2}\, \mathrm{d}x$$

$(u, v \in V)$. We note that A is a monotone, hemicontinuous and coercive gradient mapping: $A = \operatorname{grad} F$, where

$$F(u) = \int_\Omega \left[\int_0^{D_{II}(u)} a(\xi) \, d\xi \right] dx = \int_0^1 (A(su), u) \, ds, \quad u \in V.$$

Our first existence result is the following

Theorem 2.1 ($N = 2, 3$). *Let $a(\cdot)$ satisfy* (2.1), (2.2), *and let $f \in V^*$.*

Then there exists a function $u \in V$ such that

$$(Au, v - u) + (Bu, v) + \Phi(v) - \Phi(u) \geqq (f, v - u) \quad \forall\, v \in V. \tag{2.3}$$

In addition,

$$\|u\| \leqq c_0 \|f\|_* \quad (c_0 = \text{const} > 0). \tag{2.4}$$

Remark 2.1. — 1° The inequality (2.3) represents the weak formulation of (1.3) — (1.6) in the following sense:

(i) The inequality (2.3) may be obtained as limit case by an appropriate approximation argument.

(ii) Let $u \in V$ satisfy (2.3). If in addition $u_{i,kl} \in L^2_{\text{loc}}(\Omega)$ $(i, k, l = 1, 2, 3)$ and $D_{II}(u) \in C(\Omega)$ then (1.3) is satisfied a.e. in Ω_u (cf. Theorems 3.4 and 3.5 below).

2° Let $a(\cdot)$ satisfy (2.1), (2.2′), and let $f \in V^*$. Then the solution to (2.3) is unique provided that

$$\|f\|_* < a_1 (2 c_0 k_0)^{-1} \tag{2.5}$$

(a_1: according to (2.2′), c_0: according to (2.4), $k_0 = \text{const} > 0$:

$$\left| \int_\Omega u_i v_{j,i} w_j \, dx \right| \leqq k_0 \|u\| \, \|v\| \, \|w\| \quad \text{for any } (u, v, w \in V).$$

2.3. The nonstationary case

The weak formulation of (1.6)—(1.8) leads to an evolution inequality the solvability properties of which sharply differ for $N = 2$ and $N = 3$.

Theorem 2.2 ($N = 2$). *Let $a(\cdot)$ satisfy* (2.1), (2.2), *and let $u_0 \in H$, $f \in L^2(0, T; V^*)$.*

Then there exists a function $u \in L^2(0, T; V) \cap C([0, T]; H)$ such that

$$u' \in L^2(0, T; V^*), \quad u(0) = u_0; \tag{2.6}$$

$$\left.\begin{aligned} &(u'(t) + Au(t), v - u(t)) + (Bu(t), v) \\ &+ \Phi(v) - \Phi(u(t)) \geqq (f(t), v - u(t)) \\ &\textit{for a.a. } t \in (0, T) \textit{ and all } v \in V. \end{aligned}\right\} \tag{2.7}$$

Theorem 2.3 ($N = 2$). *Let $a(\cdot)$ satisfy* (2.1), (2.2′).

(*i*) *Let $u_0 \in H$, $f \in L^2(0, T; V^*)$.*

Then there exists exactly one function $u \in L^2(0, T; V) \cap C([0, T]; H)$ satisfying (2.6), (2.7).

(ii) *Let* $u_0 \in H$, $f \in L^2(0, T; V^*)$ *with* $t^\alpha f' \in L^2(0, T; V^*)$ $(1 \leqq \alpha < \infty)$. *Then, in addition to* (i), $t^\alpha u' \in L^\infty(0, T; H) \cap L^2(0, T; V)$ *(hence* $u \in C((0, T]; V)$*)*.

Remark 2.2. Under appropriate sharper conditions on u_0 and f it can be shown that $u' \in L^\infty(0, T; H) \cap L^2(0, T; V)$ (cf. [3], [4] for the case $a(\xi) \equiv \mu = \text{const} > 0$ $(\xi \in [0, \infty))$).

Theorem 2.4 ($N = 3$). *Let* $a(\cdot)$ *satisfy* (2.1), (2.2), *and let* $u_0 \in H$, $f \in L^2(0, T; V^*)$. *Then there exists a function* $u \in L^\infty(0, T; H) \cap L^2(0, T; V)$ *such that*

$$u' \in L^{4/3}(0, T; V^*), \qquad u(0) = u_0; \tag{2.8}$$

$$\left.\begin{aligned} &\int_0^T (v', v - u)\, dt + \int_0^T [F(v) + \Phi(v)]\, dt + \int_0^T (Bu, v)\, dt \\ &- \int_0^T [F(u) + \Phi(u)]\, dt \geqq \int_0^T (f, v - u)\, dt - \frac{1}{2} |v(0) - u_0|^2 \\ &\textit{for all } v \in L^3(0, T; V) \textit{ with } v' \in L^2(0, T; V^*). \end{aligned}\right\} \tag{2.9}$$

Theorem 2.5 ($N = 3$). *Let* $a(\cdot)$ *satisfy* (2.1), (2.2'), *and let* $u_0 \in H$, $f \in L^2(0, T; V^*)$. *Then there exists at most one function* $u \in L^2(0, T; V) \cap L^8(0, T; [L^4(\Omega)]^3) \cap L^\infty(0, T; H)$ *which fulfils* (2.9).

Remark 2.3. — 1⁰ Let $u \in L^3(0, T; V) \cap L^\infty(0, T; H)$ and let (2.8), (2.9) be fulfilled. Then u satisfies (2.7) (clearly, (2.7) implies (2.9)).

2⁰ Let $a(\xi) \equiv \mu = \text{const} > 0$ $(\xi \in [0, \infty))$. Then instead of (2.9) we have

$$\int_0^T (v', v - u)\, dt + \mu \int_0^T ((u, v - u))\, dt + \int_0^T (Bu, v)\, dt$$

$$+ \int_0^T \Phi(v)\, dt - \int_0^T \Phi(u)\, dt \geqq \int_0^T (f, v - u)\, dt - \frac{1}{2} |v(0) - u_0|^2$$

for all $v \in L^3(0, T; V)$ with $v' \in L^2(0, T; V^*)$ (cf. [4]).

The following result essentially coincides with [6; Theorem 2].

Theorem 2.6 ($N = 3$). *Let* $a(\cdot)$ *satisfy* (2.1), (2.2'). *In addition, suppose that*

$$|a(\xi^2)\, \xi - a(\eta^2)\, \eta| \leqq a_3\, |\xi - \eta| \ \forall\ \xi, \eta \in [0, \infty) \quad (a_3 = \text{const} > 0).$$

Further, let $f, f' \in L^1(0, T; H)$ *and*

$$\|f\|_{L^\infty(0,T;V^*)} + k_1 \left(|f(0)| + \int_0^T |f'|\, dt \right) < a_1^2 k_0^{-1}\ [1].$$

Then there exists exactly one function $u \in C([0, T]; V)$ *such that*

$u' \in L^\infty(0, T; H) \cap L^2(0, T; V)$, $u(0) = 0$;

u satisfies (2.7).

---

[1] k_1: $|v| \leqq k_1 \|v\|$ for all $v \in V$, a_1: according to (2.2'), k_0: cf. (2.5).

3. Existence and regularity of weak solutions (laminar motion through a cylindrical tube)

Let Ω be a bounded domain in $\mathbb{R}^2$ with Lipschitz boundary Γ.[1]

3.1. The stationary case

Let $g_0 = \text{const} > 0$ (fixed)[2].

Theorem 3.1. *Let* $a(\cdot)$ *satisfy* (2.1), (2.2).

Then there exists a function $u \in H_0^1(\Omega)$ *such that*

$$\int_\Omega a(|\nabla u|^2)\, \nabla u \cdot \nabla (v - u)\, dx + \beta \int_\Omega |\nabla v|\, dx - \beta \int_\Omega |\nabla u|\, dx \geqq g_0 \int_\Omega (v - u)\, dx \tag{3.1}$$

for all $v \in H_0^1(\Omega)$. *In addition:* $0 \leqq u \leqq M = \text{const} < +\infty$ *a.e. in* Ω, u *is Lipschitz continuous on each compact subset of* Ω.

If $a(\cdot)$ *satisfies* (2.1), (2.2′) *then the solution to* (3.1) *is unique.*

Let us sharpen the conditions on $a(\cdot)$ as follows:

$$\left.\begin{array}{l} a \in C^1([0, \infty)), \\ a_1 \leqq a(\xi) \leqq a_2 \quad \forall\ \xi \in [0, \infty), \quad a_i = \text{const} > 0\ (i = 1, 2); \end{array}\right\} \tag{3.2}$$

$$a_1 \leqq \frac{d}{d\xi}\,[a(\xi^2)\,\xi] \leqq a_3 \quad \forall\ \xi \in [0, \infty), \quad a_3 = \text{const} > 0. \tag{3.3}$$

We then have

Theorem 3.2 (*local regularity*). *Let* $a(\cdot)$ *satisfy* (2.2), (3.3).

Then

$$u_{,ij} \in L^2_{\text{loc}}(\Omega) \quad (i, j = 1, 2).$$

Corollary. *Let* $a(\cdot)$ *satisfy* (3.2), (3.3), *let* $u \in H_0^1(\Omega)$ *denote the* (*uniquely determined*) *solution to* (3.1) *and let* Ω_u *be open. Then: if* a' *is Lipschitz continuous on each* $[0, R]$ $(0 < R < \infty)$ *then* $u|_{\Omega_u} \in C^{2+\alpha}(\Omega_u)$ ($\alpha \in (0, 1)$ *arbitrary*); *if* $a \in C^\infty([0, \infty))$ *then* $u|_{\Omega_u} \in C^\infty(\Omega_u)$.

Theorem 3.3 (*global regularity*). *Let* $\Gamma \in C^3$, *let* $a(\cdot)$ *satisfy* (3.2), (3.3), *and let* $u \in H_0^1(\Omega)$ *denote the* (*uniquely determined*) *solution to* (3.1).

Then: $u \in H_0^1(\Omega) \cap H^2(\Omega)$.

Remark 3.1. — 1⁰ Theorem 3.3 has been proved in [2] for the case $a(\xi) \equiv \mu = \text{const} > 0$ $(\xi \in [0, \infty))$. The method of proof used in this paper easily applies to (3.1) if one of

[1] In the present section we let denote $\Omega = \Omega_0$, $u \equiv u_3$ and $a \equiv \tilde{a}$ (cf. Section 1.3).

[2] We restrict our attention to the situation considered in Section 1.3. Nevertheless, most of the following discussion continues to hold when g_0 is replaced by an appropriate function of x (analogously, when g below is replaced by a function of x, t).

the following condition is satisfied:

(*) $a(\cdot)$ *satisfies* (3.2), (3.3), Ω *is convex*;

(**) $\left\{\begin{array}{l} a(\cdot) \textit{ satisfies } (3.2),\ |a'(\xi)\,\xi| \leqq \bar{a}_3 = \text{const } \textit{for all } \xi \in [0, \infty) \textit{ where } \bar{a}_3 \textit{ is sufficiently} \\ \textit{small.} \end{array}\right.$

2⁰ If Γ is Lipschitzian and a portion $\Gamma_0 \subset \Gamma$ is in C^3 then we have the existence of $u_{,ij}$ $(i, j = 1, 2)$ in L^2 up to any part $\Gamma_0' \subset \overline{\Gamma}_0' \subset \Gamma_0$.

Remark 3.2. For technical simplicity, let $a(\cdot) \equiv \mu = \text{const} > 0$ $(\xi \in [0, \infty))$ Define

$$w_0 \in H_0^1(\Omega): \int_\Omega \nabla w_0 \cdot \nabla v \, dx = g_0 \int_\Omega v \, dx \quad \forall\, v \in H_0^1(\Omega),$$

$$K = \left\{ w \in H_0^1(\Omega): \left| \int_\Omega \nabla w \cdot \nabla v \, dx \right| \leqq \beta \int_\Omega |\nabla v| \, dx \quad \forall\, v \in H_0^1(\Omega) \right\}.$$

It is easy to check that there exists exactly one $\overline{w} \in K$ such that

$$\int_\Omega \nabla(\overline{w} + w_0) \cdot \nabla(z - \overline{w}) \, dx \geqq 0 \quad \forall\, z \in K. \tag{3.1'}$$

We then have the following

Equivalence Theorem. $u \in H_0^1(\Omega)$ *is a solution to* (3.1) *if and only if*

$$u = \overline{w} + w_0.$$

The following two theorems bring into relation the boundary value problem (1.1), (1.5), (1.6) and the variational inequality (3.1).

Theorem 3.4. *Let* $a(\cdot)$ *satisfy* (2.1), *and let* $u \in H_0^1(\Omega)$ *be a solution to* (3.1) *such that* $u_{,ij} \in L^2_{\text{loc}}(\Omega)$ $(i, j = 1, 2)$, $|\nabla u| \in C(\Omega)$.

Then

$$-\left[\left(a(|\nabla u|^2) + \beta\, |\nabla u|^{-1}\right) u_{,j}\right]_{,j} = g_0 \quad \textit{a.e. in } \Omega_u.$$

Theorem 3.5. *Let the assumptions of Theorem* 3.4 *be satisfied.*

Then there exist (*not uniquely determined*) *functions* $z_i \in L^\infty(\Omega)$ $(i = 1, 2)$ *such that:*

(i) $\int_\Omega a(|\nabla u|^2)\, \nabla u \cdot \nabla v \, dx + \int_\Omega z_i v_{,i} \, dx = g_0 \int_\Omega v \, dx \quad \forall\, v \in H_0^1(\Omega)$;

(ii) $\int_{\Omega_u} (z_i - \beta\, |\nabla u|^{-1} u_{,i})\, \varrho_{,i} \, dx = 0 \quad \forall\, \varphi \in \mathscr{D}(\Omega_u)$,

$$\int_{\Omega_u} z_i u_{,i} \, dx = \beta \int_{\Omega_u} |\nabla u| \, dx,$$

$(z_1^2 + z_2^2)^{1/2} \leqq \beta$ *a.e. on* $\Omega \setminus \Omega_u$;

(iii) *if there exists an open subset* $\tilde{\Omega} \subset \Omega \setminus \Omega_u$ *then*

$$\int_{\tilde{\Omega}} z_i \varphi_{,i} \, dx = g_0 \int_{\tilde{\Omega}} \varphi \, dx \quad \forall\, \varphi \in \mathscr{D}(\tilde{\Omega}).$$

Remark 3.3. Similar results as those stated in Theorems 3.4 and 3.5 are also true with respect to (1.3)—(1.6) and (2.3) (cf. Remark 2.1, 1⁰), (1.7)—(1.10), (1.5) and (2.7) as well as with respect to (1.12), (1.5), (1.9), (1.10) and (3.7) below. A related but less precise interpretation of (2.7) has been given in [4].

Remark 3.4. — 1° Let us consider the one-dimensional case of (1.11), (1.5), (1.6), i.e. $u = u(x_1)$ $(u \equiv u_3)$, $\Omega = \Omega_0 = (0, 1)$ (without loss of generality), $\Omega_u = \Omega_{0u} = \left\{x_1 \in (0,1): \left|\frac{du}{dx_1}(x_1)\right| > 0\right\}$. If we restrict ourselves to the case $\tilde{a}(\xi) \equiv \mu = \text{const} > 0$ $(\xi \in [0, \infty))$ then (1.11), (1.6) take the form

$$-\frac{d}{dx_1}\left[\left(\mu + \beta\left|\frac{du}{dx_1}\right|^{-1}\right)\frac{du}{dx_1}\right] = g_0 \quad in \quad \Omega_u. \tag{3.4}$$

$$u(0) = u(1) = 0. \tag{3.5}$$

If $\beta \geqq \frac{g_0}{2}$ (= critical yield value) then $\Omega_u = \emptyset$ and $u \equiv 0$. On the other hand, if $\beta < \frac{g_0}{2}$ then the solution to (3.4), (3.5) is given by

$$u(x_1) = \begin{cases} \frac{g_0}{2\mu}\, x_1(1 - x_1) - \frac{\beta}{\mu}\, x_1 & if \quad 0 \leqq x_1 < \frac{1}{2} - \frac{\beta}{g_0}, \\[2ex] \frac{g_0}{2\mu}\left(\frac{1}{2} - \frac{\beta}{g_0}\right)^2 & if \quad \frac{1}{2} - \frac{\beta}{g_0} \leqq x_1 \leqq \frac{1}{2} + \frac{\beta}{g_0}, \\[2ex] \frac{g_0}{2\mu}\, x_1(1 - x_1) - \frac{\beta}{\mu}\,(1 - x_1) & if \quad \frac{1}{2} + \frac{\beta}{g_0} < x_1 \leqq 1, \end{cases}$$

$\Omega_u = \left[0, \frac{1}{2} - \frac{\beta}{g_0}\right) \cup \left(\frac{1}{2} + \frac{\beta}{g_0}, 1\right]$; clearly, $u \in H_0^1\big((0, 1)\big)$, $\frac{d^2u}{dx_1^2} \in L^\infty(0, 1)$, $u \notin W^{3,1}\big((0, 1)\big)$. This is deduced in [7; pp. 128—130] but the reasoning seems to be not complete from the mathematical point of view. A complete proof can be given on the following pattern:

— $u \in H_0^1\big((0, 1)\big)$, $\frac{d^2u}{dx_1^2} \in L^2_{\text{loc}}(0, 1)$ (u being the solution to the corresponding variational inequality; cf. Theorem 3.2);
— representation of Ω_u as countable union of open intervals, u is concave on $(0, 1)$ ($\Rightarrow \Omega \setminus \Omega_u$ is an interval);
— calculation of the points $\frac{1}{2} - \frac{\beta}{g_0}$ and $\frac{1}{2} + \frac{\beta}{g_0}$, calculation of u on $[0, 1]$.

2° Let Ω be a circle, and let $a(\xi) \equiv \mu = \text{const} > 0$ $\big(\xi \in [0, \infty)\big)$. Then it can be proven that the solution $u \in H_0^1(\Omega)$ to (3.1) has rotational symmetry. Our preceding discussion then easily implies the explicit representation of u (cf. [5; p. 103]). As above, $u_{,ij} \in L^\infty(\Omega)$ $(i, j = 1, 2)$, $u \notin W^{3,1}(\Omega)$, while $\Omega \setminus \Omega_u$ is a concentric circle whose radius increases if β does.

3° Numerical studies on (3.1) $\big(a(\xi) \equiv \mu = \text{const} > 0$ $(\xi \in [0, \infty))\big)$ are presented in [1, 5] for the case when Ω is a square. These results give a very useful insight into position and shape of $\Omega \setminus \Omega_u$.

4° The following conjecture on the solution u to (3.1) seems to be reasonable:

(i) $u_{,ij} \in L^\infty_{\text{loc}}(\Omega)$ when Γ is Lipschitzian, and $u_{,ij} \in L^\infty(\Omega)$ when Γ is sufficiently smooth ($i, j = 1, 2$);

(ii) the interior of $\Omega \setminus \Omega_u$ is not empty; $\Omega \setminus \Omega_u$ is convex if Ω does; the yield curve Γ_u (free boundary) is smooth;

(iii) precisely along Γ_u the second order derivatives of u suffer jumps.

3.2. The nonstationary case

The following theorem presents our principal existence and uniqueness result on (1.12) (cf. footnote 4).

Theorem 3.6. — 1. *Let $a(\cdot)$ satisfy* (2.1), (2.2), *and let $g \in L^2(0, T)$.*

If $u_0 \in L^2(\Omega)$ then there exists exactly one $u \in L^2(0, T; H_0^1(\Omega)) \cap C_W((0, T]; H_0^1(\Omega)) \cap C([0, T]; L^2(\Omega))$ such that

$$\left.\begin{array}{l} u' \in L^2(0, T; H^{-1}(\Omega)), \quad t^\alpha u' \in L^2(0, T; L^2(\Omega)) \ \left(\alpha \in \left[\frac{1}{2}, \infty\right) \text{ arbitrary}\right), \\ \qquad u(0) = u_0; \end{array}\right\} \tag{3.6}$$

$$\left.\begin{array}{l} \int_0^T (u', v - u)\, dt + \int_0^T \int_\Omega a(|\nabla u|^2)\, \nabla u \cdot \nabla(v - u)\, dx\, dt \\ \quad + \beta \int_0^T \int_\Omega |\nabla v|\, dx\, dt - \beta \int_0^T \int_\Omega |\nabla u|\, dx\, dt \leqq \int_0^T g \int_\Omega (v - u)\, dx\, dt \\ \text{for all } v \in L^2(0, T; H_0^1(\Omega)). \end{array}\right\} \tag{3.7}$$

Furthermore:

(i) *if $g \geqq 0$ a.e. in $(0, T)$ and $u_0 \geqq 0$ a.e. in Ω then $u \geqq 0$ a.e. in Q;*

(ii) *if $g \in L^r(0, T)$ $(3 < r < \infty)$ and $u_0 \leqq M = \text{const} < +\infty$ a.e. in Ω then $u \leqq 2M \times (1 + \text{const}\, M^{-1} T^{1-2/r} \|g\|_{L^r(0,T)})$ a.e. in Q;*

(iii) *if $g \in L^r(0, T)$ $(3 < r < \infty)$, $g \geqq 0$ a.e. in $(0, T)$ and $0 \leqq u_0 \leqq M = \text{const} < +\infty$ a.e. in Ω then, in addition to the assertions of* (i), (ii), *$u \in H^{\gamma,\gamma/2}(Q)$ for a $\gamma \in (0, 1)$*[1].

Further, if $u_0 \in H_0^1(\Omega)$ then $u \in C_W([0, T]; H_0^1(\Omega))$, $u' \in L^2(0, T; L^2(\Omega))$.

2. *Let $a(\cdot)$ satisfy* (2.1), (2.2'), *let $u_0 \in L^2(\Omega)$ and let $g \in L^2(0, T)$, $t^\alpha g' \in L^2(0, T)$ for a real $1 \leqq \alpha < \infty$. Then*

$$u \in C((0, T]; H_0^1(\Omega)), \quad t^\alpha u' \in L^\infty(0, T; L^2(\Omega)) \cap L^2(0, T; H_0^1(\Omega)).$$

The latter result can be further sharpened by the method used in [3] (cf. also Remark 2.2).

---

[1] $Q = \Omega \times (0, T)$; $H^{\gamma,\gamma/2}(Q)$ = space of all functions $v = v(x, t)$ in Q such that $v(\cdot, t)$, $t \in (0, T)$ (resp. $v(x, \cdot)$, $x \in \Omega$), is Hölder continuous on each compact subset of Ω (resp. of $(0, T)$) with respect to the exponent γ (resp. $\gamma/2$).

Theorem 3.7. *Let $a(\cdot)$ satisfy* (3.2), (3.3), *and let u denote the solution to* (3.7) *obtained in the preceding theorem. Then:*

(i) *if* $u_0 \in L^2(\Omega)$, $g \in L^2(0, T)$ *then* $t^\alpha u \in L^2(0, T; H^2(\Omega'))$ *for any* $\frac{1}{2} \leqq \alpha < \infty$ *and* $\Omega' \subset\subset \Omega$;[1]

(ii) *if* $u_0 \in H_0^1(\Omega)$, $g \in L^2(0, T)$ *then* $u \in L^2(0, T; H^2(\Omega'))$ *for any* $\Omega' \subset\subset \Omega$;

(iii) *if* $u_0 \in L^2(\Omega)$, $g \in L^2(0, T)$ *and* $t^\alpha g' \in L^2(0, T)$ *for a real* $1 \leqq \alpha < \infty$ *then* $t^\alpha u \in L^\infty(0, T; H^2(\Omega'))$ *for any* $\Omega' \subset\subset \Omega$.

Theorem 3.8. *Let Γ be sufficiently smooth, and let one of the conditions* (*) *or* (**) *be satisfied* (*cf. Remark* 3.1). *Then*:

(i) *if* $u_0 \in L^2(\Omega)$, $g \in L^2(0, T)$ *then* $u \in L^2(\delta, T; H_0^1(\Omega) \cap H^2(\Omega))$ *for any* $0 < \delta < T$, *and* $u \in C((0, T]; H_0^1(\Omega))$;

(ii) *if* $u_0 \in H_0^1(\Omega)$, $g \in L^2(0, T)$ *then* $u \in L^2(0, T, H_0^1(\Omega) \cap H^2(\Omega)) \cap C([0, T]; H_0^1(\Omega))$.

Remark 3.5. — If $|\nabla u| \in C(Q)$ then the assertions of Theorem 3.8 continue to hold under the only assumption that $a(\cdot)$ satisfies (3.2), (3.3).

References

[1] Begis, D., Etude numérique du comportement d'un fluid de Bingham. IRIA Laboria, Rapp. Recherche no. 42 (1973).

[2] Brézis, H., Monotonicity methods in Hilbert spaces and some applications to nonlinear partial differential equations. *In:* Contributions to Nonlinear Functional Analysis (Ed. E. H. Zarantonello). Academic Press, New York, London 1971, 101–156.

[3] —, Inéquations variationnelles relatives à l'opérateur de Navier-Stokes. J. Math. Anal. Appl. **39** (1972), 159–165.

[4] Duvaut, G., et Lions, J. L., Les inéquations en mécanique et en physique. Dunod, Paris 1972.

[5] Glowinski, R., Lions, J. L., et Trémolières, R., Analyse numérique des inéquations variationnelles; tome 2. Dunod, Paris 1976.

[6] Naumann, J., On evolution inequalities of Navier-Stokes type in three dimensions. Ann. Mat. Pura Appl. (4) **74** (1980), 107–125.

[7] Prager, W., Einführung in die Kontinuumsmechanik. Birkhäuser Verlag, Basel und Stuttgart 1961.

Address of the author:

Humboldt-Universität zu Berlin
Sektion Mathematik
1080 Berlin
Unter den Linden 6
DDR

---

[1] $\Omega' \subset\subset \Omega$ means: Ω' open, $\bar{\Omega}' \subset \Omega$.

A necessary and sufficient condition for the regularity of weak solutions to nonlinear elliptic systems of partial differential equations

Jindrich Nečas, Prague

1. Introduction

We consider the second order elliptic systems

$$-\frac{\partial}{\partial x_i}[a_i^r(x, u, \nabla u)] + a^r(x, u, \nabla u) = -\frac{\partial f_i^r}{\partial x_i} + f^r, \tag{1.1}$$

$r = 1, 2, \ldots, m,$

for vector-functions $u = (u_1, u_2, \ldots, u_m)$.
The summation convention is used. Let $\Omega \subset \mathbb{R}^n$, $n \geqq 2$, be a bounded domain. We restrict ourselves to such solutions of (1.1) that $u \in [W^{1,\infty}(\Omega)]^m$, i.e. to Lipschitz-like continuous solutions. We suppose $a_i^r(x, \xi, \eta)$, $a^r(x, \xi, \eta)$ to be once continuously differentiable in $\bar{\Omega} \times \mathbb{R}^m \times \mathbb{R}^{mn}$, $f_i^r \in W^{1,p}(\Omega)$, $f^r \in W^{1,\frac{p}{2}}(\Omega)$ with $p > n$, and the strong ellipticity condition in the form

$$\frac{\partial a_i^r}{\partial \eta_j^s}(x, \xi, \eta)\, \xi_i^r \xi_j^s > 0 \quad \text{for} \quad \xi \neq 0. \tag{1.2}$$

The regularity of the weak solution u means that $u \in [C^{1,\alpha}(\Omega)]^m$ (the interior regularity (R)). The history of the 19-th Hilbert's problem, i.e. the history of the regularity question was described many times and we recommend the book by O. A. Ladyženskaja and N. N. Ural'ceva [1], and the paper by Ch. B. Morrey [2]. It is well known that the answer is positive if $n = 2$, this was proven by Ch. B. Morrey [3] in 1940, and for $m = 1$, this is the result due to E. De Giorgi [4] (1957) and to J. Nash [5] (1958). By the counter example due to the author (1979) (to appear in the lecture notes [6], see also its modification by O. John, J. Stará, J. Nečas [7]), there exists a strongly elliptic system (see also Section 4)

$$-\frac{\partial}{\partial x_i}[a_i^r(\nabla u)] = 0 \tag{1.3}$$

with analytic coefficients and with the solution, for $n \geqq 3$, of the form $u_{ij} = \frac{x_i x_j}{|x|} - \frac{1}{n}\delta_{ij}|x|$ in the whole $\mathbb{R}^n$, i.e. with the solution with bounded $\nabla u : |\nabla u| \leqq c < \infty$ and not regular. So, in general, also for systems (1.3), the 19-th Hilbert's problem is answered negatively. The mentioned example shows that there exist systems (13) with a solution u in $\mathbb{R}^n$ such that $|\nabla u| \leq c < \infty$ and that u is not a polynomial of the first degree. So we come to

Definition 1.1. The system (1.1) satisfies the *Liouville property* (L) if $\forall\, x^0 \in \bar{\Omega}$, $\forall\, \xi \in \mathbb{R}^m$, every solution v to the system

$$-\frac{\partial}{\partial x_i}\,[a_i^r(x^0, \xi, \nabla v)] = 0 \tag{1.4}$$

in $\mathbb{R}^n$ with $|\nabla v| \leq c < \infty$ is a polynomial of at most first degree.

It is proved in the paper by M. Giaquinta and J. Nečas [8] that (L) $\Rightarrow$ (R), i.e. that $u \in [C^{1,\alpha}(\Omega)]^m$ for $\alpha < \frac{1}{2} \wedge \left(1 - \frac{n}{p}\right)$; this result and other consequences (see Sections 2 and 5) are contained also in the lecture notes [6]. In this paper we prove also that (R) $\Rightarrow$ (L); for the precise statement, see Section 3.

2. The interior regularity

We denote by $B_R(x^0)$ a ball with the center x^0 and the radius R. The mean value of u is denoted by

$$u_{x^0,R} \overset{\text{def}}{=} \frac{1}{\operatorname{meas} B_R(x^0)} \int_{B_R(x_0)} u(x)\, dx\,. \tag{2.1}$$

Put

$$U(x^0, R) \overset{\text{def}}{=} R^{-n} \int_{B_R(x^0)} |u(x) - u_{x^0,R}|^2\, dx\,. \tag{2.2}$$

We have the following result due to S. Campanato [9]: Let b_{ij}^{rs} be constant coefficients and $u \in \left[W_{\text{loc}}^{1,2}(B_1(0))\right]^m \cap \left[L^2(B_1(0))\right]^m$ a solution to the strongly elliptic system

$$\int_{B_1(0)} b_{ij}^{rs} \frac{\partial \psi_r}{\partial x_i} \frac{\partial u_s}{\partial x_j}\, dx = 0\,. \tag{2.3}$$

Let $0 < \varrho \leq 1$. Then

$$U(0, \varrho) \leq K \varrho^2 (0, 1)\,, \tag{2.3}$$

where K depends on $\max |b_{ij}^{rs}|$ and the constant of strong ellipticity.

Let us further consider the quasilinear system

$$\int_\Omega \left[A_{ij}^{rs}(x, v) \frac{\partial \varphi_r}{\partial x_i} \frac{\partial v_s}{\partial x_j} + A_j^{rs}(x, v)\, \varphi_r \frac{\partial v_s}{\partial x_j}\right] dx = \int_\Omega \left[\frac{\partial \varphi_r}{\partial x_i} g_i^r + \varphi_r g^r\right] dx \tag{2.4}$$

with $v = (v_1, v_2, \ldots, v_M) \in [W^{1,2}(\Omega)]^M \times [L^\infty(\Omega)]^M$ and $\Omega \subset \mathbb{R}^n$ being a bounded domain. We suppose $A_{ij}^{rs}(x, \xi)$ and $A_j^{rs}(x, \xi)$ to be continuous in $\bar{\Omega} \times \mathbb{R}^M$, the strong ellipticity condition

$$A_{ij}^{rs}(x, \xi)\, \xi_i^r \xi_j^s > 0 \quad \text{for} \quad \xi \neq 0 \tag{2.5}$$

and $g_i^r \in L^p(\Omega)$, $g^r \in L^{\frac{p}{2}}(\Omega)$, $p < n$.

For $B_R(x^0) \subset \Omega$, put $v = v^R + w^R$, where $w^R \in \left[W_0^{1,2}(B_R(x^0))\right]^M$ is a solution to the system

$$\int\limits_{B_R(x^0)} \left[A_{ij}^{rs}(x, v) \frac{\partial \varphi_r}{\partial x_i} \frac{\partial w_s}{\partial x_j} + A_j^{rs}(x, v)\, \varphi_r \frac{\partial w_s}{\partial x_j} \right] dx$$

$$= \int\limits_{B_R(x^0)} \left[\frac{\partial \varphi_r}{\partial x_i} g_i^r + \varphi_r g^r \right] dx. \tag{2.5'}$$

We have the following lemma, a slight generalization of a result of E. Giusti [10]; for the proof see [8] or [6]:

Lemma 2.1. *Let us consider the family of v, the solutions of* (2.4) *such that*

$$\|v\|_{[L^\infty(\Omega)]^m} + \sum_{r,i} \|g_i^r\|_{L^p(\Omega)} + \sum_r \|g^r\|_{L^{p/2}(\Omega)} \leqq c_1. \tag{2.6}$$

Let K be the constant from (2.3') *for the coefficients $A_{ij}^{rs}(x^0, \xi)$, $x^0 \in \Omega$, $|\xi| \leqq c_1$. For every $0 < \tau < 1$ then there exist $\varepsilon_0 = \varepsilon_0(\tau, c_1)$ and $R_0 = R_0(\tau, c_1)$ such that if $R \leqq R_0 \wedge$ dist $(x^0, \partial\Omega)$ and if*

$$V^R(x^0, R) < \varepsilon_0^2 \tag{2.7}$$

then

$$V^R(x^0, \tau R) \leqq 2K\tau^2 V^R(x^0, R). \tag{2.8}$$

In the standard way we get as in [8] or [6]:

Lemma 2.2. *Under the conditions of Lemmma 2.1, for every point $x^0 \in \Omega$ such that $\liminf_{R\to 0} V(x^0, R) = 0$, there exists $B_{R_1}(x^0) \subset \Omega$ such that $v \in C^\alpha\left(\overline{B_{R_1}(x^0)}\right)$ with $\alpha < \frac{1}{2} \wedge \left(1 - \frac{n}{p}\right)$.*

So we are able to prove our first main theorem:

Theorem 2.3. *Let $\Omega \subset \mathbb{R}^n$ be a bounded domain, let $|\nabla u| \leqq c < \infty$ in Ω, and let u be a weak solution to the system* (1.1). *Further let the conditions on a_i^r, a^r, f_i^r, f^r, mentioned in the introduction, be satisfied. Let the condition* (L) *from Definition* 1.1 *be valid. Then the solution $u \in [C^{1,\alpha}(\Omega)]^m$, $\alpha < \frac{1}{2} \wedge \left(1 - \frac{n}{p}\right)$, and if*

$$\|u\|_{[W^{1,\infty}(\Omega)]^m} + \sum_{r,i} \|f_i^r\|_{W^{1,p}(\Omega)} + \sum_r \|f^r\|_{W^{1,p/2}(\Omega)} \leqq c_1 \tag{2.9}$$

then for every $\overline{\Omega}' \subset \Omega$,

$$\|u\|_{[C^{1,\alpha}(\overline{\Omega}')]^m} \leqq c^2(\Omega'). \tag{2.10}$$

The detailed proof is in [8] or [6] and we give here only its main idea: we claim that if $x^0 \in \overline{\Omega}'$ then $\lim_{R\to\infty} \nabla U(x^0, R) = 0$ uniformly with respect to x^0 from $\overline{\Omega}'$ and the family

(2.9). Put $u_R(y) = \frac{1}{R}\,[u(x^0 + Ry) - u(x^0)]$, $x^0 + Ry = x$, and let O_R be the image of Ω. We get

$$\int_{O_R} \Big[a_i^r(x^0 + Ry,\, Ru_R + u(x^0),\, \nabla_y u_R) \frac{\partial \psi_r}{\partial y_i} + a^r\big(x^0 + Ry,\, Ru_R + u(x^0),\, \nabla_y u_R\big) R\psi_r - f_i^r(x^0 + Ry) \frac{\partial \psi_r}{\partial y_i} - Rf^r(x^0 + Ry)\, \psi_r \Big]\, dy = 0\,. \tag{2.11}$$

But the equations in variations give the estimate

$$\int_{B_a(0)} |D_y^2 u_R|^2 \leqq c(a) \tag{2.12}$$

for $R \leqq \frac{1}{2}$ dist $(\partial\Omega', \partial\Omega)$. If our assertion were false then there would exist $x^k \to x^0 \in \bar{\Omega}'$, $R_k \to 0$, and u^k from our family (2.9) such that $\nabla U^k(x^k, R_k) \geqq \delta > 0$. But because of (2.12), we can suppose that $u^k_{R_k} \to p$ in $W^{1,2}\big(B_a(0)\big)$ $\forall\, a > 0$. From (2.11) we get that

$$\int_{\mathbb{R}^n} a_i^r(x^0, \xi, \nabla_y p) \frac{\partial \psi_r}{\partial y_i}\, dy = 0\,, \tag{2.13}$$

where $\xi = \lim_{k\to\infty} u^k(x^0)$ (which can be supposed to exist because of (2.9)). We have $|\nabla p| \leqq c < \infty$, so p is a polynomial of at most first degree. Hence

$$0 \leftarrow \int_{B_1(0)} |\nabla_y u^k_{R_k} - \nabla p|^2\, dy = R_k^{-n} \int_{B_{R_k}(x^k)} |\nabla_x u^k - \nabla p|^2\, dx \geqq \nabla U^k(x^k, R_k)\,, \tag{2.14}$$

which is a contradiction; clearly the equations in variations to the system (1.1) play the role of the system (2.4). □

3. Necessity of the condition (L) for the regularity

The condition (L) from Definition 1.1 is also necessary for the regularity. This new fact appears in this paper first time, so we give its proof in more detail.

We essentially use the "maximum principle" for systems, i.e. the result of A. Canfora [11], see also the book by B. W. Schulze, G. Wildenhain [12], where other references are given especially concerning the work of C. Miranda [13] and S. Agmon [14].

We formulate the result of A. Canfora only for second-order elliptic systems in the divergence form:

Lemma 3.1. *Let $\Omega \subset \mathbb{R}^n$, $n \geqq 2$, be a bounded domain smooth enough. Let $v = (v_1, \ldots, v_M) \in [C^{(2)}(\Omega)]^M \times [C(\bar{\Omega})]^M$ be a solution to the strongly elliptic system*

$$-\frac{\partial}{\partial x_i}\left[A_{ij}^{rs}(x) \frac{\partial v_s}{\partial x_j} \right] = 0\,, \tag{3.1}$$

where $A_{ij}^{rs} \in C^{(2)}(\bar{\Omega})$. Then there exists $L > 0$ depending on the ellipticity constant and on the $C^{(2)}$ norms of A_{ij}^{rs} such that

$$\sum_{s=1}^{M} \max_{x \in \bar{\Omega}} |v_s(x)| \leqq L \left(\sum_{s=1}^{M} \max_{x \in \partial\Omega} |v_s(x)| + \sum_{s=1}^{M} \|v_s\|_{L^2(\Omega)} \right). \tag{3.2}$$

Theorem 3.2. *Let the coefficients $a_i^r(x^0, \xi, \eta)$ of the system* (1.1) *be 3-times continuously differentiable with respect to η. Let* (2.10) *be satisfied for the system*

$$-\frac{\partial}{\partial x_i} [a_i^r(x^0, \xi, \nabla v)] = 0 \;\forall\; x^0 \in \bar{\Omega}, \quad \xi \in \mathbb{R}^m \tag{3.3}$$

and bounded $O \subset \mathbb{R}^n$. Then the condition (L) *is valid.*

Proof. Let v be a solution to the system (3.3) in the whole $\mathbb{R}^n$ and such that $|\nabla v| \leqq c < \infty$. Put $v_R(y) = \frac{1}{R} v(Ry)$. v_R is a solution to the system

$$-\frac{\partial}{\partial y_i} [a_i^r(x^0, \xi, \nabla_y v_R)] = 0 \tag{3.4}$$

and we have the system in variations

$$-\frac{\partial}{\partial y_i} \left[\frac{\partial a_i^r}{\partial \eta_i^s} (x^0, \xi, \nabla_y v_R) \frac{\partial v'_R}{\partial y_i} \right] = 0, \tag{3.5}$$

where v' is some fixed derivative. Using the known fact about linear systems especially applied to (3.5), i.e. that $\frac{\partial a_i^r}{\partial \eta_j^s} \in C^{l,\alpha}(\bar{O})$, $l \geqq 1$, $w \in [W^{1,2}(O)]^M$ being a weak solution to the strongly elliptic system (3.5), implies for $\bar{O}' \subset O$ that

$$\|w\|_{[C^{l+1,\alpha}(\bar{O}')]^M} \leqq C(O') \, \|w\|_{[W^{1,2}(O)]^M}, \tag{3.6}$$

we get from (2.10) and (3.6), applied to v'_R and $O = B_1(0)$, that

$$\|v_R\|_{C^{3,\alpha}(\overline{B_{1/2}(0)})} \leqq c_3. \tag{3.7}$$

If ϑ_1, ϑ_2 are two spherical angles we get from (3.7)

$$\int_0^{1/2} |D_y^2 v_R(r, \vartheta_i)| \, dr \leqq c_4, \quad i = 1, 2, \tag{3.8}$$

$$\int_0^{1/2} |D_y^2 v_R(r, \vartheta_1) - D_x^2 v_R(r, \vartheta_2)| \, dr \leqq c_5 |\vartheta_1 - \vartheta_2|. \tag{3.9}$$

Because

$$\int_0^{1/2} |D_y^2 v_R(r, \vartheta_i)| \, dr = \int_0^{R/2} |D_x^2 v(p, \vartheta_i)| \, d\varrho \tag{3.10}$$

we get

$$\lim_{r \to \infty} \nabla v(r, \vartheta) = l(\vartheta) \tag{3.11}$$

uniformly in ϑ. But it cannot be $l(\vartheta_1) \neq l(\vartheta_2)$ because for $\delta > 0$

$$\nabla_y v_R(\delta, \vartheta) = \nabla_x v(\delta R, \vartheta) \tag{3.12}$$

and we have (3.6). So

$$\lim_{|x|\to\infty} \nabla v = \nabla p, \tag{3.13}$$

where p is a polynomial of the first degree. But we have $v'_R \to p'$ uniformly on $\partial B_\delta(0)$, $0 < \delta \leqq \frac{1}{2}$, and in $[L^2(B_1(0))]^m$ so, applying (3.2) to $v'_R - p'$ and $B_{1/2}(0)$, we get

$$0 \leftarrow \max_{y \in \overline{B_{1/2}(0)}} |v'_R - p'| = \max_{x \in \overline{B_{R/2}(0)}} |v' - p'|, \tag{3.14}$$

hence $v' = p'$. □

4. Counterexample

The present counterexample is a variation of the counterexample given by the author in [15]; for details see [6] and [7]. Put

$$u_{ij} = \frac{x_i x_j}{|x|} - \frac{1}{n} \delta_{ij} |x|, \quad n \geqq 2,$$

for the test functions put $\varphi_{ij} = \varphi_{ji}$, and define the elementary operators:

$$\nabla_i \varphi = \frac{\partial \varphi_{ij}}{\partial x_j}, \quad \delta_i \varphi = \frac{\partial \varphi_{jj}}{\partial x_i}, \quad \nabla_{ijk} \varphi = \frac{\partial \varphi_{ij}}{\partial x_k}$$
$$+ \gamma[\nabla_i \varphi \delta_{jk} + \nabla_j \varphi \delta_{ik} + \nabla_k \varphi \delta_{ij}], \quad \gamma \leqq 0.$$

The strongly elliptic system to be considered is

$$\int_{\mathbf{R}^n} \nabla_{ijk} u \nabla_{ijk} \varphi \, dx + \delta \int_{\mathbf{R}^n} \nabla_i u \delta_i \varphi \, dx$$
$$+ \beta \int_{\mathbf{R}^n} \delta_i u \delta_i \varphi \, dx + \lambda \int_{\mathbf{R}^n} \nabla_i u \nabla_j u \nabla_k u \nabla_{ijk} \varphi \, dx$$
$$+ \nu \int_{\mathbf{R}^n} (\nabla_i u \nabla_i u)^2 \nabla_j u \nabla_j \varphi dx = 0.$$

We look for $\nu \geqq 0$, $\beta > 0$, $\gamma \leqq 0$, λ, δ such that (4.1) is satisfied and that the strong ellipticity holds. We get

$$\gamma = 0 \quad \text{for} \quad n \geqq 4, \quad \frac{-4 - \sqrt{14}}{12} < \gamma < \frac{-4 + \sqrt{14}}{12} \quad \text{for} \quad n = 3,$$

and a possible choice of ν and λ such that $\nu - \frac{3}{4} \lambda^2 > 0$. This is, of course, impossible for $n = 2$.

5. Conditions for the validity of the condition (L)

It follows from Theorem 3.1, provided that the coefficients $a_i^r(x, \xi, \eta)$ are smooth enough in η, that (L) is satisfied for $m = 1$, $n \geqq 2$, and for $n = 2$, $m \geqq 1$. We sketch the proofs of these statements; for details see [6]. In the book by D. Gilbarg, N. S. Trudinger [16] we can find the following formulation of Harnack's inequality:

Lemma 5.1. *Let $\Omega \subset \mathbb{R}^n$, let $a_{ij} \in L^\infty(\Omega)$, and let $a_{ij}\xi_i\xi_j \geqq \nu |\xi|^2$. Let $u \in W^{1,2}(\Omega)$ be a weak solution to the equation*

$$-\frac{\partial}{\partial x_i}\left[a_{ij}\,\frac{\partial u}{\partial x_j}\right] = 0. \tag{5.1}$$

Let $u \geqq 0$, $B_{4R}(y) \subset \Omega$. Then

$$\sup_{x \in B_R(y)} u(x) \leqq c \inf_{x \in B_R(y)} u(x). \tag{5.2}$$

Consequence 5.2. *If $m = 1$ then* (L) *is satisfied.*

Proof. *Let* $\int\limits_{\mathbb{R}^n} a_i(x^0, \xi, \nabla v)\,\frac{\partial \varphi}{\partial x_i}\,dx = 0$, $|\nabla v| \leqq c < \infty$ in $\mathbb{R}^n$. Let $\int\limits_{\mathbb{R}^n} \frac{\partial a_i}{\partial \eta_j}(x^0, \xi, \eta)\,\frac{\partial \varphi}{\partial x_i}\,\frac{\partial v'}{\partial x_j}\,dx$. Put $M(R) = \sup\limits_{B_R(0)} v'(x)$, $m(R) = \inf\limits_{B_R(0)} v'(x)$, $\omega(R) = M(R) - m(R)$. Apply (5.2) to $M(R) - v'$ and $v' - m(R)$. We obtain

$$\omega\left(\frac{R}{4}\right) \leqq \frac{c-1}{c+1}\,\omega(R) \leqq \frac{c-1}{c+1}\,\omega(\infty), \tag{5.3}$$

hence $\omega(\infty) \leqq \dfrac{c-1}{c+1}\,\omega(\infty) \Rightarrow \omega(\infty) = 0$. □

Proposition 5.3. *If $n = 2$ then* (L) *is satisfied.*

See [8] for the proof.

Let us consider such systems that $\dfrac{\partial a_i^r}{\partial \eta_j^s} = \dfrac{\partial a_j^s}{\partial \eta_i^r}$ and let us suppose that these derivatives are bounded in $\bar{\Omega} \times \mathbb{R}^m \times \mathbb{R}^{mn}$. Let us remark that the condition (L) is invariant under the transforms $x^* = Ax$, $v^* = Bv$, where A, B are nonsingular matrices with constant coefficients.

We have for $\forall\, x^0 \in \bar{\Omega}$, $\xi \in \mathbb{R}^m$, $\eta \in \mathbb{R}^{mn}$,

$$\lambda_1 |\xi|^2 \leqq \frac{\partial a_i^r}{\partial \eta_j^s}(x^0, \xi, \eta)\,\xi_i^r\xi_j^s \leqq \lambda_2 |\xi|^2; \tag{5.4}$$

we take the best values for λ_1 and λ_2. Put $K = \sup\limits_{A,B} \dfrac{\lambda_1}{\lambda_2}$; K is called the condition number.

Theorem 5.4. *If* $K > \dfrac{\sqrt{1 + \dfrac{(n-2)^2}{n-1}} - 1}{\sqrt{1 + \dfrac{(n-2)^2}{n-1}} + 1}$ *then* (L) *is satisfied.*

In [6] we find a modification of a lemma from the paper by the author [17].

Lemma 5.5. *Let* $v \in L^\infty(\mathbb{R}^n) \cap W^{1,2}_{\text{loc}}(\mathbb{R}^n) \cap C^1\left(\overline{B_\delta(0)}\right)$, $\delta > 0$, $g_i \in L_{\text{loc}}(\mathbb{R}^n) \cap L^\infty\left(B_\delta(0)\right)$, $n \geqq 3$, $n - 2 < \lambda < n$, $\int_{\mathbb{R}^n} |\nabla v|^2 r^{-\lambda}\, dx < \infty$, $\int_{\mathbb{R}^n} g_i g_i r^{-\lambda}\, dx < \infty$, $-\nabla v = -\frac{\partial g_i}{\partial x_i}$. *Then*

$$\int_{\mathbb{R}^n} |\nabla v|\, r^{-\lambda}\, dx \leqq \alpha(\lambda) \left(1 + \frac{(n-2)^2}{n-1}\right) \int_{\mathbb{R}^n} g_i g_i r^{-\lambda}\, dx, \tag{5.5}$$

where $\alpha(\lambda) \to 1$ *as* $\lambda \to (n-2)$.

Sketch of the proof of Theorem 5.4: Because ∇v is bounded in $\mathbb{R}^n$ we have

$$\int_{\mathbb{R}^n} |\nabla v'|^2 r^{-\lambda}\, dx < \infty.$$

Put $\gamma = \frac{2}{\lambda_1 + \lambda_2}$. We get

$$\int_{\mathbb{R}^n} \frac{\partial v'_r}{\partial x_i} \frac{\partial \varphi_r}{\partial x_i}\, dx = \int_{\mathbb{R}^n} \left(\frac{\partial v'_r}{\partial x_i} - \gamma \frac{\partial a_i^r}{\partial \eta_j^s} \frac{\partial v'_s}{\partial x_j}\right) \frac{\partial \varphi_r}{\partial x_i}\, dx. \tag{5.6}$$

Put $g_i^r \overset{\text{def}}{=} \frac{\partial v'_r}{\partial x_i} - \gamma \frac{\partial a_i^r}{\partial \eta_j^s} \frac{\partial v'_s}{\partial x_j}$. We have

$$\int_{\mathbb{R}^n} g_i^r g_i^r r^{-\lambda}\, dx \leqq \left(\frac{\lambda_2 - \lambda_1}{\lambda_2 + \lambda_1}\right)^2 \int_{\mathbb{R}^n} |\nabla v'|^2 r^{-\lambda}\, dx, \tag{5.7}$$

so from (5.5) it follows

$$\int_{\mathbb{R}^n} |\nabla v'|^2\, r^{-\lambda}\, dx \leqq \alpha(\lambda) \left(1 + \frac{(n-2)^2}{n-1}\right) \left(\frac{\lambda_2 - \lambda_1}{\lambda_2 + \lambda_1}\right)^2 \int_{\mathbb{R}^n} |\nabla v'|^2\, r^{-\lambda}\, dx,$$

and because

$$\left(1 + \frac{(n-2)^2}{n-1}\right) \left(\frac{\lambda_2 - \lambda_1}{\lambda_2 + \lambda_1}\right)^2 < 1,$$

we have the result. □

References

[1] О. А. Ладыженская, Н. Н. Уральцева, Линейные и квазилинейные уравнения элиптического типа. Москва **1973** (второе издание).

[2] Ch. B. Morrey, Differentiability theorems for weak solutions of nonlinear elliptic differential equations. Bull. Amer. Math. Soc. **75** (1969), 684–705.

[3] Ch. B. Morrey, Existence and differentiability theorems for the solutions of variational problems for multiple integrals. Bull. Amer. Math. Soc. **46** (1940), 439–458.

[4] E. De Giorgi, Sulla differenziabilità e analiticità delle estramali degli integrali multipli regolari. Mem. Acad. Sci. Torino, Cl. Sci. Fis. Mat. Nat. **3** (1957), 25–43.

[5] J. Nash, Continuity of solutions of parabolic and elliptic equations. Comm. Pure Appl. Math. **80** (1958), 931–954.

[6] J. Nečas, On the regularity of weak solutions to nonlinear elliptic systems of partial differential equations. Editrice Tecnico Scientifica, Pisa (to appear).

[7] O. John, J. Nečas, J. Stará, Counterexample to the regularity of weak solution of elliptic systems, Comment. Math. Univ. Carolinae (to appear).

[8] M. Giaquinta, J. Nečas, On the regularity of weak solutions to nonlinear elliptic systems via Liouville's type property. Comment. Math. Univ. Carolinae **20** (1979) 1, 111–121.

[9] S. Campanato, Equazioni ellitiche del II ordine e spazi $L^{2,\lambda}$. Ann. Mat. Pura Appl. **69** (1965), 321–381.

[10] E. Giusti, Regolarità parziale delle soluzioni di sistemi ellitici quasilineari di ordine arbitrario. Ann. Scuola Norm. Sup. Pisa **23** (1969), 115–141.

[11] A. Canfora, Teorema del massimo modulo e teorema di esistenza per il problema di Dirichlet relativo ai sistemi fortemente ellitici. Ricerche Mat. **15** (1966), 249–222.

[12] B. W. Schulze, G. W. Wildenhain, Methoden der Potentialtheorie für elliptische Differentialgleichungen beliebiger Ordnung. Akademie-Verlag, Berlin 1977.

[13] C. Miranda, Teorema del massimo modulo e teorema di esistenza e di unicità per il problema di Dirichlet relativo alle equazioni ellitiche in due variabili. Ann. Math. Pura Appl. **46** (1958), 265–311.

[14] S. Agmon, Maximum theorems for solutions of higher order elliptic equations. Bull. Amer. Math. Soc. **66** (1960), 77–80.

[15] J. Nečas, Example of an irregular solution to a nonlinear elliptic system with analytic coefficients and conditions for regularity. *In:* Theory of Nonlinear Operators (Proc. Internat. Summer School Berlin 1975). Abh. Akad. Wiss. DDR, Jahrg. 1977 Nr. 1N, Akademie-Verlag Berlin 1977, 197–206.

[16] D. Gilbarg, N. S. Trudinger, Elliptic Partial Differential Equations of Second Order. Springer-Verlag, Berlin–Heidelberg–New York 1977.

[17] J. Nečas, On the regularity of weak solutions to variational equations and inequalities for nonlinear second order elliptic systems. Proceedings Equadiff IV, Springer-Verlag, Berlin–Heidelberg–New York 1979.

Address of the author:

Faculty of Mathematics and Physics
Charles University
Malostranské nám. 2/25
11800 Praha 1
Czechoslovakia

Problems in connection with the Hahn-Banach Theorem

Reinhard Nehse, Ilmenau

We consider two aspects of the Hahn-Banach extension property. Firstly, we summarize some well-known facts about ext⁻⁻sions of linear continuous mappings preserving the norm of the initial mapping. Seconuly, we give some new results for conditionally complete vector spaces including separation theorems and applications to nonlinear optimization problems.

1. Preliminaries

Throughout this paper we consider only *real* vector spaces. If E and F are (topological) vector spaces, then $\mathscr{L}(E, F)$ and $\mathscr{L}'(E, F)$ denote the real vector spaces of all linear operators and all linear continuous operators $L: E \to F$, respectively. If $F = R$ (the space of the real numbers), we write $E^* := \mathscr{L}(E, R)$ and $E' := \mathscr{L}'(E, R)$.

For a non-empty subset A of a vector space we define ${}^iA$ — the *algebraical relative interior* of A, i.e.

${}^iA := \{x \in A / \forall\, y \in {}^lA\ \exists\, \varepsilon := \varepsilon(x, y) > 0 \quad \forall\, \lambda \in (-\varepsilon, \varepsilon): x + \lambda(y - x) \in A\}$

where ${}^lA$ denotes the *affine manifold generated by* A;

${}^cA$ — the *convex hull* of A;

$K(A)$ — the *cone generated by* A, i.e. $K(A) := \bigcup\limits_{\lambda \in R_+} \lambda A$

where R_+ denotes the non-negative real numbers.

For a non-empty subset A of a product of vector spaces $E \times F$ the *projection* $P_E(A)$ of A on E is used:

$$P_E(A) := \{x \in E / \exists\, y \in F: (x, y) \in A\}.$$

In connection with ordered sets we shall say that the vector space F is *quasiordered* if the order-relation "$\lesssim$" in F is reflexive, transitive and compatible with the structure of the vector space F, i.e. the *order cone* $K_F := \{y \in F / y \gtrsim 0\}$ is a convex cone with $0 \in K_F$. If $y^1, y^2 \in F$ are indifferent (that means $y^1, y^2 \in K_F \cap (-K_F)$), we write $y^1 \sim y^2$.

2. Hahn-Banach extension properties

In this section we summarize some well-known results about extensions of linear (continuous) operators (see [4], [7], [9]).

Definition 2.1. A normed vector space Y has the *first Hahn-Banach extension property* if for any normed space X, for any (closed) vector subspace $X_0 \subset X$ and for every $L_0 \in \mathscr{L}'(X_0, Y)$ there is an $L \in \mathscr{L}'(X, Y)$ such that $L_0(x) = L(x)$ for all $x \in X_0$ and $\|L_0\| = \|L\|$ where $\|\cdot\|$ denotes the usual operator norm.

Definition 2.2. A normed vector space Y has the *binary intersection property* (or Y is a *space of $\mathfrak{M}$-type*) if each family $\{M_i/i \in I\}$ of closed balls in Y satisfying $M_i \cap M_j \neq \emptyset$ for all $i, j \in I$ has a non-empty intersection $\bigcap_{i \in I} M_i \neq \emptyset$.

Clearly, if Y has the binary intersection property, then Y is complete, i.e. Y is a Banach space. Moreover, as an essential result we know (see [4, 6, 7, 9])

Theorem 2.1. *Let Y be a normed vector space. The following conditions are equivalent:*

(1) *Y has the binary intersection property;*
(2) *Y has the first Hahn-Banach extension property;*
(3) *Y is isometric isomorphic to a space $C(S)$ where S is a compact and extremally disconnected Hausdorff space.*

Here $C(S)$ denotes the real normed vector space of all real-valued bounded continuous functions x on S with the norm

$$\|x\| := \sup \{|x(s)|/s \in S\}.$$

S is said to be *extremally disconnected* if the closure of every open set is open.

In order to give a more general assertion of Hahn-Banach type we will use

Definition 2.3. A quasiordered vector space F is called *conditionally complete* if every non-empty subset A of F has a least upper bound in F whenever there exists an upper bound of A in F (in other words: F has the *least upper bound property*).

Definition 2.4. A quasiordered vector space F has the *second Hahn-Banach extension property* if for any vector space E, for any vector subspace $E_0 \subset E$, for any sublinear mapping $T: E \to F$, and for any $L_0 \in \mathscr{L}(E_0, F)$ with $L_0(x) \lesssim T(x)$ for all $x \in E_0$ there is $L \in \mathscr{L}(E, F)$ such that $L(x) = L_0(x)$ for all $x \in E_0$ and $L(x) \lesssim T(x)$ for all $x \in E$.

If the quasiordered vector space F is conditionally complete, then it is possible to develop fruitful necessary and sufficient conditions under which such extensions exist (see [3, 5, 10, 12, 14]; cf. also [8]). Here we give a fundamental result proved by To [15] (see also [1]).

Theorem 2.2. *Let F be a quasiordered vector space. The following conditions are equivalent:*

(1) *F is conditionally complete;*
(2) *F has the second Hahn-Banach extension property.*

Using the fact that every $C(S)$ (let S be a compact Hausdorff space, let $C(S)$ be equipped with the pointwise ordering) is conditionally complete if and only if S is extremally disconnected, we get in connection with Theorem 2.1:

Each normed vector space Y possessing the first Hahn-Banach extension property is isometric isomorphic to a quasiordered vector lattice having the second Hahn-Banach extension property.

The converse assertion is not true (see Nachbin [9]).

While the binary intersection property is a condition of the same type as the compactness condition, the condition of conditional completeness is more geometric. This we get from the following simple property.

Theorem 2.3. *Let F be a quasiordered vector space. F is conditionally complete if and only if for every non-empty subset A of F there is a $y_A \in F$ such that*

$$\bigcap_{a \in A} (a + K_F) \neq \emptyset \Rightarrow \bigcap_{a \in A} (a + K_F) = y_A + K_F .$$

Therefore, with respect to applications to optimization problems it is more useful to consider quasiordered and conditionally complete vector spaces.

3. Separation of two sets

In order to study the second Hahn-Banach extension property and equivalent conditions we give (see [11, 12, 13])

Theorem 3.1. *Let F be a quasiordered vector space. If F is conditionally complete, then F has the following separation property: For any vector space E, for any subsets A and B of $E \times F$ with*

$$(0, y) \in {}^cK(A - B) \Rightarrow y \gtrsim 0, \tag{3.1}$$

$${}^iP_E[{}^cK(A - B)] \subset P_E[{}^cK(A - B)] \tag{3.2}$$

there exist $L \in \mathscr{L}(E, F)$ and $y^0 \in F$ such that

$$L(x^1) - y^1 \lesssim y^0 \lesssim L(x^2) - y^2 \qquad \forall\, (x^1, y^1) \in A, \qquad \forall\, (x^2, y^2) \in B. \tag{3.3}$$

It is easy to prove that (3.1) is a necessary condition for (3.3). Sufficient conditions for (3.1) and (3.2) are contained in [11], [12] and in

Lemma 3.2. *Let E be a vector space, let F be a quasiordered vector space, and let A and B be subsets of $E \times F$.*

(1) *If $K(A - B)$ is convex and if*

$$\left.\begin{matrix}(x, y^1) \in A\\ (x, y^2) \in B\end{matrix}\right\} \Rightarrow y^1 \gtrsim y^2,$$

then (3.1) *is satisfied.*

(2) *If $0 \in {}^iP_E[{}^cK(A - B)]$, then* (3.2) *holds.*

From this we can see that (3.1) is a certain "convexity condition" for A and B, (3.2) is a certain "interior point condition". While (3.1) is a necessary condition for (3.3), (3.2) is a more technical assumption. But this condition cannot be dropped, as the following example shows.

Consider the convex sets

$$A = \{(x, y) \in R^2 / -\sqrt{x} \leqq y\}, \quad B = \{(x, y) \in R^2 / x = 0, y < 0\}.$$

If we use the following representation $E \times F$ for R^2: E is spanned by $(1, 0) \in R^2$, F is generated by $(0, 1) \in R^2$ and ordered by $(0, y_1) \gtrsim (0, y_2)$ iff $y_1 \geqq y_2$, then (3.1) is satisfied, but (3.2) is not fulfilled. There are not any $L \in \mathscr{L}(R, R)$ and $(0, y_0) \in F$ that (3.3) holds.

Using another representation for R^2, for example E as above, F is spanned by $(1, 1) \in R^2$ and ordered by $(y_1, y_1) \gtrsim (y_2, y_2)$ iff $y_1 \geqq y_2$, both conditions (3.1) and (3.2) are satisfied for A and B and a separation of these sets is possible in the sense of (3.3).

These examples show that the separation of sets A and B in the sense of (3.3) depends on the representation of the space containing these sets. Therefore, it is useful to have the following interpretation for Theorem 3.1.

Theorem 3.3. *Let A and B be subsets of a vector space $\bar{E}$ having a representation as a product of vector spaces $E \times F$ such that the following conditions are satisfied:*

(a) *F is quasiordered and conditionally complete,*
(b) *see* (3.1),
(c) *see* (3.2).

Then there are $L \in \mathscr{L}(E, F)$ and $y^0 \in F$ such that (3.3) *holds.*

For a separation theorem of sets in topological spaces we refer to

Theorem 3.4 (see [11], [12]). *Let A and B be subsets of a topological vector space $\bar{E}$ possessing a representation as a product of topological vector spaces $E \times F$ such that the following conditions are satisfied:*

(a) *F is quasiordered and conditionally complete,*
(b) *the order-cone K_F is closed in the topology $\mathfrak{T}$ of F,*
(c) *the topology $\mathfrak{T}$ has a basis $\mathfrak{B}$ of neighbourhoods of zero such that every $V \in \mathfrak{B}$ has a representation of the following form: $V = (V + K_F) \cap (V - K_F)$,*
(d) *see* (3.1),
(e) $0 \in \operatorname{int} P_E[{}^cK(A - B)]$,
(f) $T(x) := \inf \{y/(x, y) \in P_{E \times F_0}[{}^cK(A - B)]\}, \quad x \in E,$

is continuous at the origin (F_0 denotes a partially ordered subspace of F; an algebraical complementary space of $K_F \cap (-K_F)$ ordered by $F_0 \cap K_F$ can be used).

Then there are $L \in \mathscr{L}'(E, F)$ and $y^0 \in F$ such that (3.3) *holds.*

As a special case including the usual separation by hyperplanes we get (see [11, 12, 13])

Corollary 3.5. *Let A and B be subsets of a vector space $\bar{E}$ with ${}^{ic}K(A - B) \neq \emptyset$. A and B can be separated properly by a hyperplane if and only if $0 \notin {}^{ic}K(A - B)$.*

Corollary 3.6. *Let A and B be subsets of a topological vector space $\bar{E}$ with* $\operatorname{int} {}^cK(A - B) \neq \emptyset$. *$A$ and B can be separated properly by a closed hyperplane if and only if* $0 \notin \operatorname{int} {}^cK(A - B)$.

4. Related results to the second Hahn-Banach extension property

We know several conditions which are equivalent to the second Hahn-Banach extension property (see [4, 5, 10, 12]). Here we consider a special group of them. Simultaneously, the assumptions in these conditions are weaker than in previous papers. The proofs of the following theorems are straightforward, if we use Theorem 3.1.

Theorem 4.1. *Let F be a quasiordered vector space. If F is conditionally complete, then F has the geometrical Hahn-Banach property: For any vector space E, for any subset A of $E \times F$ and any affine manifold B of $E \times F$ with* (3.1) *and* (3.2) *there are $L \in \mathscr{L}(E, F)$ and $y^0 \in F$ such that*

$$L(x^1) - y^1 \lesssim y^0 \sim L(x^2) - y^2 \qquad \forall\, (x^1, y^1) \in A, \qquad \forall\, (x^2, y^2) \in B. \tag{4.1}$$

If F is partially ordered and conditionally complete, then we get

$$L(x^1) - y^1 \lesssim y^0 = L(x^2) - y^2 \qquad \forall\, (x^1, y^1) \in A, \qquad \forall\, (x^2, y^2) \in B$$

instead of (4.1). In this case B is contained in the affine manifold M defined by

$$M := \{(x, y) \in E \times F / L(x) - y = y^0\};$$

this is the usual meaning of the geometrical form of the Hahn-Banach theorem.

Theorem 4.2. *Let F be a quasiordered vector space. If F is conditionally complete, then F has the Krein-Rutman extension property: For any quasiordered vector space F_1, for any subspace A of F_1 quasiordered by $K_A := A \cap K_{F_1}$, for any convex subset D of F_1 with $0 \in {}^i[(D \cap A) - K_{F_1}]$, and for any $L_0 \in \mathscr{L}(A, F)$ with $L_0(x) \gtrsim 0$ for all $x \in D \cap K_A$ there is an $L \in \mathscr{L}(F_1, F)$ such that $L(x) = L_0(x)$ for all x belonging to A and*

$$L(x) \gtrsim 0 \qquad \forall\, x \in [(D \cap A) + A_1] \cap K_{F_1}$$

where A_1 denotes an algebraical complementary space of A in F_1.

Clearly, several well-known theorems about extension of positive linear operators are contained in Theorem 4.2 (for example $D = A$, $A \cap {}^iK_{F_1} \neq \emptyset$ or $D = A$, $A + K_{F_1}$ is a subspace of F_1).

Theorem 4.3. *Let F be a quasiordered vector space. If F is conditionally complete, then F has the sandwich property: For any vector space E, for any non-empty subsets $D(U)$ and $D(V)$ of E, for any mappings $U : D(U) \to F$, $V : D(V) \to F$ satisfying condition* (3.1) *and condition* (3.2) *of Theorem* 3.1 *with*

$$A := \text{epi } U := \{(x, y) \in E \times F / U(x) \lesssim y, \;\; x \in D(U)\},$$

$$B := \text{hypo } V := \{(x, y) \in E \times F / V(x) \gtrsim y, \;\; x \in D(V)\}$$

there is an "affine" operator $x \to L(x) - y^0$, $L \in \mathscr{L}(E, F)$, $y^0 \in F$ such that

$$L(x) - y_0 \lesssim U(x) \qquad \forall\, x \in D(U), \quad V(x) \lesssim L(x) - y^0 \qquad \forall\, x \in D(V).$$

In order to show that the conditional completeness of F is also a necessary condition in the theorems of Section 3 and Section 4 we may prove that the second Hahn-Banach extension property follows from the geometrical Hahn-Banach property, from the Krein-Rutman extension property and from the sandwich property (see [5, 10, 12, 13]). The proofs are finished by use of Theorem 2.2.

5. An application to non-convex optimization

By Theorem 3.1 (Theorem 3.3 or Corollary 3.5) we may extend some results in optimization theory for $F = R$, too. As an example we give an optimality condition (see [11, 12]).

In this section let E and W be (topological) vector spaces, let K_W be a (closed) convex cone in W with $0 \in K_W$, let $D(f)$ and $D(g)$ be subsets of E with $P_0 := D(f) \cap D(g) \neq \emptyset$ and let be $f : D(f) \to R$, $g : D(g) \to W$.

Consider the problem

$$\text{(P)} \qquad f(x) \to \min,$$

$$x \in G, \quad G := \{y \in P_0 / -g(y) \in K_W\}.$$

Theorem 5.1.

(1) *Let $x^0 \in G$ be a solution of (P). If the conditions* (3.1) *and* (3.2) *of Theorem* 3.1 *are satisfied for*

$$A := \{(g(x) + k, f(x) - f(x^0) + \lambda)/x \in P_0, k \in K_W, \lambda \in R_+\},$$

$$B := \{(0, 0)\} \subset W \times R,$$

then there is a linear functional $u^0 \in W^*$ with

$$\langle u^0, y \rangle \geqq 0 \qquad \forall\, y \in K_W \quad (\textit{that is} \quad u^0 \in K_W^*)$$

such that (x^0, u^0) is a saddle-point of the Lagrangian of (P) with respect to $P_0 \times K_W^*$.

(2) *Let K_W be algebraically closed. Let ${}^iK_W \neq \emptyset$ or let exist a countable algebraical basis of W. If (x^0, u^0) is a saddle-point of the Lagrangian of (P), then x^0 is a solution of (P).*

Since (3.1) is a necessary condition for (3.3), it is easy to show that (3.1) for the sets A and B given in Theorem 5.1 is a necessary condition for the existence of a saddle-point, too.

Now, it is an essential problem to get more practical conditions for the mappings f and g such that (3.1) is valid for sets A and B given above. As a sufficient condition for (3.2) we may use

$$0 \in {}^{ic}[g(P_0) + K_W],$$

a so-called *generalized Slater-condition.*

We will later on need the following notion. Let (f, g) be a mapping defined by $(f, g)(x) := (f(x), g(x))$, $x \in P_0$.

Definition 5.1. (1) (f, g) is said to be *convex-like* on P_0 if for every $x^1, x^2 \in P_0$, $\lambda \in (0, 1)$ there is $x^3 := x^3(x^1, x^2, \lambda) \in P_0$ such that

$$\lambda f(x^1) + (1 - \lambda) f(x^2) - f(x^3) \geqq 0,$$

$$\lambda g(x^1) + (1 - \lambda) g(x^2) - g(x^3) \in K_W.$$

(2) Let P_0 be convex. (f, g) is said to be *strongly v-pseudoconvex* on P_0 if for every $x^2 \in P_0$ there are $u_f \in E^*$ and $L_g \in \mathcal{L}(E, W)$ such that

$$\left.\begin{array}{ll} p_f(x^1, x^2)\,[f(x^1) - f(x^2)] - \langle u_f, x^1 - x^2\rangle \geqq 0 & \forall\, x^1 \in P_0, \\ p_g(x^1, x^2)\,[g(x^1) - g(x^2)] - L_g(x^1 - x^2) \in K_W & \forall\, x^1 \in P_0 \end{array}\right\} \tag{5.1}$$

for certain functions

$$p_f : P_0 \times P_0 \to R_+ \setminus \{0\}, \quad p_g : P_0 \times P_0 \to R_+ \setminus \{0\} \quad \text{with} \quad p_f \equiv p_g.$$

(5.1) means that the mappings f and g are subdifferentiable in a certain generalized sense.

Theorem 5.2. *If (f, g) is convex-like on P_0, then* (3.1) *holds for the sets A and B given in Theorem* 5.1.

This assertion easily follows from a result proved by Vogel [16] in connection with the assumption that x^0 is a solution of (P).

In order to give more analytical conditions for the convex-likeness of (f, g) let us refer to

Theorem 5.3. *Let (f, g) be strongly v-pseudoconvex on P_0 and let $p := p_f = p_g$ on $P_0 \times P_0$.*

(1) *If E is a topological vector space and if $p(x, \cdot)$ is continuous on P_0 for every fixed $x \in P_0$, then (f, g) is convex-like on P_0.*

(2) *If E is a vector space and if the equation*

$$[(1 - \lambda)\, p(x^1, x) + \lambda p(x^2, x)]\, x = \lambda p(x^2, x)\, x^1 + (1 - \lambda)\, p(x^1, x)\, x^2 \tag{5.2}$$

has a solution $\bar{x} := \bar{x}(x^1, x^2, \lambda)$ for every $x^1, x^2 \in P_0$ and every $\lambda \in (0, 1)$, then (f, g) is convex-like on P_0.

Proof. (1) Let $x^1, x^2 \in P_0$ be fixed, let $\lambda \in [0, 1]$, $h := (f, g)$, $K := R_+ \times K_W$, $L := (u_f, L_g)$, $p_1(\lambda) := p\big(x^1, \lambda x^1 + (1 - \lambda)\, x^2\big)$ and $p_2(\lambda) := p\big(x^2, \lambda x^1 + (1 - \lambda)\, x^2\big)$. By definition there is a linear operator $L\colon E \to R \times W$ connected with $\lambda x^1 + (1 - \lambda) \times x^2 \in P_0$ such that

$$p_1(\lambda)\,\big[h(x^1) - h\big(\lambda x^1 + (1 - \lambda)\, x^2\big)\big] - (1 - \lambda)\, L(x^1 - x^2) \in K,$$

$$p_2(\lambda)\,\big[h(x^2) - h\big(\lambda x^1 + (1 - \lambda)\, x^2\big)\big] + \lambda L(x^1 - x^2) \in K.$$

Hence

$$\mu h(x^1) + (1 - \mu)\, h(x^2) - h\big(\lambda x^1 + (1 - \lambda)\, x^2\big) - \left[\frac{\mu(1 - \lambda)}{p_1(\lambda)} - \frac{\lambda(1 - \mu)}{p_2(\lambda)}\right] L(x^1 - x^2) \in K \tag{5.3}$$

follows for any fixed $\mu \in (0, 1)$. Since p_1 and p_2 are positive-valued and continuous on $[0, 1]$, the function $\bar{p}$ with

$$\bar{p}(\lambda) := \frac{\mu(1 - \lambda)}{p_1(\lambda)} - \frac{\lambda(1 - \mu)}{p_2(\lambda)}, \quad \lambda \in [0, 1],$$

is continuous on [0, 1] and we have $\overline{p}(0) > 0$, $\overline{p}(1) < 0$. Applying Bolzano's theorem there is $\overline{\lambda} \in (0, 1)$ such that $\overline{p}(\overline{\lambda}) = 0$. Therefore, we obtain from (5.3)

$$\mu h(x^1) + (1 - \mu)\, h(x^2) - h\big(\overline{\lambda} x^1 + (1 - \overline{\lambda})\, x^2\big) \in K$$

for this $\overline{\lambda} := \overline{\lambda}(x^1, x^2, \mu)$. Then we can choose $x^3 := \overline{\lambda} x^1 + (1 - \overline{\lambda})\, x^2$.

(2) Let $x^1, x^2 \in P_0$, $\lambda \in (0, 1)$, let $\overline{x}$ be a solution of (5.2), let $p_1 := p\,(x^1, \overline{x})$, $p_2 := p(x^2, \overline{x})$. Then

$$\overline{x} = \frac{\lambda p_2}{(1 - \lambda)\, p_1 + \lambda p_2}\, x^1 + \frac{(1 - \lambda)\, p_1}{(1 - \lambda)\, p_1 + \lambda p_2}\, x^2 \tag{5.4}$$

and $\overline{x} \in P_0$ follows because of the convexity of P_0. Using $h := (f, g)$, $K := R_+ \times K_W$, $L := (u_f, L_g)$ as a linear operator connected with $\overline{x}$ we get

$$p_1[h(x^1) - h(\overline{x})] - L(x^1 - \overline{x}) \in K,$$

$$p_2[h(x^2) - h(\overline{x})] - L(x^2 - \overline{x}) \in K$$

and, therefore,

$$\lambda h(x^1) + (1 - \lambda)\, h(x^2) - h(\overline{x}) - L\left(\frac{\lambda}{p_1}\,(x^1 - \overline{x}) + \frac{1 - \lambda}{p_2}\,(x^2 - \overline{x})\right) \in K.$$

In connection with (5.4)

$$\lambda h(x^1) + (1 - \lambda)\, h(x^2) - h(\overline{x}) \in K$$

follows.

Theorem 5.4. *Let f and g be strongly v-pseudoconvex on P_0 (that means $p_f = p_g$ is not necessary* (see Definition 5.1 above)). *If there is $\overline{\mu} := \overline{\mu}(x^1, x^2, \lambda_1, \lambda_2, \lambda) \in [0, 1]$ such that $h_1(\overline{\mu}) = h_2(\overline{\mu}) = 0$, where $x^1, x^2 \in P_0$, $\lambda_1, \lambda_2 \in R_+$, $\lambda \in (0, 1)$ are any fixed elements and*

$$h_1(\mu) := \frac{(1 - \mu)\, \lambda\lambda_1}{p_{f1}(\mu)} - \frac{\mu(1 - \lambda)\, \lambda_2}{p_{f2}(\mu)},$$

$$h_2(\mu) := \frac{(1 - \mu)\, \lambda\lambda_1}{p_{g1}(\mu)} - \frac{\mu(1 - \lambda)\, \lambda_2}{p_{g2}(\mu)}, \quad \mu \in [0, 1],$$

$$p_{fj}(\mu) := p_f\big(x^j, \mu x^1 + (1 - \mu)\, x^2\big),$$

$$p_{gj}(\mu) := p_g\big(x^j, \mu x^1 + (1 - \mu)\, x^2\big), \quad j = 1, 2,$$

then (3.1) *holds for the sets A and B given in Theorem* 5.1.

An easy calculation proves this assertion. As a sufficient condition for the complicated assumption in Theorem 5.4 let us refer to the following fact: If for each $x' \in P_0$

$$p_f(x', x) = p_g(x', x) = c_{x'} \qquad \forall\, x \in P_0$$

holds, where $c_{x'} \in R_+ \setminus \{0\}$ is a constant depending only on x', then there is $\overline{\mu} \in [0, 1]$ such that $h_1(\overline{\mu}) = h_2(\overline{\mu}) = 0$.

While Theorem 5.2 guarantees that A (see Theorem 5.1) is convex, from Theorem 5.4 the convexity of $K(A)$ follows only.

References

[1] Akilov, G. P., Kutateladze, S. S., Ordered vector spaces (Russian). Nauka, Novosibirsk 1978.
[2] Bair, J., Fourneau, R., Étude geometrique des espaces vectoriels. Springer-Verlag, Berlin—Heidelberg—New York 1975.
[3] Bittner, L., Extension theorems of Hahn-Banach type and monotone operators. *In:* Theory of Nonlinear Operators (Proc. Summer School Berlin 1977), Abh. Akad. Wiss. DDR, Jahrg. 1978, Nr. 6N, Akademie-Verlag, Berlin 1978, 41—50.
[4] Day, M. M., Normed linear spaces. Springer-Verlag, Berlin—Heidelberg—New York 1973 (third ed.).
[5] Elster, K.-H., Nehse, R., Necessary and sufficient conditions for the order-completeness of partially ordered vector spaces. Math. Nachr. **81** (1978), 301—311.
[6] de Jonge, E., van Rooij, A. C. M., Introduction to Riesz spaces. Mathematisch Centrum, Amsterdam 1977.
[7] Kantorovich, L. W., Akilov, G. P., Functional analysis (Russian). Nauka, Moscow 1977.
[8] Krauß, E., A general Hahn-Banach type theorem. *In:* Theory of Nonlinear Operators (Proc. Summer School Berlin 1977), Abh. Akad. Wiss. DDR, Jahrg. 1978, Nr. 6N, Akademie-Verlag, Berlin 1978, 323—333.
[9] Nachbin, L., A theorem of the Hahn-Banach type for linear transformations. Trans. Amer. Math. Soc. **68** (1950), 28—46.
[10] Nehse, R., The Hahn-Banach property and equivalent conditions. Comment. Math. Univ. Carolinae **19** (1978), 165—177.
[11] Nehse, R., Some general separation theorems. Math. Nachr. **84** (1978), 319—327.
[12] Nehse, R., Ordnungstheoretische Untersuchungen zu nichtkonvexen Optimierungsproblemen, Dissertation B, TH Ilmenau 1978.
[13] Nehse, R., Separation of two sets in a product space. Math. Nachr. **97** (1980), 179—187.
[14] Thera, M., Étude des fonctions convexes vectorielles semicontinues. Thèse, Univ. de Pau 1978.
[15] To, T. O., The equivalence of the least upper bound property and the Hahn-Banach extension property in ordered linear spaces. Proc. Amer. Math. Soc. **30** (1972), 287—295.
[16] Vogel, W., Ein Maximumprinzip für Vektor-Optimierungsaufgaben. Oper. Res. Verf. XIX (1975), 161—184.

Address of the author:

Technische Hochschule Ilmenau
Sektion Mathematik, Rechentechnik und ökonomische Kybernetik
6300 Ilmenau
Am Ehrenberg
DDR

Parabolic problems with free boundaries: Existence and properties of solutions; optimal control problems

Irena Pawłow, Warszawa

Abstract. The paper is concerned with multidimensional parabolic free boundary value problems and some related optimal control problems. The multidimensional two-phase Stefan type problem including both nonlinear parabolic equations and boundary conditions is considered. Existence, uniqueness and regularity of a weak solution as well as its continuous and monotone dependence upon the data are shown. Some typical optimal control problems associated with the Stefan type processes are considered. Existence of optimal controls for boundary control problems with distributed, terminal or boundary observation is discussed.

Some open questions concerning various parabolic free boundary problems are formulated.

1. Introduction

The study of many physical processes of heat flow and diffusion involving change of phase gives rise to free boundary value problems for parabolic equations [18, 51, 54].

In this paper we are concerned with multidimensional parabolic free boundary value problems of the Stefan type [20, 51] and also with some related optimal control problems. A number of results briefly discussed here was extensively studied in [46, 47]. Let $\Omega \subset \mathscr{R}^N$ (N denotes a given positive integer) be a domain satisfying assumption (A1) (see Section 2), $\Gamma = \partial\Omega$. Given $0 < T < \infty$, denote $Q = \Omega \times (0, T)$, $\Sigma = \Gamma \times (0,T)$, $S \subset Q$ — free boundary. Then $Q = Q_1 \cup S \cup Q_2$ where Q_1 and Q_2 are separated by S. We will consider the following two-phase multidimensional problem of the Stefan type including both nonlinear parabolic equation and boundary conditions (called further (SP))

$$c(x, t, \theta)\,\frac{\partial\theta}{\partial t} - \sum_{j=1}^{N} \frac{\partial}{\partial x_j}\left[k(x, t, \theta)\,\frac{\partial\theta}{\partial x_j}\right] = \bar{F}(x, t, \theta) \quad \text{in} \quad Q \setminus S, \tag{1.1}$$

$$k(x, t, \theta)\,\frac{\partial\theta}{\partial\nu} = \bar{\psi}(x, t, \theta), \quad (x, t) \in \Sigma, \tag{1.2}$$

$$\theta(x, t) = \theta^* \quad \text{on} \quad S, \tag{1.3}$$

$$\varkappa\,\frac{\partial\Phi}{\partial t} = \left[\sum_{j=1}^{N} k(x, t, \theta)\,\frac{\partial\theta}{\partial x_j}\,\frac{\partial\Phi}{\partial x_j}\right]\Bigg|_{\theta^*-}^{\theta^*+}, \quad (x, t) \in S, \tag{1.4}$$

$$\theta(x, 0) = \theta_0(x), \quad x \in \Omega, \tag{1.5}$$

where $S \triangleq \{(x, t) \in Q \mid \Phi(x, t) = 0\}$; c, k are given strictly positive functions with discontinuities of the first kind on S; $\bar{F}$, $\bar{\psi}$, θ_0 are given functions and $\varkappa$, θ^* are given finite positive constants. Complete assumptions concerning the data will be specified in Section 2.

By a classical solution of problem (SP) we understand a pair $\{\theta, \Phi\}$ satisfying (1.1) to (1.5), such that θ, $\nabla\theta \in C(\bar{Q}_i)$, $\Delta\theta$, $\frac{\partial\theta}{\partial t} \in C(Q_i)$, $i = 1, 2$, $\Phi \in C^1(\bar{Q})$, and S is an N-dimensional C_1-manifold.

The question about existence of such a solution stays open except for some special cases. In the case of space dimension one there are numerous results concerning existence, uniqueness, stability, regularity and asymptotic behaviour of the solution. We refer to [48, 51, 54] and the literature quoted therein, as well as to [5, 6, 9, 17, 19, 21, 23, 44]. There are also some results on regularity of "almost classical" solutions of multidimensional one-phase problems [4, 34, 35]. Two-phase and multi-phase Stefan problems in several space dimensions on the contrary to those one-dimensional may show lack of regularity even if the data are arbitrarily smooth [20, 48]. For this reason it is a common practice to reformulate these problems in such a way as to absorbe the free boundary into the differential equation.

The concept of a weak solution to problems of Stefan type was introduced firstly by Kamenomotskaya [30, 31] and next extended by Oleinik [49]. It was based on applying the so called Kirchhoff transformation and introducing enthalpy function (see Section 3).

Kamenomotskaya [30, 31] considered the Stefan problem (SP) in the case $c(x, t, \theta) \equiv c(\theta)$, $k(x, t, \theta) \equiv k(\theta)$, $\bar{F}(x, t, \theta) \equiv 0$, including the Dirichlet boundary condition instead of (1.2). She proved existence and uniqueness of the weak solution in the multidimensional case by means of approximations using an explicit finite-difference scheme. Oleinik [49] considered a more general Stefan problem. She proved existence and uniqueness of the weak solution in the case of $N = 1$ via a smoothing technique. The results of Kamenomotskaya and Oleinik were next refined and extended by Friedman [20].

The Stefan problem in the form (1.1)–(1.5) was investigated only by Budak and Gaponenko [3]. In principle they considered the case $N = 1$. Under some assumptions concerning the coefficients and the data they proved existence and uniqueness of a weak solution, exploiting the smoothing technique of Oleinik.

Another approach to the analysis of Stefan problems, based on the employing of monotonicity methods, was proposed by Brezis [2] and extended by Lions [37]. That approach was also developed by Jerome [28, 29] and Damlamian [11, 12, 13].

Stefan problems were also extensively studied via transforming them to variational inequalities. Such an approach was introduced by Duvaut [14] and independently by Frémond [18]. This technique was applied by several authors, among them Lions [38, 39, 40], Friedman [22], Magenes [42], Moreno and Saguez [43], Saguez [52, 53]. Besides, in [43, 52, 53] some optimal control problems for variational inequalities were considered.

Our paper [47] extends results of Budak and Gaponenko [3] to the multidimensional case. In most of the papers quoted above only regular domains (i.e. domains with smooth boundaries) were considered, whereas from the practical point of view it is often necessary to have the possibility of taking into account also domains with less regular boundaries, in particular piecewise smooth (having angle points). Apart from Jerome [29],

Budak and Gaponenko [3] suggested the possibility of taking into consideration such domains. We also consider domains having only piecewise regular boundary.

Now we give a short outline of our paper. In Section 2 the basic assumptions concerning the domain Ω and data are formulated. In Section 3 we introduce a weak statement of the Stefan problem (SP). In the next section we are concerned with the existence and uniqueness of the weak solution, its regularity as well as some a priori estimates for it. In Section 5 stability of the solution and its monotone dependence upon the data are discussed. Section 6 is devoted to some optimal control problems formulated for processes of Stefan type. Problems with boundary controls and distributed, terminal or boundary observation are considered. Some results concerning continuity of mappings control $\rightarrow$ state are presented. For all the problems the existence of optimal controls is discussed. Finally. in Section 7 we formulate some open questions concerning various parabolic free boundary problems.

We start with introducing basic notations:

$y|_\Sigma$ — trace of the function y at the boundary hypersurface Σ;
$y|_T$ — trace of y at the hyperplane $t = T$;
$\vec{\nu}$ — vector normal to Σ, external with respect to Q;
μ_j, $j \in \mathcal{N}$ — j-dimensional Lebesgue measure;
μ_Σ — Lebesgue measure defined over the hypersurface Σ;
Δ — Laplace operator in $\mathcal{R}^N$, $\Delta = \sum_{j=1}^{N} \frac{\partial^2}{\partial x_j^2}$;
$\nabla = \left(\frac{\partial}{\partial x_1}, \ldots, \frac{\partial}{\partial x_N}\right)$;
$\underset{X}{\rightarrow}$, $\underset{X}{\rightharpoonup}$ — denote respectively strong and weak convergence in the normed space X;

$H^1(Q)$, $H^1(0, T; L^2(\Gamma))$ denote appropriate Sobolev spaces [1, 36, 41]. For simplicity we will often denote functions f depending on x, t and y, which in turn depends also on x and t, by $f(y)$ instead of $f(x, t, y)$.

2. Assumptions (A)

(A 1) Ω is a bounded, open and connected domain satisfying a strong cone condition [1]. Γ is piecewise of the Hölder class $\mathcal{H}^{2+\beta}$, $0 < \beta < 1$ [36].

(A 2) c, k are continuous with respect to $\theta \in \mathcal{R} \setminus \{\theta^*\}$, strictly positive and bounded from above by finite constants (dependent upon M, where $\theta \in [-M, M]$); moreover $c(\cdot, \cdot, \theta) \in C^{0,1}(\bar{Q})$, $k(\cdot, \cdot, \theta) \in C^{2,1}(\bar{Q})$ $\quad \forall\, \theta \in \mathcal{R} \setminus \{\theta^*\}$.

(A 3) $\frac{\partial k}{\partial x_j}(x, t, \theta)/k(x, t, \theta)$ does not depend on θ.

(A 4) $\bar{F} \in C(\bar{Q} \times \mathcal{R})$, moreover $\bar{F}(x, t, \cdot)$ is θ-Lipschitz continuous uniformly for $(x, t) \in Q$ and satisfies the growth condition $\theta \bar{F}(x, t, \theta) \leq \alpha_1 \theta^2 + \alpha_2$, $(x, t, \theta) \in Q \times \mathcal{R}$, with constants $0 < \alpha_1, \alpha_2 < \infty$.

(A 5) $\bar{\psi} \in C(\bar{\Sigma} \times \mathcal{R})$, moreover $\bar{\psi}(x, t, \cdot)$ is θ-Lipschitz continuous uniformly for $(x, t) \in \Sigma$, satisfies the growth condition $\theta \bar{\psi}(x, t, \theta) \leq \alpha_3 \theta^2 + \alpha_4$, $(x, t, \theta) \in \Sigma \times \mathcal{R}$, with constants $0 < \alpha_3, \alpha_4 < \infty$, and $\bar{\psi}(x, \cdot, \theta)$ is t-Lipschitz continuous uniformly for $(x, \theta) \in \Gamma \times (-M, M)$ with any $0 < M < \infty$.

(A 6) $\theta_0 \in H^1(\Omega) \cap L^\infty(\Omega)$ and $\mu_N(S_0) = 0$ where $S_0 = \{x \in \Omega \mid \theta_0(x) = \theta^*\}$.

(A 7) $\varkappa, \theta^*$ are given positive finite constants.

Remark. In Section 6 we will assume that $\bar{\psi}(x, t, \theta) \equiv \bar{\psi}_1(x, t, \theta) + \bar{\psi}_2(x, t)$. Then we will postulate that

(A 5′) $\bar{\psi}_1$ satisfies (A 5), $\bar{\psi}_2 \in L^\infty(\Sigma)$, $\partial\bar{\psi}_2/\partial t \in L^1(\Sigma)$.

3. Weak formulation of Stefan problem

We give here a weak formulation of the two-phase Stefan problem (SP) nevertheless it could be extended onto m-phase Stefan problems $(m \geqq 2)$.

Following [31, 49] we transform problem (SP) via the Kirchhoff transformation

$$y = K(x, t, \theta) \triangleq \int_0^\theta k(x, t, \xi)\, d\xi ,$$

introducing also so called enthalpy

$$\gamma(x, t, y) \triangleq \int_0^y \frac{c\big((x, t, K^{-1}(x, t, \eta)\big)}{k\big(x, t, K^{-1}(x, t, \eta)\big)}\, d\eta + \lambda(x, t, y) ,$$

where $$\lambda(x, t, y) = \begin{cases} 0 & \text{for} \quad y < y^*(x, t) \\ \varkappa & \text{for} \quad y > y^*(x, t) \\ \varkappa^*(x, t) & \text{for} \quad y = y^*(x, t) \end{cases}$$

with any measurable function $\varkappa^*$ satisfying the condition

$$0 \leqq \varkappa^*(x, t) \leqq \varkappa \quad \text{a.e. on } S .$$

Next, multiplying the transformed parabolic equation by any function $\eta \in C^{2,1}(\bar{Q})$, $\eta(\cdot, T) \equiv 0$, integrating by parts over Q_i, $i = 1, 2$ and taking into account (1.5), we come to the integral identity

$$\begin{aligned} &\iint_Q \left[\gamma(y) \frac{\partial \eta}{\partial t} + y \Delta \eta + \sum_{j=1}^N b_j^1(y) \frac{\partial \eta}{\partial x_j} + b^1(y)\, \eta\right] dQ \\ &+ \int_\Sigma \left[\psi(y)\, \eta - y \frac{\partial \eta}{\partial \nu}\right] d\Sigma + \int_\Omega \gamma\big(y_0(x)\big)\, \eta(x, 0)\, d\Omega = 0, \end{aligned} \tag{3.1}$$

where $y_0 = K(\theta_0)$, b_j^1, b^1 correspond to lower order terms in (1.1) and

$$\psi(y) = \bar{\psi}\big(K^{-1}(y)\big) + \sum_{j=1}^N \cos(\vec{\nu}, x_j) \left[-y \frac{\dfrac{\partial k}{\partial x_j}\big(K^{-1}(y)\big)}{k\big(K^{-1}(y)\big)} + \int_0^{K^{-1}(y)} \frac{\partial k}{\partial x_j}(\xi)\, d\xi \right].$$

Definition 3.1. A function $y \in L^\infty(Q)$, satisfying (3.1) for every $\eta \in C^{2,1}(\bar{Q})$, $\eta(\cdot, T) \equiv 0$, is called *weak solution of problem* (SP).

4. Exitence and uniqueness of weak solution

Theorem 4.1. *Let the assumptions* (A) *hold. Then:* (i) *there exists a weak solution* y *of problem* (SP); (ii) $y \in H^1(Q)$ *and satisfies a priori estimates*

$$\|y\|_{L^\infty(Q)} \leqq M_1, \quad \|y\|_{H^1(Q)} \leqq M_2, \tag{4.1}$$

where M_1, M_2 *are finite constants dependent only upon the bounds on the data of problem* (SP).

We present here only the idea of the proof of this theorem, referring to [47] for details. The proof consists of two stages.

First we introduce a family of auxiliary nonlinear parabolic boundary value problems including smooth coefficients, boundary and initial data. These problems are formulated in the domains $Q^n = \Omega^n \times (0, T)$, where $\Omega^n \subset \mathscr{R}^N$, $n \in \mathscr{N}$ is a decreasing family of domains having boundaries Γ^n of the Hölder class $\mathscr{H}^{2+\beta}$, $0 < \beta < 1$, and approximating Ω in the following sense: $\Omega \subset \Omega^n$ for every $n \in \mathscr{N}$, $\mu_N\{\Omega^n \setminus \Omega\} \xrightarrow{n} 0$, $\mu_{N-1}\{\Gamma^n \cap \Gamma\} \xrightarrow{n} \mu_{N-1}(\Gamma)$. Let us formulate the auxiliary problem (SP^n):

$$(SP^n) \begin{cases} \dfrac{\partial y^n}{\partial t} = a^n(x, t, y^n)\,\Delta y^n + \bar{b}^n(x, t, y^n, \nabla y^n) \quad \text{in} \quad Q^n, \\[2mm] \dfrac{\partial y^n}{\partial \nu} = \psi^n(x, t, y^n) \quad \text{on} \quad \Sigma^n = \Gamma^n \times (0, T), \\[2mm] y^n(x, 0) = y_0^n(x) \quad \text{for} \quad x \in \Omega^n, \end{cases}$$

where a^n, $\bar{b}^n$, ψ^n, y_0^n are smooth functions approximating in an appropriate sense a, $\bar{b}$, ψ and y_0, respectively. Moreover we assume that the compatibility condition

$$\frac{\partial y_0^n}{\partial \nu}(x) = \psi^n\big(x, 0, y_0^n(x)\big) \quad \text{for} \quad x \in \Gamma^n$$

holds. Then there exists a unique classical solution $y^n \in C^{2,1}\big(\overline{Q^n}\big)$ of problem (SP^n). Next, we obtain the following uniform a priori estimates for the solutions y^n

$$\|y^n\|_{C(\overline{Q^n})} \leqq M_1, \tag{4.1}$$

$$\|y^n\|_{H^1(Q^n)} \leqq M_2, \tag{4.2}$$

where the finite constants M_1, M_2 depend only upon the bounds of the data of problem (SP). From (4.3) it follows that there exists a subsequence $\{y^{n_1}\} \subset \{y^n\}$, such that $y^{n_1} \rightharpoonup \bar{y}$ in $H^1(Q)$. Hence by weak lower semicontinuity of the norm we get for the element $\bar{y}$ the second estimate (4.1). Taking into account the compactness of the embedding $H^1(Q)$ into $L^2(Q)$ we get $y^{n_1} \to \bar{y}$ in $L^2(Q)$. Therefore, there exists a subsequence $\{y^{n_2}\} \subset \{y^{n_1}\}$ such that $y^{n_2}(x, t) \to \bar{y}(x, t)$ a.e. in Q. Hence by virtue of (4.2) we get the estimate of $\bar{y}$ in the L^∞-norm.

The second stage of the proof consists in demonstrating that the obtained element $\bar{y}$ satisfies the integral identity (3.1).

Theorem 4.2. *If assumptions (A) are satisfied then the weak solution of problem (SP) is unique.*

Proof of this theorem is an extension of those carried out by Kamenomostskaya [31] and Friedman [20] (see [47] for details).

5. Dependence of weak solution upon the data

The weak solution of problem (SP) is in some sense continuously and monotone dependent upon the perturbations of the data.

Theorem 5.1 [47]. *Denote* $F(y) \triangleq \bar{F}(K^{-1}(y))$. *Let* y *and* $\tilde{y}$ *be the weak solutions of problem (SP) corresponding to given* F, ψ, y_0, S_0 *and* $\tilde{F}, \tilde{\psi}, \tilde{y}_0, \tilde{S}_0$*, respectively. If all the data satisfy the assumptions (A) then there exists a constant* B *independent of the data, such that*

$$\|y - \tilde{y}\|_{L^2(Q)} \leqq B\big[\|F(y) - \tilde{F}(y)\|_{L^2(Q)} + \|\psi(y) - \tilde{\psi}(y)\|_{L^2(\Sigma)}$$
$$+ \|y_0 - \tilde{y}_0\|_{L^2(\Omega)} + \mu_N\{x \in \Omega \mid (y_0(x) - y^*(x, 0))\,(\tilde{y}_0(x) - y^*(x, 0)) < 0\}\big]^{1/2}.$$

Theorem 5.2 [47]. *Under monotonicity assumptions concerning the functions* $F(z)$, $\tilde{F}(z)$ *and* $\psi(z)$, $\tilde{\psi}(z)$, *if* $F(z) \geqq \tilde{F}(z)$, $\psi(z) \geqq \tilde{\psi}(z)$, $y_0 \geqq \tilde{y}_0$ *then* $y \geqq \tilde{y}$, *where all the inequalities are understood as fulfilled almost everywhere in the appropriate domains.*

6. Optimal control problems associated with process (SP)

Let the control appear in the nonlinear Neumann condition (1.2). More precisely, the system to be controlled is described by (1.1)—(1.5) with boundary condition (1.2) in the form

$$k(x, t, \theta)\,\frac{\partial\theta}{\partial\nu} = \bar{\psi}_1(x, t, \theta) + \bar{\psi}_2(x, t), \quad x, t \in \Sigma.$$

The functions $\bar{\psi}_1$ and $\bar{\psi}_2$ are assumed to satisfy (A 5′) with $\bar{\psi}_2 = \mathcal{A}u$, $\mathcal{A} \in \mathcal{L}(\mathcal{U}, \mathcal{U})$, where u denotes the control variable and $\mathcal{U}$ is the control space. We take $\mathcal{U} = H^1(0, T; L^2(\Gamma))$ and a bounded, convex, closed subset $\mathcal{U}_{\mathrm{ad}} \subset \mathcal{U}$ as the set of admissible controls. Moreover we assume that $\mathcal{U}_{\mathrm{ad}}$ is bounded in $L^\infty(\Sigma)$.

Remark 6.1. $\mathcal{A}$ may be defined in particular as a truncation operator on Σ (when the control acts only on a part of the boundary).

Definition 6.1. A function $y(u) \in H^1(Q)$ is said to be an *admissible trajectory of the process* (SP) iff there exists $u \in \mathcal{U}_{\mathrm{ad}}$ such that the integral identity (3.1) is satisfied by $y(u)$.

By $\mathcal{Z}$ we will denote the set of all admissible trajectories of the process (SP). We also introduce the sets of all states attainable at the terminal time moment T and at the boundary Σ:

$$\mathscr{Z}_T = \{\mathfrak{z} \in L^2(\Omega) \mid \exists\, y \in \mathscr{Z},\, \mathfrak{z} = y|_T\},$$

$$\mathscr{Z}_\Sigma = \{\mathfrak{z} \in L^2(\Sigma) \mid \exists\, y \in \mathscr{Z},\, \mathfrak{z} = y|_\Sigma\}.$$

Theorem 6.1 [46, 47]. *If assumptions (A) are satisfied then $\mathscr{Z}$ is bounded and weakly closed in $H^1(Q)$.*

In view of the embedding and trace theorems [36, 41], from Thm. 6.1 it follows immediately

Corollary 6.1. *$\mathscr{Z}$ is compact in $L^2(Q)$, $\mathscr{Z}_T$ is compact in $L^2(Q)$, $\mathscr{Z}_\Sigma$ is compact in $L^2(\Sigma)$.*

Besides, from the proof of Thm. 6.1 we can deduce more about the continuity of the operations control $\to$ state.

Corollary 6.2. *Let $\{u^n\} \subset \mathscr{U}_{\text{ad}}$ and $u^n \rightharpoonup u \in \mathscr{U}_{\text{ad}}$. Then*

$$y(u^n) \xrightarrow[L^2(Q)]{} y(u), \quad y(u^n)|_T \xrightarrow[L^2(\Omega)]{} |\, y(u)|_T \quad y(u^n)|_\Sigma \xrightarrow[L^2(\Sigma)]{} y(u).$$

Remark 6.2. From Thm. 5.1 it follows that the mapping control $\to$ state from $L^2(\Sigma)$ into $L^2(Q)$ is Hölder continuous with the Hölder index 1/2.

Let us consider the following optimal control problems with distributed, terminal and boundary observation:

(P_Q) $$\inf_{u \in \mathscr{U}_{\text{ad}}} \{J_Q(u) \equiv I_Q[y(u)]\}$$

(P_T) $$\inf_{u \in \mathscr{U}_{\text{ad}}} \{J_T(u) \equiv I_T[y(u)|_T]\}$$

(P_Σ) $$\inf_{u \in \mathscr{U}_{\text{ad}}} \{J_\Sigma(u) \equiv I_\Sigma[\mathscr{B}(y(u)|_\Sigma)]\}$$

In these problems the functionals $I_Q: L^2(Q) \to \mathscr{R}$, $I_T: L^2(\Omega) \to \mathscr{R}$, $I_\Sigma: L^2(\Sigma) \to \mathscr{R}$ are assumed to be lower semicontinuous and bounded from below on every bounded subset of the appropriate space. The operator $\mathscr{B} \in \mathscr{L}(L^2(\Sigma), L^2(\Sigma))$. In particular $\mathscr{B}$ may be defined as a truncation operator on Σ (when observation is realized only on a part of the boundary Σ).

The above problems have the following physical interpretations. Problem (P_Q) with distributed observation may represent stabilization of the process. The control problem with terminal observation is typical for technological processes where the aim of control is to obtain a desired final product [26, 45]. The problem with boundary observation is charateristic for many diffusion and reaction processes as well as for controlled combustion [48, 51].

By the Weierstrass theorem the following result is clear from Corollary 6.1.

Theorem 6.2. *If assumptions (A) are satisfied then there exist optimal controls for problems (P_Q), (P_T), (P_Σ).*

Remark 6.3. Constrained optimal control problems may be also taken into consideration.

Let $\mathscr{W}_Q$ be a nonempty weakly closed subset of $H^1(Q)$. Consider problems (P_T), (P_Σ) subject to the constraints that $y(u) \in \mathscr{W}_Q$. Using arguments as above and assuming that $\mathscr{W}_Q \cap \mathscr{Z} \neq \emptyset$ we get existence of optimal controls for the constrained optimal control problems (P_T) and (P_Σ).

Of course, instead of the constraint on the whole trajectory $y(u)$ there can be imposed erminal or boundary constraints.

7. Remarks concerning some open questions in parabolic free boundary value problems

All the phase-change phenomena such as melting of solids, solidification of liquids and others may be described by the mathematical model (1.1)—(1.5). We refer to the monograph [51] and the proceedings [48, 54] where many physical processes giving rise to parabolic problems with free boundaries are discussed and on the other hand various mathematical aspects concerning these problems are considered.

An open question associated with multi-phase Stefan problems in several space dimensions is the characterization of the free boundary. There are some general results on regularity of weak solutions to Stefan type problems, obtained recently by Damlamian [11, 12, 13]. However, the results obtained there are not enough to recover any regularity of the free boundary (not even existence as a closed set).

Another question, which seems to be unsolved, is concerned with the multidimensional so called generalized Muskat's problem [5, 8, 25], known also from Russian literature as Verigin's problem [32, 33, 51]. Such parabolic free boundary problems arise in modelling of the behaviour of two slightly compressible, immiscible fluids which are in contact with one another in a porous medium. In the one-dimensional case Muskat's problem consists in determining the triple $\{p_1, p_2, g\}$, satisfying:

— parabolic equations

$$\frac{\partial p_i}{\partial t} = \alpha_i \frac{\partial^2 p_i}{\partial x^2} \quad \text{in} \quad Q_i, \quad i = 1, 2,$$

where $Q_1 \triangleq \{(x, t) \mid x \in (0, g(t)),\ t \in (0, T)\}$, $Q_2 \triangleq \{(x, t) \mid x \in (g(t), 1),\ t \in (0, T)\}$, $\alpha_i > 0$;

— boundary conditions of specified type at $x = 0$ and $x = 1$;

— initial conditions: $g(0) = g_0$, $0 < g_0 < 1$, $p_i(x, 0) = p_{i0}(x)$;

— conditions at the free boundary g:

$$p_1(g(t), t) = p_2(g(t), t),$$

$$a_1 \frac{\partial p_1}{\partial x}(g(t), t) = a_2 \frac{\partial p_2}{\partial x}(g(t), t), \qquad t \in (0, T),\ a_i > 0,$$

$$\frac{dg}{dt}(t) = -\beta \frac{\partial p_1}{\partial x}(g(t), t), \quad t \in (0, T), \quad \beta > 0.$$

Note the essential difference of the form of the above conditions at the free boundary in comparison with the appropriate conditions in the Stefan problem (values of the functions p_i are not given at the free boundary).

To the author's knowledge Muskat's problems as formulated above and their generalizations have been studied only in the one-dimensional case [51, 32, 33, 25, 50, 8, 24]. It is of great interest to study Muskat's type problems for any space dimension greater than one.

References

[1] O. B. Besov, V. P. Il'in and S. M. Nikolskii, Integral representations of functions and embedding theorems, Nauka, Moskva 1975 (in Russian).

[2] H. Brezis, On some degenerate nonlinear parabolic equations, Nonlinear Functional Analysis, part I, F. E. Browder, ed., Proc. Symposia in Pure Math., Vol. 18, AMS, Providence, R. I., 1970, 28—38.

[3] B. M. Budak and Yu. L. Gaponenko, On the solution of Stefan problem involving quasilinear parabolic equation and quasilinear boundary conditions, Trudy Vyčisl. Centra Moskov. Gos. Univ., 1971, 235—313 (in Russian).

[4] L. A. Caffarelli, The regularity of free boundaries in higher dimensions, Acta Math. **137** (1978), 155—184.

[5] J. R. Cannon, Multiphase parabolic free boundary value problems, Moving Boundary Problems, D. G. Wilson et al., eds., Academic Press, New York 1978, 3—24.

[6] J. R. Cannon, J. Douglas, Jr. and C. D. Hill, A multi-boundary Stefan problem and the disappearance of phases, J. Math. Mech. **17** (1967), 21—34.

[7] J. R. Cannon and A. Fasano, Boundary value multidimensional problems in fast chemical reactions, Arch. Rat. Mech. Anal. **53** (1973/74), 1—13.

[8] J. R. Cannon and A. Fasano, A nonlinear parabolic free boundary problem, Ann. Mat. Pura Appl. (IV) **112** (1977), 119—148.

[9] J. R. Cannon, D. B. Henry and D. B. Kotlow, Classical solutions of the one-dimensional two-phase Stefan problem, Ibid. **107** (1976), 311—341.

[10] J. R. Cannon and C. D. Hill, On the movement of a chemical reaction interface, Indiana Univ. Math. J. **20** (1970), 429—454.

[11] A. Damlamian, Problèmes aux limites non-linéaires du type du problème de Stefan et inéquations variationelles d'évolution, Thése, Univ. Paris VI, 1976.

[12] A. Damlamian, Some results on the multi-phase Stefan problem, Comm. Partial Differential Equations **2** (1977), 1017—1044.

[13] A. Damlamian, Une generalisation concernant le problème de Stefan, Recent Methods in Nonlinear Analysis, E. De Giorgi et al., eds., Pitagora Editrice, Bologna 1979, 25—28.

[14] G. Duvaut, The solution of a two-phase Stefan problem by a variational inequality, Moving Boundary Problems in Heat Flow and Diffusion, J. R. Ockendon and W. R. Hodgkins, eds., Clarendon Press, Oxford 1975, 173—181.

[15] L. C. Evans, A free boundary problem: The flow of two immiscible fluids in a one-dimensional porous medium, I, Indiana Univ. Math. J. **26** (1977), 915—932.

[16] L. C. Evans, A free boundary problem: The flow of two immiscible fluids in a one-dimensional porous medium, II, Ibid. **27** (1978), 93—111.

[17] A. Fasano, M. Primicerio and S. Kamin, Regularity of one-dimensional two-phase Stefan problems, Ann. Mat. Pura Appl. (IV) **115** (1978), 341—348.

[18] M. Frémond, Frost propagation in porous media, International Conference on Computational Methods in Nonlinear Mechanics, Austin 1974.

[19] A. Friedman, Partial differential equations of parabolic type, Prentice-Hall, Englewood Cliffs, N. J., 1964.

[20] A. Friedman, The Stefan problem in several space dimensions, Trans. Amer. Math. Soc. **133** (1968), 51—87.

[21] A. Friedman, One dimensional Stefan problems with nonmonotone free boundary, Ibid. 89–114.

[22] A. Friedman, Parabolic variational inequalities in one space dimension and smoothness of the free boundary, J. Functional Analysis **18** (1975), 151–176.

[23] A. Friedman, Analyticity of the free boundary for the Stefan problem, Arch. Rat. Mech. Anal. **61** (1976), 97–125.

[24] A. Friedman, Time dependent free boundary problems, SIAM Review **21** (1979), 213–221.

[25] W. Fulks and R. B. Guenther, A free boundary problem and an extension of Muskat's model, Acta Math. **122** (1969), 273–300.

[26] A. Gosiewski and M. Niezgódka, Mathematical models of some controlled crystallization process, Arch. Automatyki Telemech. **23** (1978), 255–265 (in Polish).

[27] A. Gosiewski and I. Pawłow, Models of dynamics of an underground gas reservior, Ibid. 267–277 (in Polish).

[28] J. W. Jerome, Existence and approximation of weak solutions of the Stefan problem with nonmonotone nonlinearities, Dundee Conference on Numerical Analysis, G. A. Watson, ed., Lecture Notes in Mathematics, Vol. 506, Springer-Verlag, 1976, Berlin—Heidelberg—New York 1976.

[29] J. W. Jerome, Nonlinear equations of evolution and a generalized Stefan problem, J. Differential Equations **26** (1977), 240–261.

[30] S. Kamenomotskaya, On the Stefan problem, Naučnye Doklady Vysšej Školy **1** (1958), 60–62 (in Russian).

[31] S. Kamenomotskaya, On the Stefan problem, Mat. Sbornik **53** **(95)** (1961), 488–514 (in Russian).

[32] L. I. Kamynin, On the existence of a solution of Verigin's problem, Žurnal Vyčisl. Mat. Mat. Fiz. **2** (1962), 833–858 (in Russian).

[33] L. I. Kamynin, On the linear Verigin's problem, DAN SSSR **150** (1963), 1210–1213 (in Russian).

[34] D. Kinderlehrer and L. Nirenberg, Hodograph methods and the smoothness of the free boundary in the one phase Stefan problem, Moving Boundary Problems, D. G. Wilson et al., eds., Academic Press, New York 1978, 57–69.

[35] D. Kinderlehrer and L. Nirenberg, Regularity in free boundary problems, Ann. Scu. Norm. Super. Pisa (IV) **4** (1977), 373–391.

[36] O. A. Ladyženskaya, V. A. Solonnikov and N. N. Ural'ceva, Linear and quasilinear equations of parabolic type, Nauka, Moskva 1967 (in Russian).

[37] J. L. Lions, Quelques méthodes de résolution des problèmes aux limites non linéaires, Dunod, Paris 1969.

[38] J. L. Lions, Introduction to some aspects of free surface problems, SYNSPADE, Maryland 1975.

[39] J. L. Lions, On free surface problems: methods of variational and quasi-variational inequalities, Computational Mechanics, J. T. Oden, ed., Lecture Notes in Mathematics, Vol. 461, Spinger-Verlag, Berlin—Heidelberg—New York 1975.

[40] J. L. Lions, Sur quelques questions d'analyse, de mécanique et de contrôle optimal, Les Presses de l'Université de Montreal 1976.

[41] J. L. Lions and E. Magenes, Problèmes aux limites non homogènes et applications, Vol. 1 et 2, Dunod, Paris 1968.

[42] E. Magenes, Topics in parabolic equations: some typical free boundary problems, Boundary Value Problems for Linear Evolution Partial Differential Equations, H. G. Garnir, ed., D. Reidel, Dordrecht/Boston 1977, 239–312.

[43] C. Moreno and C. Saguez, Dépendance, par rapport aux données de la fontière libre associée

à certaines inéquations variationnelles d'évolution, IRIA Laboria, Rapport de Recherche, 1978.
[44] M. Niezgódka, On some properties of two-phase parabolic free boundary value control problems, Control and Cybernetics 8 (1979), 23—42.
[45] M. Niezgódka, Control of parabolic systems with free boundaries — application of inverse formulations, Ibid. 8 (1979) 3, 213—226.
[46] M. Niezgódka and I. Pawłow, Optimal control for parabolic systems with free boundaries: existence of optimal controls, approximation results, Proc. 9th IFIP Conference on Optimization Techniques, Part 1, 412—420. Lecture Notes in Control and Information Sciences, Vol. 22, Springer-Verlag, Berlin—Heidelberg—New York 1980, 412—420.
[47] M. Niezgódka and I. Pawłow, A generalized Stefan problem in several space variables, Applied Mathematics and Optimization (to appear).
[48] J. R. Ockendon and W. R. Hodgkins, eds., Moving boundary problems in heat flow and diffusion, Clarendon Press, Oxford 1975.
[49] O. A. Oleinik, A method of solution of the general Stefan problem, DAN SSSR **135** (1960), 1054—1057 (in Russian).
[50] I. Pawłow, On some properties of two-layer parabolic free boundary value problems, Control and Cybernetics 7 (1978), 19—37.
[51] L. I. Rubinstein, The Stefan problem, Zvajgzne, Riga 1967 (in Russian).
[52] C. Saguez, Contrôle optimal de systèmes gouvernés par des inéquations variationnelles; applications à des problèmes de frontière libre, IRIA Laboria, Rapport de Recherche, 1976.
[53] C. Saguez, Contrôle optimal d'inéquations variationnelles avec observation de domaines, IRIA Laboria, Rapport de Recherche, 1978.
[54] D. G. Wilson, A. D. Solomon and P. T. Boggs, eds., Moving boundary problems, Academic Press, New York 1978.

Address of the author:

Systems Research Institute
Polish Academy of Sciences
Newelska 6
01-447 Warszawa
Poland

Optimization methods for non-linear partial differential equations

Olivier Pironneau, Paris

Abstract. In this paper we discuss the various conjugate gradient methods available for the solution of non linear partial differential equations discretized by the finite element method. Speed and memory required will be our criterion for selection.

0. Introduction

Many partial differential equations of physics are still difficult to solve numerically on digital computers. For example some simulation of flows requires the biggest computers and the special machines such as ILLIAC IV were built mostly for the numerical solutions of the Navier-Stokes equation.

After discretization non linear PDE yield non linear systems, and one of the difficulties of such simulations is that of a fast, reliable, and corewise economical non linear system solver. Minimum energy principles have been used widely mostly with the finite element method, to convert the non linear systems into optimization problems, the latter can be solved numerically by efficient conjugate gradient methods with or without conditioning matrices (Axelson [1]) depending upon the memory available.

In this lecture we wish to review some of these methods and compare their odds and assets on a typical non linear partial differential equation taken from fluid mechanics with industrial applications into aeronautics. The material of this lecture is an updated version of a team work done at Laboria and AMD-BA for the construction of an industrial code for the solution of the transonic potential flow equation [2, 3, 4].

1. Position of the problem

Instead of trying to place the discussion in a general framework to survey optimization methods for PDE we shall concentrate on a example taken from fluid mechanics.

$$\nabla \cdot \varrho \nabla \phi = 0 \quad \text{in } \Omega, \quad \varrho = (1 - |\nabla \phi|^2)^{1/\gamma - 1}. \tag{1.1}$$

This equation describes the velocity $\nabla \Phi$ of a compressible perfect adiabatic fuid ($\gamma = 1.4$); ϱ is the density of the fluid and Φ is the potential, $\Omega \subset R^n$ is the region by occupied the flow.

Typical boundary condition may be

$$\varrho \left. \frac{\partial \varrho}{\partial n} \right|_{\Gamma} = g; \qquad \Gamma = \text{boundary of } \Omega. \tag{1.2}$$

In variational form this problem may be written.

$$\left((1 - |\nabla\phi|^2)^{\frac{1}{\gamma-1}} \nabla\phi, \nabla\varphi\right) = (g, \psi)_\Gamma \quad \forall\, \psi \in H^1(\Omega), \tag{1.3}$$

where

$$(a, b) = \int_\Omega a(x)\, b(x)\, dx, \tag{1.4}$$

$$(a, b)_\Gamma = \int_\Gamma ab d\Gamma, \tag{1.5}$$

$$H^1(\Omega) = \{\psi \in L^2(\Omega) : \nabla\psi \in (L^2(\Omega))^n\}. \tag{1.6}$$

Theorem 1. *If ϱ is a continuous function from R to R such that $\varrho(\lambda) + 2\lambda \frac{d\varrho}{d\lambda}(\lambda) \geqq \alpha$ $\forall\, \lambda \geqq 0$, for some $\alpha > 0$; if Ω is a bounded open set of R^n and if $g \in L^2(\Gamma)$ then*

$$\bar{\phi} = \arg\min_{\phi \in H^1(\Omega)} \left\{ \int_\Omega dx \int_0^{|\nabla\phi|^2} \varrho(\lambda)\, d\lambda - 2 \int_\Gamma g\varphi\, d\Gamma \right\} \tag{1.7}$$

is unique in $H^1(\Omega)/R$ and it is the solution of

$$\nabla \cdot \varrho(|\nabla\phi|^2)\, \nabla\phi = 0 \quad \text{in} \quad \Omega; \quad \varrho \left. \frac{\partial\phi}{\partial n} \right|_\Gamma = g. \tag{1.8}$$

Proof. Call

$$E(|\nabla\phi|^2) = \int_\Omega dx \int_0^{|\nabla\phi|^2} \varrho(\lambda)\, d\lambda - 2 \int_\Gamma g\phi\, d\Gamma, \tag{1.9}$$

then

$$\frac{d}{d\mu} E\left(|\nabla(\bar{\phi} + \mu\psi)|^2\right)\Big|_{\mu=0} = 2 \int_\Omega \varrho(|\nabla\bar{\phi}|^2)\, \nabla\bar{\phi} \nabla\psi\, dx - 2 \int_\Gamma g\psi\, d\Gamma$$

but by definition of $\bar{\phi}$:

$$\frac{1}{\mu} \left[E\left(|\nabla(\bar{\phi} + \mu\psi)|^2\right) - E(|\nabla\bar{\phi}|^2)\right] \geqq 0 \quad \forall\, \psi \in H^1(\Omega) \quad \forall\, \mu > 0,$$

therefore

$$(\bar{\varrho} \nabla\bar{\phi}, \nabla\psi) - (g, \psi)_\Gamma \geqq 0 \quad \forall\, \psi \in H^1(\Omega);$$

changing ψ into $-\psi$ yields

$$\left(\varrho(|\nabla\bar{\phi}|^2)\, \nabla\phi, \nabla\psi\right) = (g, \psi)_\Gamma \quad \forall\, \psi \in H^1(\Omega), \tag{1.10}$$

therefore $\bar{\phi}$ satisfies (1.8).

To prove that $\bar{\phi}$ exists and is unique one shows that the minimization problem has a unique solution.

Since

$$\left. \frac{d^2E}{d\mu^2} \right|_{\mu=0} = 2 \int_\Omega \{\varrho |\nabla\psi|^2 + 2\dot{\varrho}[\nabla\phi\nabla\psi]^2\}\, dx \geqq 2\alpha \int_\Omega |\nabla\psi|^2\, dx,$$

E is convex; therefore it is lower semi continuous and tends to $+\infty$ when $|\nabla\phi|$ tends to $+\infty$. From Lax-Milgram lemma the solution exists and is unique. ■

In the case of (1.2)

$$\varrho + \lambda\dot{\varrho} = \left(1 - \frac{\gamma+1}{\gamma-1}\lambda\right)(1-\lambda)^{\frac{2-\gamma}{\gamma-1}},$$

therefore the theorem applies only if

$$|\nabla\tilde{\phi}|^2 \leqq \frac{\gamma-1}{\gamma+1} \tag{1.11}$$

which means physically that when the velocity of the flow is everywhere smaller than the speed of sound one can solve (1.7) instead of (1.8).

In the case of transonic flows there is no such variational principle and one may have to use a least square method.

It was shown in [2] that the proper formulations of a least square for (1.1) (1.2) is:

$$\min_{\phi\in Hg} \|\nabla\cdot(\varrho\nabla\phi)\|_{-1}^2, \tag{1.12}$$

where

$$\|u\|_{-1} = \sup_{\psi\in H^1(\Omega)/R} \{(u,\psi)/|\nabla\psi|_0\}, \tag{1.13}$$

$$|v|_0 = (v, v)^{1/2}, \tag{1.14}$$

$$Hg = \left\{\phi \in H^1(\Omega) : \varrho(|\nabla\phi|^2)\left.\frac{\partial\phi}{\partial n}\right|_\Gamma = g\right\}. \tag{1.15}$$

We recall the following simple results:

Theorem 2. *Problem* (1.12) *and problem* (1.1) (1.2) *are equivalent to the optimal control problem*

$$\min_{\phi\in H^1(\Omega)} \left\{|\nabla\varepsilon|_0^2 : \Delta\varepsilon = \nabla\cdot[\varrho(|\nabla\phi|^2)\nabla\phi] \text{ in } \Omega;\ \left.\frac{\partial\varepsilon}{\partial n}\right|_\Gamma = \varrho\left.\frac{\partial\phi}{\varkappa n}\right|_\Gamma - g\right\}. \tag{1.16}$$

Proof. Define $\tilde{\varepsilon} \in H^1(\Omega)$ by

$$(\nabla\tilde{\varepsilon}, \nabla\psi) = \left(\varrho(|\nabla\phi|^2)\nabla\phi, \nabla\psi\right) \quad \forall\, \psi \in H^1(\Omega),$$

then

$$\|\nabla\cdot\varrho\nabla\phi\|_{-1} = \sup_{\phi\in H^1(\Omega)/R} (\varrho\nabla\phi, \nabla\psi)/|\nabla\psi|_0 = |\nabla\tilde{\varepsilon}|_0.$$

Therefore (1.12) is equivalent to

$$\min_{\phi\in Hg}\left\{|\nabla\tilde{\varepsilon}|_0^2 : \Delta\tilde{\varepsilon} = \nabla\cdot[\varrho(|\nabla\phi|^2)\nabla\phi],\ \left.\frac{\partial\varepsilon}{\partial n}\right|_\Gamma = 0\right\}, \tag{1.17}$$

but (1.16) and (1.17) are equivalent.

Remark. Problem (1.1)—(1.2) is not equivalent to

$$\min_{\phi\in Hg} |\nabla\cdot(\varrho\nabla\phi)|_0^2. \tag{1.18}$$

There are indeed instances where $\nabla \cdot (\varrho \nabla \phi)$ is not square integrable so that problem (1.18) does not make sense. Problem (1.18) is in fact a disguised 4th order PDE and numerically it is found that the solution of (1.18) may have unphysical boundary layers near Γ.

2. Optimization in Hilbert spaces and discretization

2.1. The conjugate gradient method

$$\min_{\phi \in H} E(\phi), \tag{2.1}$$

where H is some Hilbert space with scalar product $\langle , \rangle$ and E is a Fréchet differentiable function.

The gradient of E in H, $\mathrm{Grad}_H E$, is defined by

$$\frac{d}{d\mu} E(\phi + \mu\psi)|_{\mu=0} = \langle \mathrm{Grad}_H\, E, \psi \rangle \quad \forall\ \psi \in H; \quad \mathrm{Grad}_H\, E \in H.$$

The method of steepest descent with fixed step size μ is the following algorithm:

0. Choose ϕ^0, μ; set $n = 0$.
1. Compute $\mathrm{Grad}_H\, E(\phi^n)$.
2. Set $\phi^{n+1} = \phi^n - \mu\, \mathrm{Grad}_H\, E(\phi^n)$, $n = n + 1$ and go to 1.

Example 1. If

$$\inf_{\substack{\phi \in L^2(\Omega) \\ \phi|_\Gamma = 0}} \frac{1}{2} |\nabla \phi|_0^2 - (f, \phi) \tag{2.2}$$

is solved by the method of steepest descent it goes like: $\phi^{n+1} = \phi^n - \mu(\Delta\phi^n + f)$ in Ω, which is an explicit scheme to solve

$$\frac{\partial \phi}{\partial t} - \Delta \phi = f \quad \text{in } \Omega, \quad \phi_{|\Gamma} = 0.$$

This method is naturally quite slow.

Example 2. If the same technique is applied to

$$\min_{\phi \in H_0^1(\Omega)} \frac{1}{2} |\nabla \phi|_0^2 - (f, \phi), \tag{2.3}$$

where

$$H_0^1(\Omega) = \{\phi \in H^1(\Omega) : \phi_{|\Gamma} = 0\}$$

then with scalar product

$$\langle a, b \rangle = (\nabla a, \nabla b) \tag{2.4}$$

one finds that ϕ^{n+1} is computed from ϕ^n by

$$\Delta\phi^{n+1} = \Delta\phi^n - \mu(\Delta\phi^n - f) \quad \text{in } \Omega \quad \phi^{n+1}{}_{|\Gamma} = 0.$$

If $\mu = 1$ then the solution of (2.3) is found in one iteration.

Examples 1 and 2 show that the proper choice of the space of optimization is numerically very important.

In practice one does not use the method of steepest descent; one uses the method of conjugate gradient:

0. Choose φ^0, set $n = 0$ and $\gamma^0 = 0$.
1. Compute $g^n = -\operatorname{Grad}_H E(\phi^n)$
2. Compute $\gamma^n = \langle g^n - g^{n-1}, g^n\rangle / \langle g^{n-1}, g^{n-1}\rangle$ and set $h^n = g^n + \gamma^n h^{n-1}$
3. Compute $\mu^n = \arg\min E(\phi^n + \mu h^n)$
4. Set $\phi^{n+1} = \phi^n + \mu^n h^n$; $n = n + 1$ and go back to 1.

Remark. Step 3 is usually the most costly but it is not necessary to compute μ^n exactly. A good method is to find by dichotomy a point $\tilde{\mu}^n$ such that $E(\tilde{\mu}^n) \leq \min\{E(\tilde{\mu}^n/2), E(2\tilde{\mu}^n)\}$ and then to compute the minimum of the parabolic fit of E with these 3 points.

2.2. Discretization

For our purpose we shall work with only two Hilbert spaces, namely $L^2(\Omega)$ and $H^1(\Omega)$ with the scalar products (1.4) and (2.4) These spaces will be discretized as follows:

Ω is divided into non overlapping triangles which intersect into a vertex, or a whole side, or the empty set; h will denote the length of the largest side of the triangulation $\mathfrak{T}_h$. Then $L^2(\Omega)$ will be approximated by

$$L_h = \{\phi_h : \phi_{h|T} = \text{const} \quad \forall\, T \in \mathfrak{T}_h\} \tag{2.5}$$

and $H^1(\Omega)$ is approximated by the space of continuous piecewise linear functions

$$H_h = \{\phi_h \in C^0(\Omega) : \phi_{h|T} \quad \text{is affine} \quad \forall\, T \in \mathfrak{T}_h\}. \tag{2.6}$$

2.3. Implementation on the discrete problem

Now let us look at the last two examples after discretization. Naturally (2.2) cannot be discretized by replacing $L^2(\Omega)$ by L_h because $\nabla\Phi$ cannot be computed easily in L_h. Therefore we would still discretize (2.2) by

$$\min_{\substack{\phi \in H_h \\ \phi_{|\Gamma} = 0}} \frac{1}{2} |\nabla\phi|_0^2 - (f, \phi), \tag{2.7}$$

but we may retain (1.4) for the scalar product in which the method of steepest descent is applied. Then one finds that ϕ^{n+1} is computed from ϕ^n by solving

$$(\phi^{n+1}, \psi) = (\phi^n, \psi) - \mu(\nabla\phi^n, \nabla\psi) + (f, \psi)\,\mu \quad \forall\, \psi \in H_h, \quad \psi_{|\Gamma} = 0 \tag{2.8}$$

or equivalently

$$B\Phi^{n+1} = B\Phi^n - \mu(A\Phi^n - F), \tag{2.9}$$

where as usual with the finite element method B and A are the matrices of size $N \times N$ (N = number of internal vertices):

$$b_{ij} = (w_i, w_j)\,; \quad a_{ij} = (\nabla w_i, \nabla w_j) \tag{2.10}$$

and w_i are the form functions attached to vertex x^i:

$$w_i(x^j) = \delta_{ij}, \quad w_i \in H_h\,; \tag{2.11}$$

Φ and F are the vectors of components:

$$\Phi_i = \phi(x^i)\,; \quad F_j = (f, w_j). \tag{2.12}$$

Similarly if (2.3) is discretized by (2.7) but the scalar product (2.4) is used, then one finds:

$$A\Phi^{n+1} = A\Phi^n - \mu(A\Phi^n - F). \tag{2.13}$$

In the vocabulary of optimization to replace (2.9) by (2.13) is called *preconditioning*; here the matrix of preconditioning is AB^{-1}. The idea is quite old and natural once it is noted that the scheme

$$M\Phi^{n+1} = M\Phi^n - \mu E^+(\Phi^n)$$

(to minimize $E(\Phi)$) is fastest when M is "close" to $\ddot{E}(\Phi^n)$. By "close" one usually means that the spectrums are close.

2.4. A good preconditioning scheme for elliptic PDE

Very useful preconditioning matrices have been given by Meijering-Van der Worst [5] known as incomplete Choleski factorization. This method was further improved in [4] for our purpose.

The construction of the preconditioning matrix is the following:

0. Construct A
1. Factorize A into L^tL by the Choleski algorithm
2. Construct $\tilde{L}$ by putting to 0 all the small non influential elements of L until only $\frac{5}{100}$ of the original non zero elements are left
3. Set $\tilde{A} = \tilde{L}^tL$.

An influential element L_{ij} is an element such that $a_{ij} \neq 0$.

Then an intermediate method between (2.9) and (2.13) is:

$$\tilde{A}\Phi^{n+1} = \tilde{A}\Phi^n - \mu(A\Phi^n - F)$$

or its conjugate gradient equivalent:

0. Choose Φ^0, set $n = 0$, $\gamma^0 = 0$.
1. Compute $G^n = \tilde{A}^{-1}(A\Phi^n - F)$
2. Compute $\gamma^n = (G^n - G^{n-1})^t \tilde{A} G^n / G^{nt} \tilde{A} G^n$ and set $H^n = G^n + \gamma^n H^{n-1}$

3. Compute $\mu^n = \arg\min \frac{1}{2} (\Phi^n + \mu H^n)^t A(\Phi^n + \mu H^n) - F(\Phi^n + \mu H^n)$
4. Set $\Phi^{n+1} = \phi^n + \mu^n H^n$, $n = n + 1$ and go back to 1.

Remark. Preconditioning has been introduced also directly in the context of PDE (Jameson [6], see also [5]), but then there is no provision for variable μ.

3. Application to the case study

3.1. Solution of (1.1), (1.2) by (1.7)

After discretization (1.7) is approximated by

$$\tilde{\phi}_h = \arg\min_{\phi \in H_h} \left\{ -\frac{\gamma - 1}{\gamma} \int_\Omega (1 - |\nabla\phi|^2)^{\frac{\gamma}{\gamma-1}} dx - 2 \int_\Gamma g\varphi \, d\Gamma \right\}. \tag{3.1}$$

Denote by H_h^1 the space H_h structured by the scalar product (2.4) and H_h^0 if (1.4) is used. Since, with the notation (1.9):

$$\frac{d}{d\mu} E(|\nabla(\phi + \mu\psi)|^2)\Big|_{\mu=0} = 2 \int_\Omega (1 - |\nabla\Phi|^2)^{\frac{1}{\gamma-1}} \nabla\phi \nabla\psi \, dx - 2 \int_\Gamma g\psi \, d\Gamma, \tag{3.2}$$

$\mathrm{Grad}_{H_h^1} E$ is defined by

$$\left.\begin{aligned} &(\nabla \mathrm{Grad}_{H_h^1} E, \nabla\psi) = 2\left((1 - |\nabla\phi|^2)^{\frac{1}{\gamma-1}} \nabla\phi, \nabla\psi\right) - 2(g, \psi)_\Gamma \quad \forall \psi \in H_h, \\ &\mathrm{Grad}_{H_h^1} E \in H_h, \end{aligned}\right\} \tag{3.3}$$

while $\mathrm{Grad}_{H_h^0} E$ is defined by

$$\left.\begin{aligned} &(\mathrm{Grad}_{H_h^0} E, \psi) = 2\left((1 - |\nabla\phi|^2)^{\frac{1}{\gamma-1}} \nabla\phi, \nabla\psi\right) - 2(g, \psi)_\Gamma \quad \forall \psi \in H_h, \\ &\mathrm{Grad}_{H_h^0} E \in H_h. \end{aligned}\right\} \tag{3.4}$$

Method 1 (Conjugate gradient in L^2). In the latter case (3.4) we may use mass lumping to approximate B by

$$b_{ij} = \frac{\delta_{ij}}{6} \sum_{T_k \supset \{x^i\}} \text{area}\,(T_k) \tag{3.5}$$

and then

$$\mathrm{Grad}_{H_h^0} E(x^i) = \frac{2}{b_{ii}} \left[\left((1 - |\nabla\phi|^2)^{\frac{1}{\gamma-1}} \nabla\phi, \nabla w_i\right) - (g, w_i)_\Gamma\right], \tag{3.6}$$

so that with this metric, the conjugate gradient algorithm built Φ^{n+1} from Φ^n by:

$$\Phi^n = \sum_1^{N'} w^i \varphi_i^n \qquad N' = \text{number of vertices}, \tag{3.7}$$

$$\varphi_i^{n+1} = \varphi_i^n - 2 \frac{\mu^n}{b_{ii}} \left((1 - |\nabla\phi^n|^2)^{\frac{1}{\gamma-1}} \nabla\phi, \nabla w_i\right) - (g, w_i)_\Gamma + \gamma^n \mu^n H^n, \tag{3.8}$$

where γ, μ, H are defined in § 2; in particular

$$\gamma^n = \sum_1^{N'} b_{ii}(G_i^{n+1} - G_i^n)\, G_i^n \Big/ \sum_1^{N'} b_{ii} G_i^{n^2}. \tag{3.9}$$

Method 2 (Conjugate gradient in H^1). If H_h^1 is used then (3.3) implies that

$$A[\operatorname{Grad}_{H_h^1} E(x^i)] = 2\left((1 - |\nabla\phi|^2)^{\frac{1}{\gamma-1}}\, \nabla\phi,\, \nabla w^i\right) - 2(g, w^i) \tag{3.10}$$

and then in the conjugate gradient algorithm:

$$\gamma^n = \sum_{i,j=1}^{N'} (G_i^{n+1} - G_i^n)\, a_{ij} G_j^n \Big/ \sum_{i,j=1}^{N'} G_i^n a_{ij} G_j^n . \tag{3.11}$$

Method 3 (precondition of method 1). Denote by $\tilde{H}_h^1$ the space H_h with the scalar product: $\langle\phi, \psi\rangle = \sum \tilde{a}_{ij}\phi(x^i)\, \psi(x^j)$, where $\tilde{A}$ is defined in § 2. Then the reader will check that the formulas are as in method 2 where A is replaced by $\tilde{A}$.

3.2. Solution of (1.1), (1.2) by (1.16)

Problem (1.16) is approximated by

$$\min_{\phi\in H_h} \left\{|\nabla\varepsilon|_0^2 : (\nabla\varepsilon, \nabla\psi) = \left((1 - |\nabla\phi|^2)^{\frac{1}{\gamma-1}}\, \nabla\phi, \nabla\psi\right) - (g, \psi)_\Gamma \quad \forall\, \psi \in H_h, \quad \varepsilon \in H_h\right\}. \tag{3.12}$$

If we denote by

$$E(\phi) = |\nabla\varepsilon(\phi)|_0^2, \tag{3.13}$$

where $\varepsilon(\phi)$ is the solution of the right member of (3.12), then

$$\frac{d}{d\mu} E(\phi + \mu\psi)|_{\mu=0} = 2\left(\nabla \frac{d\varepsilon}{d\mu}(\phi + \mu\psi)|_{\mu=0}, \nabla\varepsilon\right)$$

$$= 2\left(\frac{d}{d\mu}(1 - |\nabla\phi + \mu\psi|^2)^{\frac{1}{\gamma-1}}\, \nabla(\phi + \mu\psi), \nabla\varepsilon\right)_{\mu=0}$$

$$= \int_\Omega (1 - |\nabla\phi|^2)^{\frac{1}{\gamma-1}}\, \nabla\psi\nabla\varepsilon - \frac{2}{\gamma - 1}(1 - |\nabla\phi|^2)^{\frac{2-\gamma}{\gamma-1}}\, \nabla\phi\nabla\psi\nabla\phi\nabla\varepsilon\, dx. \tag{3.14}$$

Hence if H_h is used with (2.4) to find the gradient of E, we must solve:

$$\left.\begin{aligned} &(\nabla \operatorname{Grad}_{H_h^1} E, \nabla\psi) = \Big((1 - |\nabla\phi|^2)^{\frac{1}{\gamma-1}}\, \nabla\varepsilon + \frac{2}{\gamma - 1}(1 - |\nabla\phi|^2)^{\frac{2-\gamma}{\gamma-1}} \\ &\quad \times \nabla\phi\nabla\varepsilon\nabla\phi, \nabla\psi\Big) \quad \forall\, \psi \in H_h^1, \\ &\operatorname{Grad}_{H_h^1} E \in H_h^1. \end{aligned}\right\} \tag{3.15}$$

If we denote by D_i the right member of (3.15) for $\psi = w^i$ then (3.15) gives

$$c_{ij} \operatorname{Grad}_{H_h^1} E(x^i) = D_j, \qquad j = 1, \ldots, N', \tag{3.16}$$

where C is the matrix of the Laplacian operator.

So we can also construct 3 methods:

Method 4: $C = B$
Method 5: $C = A$
Method 6: $C = \tilde{A}$

For all 3 methods the algorithm will be

0. Choose C, ϕ^0; set $i = 0, \gamma^0 = 0$.
1. Compute ε^n by solving the right member of (3.12) with $\phi = \phi^n$
2. Compute $G^n = \text{Grad}_{H_h^1} E$ by (3.16); $\gamma^n = (G^{n+1} - G^n)^t \, CG^n / G^{nt} CG^n$; set $H^n = G^n + \gamma^n H^{n-1}$
4. Compute $\mu^n = \arg\min |\nabla \varepsilon(\phi^n + \mu H^n)|_0^2$; set $\phi^{n+1} = \phi^n + \mu^n H^n$; $n = n + 1$ and go back to 1.

Remark. A seventh method can be obtained if we do not minimize $|\nabla \varepsilon|_0^2$ but $\varepsilon^t \tilde{A} \varepsilon$ instead and the Laplacian is replaced by $\tilde{A}$ (see [4]).

3.3. Comparison of the 6 methods

The simplest method to program is method 1; it is also the one that requires the least core memory because there are no matrices to store. To speed up the calculations it is however wise to store a $6 \times M$ array Q containing the derivatives $\nabla w_i|_{T_k}$, $k = 1, \ldots, M$; an array $S(M)$ containing the area of each triangle, and an array $B(M)$ for b_{ii}. Then (3.6) is computed in approximately $14M$ multiplications. The computation of μ^n costs approximately 3 evaluations of $E(\phi)$; thus for L_i iterations: $56ML_i$ multiplications are needed. Our experience shows that $L_i \sim 10\sqrt{N}$.

For methods 2 and 3 one has to add to these numbers the construction of A and $\tilde{A}$, the factorization of A and the solutions of the linear systems. It sums up to $(\nu + 2)\,\nu N + \left(2(\gamma + 2)\,N + 56M\right)$ for method 2, where ν is the bandwidth of the triangulation:

$$\nu = \sup_{T_k \supset \{x_i, x_j\}} |i - j| .$$

For method 2 $L_i = 5$ but for method 3 $L_i \sim 10$. Method 2 requires an additional memory of νN, approximately, for the storage of L while method 3 will need only an additional storage of $5\nu N/100$ over method 1.

Due to the sparcity of $\tilde{A}, \tilde{L}$, method 3 requires $\left((\nu + 2)\,\nu N + \frac{1}{10}(\nu + 2)\,N + 56M\right) \times 50$ multiplications ($L_i \sim 50$).

Method 4 is not efficient as it turned out that up to 5000 iterations may be necessary to solve a problem with $N \sim 1000$.

Methods 5 and 6 are of course longer to program than methods 1, 2, 3 and they should be used only for transonic flows. The computation of ε can itself be done by factorization and direct solution: $2(\nu + 2)\,N + 56N$ multiplications, or by method 3: $2(\nu + 2)\,N + 56N + \nu N$ multiplications; the computation of the gradient (3.15) costs also $2(\nu + 2)\,N + 112M(+\nu N)$ where the last number it taken if method 3 is used to solve (3.15); and the computation of $E(\Phi)$ costs $n_\nu N$ multiplications, where n_ν is the average of neighbour vertices for each node.

For each iterations it appears that the computation of the gradient Grad E account for 1/4 of the number of multiplications, therefore method 6 is not feasible and the difference will be essentially between method 5 and method 5′: the linear system for ε and $\text{Grad}_{H_n^1} E$ are solved by method 3.

Then the total number of multiplications is

$$(\nu + 2)\,\nu N + L_i\big(8(\nu + 2)\,N + 504M + 4\nu N + 3n_v N\big).$$

In practice $L_i \sim 20$.

The storage required is:

— the matrix A (for fast computation of $|\nabla\varepsilon|_0^2$): $n_0 N + (\nu N)_i$, where i refers to the fact that integer arrays can be packed.

— the matrix $\tilde{L}$ $\quad \frac{5}{100}\,\nu N + (\nu N)_i$

— the vectors ε, Φ, Grad E $\quad 3N$

— The triangulation and $S(M)$, $Q(6M)$ $\quad 6N + (9M + N)_i$

The results are summarized in Table 1.

Table 1

Methods	1	2	3	5	5′
Storage required	$12N + (3M + N)_i$	$12N + (3M + N)_i + \nu N$	$12N + (3M + N)_i + \frac{5}{100}\,\nu N + (\nu N)_i$	$12N + (3M + N)_i + \nu N + n_v N + (\nu N)_i$	← id but $\frac{5}{100}\,\nu N$
number of multiplications	$560M\sqrt{N}$	$280N + 10 \times (\nu + 2)\,N + (\nu + 2)\,\nu N$	$560N + (\nu + 2)\,\nu N$	$(\nu + 2)\,\nu N + 60n_v N + 160(\nu + 2) \times N + 10160N$	$(\nu + 2)\,\nu N + 60n_v N + 160(\nu + 2) \times N + 10160N + 80\nu N$
programming complexity	1	2	2	3	3
restriction on reach number	$M < 1$	$M < 1$	$M < 1$		

Remark. For a treatment of the entropy condition see [3].

References

[1] Axelson, O., A class of iterative methods for finite element equations, Comput. Methods Applied Mech. Eng. **9** (1976), 123—138.

[2] Glowinski, R., Periaux, J., Pironneau, O., Abstract least squares-Bat-Sheva Seminars on finite elements for non elliptic problems. Tel Aviv 77 Geffen-Bercovier ed. (to appear).

[3] Bristeau, M. O., Glowinski, R., Periaux, J., Perrier, P., Poirier, G., Pironneau, O., Rapport de contrat DRME 77/011, Calcul des ecoulements transsoniques par le controle optimal. Laboria 294.
[4] Glowinski, R., Periaux, J., Pironneau, O., Applied Math. Modelling **4** (1980) 3.
[5] Meijerin Van der Worst, An iterative solution method for linear systems of which the coefficient matrix is a symmetric M-matrix, Math. of Comp. **31** (1977), 148—162.
[6] Jameson, A., Transonic flow calculations, V.K.I. Lecture series: Computational Fluid Dynamics, Von Kármán Institute for Fluid Dynamics, Rhode-St. Genese, Belgium, March 15—19, 1976.

Address of the author:

Centre Scientifique et Polytechnique
Université de Paris 13
Av. J.-B. Clément
93430 Villetaneuse

and

IRIA Laboria
Domaine de Voluceau
Rocquencourt — B.P. 5
78150 Le Chesnay
France

Design sensitivity analysis in structural mechanics[1]

Bernard Rousselet, Nice

1. Introduction

Many papers have investigated optimization problems in structural mechanics; a synthesis on recent results as well as many references may be found in [8]. These topics have been mostly studied by engineers for discretized structures. Works on continuous structures lacked theoretical proof [9].

The aim of joint work with E. J. Haug[2] is to give proof to some previously introduced methods. Design sensitivity analysis has been rigorously proved in a sequence of three papers [10, 11, 19]. Static response as well as eigenvalues of beams, plates and planar elasticity have been studied, as also buckling of beams. These structures are of variable thickness and from a mathematical point of view the control u, here, is the thickness of the structure; the case where the control is the geometrical domain occupied by the structure (shape optimal design) has also been considered in the third paper of the sequence [18], following the lines of [15, 16] (see [17] for a compact report).

Here, we are only dealing with thickness optimization.

2. Setting of the problems

Let us first formulate some examples which have been selected to illustrate classes of mechanical systems in which design dependence arises in a consistent way.

2.1 State equations

Here, we give equations of the structures in variational form (see [10, 11] for the formal equations) which are adequate here; we do not use any further regularity of the variational solution in our analysis; and this is impossible, moreover, in our examples, as the coefficients involved in the highest derivative belong only to L^∞. The static problems may be expressed to find $z \in V$ such that for all $v \in V$

$$a_u(z, v) = (f, v)_{L^2} \tag{2.1}$$

where $f \in L^2(\Omega)$ and a_u is a bilinear form on V.

For simplicity we only consider clamped structures (i.e., Dirichlet condition for fourth order problems: u and $\partial u/\partial v$ equal zero on the boundary).

---

[1] Research supported by U.S. National Science Foundation Grant, E.N.G. 77-19967.

[2] College of Engineering, The University of Iowa, Iowa City, Iowa U.S.A.

For the beam, $V = H_0^2(0, 1)$, $\Omega =]0, 1[$,

$$a_u(z, v) = \int_0^1 E\alpha u^2 z_{xx} v_{xx}\, dx \tag{2.2}$$

where E is Young's modulus (supposed to be fixed here), α is a positive constant depending on the shape of the cross-section (it is presumed that all its dimensions vary with the same ratio); u is the variable cross-sectional area; and it is presumed that $u \in L^\infty(0, 1)$.

For the plate, Ω is a bounded open subset of R^2, $V = H_0^2(\Omega)$ and

$$a_u(z, v) = \int_\Omega \hat{D}(u)\, [z_{11}v_{11} + z_{22}v_{22} + \nu(z_{11}v_{22} + z_{22}v_{11}) + 2(1 - \nu)\, z_{12}v_{12}]\, dx \tag{2.3}$$

where $\hat{D}(u) = Eu^2/[12(1 - \nu^2)]$ with the constant ν: Poisson's ratio; $u \in L^\infty(\Omega)$ is the variable thickness of the plate here.

For the plane elasticity Ω is still a bounded open subset of R^2, $V = H_0^1(\Omega) \times H_0^1(\Omega)$, $z = (z^1, z^2)$, $v = (v^1, v^2)$, and

$$a_u(z, v) = \frac{\alpha}{2(1 + \nu)} \int_\Omega u\{[4\nu(z_1^1 + z_2^2)]\, [v_1^1 + v_2^2] + 2(z_1^1 v_1^1 + z_2^2 v_2^2) + (z_2^1 + z_1^2)\, (v_2^1 + v_1^2)\}\, dx, \tag{2.4}$$

while

$$(f_1 v) = \int_\Omega (f^1 v^1 + f^2 v^2)\, dx.$$

In the three cases the bilinear form is V-continuous and V-coercive which means that

$$\forall\, z \in V, \quad \forall\, v \in V \quad |a_u(z, v)| \leqq K\, \|z\|_V\, \|v\|_v, \tag{2.5}$$

$$\forall\, z \in V \quad a_u(z, z) \geqq \gamma\, \|z\|_V^2 \quad (\gamma > 0) \quad \text{(see [7])}. \tag{2.6}$$

One may associate these bilinear forms with unbounded operators A_u in $L^2(\Omega)$ (or $L^2(\Omega) \times L^2(\Omega)$ for the plane elasticity) such that

$$\forall\, v \in V \quad (A_u z, v)_{L^2} = a_u(z, v), \tag{2.7}$$

the domain $D(A_u)$ being the $z \in V$ such that $v \to a_u(z, v)$ be L^2-continous.

It is shown in [3, 7] that the equation $A_u z_u = f$ has a unique solution in $D(A_u)$ for every $f \in L^2$.

The eigenvalue problems may be expressed so as to find $(\zeta, y) \in \mathbb{C} \times (V \setminus \{0\})$ such that

$$\forall\, v \in V \quad a_u(y, v) = \zeta b_u(y, v). \tag{2.8}$$

In the vibration of beam, plate, and plane elasticity the bilinear form a_u is the same as in the static response.

The bilinear form b_u is

$$b_u(y, v) = \varrho \int_0^1 uyv \, dx \qquad \text{(for the beam)}, \tag{2.9}$$

$$b_u(y, v) = \varrho \int_\Omega uyv \, dx \qquad \text{(for the plate)}, \tag{2.10}$$

$$b_u(y, v) = \varrho \int_\Omega u(y^1v^1 + y^2v^2) \, dx \quad \text{(in plane vibration)}. \tag{2.11}$$

They are all bounded in a L^2-space. However, for the buckling of a beam $b_u(y, v) = \int_0^1 y_x v_x \, dx$ which is bounded in H^1, but unbounded in L^2 (note that b_u does not depend on u). Let us denote by B_u the operator associated with b_u in the same way as for A_u.

Due to the Sobolev imbedding lemma [1], the identity map from $D(A_u) \subset H_0^2(\Omega)$ to $L^2(\Omega)$ is compact; this fact, the V-ellipticity, and the self-adjointness of the bilinear forms a_u and b_u allows one to show that the operator equation

$$A_u y = \zeta B_u y \tag{2.12}$$

in all the considered examples, has a purely discrete and real positive spectrum; i.e., a denumerable set of isolated eigenvalues of finite multiplicity (with are real positive numbers), as one expects for physical reasons. This result rests on F. Riesz' spectral theory [6].

To apply it one introduces the operator $C_u = A_u^{-1} B_u$, and one shows (see [10], Prop. 2.1) that if A_u^{-1} is compact, B_u bounded, and zero is neither an eigenvalue of A_u or B_u, then the eigenvalues of $A_u^{-1} B_u$ are exactly the inverses of the numbers ζ, such that there exists $y \in D(A_u)$, $y \neq 0$, satisfying $A_u y - \zeta B_u y = 0$. Note that all the previous assumptions are verified in all the examples except for the buckling column. However, a separate analysis (see [11] para. 2) shows that the same results hold also in this case. Then it is easy to see that $A_u^{-1} B_u$ is a self-adjoint operator with respect to the scalar product defined by b_u; as it is compact the F. Riesz' theory applies straightforwardly. In particular, we deduce that all the eigenvalues are semi-simple. This result will be of the uppermost importance for our perturbation study.

2.2 The optimization problems

Typical optimization problems consist of minimizing the weight of the structure which amounts to the very simple functional $J(u) = \varrho \int_\Omega u(x) \, dx$ under various constraints.

To prevent the degeneracy of the structure (and hence of the equations) one imposes

$$u(x) \geqq u_0(x) > 0. \tag{2.13}$$

For the static response one asks that the deflection be bounded:

$$\forall \, x \in [0, 1] \quad |z(x; u)| < c \tag{2.14}$$

For the vibration one considers the constraint

$$\zeta(u) \geqq \zeta_0 > 0 \tag{2.15}$$

where $\zeta(u)$ is the smallest eigenvalue (see [8] for many other examples).

The constraints (2.14) and (2.15) are implicit with respect to the control u. To write out necessary optimality conditions, or to apply a gradient projection-like algorithm, one has to derivate these constraints (in other words to make a sensitivity analysis).

This derivability is now rigorously proved in [10, 11], although it has been used previously in many numerical applications [2, 9].

We sketch the methods in the following two paragraphs.

3. Design sensitivity analysis

It is based on perturbation theory. One could try to deal with these problems by using an implicit function theorem as for example in [4, 5], but this is not tractable for multiple eigenvalues (see [19] for a weaker derivability result obtained by a different approach).

3.1 Static problems

The method consists of first proving that the bilinear form is Fréchet differentiable with respect to u, in the sense of relatively bounded perturbations; i.e., there exists a bilinear form $a^{(1)}_{u,\delta u}(z, v)$ such that $\delta u \to a^{(1)}_{u,\delta u}$ is bilinear with the bound

$$|a^{(1)}_{u,\delta u}(z, z)| \leqq e_1(u) \, \|\delta u\| \, a_u(z, z) \tag{3.1}$$

and such that for

$$a^{(2)}_{u,\delta u} = a_{u+\delta u} - a_u - a^{(1)}_{u,\delta u} \tag{3.2}$$

the following bound is valid

$$|a^{(2)}_{u,\delta u}(z, z)| \leqq e_2(u, \delta u) \, \|\delta u\| \, a_u(z, z) \tag{3.3}$$

with $e_2(u, \delta u) \to 0$, as $\|\delta u\| \to 0$. This differentiability has to be proved for every example; we are giving the complete proof for the beam equation.

A formal variational calculation from Eq. (2.2) yields

$$a^{(1)}_{u,\delta u}(z, v) = 2E\alpha \int_0^1 2u\delta u z_{xx} v_{xx} \, dx. \tag{3.4}$$

One has

$$|a^{(1)}_{u,\delta u}(z, z)| \leqq \frac{2E\alpha}{u_0} \left[\int_0^1 u^2 z_{xx}^2 \, dx \right] \|\delta u\|_{L^\infty},$$

$$|a^{(1)}_{u,\delta u}(z, z)| \leqq \frac{2}{u_0} \|\delta u\|_{L^\infty} \, a_u(z, z), \tag{3.5}$$

$$|a^{(2)}_{u,\delta u}(z, z)| = \left| \alpha E \int_0^1 (\delta u)^2 \, z_{xx}^2 \, dx \right|,$$

$$|a^{(2)}_{u,\delta u}(z, z)| \leqq \frac{\|(\delta u)^2\|_{L^\infty}}{u_0^2} \, a_u(z, z). \tag{3.6}$$

This establishes Eqs. (3.2) and (3.3) when $\|\delta u\| \to 0$. In the same way one can show that for the plate and the plane elasticity the Fréchet differentials of the bilinear forms are:

Plate

$$a^{(1)}_{u,\delta u}(z, v) = \int\limits_{\Omega} \left(\delta u \frac{3Eu^2}{12(1-\nu^2)}\right) [z_{11}v_{11} + \nu(z_{22}v_{11} + v_{22}z_{11}) + z_{22}v_{22} + 2(1-\nu)\, z_{12}v_{12}]\, dx, \tag{3.7}$$

Plane elasticity

$$a^{(1)}_{u,\delta u}(z, v) = \frac{1}{2(1+\nu)} \int\limits_{\Omega} \alpha \delta u \{v^1_1[4\nu(z^1_1 + z^2_2) + 2z^1_1] + v^1_2(z^1_2 + z^2_1) + v^2_1(z^1_2 + z^2_1) + v^2_2[4\nu(z^1_1 + z^2_2) + 2z^2_2]\}\, dx. \tag{3.8}$$

Let us state a perturbation theory result.

Theorem 3.1. *Under the hypotheses* (3.1), (3.3), *the operator* A_u^{-1} *(which exists because* a_u *is coercive) is a bounded operator in* $L^2(0, 1)$ *which is Fréchet differentiable with respect to* u.

Here is the idea of the proof (see [10] for a complete treatment). From Eq. (3.10) of [12] (p. 281) follows the existence of a self-adjoint, invertible operator G_u, such that $G_uG_u = A_u$ (square root of A_u); the domain of G_u is the one of the form a_u and one has

$$a_u(z, v) = (G_uz, G_uv). \tag{3.9}$$

Using Eqs. (3.11), (3.12) and lemma VI. 3.1. of [12] (p. 336), and Eqs. (3.1), and (3.3) one gets $C_1(u, \delta u)$, $C_2(u, \delta u)$, bounded operators in L_2, such that

$$a^{(1)}_{u,\delta u}(z, v) = \big(C_1(u, \delta u)\, G_uz, G_uv\big), \tag{3.10}$$

$$a^{(2)}_{\delta u,u}(z, v) = \big(C_2(u, \delta u)\, G_uz, G_uv\big)\,; \tag{3.11}$$

$C_1(u, \delta u)$ is linear in δu as also $a^{(1)}_{u,\delta u}$ and one has the following bounds

$$\|C_1(u, \delta u)\| \leqq e_1(u)\, \|\delta u\|, \tag{3.12}$$

$$\|C_2(u, \delta u)\| \leqq e_2(u, \delta u)\, \|\delta u\| \tag{3.13}$$

where e_1 and e_2 have already appeared in Eqs. (3.1) and (3.3).

One deduces from Eqs. (3.2), (3.10), (3.11), (3.12), and (3.13) the expansion (see [10])

$$A_{u+\delta u} = G_u(I + C_1 + C_2)\, G_u,$$

hence,

$$A^{-1}_{u+\delta u} = G_u^{-1}(I + C_1 + C_2)^{-1}\, G_u^{-1},$$

which enables one to prove the bound (see [10])

$$\|A^{-1}_{u+\delta u} - A_u^{-1} - G_u^{-1}C_1(u, \delta u)\, G_u^{-1}\| \leqq e_3(u, \delta u)\, \|\delta u\| \tag{3.14}$$

where $e_3(u, \delta u) \to 0$ when $\delta u \to 0$, which proves the Fréchet differentiability of $u \to A_u^{-1}$ and gives the following formula for the derivative

$$\delta_u A_u^{-1} = -G_u^{-1}C_1(u, \delta u)\, G_u^{-1}. \tag{3.15}$$

3.2 Linearization of the constraint (2.14) $|z(u; x)| \leq C$

It is transformed in an integral constraint:

$$\int_\Omega (|z| - C)^{+2}\, dx = 0$$

where $x^+ = \dfrac{x + |x|}{2}$; its derivative is

$$2 \int_\Omega (|z| - C)^+ \operatorname{sgn} z_u \delta_u z_u\, dx = 0. \tag{3.16}$$

One deduces from Eq. (3.15) $\delta_u z_u = (\delta_u A_u^{-1})\, f$,

$$\delta_u z_u = -G_u^{-1} C_1(u, \delta u)\, G_u^{-1} f$$

now

$$+ \int_\Omega \psi \delta_u z_u = -(\psi, G_u^{-1} C_1(u, \delta u)\, G_u^{-1} f)_{L^2}.$$

Manipulating and using the self-adjointness of A_u^{-1}, one gets:

$$\int_\Omega \psi \delta_u z_u = -\left(A_u^{-1}\psi, G_u C_1(u, \delta u)\, G_u A_u^{-1} f\right),$$

$$\int_\Omega \psi \delta_u z_u = -\left(A_u^{-1}\psi, G_u C_1(u, \delta u)\, G_u z_u\right),$$

$$\int_\Omega \psi \delta_u z_u = -a^{(1)}_{u,\delta u}(A_u^{-1}\psi, z_u).$$

If one introduces p_u (adjoint state) solution of

$$a_u(p_u, \phi) = (\psi, \phi) \tag{3.17}$$

for every $\phi \in H_0^2(]0, 1[)$, one gets

$$\int_\Omega \psi \delta_u z_u = -a^{(1)}_{u,\delta u}(p_u, z_u), \tag{3.18}$$

so that Eq. (3.16) may be written as

$$a^{(1)}_{u,\delta u}(p_u, z_u) = 0 \tag{3.19}$$

where p_u is the solution of

$$\forall \phi \in H_0^2(]0, 1[), \quad a_u(p_u, \phi) = \left(2\,(|z| - C)^+ \operatorname{sgn} z_u, \phi\right)_{L^2}. \tag{3.20}$$

3.3 Eigenvalue problems

To be tractable the eigenvalue problems are brought to the standard form $C_u y_u = \lambda_u y_u$, as was explained at the end of para. 2.2. However, before proceeding any further let us make a heuristic computation.

Let us linearize the equation

$$a_u(y_u, y_u) = \zeta_u b_u(y_u, y_u), \tag{3.21}$$

we get

$$a^{(1)}_{u,\delta u}(y_u, y_u) + 2a_u(y_u, \delta_u y_u) = \delta_u \zeta_u b_u(y_u, y_u)$$
$$+ \zeta_u b^{(1)}_{u,\delta u}(y_u, y_u) + 2\zeta_u b_u(y_u, \delta_u y_u).$$

Then, using Eq. (2.8) and y_u being an eigenvector, we get:

$$a^{(1)}_{u,\delta u}(y_u, y_u) - \zeta_u b^{(1)}_{u,\delta u}(y_u, y_u) = \delta_u \zeta_u b_u(y_u, y_u)$$

which yields, if y_u is normalized by $b_u(y_u, y_u) = 1$

$$\delta_u \zeta_u = a^{(1)}_{u,\delta u}(y_u, y_u) - \zeta_u b^{(1)}_{u,\delta u}(y_u, y_u). \tag{3.22}$$

This derivation is valid under the assumption that the eigenvalue is simple and that it is differentiable as well as the eigenvector, with respect to design. This is exactly the result to be proved in this section. Such a result has been presented in [16] in a related class of problems.

In the previous paragraph, the Fréchet differentiability of A_u^{-1}, with respect to u, has been proved. The same result is easily proved for B_u in every example except for the buckling which requires a separate analysis (see [11]).

So it remains to study the regularity of λ_u, with respect to u, where λ_u is an eigenvalue of the Fréchet differentiable bounded operator $\hat{C}(u)$.

This is not so obvious as one would except. In fact the 2×2 matrix

$$A(x_1, x_2) = \begin{pmatrix} x_1 & x_2 \\ x_2 & -x_1 \end{pmatrix} \tag{3.23}$$

where x_1 and x_2 are real parameters, shows that the operator may be analytic with respect to the parameters and may have eigenvalues which are only Gâteaux differentiable but not Fréchet differentiable, as one checks easily.

The continuity of the whole spectrum is also wrong, in general, for unbounded operators (see [12]).

However, the continuity of a finite system of eigenvalues is, of course, right for our examples (see theorem 3.1 of [12]).

The derivability result is stated in the following Theorem:

Theorem 3.2. *Let the operator A_u of Eq. (2.12) have a bounded inverse that is Fréchet differentiable with respect to $u \in U$, and let B_u be either bounded and Fréchet differentiable with respect to $u \in U$, or be unbounded ($B = B_u$ independent of u) but $A_u^{-1}B$ be bounded. Let $\zeta(u)$ be a generalized eigenvalue of (A_u, B_u) of Eq. (2.12) of multiplicity m_u. Then, the neighboring group of $\zeta(u)$ of m_u eigenvalues of Eq. (2.12) associated with operators $(A_{u+\delta u}, B_{u+\delta u})$ for $\|\delta u\|$ small enough (as stated in the previous paragraph) is Gâteaux differentiable at u; i e .*

$$\zeta_j(u + t\delta u) = \zeta(u) + t\zeta_j^{(1)}(u, \delta u) + O(t) \qquad (j = 1, \ldots, m_u), \tag{3.24}$$

where

$$\zeta_j^{(1)}(u, \delta u) = \zeta(u)^2 \, \alpha_j^{(1)}(u, \delta u) \tag{3.25}$$

and $\alpha_j^{(1)}(u, \delta u)$ *are the eigenvalues of*

$$P_u \left(\left(\frac{d}{ds} C_{u+s\delta u} \right)_{s=0} \right) P_u \tag{3.26}$$

where $C_u = A_u^{-1} B_u$ *and* P_u *is its spectral projector*[1] *associated with* $\alpha(u) = \frac{1}{\zeta(u)}$.

The proof of this theorem is rather technical and involves the use of the resolvent $R(\zeta, C) = (C - \zeta)^{-1}$ for $\zeta \in \mathbb{C}$ and the Dumford integral representation of the spectral projector (see [11] Appendix).

On the other hand, the following corollary gives a explicit computation of $\zeta_j^{(1)}(u, \delta u)$.

Corollary. *Under the hypotheses of the previous theorem, if* $\zeta(u)$ *is an m-fold eigenvalue of* (A_u, B_u), *its derivatives are the* m_u *eigenvalues of the matrix M of general term*

$$M_{ij} = a_{u,\delta u}^{(1)}(y_i, y_j) - \zeta(u)\, b_{u,\delta u}^{(1)}(y_i, y_j), \quad i = 1, \ldots, m;\ j = 1, \ldots, m \tag{3.27}$$

where y_i *(short for* y_{ui}*) satisfies* $(i = 1, \ldots, m)$

$$\begin{aligned} &\text{(i)}\ \ A_u y_i = \zeta(u)\, B_u y_i, \\ &\text{(ii)}\ \ (B_u y_i, y_j) = \delta_{ij}. \end{aligned} \tag{3.28}$$

Proof. As noted in the footnote of the previous theorem, P_u is the orthogonal projector for the scalar product $b_u(y, v) = (B_u y, v)$ on the eigenspace associated with $\zeta(u)$ of multiplicity m_u; $y_{u,i}$ $(i = 1, \ldots, m)$ denotes an orthonormal (for b_u) basis of this eigenspace. Thus, for any $y \in V$

$$P_u y = \sum_{j=1}^{m} (B_u y, y_{u,i})\, y_{u,i} \quad \text{(using Eq. (3.15))}.$$

Now,

$$\delta_u C_u = \left[\frac{d}{ds} C_{u+s\delta u} \right]_{s=0}$$

is equal to

$$\delta_u C_u = -G_u^{-1} C_1(u, \delta u)\, G_u^{-1} B_u + A_u^{-1} B_{u,\delta u}^{(1)}$$

so that if $N_{ij} = (P_u \delta_u C_u P_u)_{i,j}$, one gets

$$N_{ij} = \left(B_u P_u [-G_u^{-1} C_1(u, \delta u)\, G_u^{-1} B_u + A_u^{-1} B_{u,\delta u}^{(1)}]\, P_u y_{u,j}, y_{u,i} \right).$$

Now,

$$(P_u^* B_u y_{u,i}, v) = (B_u y_{u,i}, P_u v) = \sum_{j=1}^{m} (B_u y_{u,i}, y_{u,j})\, (B_u v, y_{u,j}) = (B_u v, y_{u,i}) = (v, B_u y_{u,i})$$

so that $P_u^* B_u y_{u,i} = B_u y_{u,i}$; which is obvious if one notes that P_u is self-adjoint with respect to b_u.

---

[1] The operator projecting elements of V onto the subspace of V spanned by the eigenvectors of $\hat{C}_u$ associated with the eigenvalue $\alpha(u)$; the projection being parallel to the supplementary invariant subspace. $\hat{C}_u$ being self-adjoint (see end of para. 2.1) for the scalar product b_u, this projection is also orthogonal with respect to b_u.

Thus,

$$N_{ij} = ([-G_u^{-1}C_1(u, \delta u)\, G_u^{-1}B_u + A_u^{-1}B^{(1)}_{u,\delta u}]\, y_{u,j},\, B_u y_{u,i}),$$

$$N_{ij} = (-A_u^{-1}G_u C_1(u; \delta u)\, G_u A_u^{-1} B_u y_{u,j},\, B_u y_{u,i}) + (A_u^{-1}B^{(1)}_{u,\delta u} y_{u,j},\, B_u y_{u,i})$$

which yields using the self-adjointness of A_u^{-1} and the equation

$$A_u^{-1}B_u y_{u,j} = \frac{1}{\zeta(u)}\, y_{u,i},$$

$$N_{ij} = \frac{1}{\zeta(u)^2}\left(-G_u C_1(u, \delta u)\, G_u y_{u,j},\, y_{u,i}\right) + \frac{1}{\zeta(u)}\,(B^{(1)}_{u,\delta u} y_{u,j},\, y_{u,i})$$

which may be expressed in terms of design derivatives of the bilinear form:

$$N_{ij} = \frac{-1}{\zeta(u)^2}\, a^{(1)}_{u,\delta u}(y_{u,j}, y_{u,i}) + \frac{1}{\zeta(u)}\, b^{(1)}_{u,\delta u}(y_{u,j}, y_{u,i})$$

and one concludes by noting $\zeta^{(1)}_{u,\delta u} = -\zeta(u)^2\, \alpha^{(1)}_{u,\delta u}$, that $\zeta^{(1)}_{u,\delta u}$ are the eigenvalues of $M_{ij} = -\zeta(u)^2\, N_{ij}$, which gives Eq. (3.27)

4. Conclusion

The method presented here has a very wide range of applicability in structural mechanics as well as in many other optimal design problems; for example, it can be used as a basis for shape optimal design [18].

Although many authors argue that multiple eigenvalues do not occur in practice, several authors have pointed out that they do appear in optimum designs, satisfying constraints on eigenvalues [13, 14]. Detailed numerical experiments are in process.

References

[1] Adams, R. A., Sobolev spaces, Academic Press, New York 1975.

[2] Armand, J. L., and Lodier, B., Optimal design of bending elements, Internat. J. Numer. Meth. Engrg. **13** (1978), 373–381.

[3] Aubin, J. P., Approximation of elliptic boundary-value problems, Wiley Interscience, New York 1972.

[4] Chavent, G., Analyse fonctionelle et identifications de coefficients répartis dans les équations aux dérivées partielles, Thèse d'Etat, Faculté des Sciences de Paris, 1971.

[5] Dervieux, A., et Palmerio, B., Une formule de Hadamard dans des problèmes d'identification de domaines, I et II, C.R. Acad. Sci. Paris Sér. A, **280, 281** (1975).

[6] Dieudonné, J. A., Foundations of modern analysis, New York, Acad. Press, 1969.

[7] Fichera, G., Existence theorems in elasticity, Handbuch der Physik, Vol. VI, a/2, 1972, 347–389.

[8] Haug, E. J., and Arora, J. S., Applied optimal design, Wiley, New York 1979.

[9] Haug, E. J., and Arora, J. S., Design sensitivity analysis of elastic mechanical systems, Comput. Methods Appl. Mech. Eng. **15** (1978), 35–62.

[10] Haug, E. J., and Rousselet, B., Design sensitivity analysis in structural mechanics I: Static response variations, J. Structural Mech. 8 (1980), 14–41.

[11] Haug, E. J., and Rousselet, B., Design sensitivity analysis in structural mechanics II: Eigenvalue variations, J. Structural Mech. 8 (1980), 161—186.
[12] Kato, T., Perturbation theory for linear operators, Springer-Verlag, Berlin—Heidelberg—New York 1966.
[13] Masur, E. F., and Mroz, Z., Non stationary optimality conditions in structural design. Internat. J. Solids and Structures 15 (1979), 503—512.
[14] Olhoff, H., and Rasmussen, S. H., On single and bimodal optimum buckling modes of clamped columns, Internat. J. Solids and Structures 13 (1977), 605—614.
[15] Rousselet, B., Etude de la régularité des valeurs propres par rapport à des déformations bilipschitziennes du domaine, C.R.Acad.Sci. Paris, Sér. A, 283 (1976).
[16] Rousselet, B., Identification de domaines et problèmes de valeurs propres, Thèse de Spécialité, Université de Nice, France, 1977.
[17] Rousselet, B., Optimal design and eigenvalue problems. *In:* Optimization Techniques, Proceedings of the 8th I.F.I.P. Conference, Lecture Notes in Control and Information Sciences Vol. 7, Springer-Verlag, Berlin—Heidelberg—New York 1978, 343—352.
[18] Rousselet, B., and Haug, E. J., Design sensitivity analysis in structural mechanics III: Effects of shape variation (to appear).
[19] Zolezio, J. P., Identification de domaines par déformation, Thèse d'Etat, Université de Nice, 1979.

Address of the author:

Département de Mathématiques
Université de Nice
Parc Valrose
06034 Nice
France

and

College of Engineering
University of Iowa
Iowa City, Iowa 52242
U.S.A.

On optimal control problems in civil-engineering and technology

Karl-Heinz Schlüßler, Cottbus

Abstract. This paper deals with optimal control problems and their application in civil-engineering and technology that can be described by ordinary or partial differential equations. Taking as model a beam and a slab as problems of practical interest the thermal prestressing and optimal design are discussed. The former is leading to an optimal control problem for the heat equation, the latter to the optimal distribution of flexural rigidity, for which a necessary condition can be given.

Introduction

Problems of optimal control in continuum mechanics have already a certain tradition, whenever the treatment has often a somewhat academic feature and application suffers under the lack of numerical standard routines. A lot of papers could be cited respecting this topic (see for instance Prager [1], Distefano [2], Klosowicz and Lurie [3], Armand [4] and Banitschuk [5]). And the number of problems could easily be extended if you think of Kármán's plate problem, of shell problems or of problems of plasticity brought with a reasonable objective functional in form of control theory. Another question is still how to overcome the numerical difficulties.

In the field of technics and technology, it seems, a more realistic point of view must be accepted to bring control theory to an adequate level of efficiency. Especially in civil-engineering that means

(i) to be more aware of the physical properties of material (e.g. concrete as the main building material in all industrial developed countries) and to exploit them in a more creative way in the sense of distributed parameter control theory,

(ii) to develop the ability of the scientific worker to recognize in real technical or technological processes those phaenomena that give the possibility to controllability,

(iii) to give with control problems a convincing contribution to production on a higher standard of quality, decreased consume of time or better continuity of the running process.

The following contribution discusses two problems that can be treated by the first glance in a quite familiar manner as common in continuum mechanics, though by the second glance it must be stated that the full truth in the view of application was not yet quite reached and needs further accomplishment. But this may be fate of optimization at all. Perhaps it is better first to get approximate results and recorrect a simpler model than to start at once with a perfect one.

The pronounced intention here is to find out those control phaenomena being of interest in the sphere of real application for elements having large production numbers. Elements of this character are in civil-engineering up to now the beam and the slab. Dealing with those objects one can not expect exciting news. Nevertheless some general demands in the field of practice can be outlined. In the opinion of the author there should nothing be to simple to be not worth of a new attack with a better scientific weapon. Besides there are a lot of places where no scientific battle took place and that are despite in a surprisingly well-posed government of experience. Them too must be paid more attention. Merely to have an idea remember concrete, a stuff from simple composition but from theoretical point of view yet undefined description, being a randomly packed multiphase granular medium that must be brought by vibration to a viscous state and minimal volume. And now answer the question: "How reads a control problem to determine the optimal frequency of vibration of concrete in a given steel matrix?"

1. Thermal prestressing

We regard as an element of a structure a beam being the simplest case of continuum mechanics. From the viewpoint of technical application we are interested in such a design that the load can be carried by a state of internal stresses without leaving the field of admissible stresses. In steel construction systems are preferred having a constant cross-section for the whole length. Consequently more material is supplied than would be necessary if another technology than hot-rolling were available. The question is whether a better exploitation of the material is possible despite of this technical restriction of constant cross-section.

Commonly it is presupposed that a system

(i) is unstressed, when it is unloaded.

(ii) is physically homogeneous, that means, it consists of only one unitary part.

We will give up this demands and think of a beam being of two physically different or indifferent parts, brought continuously into contact by a suitable technological process (like welding or glueing) explicitly with the aim to bring the system to a state of controlled self-stress. As is well known material changes its length if its state of temperature changes and gets in a state of stress if deformations are hindered. Using the fact that the two different parts of the beam being in contact constrain themselves if having initially different thermal states we have a medium to produce a prestressed beam.

1.1 The state of stress

To answer the question what temperature fields should be induced to the bars, we must find the dependence between the temperature field and the stress field of the system. These fundamentals can be studied at hand of the one-dimensional problem.

We have in mind a system as shown in fig. 1 and make the following assumptions:

A 1: the z-axis coincides with the line of zero bending stresses of the flitched beam

A 2: the contact of bar 1 and bar 2 is produced at time $t = 0$.

A 3: for both of the materials for bar 1 and 2 Hooke's law is valid.

A 4: Bernoulli's hypothesis of plane deformation of the cross-section is valid for the flitched beam.

A 5: thermal expansion of the material follows a linear law.

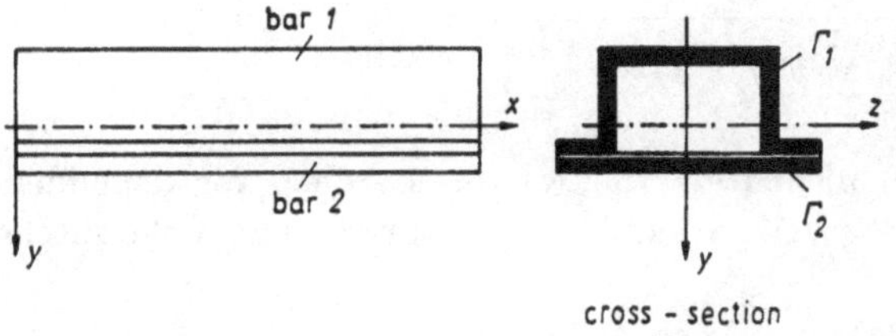

Fig. 1

For any fixed time $t > 0$ the stress-strain law is given by

$$\varepsilon_k(x, y) = \frac{1}{E_k} \sigma_k(x, y) + \alpha_k \Delta T_k(x) \quad (k = 1, 2) \tag{1.1}$$

(ε strain, σ stress, E modul of elasticity,
α coefficient of linear thermal-expansion,
$\Delta T(x) = T(x, t) - T(x, 0)$ change of temperature in the time $[0, t]$).

The strain being a linear function in y (see A 4)

$$\varepsilon_k(x, y) = \varepsilon(x, 0) + \varkappa(x)\, y, \tag{1.2}$$

compatibility at the line of contact $y = \hat{y}$ is at once satisfied. From (1.1) and (1.2) we have

$$\sigma_k(x, y) = E_k\big(\varepsilon(x, 0) + \varkappa(x)\, y\big) - \alpha_k E_k \Delta T(x) \quad (k = 1, 2). \tag{1.3}$$

The yet undetermined axial strain $\varepsilon(x, 0)$ and curvature $\varkappa(x)$ can be lead back to the axial force $N(x)$ and the moment $M_z(x)$ using the equilibrium conditions

$$\int_{\Gamma} \sigma(x, y)\, dA = \int_{\Gamma_1} \sigma_1(x, y)\, dA + \int_{\Gamma_2} \sigma_2(x, y)\, dA = N(x), \tag{1.4}$$

$$\int_{\Gamma} y\sigma(x, y)\, dA = \int_{\Gamma_1} y\sigma_1(x, y)\, dA + \int_{\Gamma_2} y\sigma_2(x, y)\, dA = M_z(x). \tag{1.5}$$

With the abbreviations

$$\int_{\Gamma_k} dA = A_k, \quad \int_{\Gamma_k} y\, dA = S_k, \quad \int_{\Gamma_k} y^2\, dA = J_k \tag{1.6}$$

and with respect to A 1 reminding

$$E_1 S_1 + E_2 S_2 = 0, \tag{1.7}$$

from (1.4) and (1.5) we have two linear equations giving the solution

$$\varepsilon(x, 0) = \frac{N(x) + \sum_{k=1}^{2} \alpha_k E_k A_k \Delta T_k(x)}{\sum_{k=1}^{2} E_k A_k}, \tag{1.8}$$

$$\varkappa(x) = \frac{M_z(x) + \sum_{k=1}^{2} \alpha_k E_k S_k \Delta T_k(x)}{\sum_{k=1}^{2} E_k J_k}. \tag{1.9}$$

From (1.3), (1.8) and (1.9) ultimately follows the searched for dependence between stress and loading (giving $N(x)$, $M_z(x)$) and temperatur-change of the flitched beam

$$\sigma_k(x, y) = E_k \frac{N(x) + \sum_{i=1}^{2} \alpha_i E_i A_i \Delta T_i(x)}{\sum_{i=1}^{2} E_i A_i}$$

$$+ E_k y \frac{M_z(x) + \sum_{i=1}^{2} \alpha_i E_i S_i \Delta T_i(x)}{\sum_{i=1}^{2} E_i J_i} - \alpha_k E_k \Delta T_k(x) \quad (k = 1, 2). \tag{1.10}$$

It must be noticed that the stresses of the system must always be kept in an admissible domain

$$|\sigma_k(x, y)| \leqq \sigma_{k,\mathrm{ad}}. \tag{1.11}$$

To get a practical advantage in the exploitation of the material the stresses caused by temperature-change and the stresses caused by loading of the system should partly equilibrate themselves.

(1.11) in connection with (1.10) means that $\Delta T_k(x)$ is a bounded function and must be chosen in a way a maximal increase of load carrying capacity of the system is obtained and the stresses in the unloaded state do not exceed the constraint (1.11).

The possible effects may be shortly illustrated with two examples of pure bending of a beam. We respectively regard of the whole beam only the cross-section $x = \mathring{x}$ that is maximally stressed. For the sake of simplicity the material parameters may be equal ($\alpha_1 = \alpha_2 = \alpha$, $E_1 = E_2 = E$), a temperature change in part 1 does not occur $\big(\Delta T_1(x) = 0\big)$ and an axial load is not present $\big(N(x) = 0\big)$. The parameter t denotes the thickness of the sheet metal!

Example 1 (s. fig. 2)

$$A_1 = 64t^2, \quad A_2 = 32t^2, \quad A = 96t^2, \quad \sigma_{1,\mathrm{ad}} = \sigma_{\mathrm{ad}}$$

$$S_1 = -192t^3, \quad S_2 = 192t^3, \quad J = 4472t^4, \quad \sigma_{2,\mathrm{ad}} = 1{,}4\,\sigma_{\mathrm{ad}}$$

$$\mathring{y} = 5{,}5t, \quad -11{,}5t \leqq y \leqq 6{,}5t$$

Example 2 (s. fig. 3)

$$A_1 = 48t^2, \qquad A_2 = 48t^2, \qquad A = 96t^2$$

$$S_1 = -216t^3, \qquad S_2 = 216t^3, \qquad J = 31524t^4$$

$$\mathring{y} = 0, \qquad -9t \leqq y \leqq 9t, \; \sigma_{1,\mathrm{ad}} = \sigma_{2,\mathrm{ad}} = \sigma_{\mathrm{ad}}$$

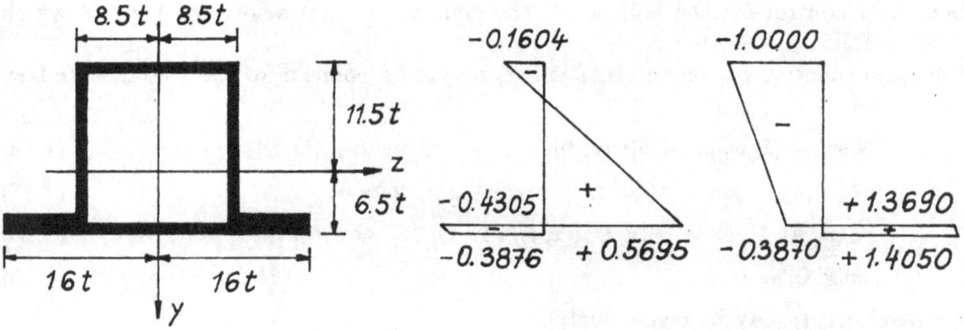

a) b) c)

Fig. 2a. Cross-section of example 1
Fig. 2b. $\sigma/\alpha E\Delta T_2$: stress distribution of the unloaded system
Fig. 2c. $\sigma/\sigma_{\mathrm{ad}}$: total stress (optimal $\alpha E\Delta T_2 = -1.756\sigma_{\mathrm{ad}}$, max $M_z = 1.2817 W_0\sigma_{\mathrm{ad}} = 1.2817\mathring{M}_z$)

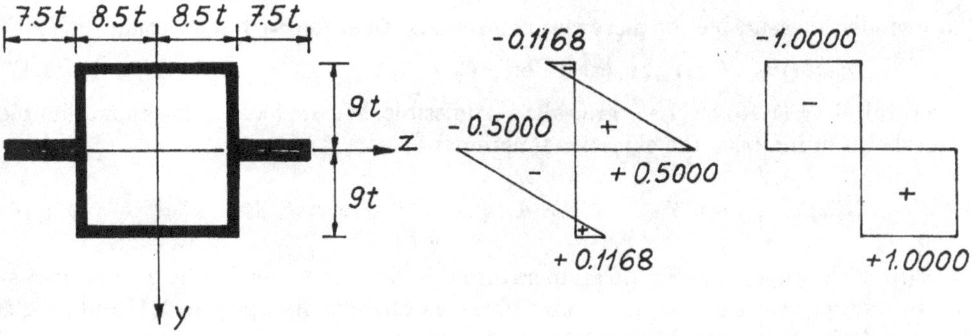

a) b) c)

Fig. 3a. Cross-section of example 2
Fig. 3b. $\sigma/\alpha E\Delta T_2$: stress distribution of the unloaded system
Fig. 3c. $\sigma/\sigma_{\mathrm{ad}}$: total stress (optimal $\alpha E\Delta T_2 = -2.0\sigma_{\mathrm{ad}}$, max $M_z = 1.2335 W\sigma_{\mathrm{ad}} = 1.2335\mathring{M}_z$)

The results from the examples 1 and 2 see figures 2 and 3. As result we have

a) the load carrying capacity in comparison with a beam of the same cross-section without prestress can be raised to 128% resp. 123% in the first resp. the second case.

b) information of the maximal temperature $\Delta T_2(\mathring{x})$ at $x = \mathring{x}$ that is to be produced to get the wished prestress (for examples 1 and 2, $\Delta T_2(\mathring{x}) = -97.5$ K resp. -111 K).

1.2 The state of temperature

We bring the beam into the wished state of elongation by a suitable field of temperature. This may be produced by means of heating respectively of cooling of e.g. three points of the system (the endpoints and the midpoint). In the cases of symmetry of the state of temperature the problem can be brought to a one-dimensional problem of boundary control for the half-beam, the general ($x \in R^n$) aelotropic type of which is given as follows.

The state function of temperature $w(x, t)$ may be a solution of the initial boundary value problem

$$bw_{|t} - (\lambda_{ij}w_{|j})_{|i} = \Phi(w(x, t)) \qquad \forall\ (x, t) \in B \times [0, T], \tag{1.12}$$

$$\lambda_{ij}w_{|j}n_i + \alpha w = \alpha\varphi(x, t) \qquad \forall\ x \in \Gamma_1, \quad t > 0, \tag{1.13}$$

$$\lambda_{ij}w_{|j}n_i + \alpha w = \Psi(x, t, w(x, t), u(x, t)) \qquad \forall\ x \in \Gamma_2, \quad t > 0, \tag{1.14}$$

$$w(x, 0) = \mathring{w}(x). \tag{1.15}$$

The control $u(x, t)$ may be respectively

(i) a square integrable function on $\Gamma_2 \times [0, T]$,

$$\int_0^T \int_{\Gamma_2} u^2(x, t)\, dx\, dt < \infty, \tag{1.16}$$

(ii) a bounded measurable (or piecewise continuous) function with restriction

$$u_u(x, t) \leqq u(x, t) \leqq u_0(x, t) \quad \text{on} \quad \Gamma_2 \times [0, T]. \tag{1.17}$$

Controls fulfilling (1.16) or (1.17) are called admissible. We are looking for an admissible control that is minimizing the objective functional

$$J[u] = \int_B [w(x, T) - \overline{w}(x)]^2\, dx + \lambda_0 \int_0^T \int_{\Gamma_2} u^2(x, t)\, dx\, dt. \tag{1.18}$$

Here $\overline{w}(x)$ is the tape function of temperature (heating) at the end time T that corresponds to $\Delta T(x)$ as the succeeding cooling phase as already discussed in 1.1. and $\lambda_0 \geqq 0$ may be a suitable factor for the cost of control.

It may be presupposed that there exists for each control $u(x, t)$ a unique (square integrable) solution $w(x, t)$ of (1.12)—(1.15),

$$w(x, t) = \chi[u]\,(x, t). \tag{1.19}$$

We introduce the Hilbert spaces H and U of states and controls respectively,

$$H = L_2(B \times (0, T)), \quad U = L_2(\Gamma_2 \times [0, T]). \tag{1.20}$$

The adjoint state $z(x, t)$ must fulfil the adjoint initial boundary value problem

$$bz_{|t} + (z_{|i}\lambda_{ij})_{|j} + z\Phi_W(w) = 0 \quad \forall (x, t) \in B \times [0, T], \tag{1.21}$$

$$z_{|i}\lambda_{ij}n_j + \alpha z = 0 \quad \forall\ x \in \Gamma_1, \quad 0 \leqq t < T, \tag{1.22}$$

$$z_{|i}\lambda_{ij}n_j + \alpha z = z\Psi \quad \forall\ x \in \Gamma_2, \quad 0 \leqq t < T, \tag{1.23}$$

$$bz(x, T) = 2[\overline{w}(x) - w(x, T)] \quad \forall\ x \in B. \tag{1.24}$$

The gradient $P = J'[u] \in L_2(\Gamma_2 \times [0, T])$ of the functional (1.18) is given by

$$P(x, t) = 2\lambda_0 u(x, t) - \Psi_u(x, t, w, u)\, z(x, t) \quad \text{for} \quad (x, t) \in \Gamma_2 \times (0, T). \tag{1.25}$$

So we have the weak minimum principle

$$\underset{\substack{v \in \Omega \\ (x,t) \in \Gamma_2 \times (0,T)}}{\mathrm{Min}} \; [2\lambda_0 u(x, t) - \Psi_u(x, t, w, u)\, z(x, t)]\, v(x, t) \tag{1.26}$$

or the strong minimum principle

$$\underset{\substack{v \in \Omega \\ (x,t) \in \Gamma_2 \times (0,T)}}{\mathrm{Min}} \; [\lambda_0 v^2(x, t) - \Psi(x, t, w, v)\, z(x, t)], \tag{1.27}$$

the minimum being attained for $v = u$, respectively.

In the case (i) of unbounded control of type

$$\Psi \equiv \alpha u(x, t) \quad \forall\, (x, t) \in \Gamma_2 \times (0, T) \tag{1.28}$$

from (1.26) we have

$$z(x, t) = \frac{2\lambda_0}{\alpha}\, u(x, t), \tag{1.29}$$

that means (with continuation to $B \times (0, T)$) the optimal control $u(x, t)$ according to (1.29) must solve the adjoint problem (1.21)—(1.24).

Whereas in the case (ii) with a control function according to (1.28) we derive the necessary condition

$$2\lambda_0 u(x, t) - \alpha z(x, t) \begin{cases} \geqq 0 & \text{for} \quad u = u_u, \\ = 0 & \text{for} \quad u_u < u < u_0, \\ \leqq 0 & \text{for} \quad u = u_0. \end{cases} \tag{1.30}$$

(For more details see for instance A. T. Egorov [6]). Because of the uncertainty how the process of heating can be realized technically in the best way we do not be jealous to give numerical solutions here. What is asked for are controls that furnish triangular or parabolic tape functions for the end temperature corresponding to single force loading resp. loading by the dead weight of the structure.

2. The aelotropic slab with optimal distribution of bending stiffness

The problem of an isotropic rectangular slab of Kirchhoff type of optimal distribution of a given volume with a minimal value of occuring maximal deflection was treated by Banitschuk, Kartvelishvili, Mironow [5] in 1977. Here we will look for the necessary condition for the optimal distribution of the bending stiffness of an aelotropic slab. The fundamental equations are written down using the symbolism of Ricci's tensor calculus.

For the sectional forces we have the well-known relations between moments M_{ik}, shearing forces Q_k and the loading $q(x_1, x_2)$

$$M_{ik/i} = Q_k, \tag{2.1}$$

$$Q_{k/k} = -q(x_1, x_2), \tag{2.2}$$

$$M_{ik/ik} + q(x_1, x_2) = 0. \tag{2.3}$$

The connection between the internal moments and the deflection $w(x_1, x_2)$ of the slab may be defined by an elasticity law of the form

$$M_{ik} = -s(x_1, x_2)\, D_{ikpq} w_{|pq}, \tag{2.4}$$

$s(x_1, x_2)$ being a function describing the distribution of bending stiffness across the domain $(x_1, x_2) \in A$ of the slab and D_{ikpq} an elasticity tensor with constant coordinates, fulfilling appropriate presuppositions of positive definiteness and the symmetry properties

$$D_{ikpq} = D_{pqik} = D_{kipq} = D_{ikqp}. \tag{2.5}$$

We now consider for the pair $\{w(x_1, x_2), s(x_1, x_2)\}$ of state and control variables the following problem

$$D_{ikpq}\big(s(x_1, x_2)\, w_{|pq}\big)_{|ik} = q(x_1, x_2) \tag{2.6}$$

with boundary conditions (on Γ_1 supported, on Γ_2 clamped)

$$w = 0, \quad M_{ik}n_i n_k = 0 \quad \text{for} \quad (x_1, x_2) \in \Gamma_1, \tag{2.7}$$

$$w = 0, \quad w_{|i}n_i = 0 \quad \text{for} \quad (x_1, x_2) \in \Gamma_2, \tag{2.8}$$

n_i being the direction vector of the outer normal of the contour Γ of the slab. Notice that on account of $w = 0$ on $(x_1, x_2) \in \Gamma$ we have always (e_i tangential vector to Γ)

$$w_{|i} = w_{|k}n_k n_i + w_{|k}e_k e_i = w_{|k}n_k n_i \quad \text{for} \quad (x_1, x_2) \in \Gamma. \tag{2.9}$$

The control $s = s(\cdot, \cdot) \in U$ with

$$U = \{s = s(\cdot,\cdot) \text{ bounded measurable with } s(x_1, x_2) \in \Omega \ \forall\ (x_1, x_2) \in A\}, \tag{2.10}$$

$$\Omega = \{s(x_1, x_2)|\ s_u(x_1, x_2) \leqq s(x_1, x_2) \leqq s_0(x_1, x_2)\} \tag{2.11}$$

may minimize the functional

$$\int_A s(x_1, x_2)\, dA = \underset{s \in U}{\text{Min}}, \tag{2.12}$$

while the compliance of the system is restricted by

$$\int_A \tilde{q}(x_1, x_2)\, w(x_1, x_2)\, dA = \mathring{C} = \text{const.} \tag{2.13}$$

Meantime $\tilde{q}(x_1, x_2)$ shall be any statically admissible loading and under circumstances different from the given external loading $q(x_1, x_2)$. Taking $\omega(x_1, x_2)$ as adjoint function to $w(x_1, x_2)$ we bring (2.12), (2.13) with the restriction (2.6) to the form of a variational principle, (2.7), (2.8) being its natural boundary conditions:

$$J[w, s] = \int_A s(x_1, x_2)\, dA + \lambda_0 \int_A \tilde{q}(x_1, x_2)\, w(x_1, x_2)\, dA$$
$$- \lambda_0 \int_A [D_{ikpq}(s w_{|pq})_{|ik} - q]\, \omega(x_1, x_2)\, dA = \text{Min}. \tag{2.14}$$

λ_0 plays here the role of a Lagrange multiplier that must be calculated in connection with (2.13). For convenience, to get the adjoint function $\omega(x_1, x_2)$ physically as a deflection, we take it as a scalar multiplier of $\omega(x_1, x_2)$ too. If we assume the boundary conditions (2.7), (2.8) to be valid also for $\omega(x_1, x_2)$ we can prove using the integral-theorem of Gauss the selfadjointness of the last integral term in (2.14).

$$J[w, s] = \int_{A'} \{s(x_1, x_2) + \lambda_0[\bar{q}(x_1, x_2)\, w(x_1, x_2) + q(x_1, x_2)\, \omega(x_1, x_2)]\}\, dA$$
$$- \lambda_0 \int_A D_{ikpq}(s\omega_{|pq})_{|ik}\, w(x_1, x_2)\, dA\,. \tag{2.15}$$

The term for the first variation becomes

$$\delta J = \int_A [\delta s - \lambda_0 D_{ikpq}(\delta s \omega_{|pq})_{|ik}\, w]\, dA$$
$$+ \lambda_0 \int_A [\bar{q} - D_{ikpq}(s\omega_{|pq})_{|ik}]\, \delta w\, dA\,. \tag{2.16}$$

Applying the integral theorem two times to the second term in the first integrand we get

$$\delta J = \int_A [1 - \lambda_0 D_{ikpq}\omega_{|pq} w_{|ik}]\, \delta s\, dA$$
$$+ \lambda_0 \int_A [\bar{q} - D_{ikpq}(s\omega_{|pq})_{|ik}]\, \delta w\, dA\,. \tag{2.17}$$

From (2.17) we conclude the necessary condition for optimality $\delta J \geqq 0$. So we can see, if the adjoint function $\omega(x_1, x_2)$ fulfils the boundary value problem

$$D_{ikpq}(s\omega_{|pq})_{|ik} = \bar{q}(x_1, x_2)\,, \tag{2.18}$$

$$\omega = 0, \quad M_{ik} n_i n_k = 0 \quad \text{for} \quad (x_1, x_2) \in \Gamma_1\,, \tag{2.19}$$

$$\omega = 0, \quad \omega_{|i} n_i = 0 \quad \text{for} \quad (x_1, x_2) \in \Gamma_2\,, \tag{2.20}$$

then the optimal distribution of bending stiffnes is described by the necessary condition

$$\lambda_0 D_{ikpq} w_{|ik} \omega_{|pq} \begin{cases} \leqq 1 & \text{for} \quad s(x_1, x)_2 = s_u(x_1, x_2)\,, \\ = 1 & \text{for} \quad s_u(x_1, x_2) < s(x_1, x_2) < s_0(x_1, x_2)\,, \\ \geqq 1 & \text{for} \quad s(x_1, x_2) = s_0(x_1, x_2)\,. \end{cases} \tag{2.21}$$

Generally the control problem can only be solved in an iterative way. (See remarks in [7]). As an exception one can expect analytically strong solutions merely for simple problems where the control is unbounded. For illustration we regard the slab spanning only in one direction.

Example. We study as special case the one dimensional problem (independence of x_2, $x_1 \to x$) and assume the control to be not bounded. Taking $\bar{q}(x) \equiv q(x) = \mathring{q}$ (constant line load) it follows $w(x) = \omega(x)$. (2.6) will get the simpler form

$$K\big(s(x)\, w''(x)\big)'' = q(x) \tag{2.6'}$$

with K being the so called bending stiffness,

$$K = \frac{EJ}{1 - \nu^2} \tag{2.22}$$

(ν Poisson's ratio). To have a solution to compare with we give in advance the well-known results for the simply supported slab with constant bending stiffness K.

$$M(x) = \frac{\mathring{q}l^2}{2}\frac{x}{l}\left(1 - \frac{x}{l}\right), \tag{2.23}$$

$$\max_x M(x) = M\left(\frac{l}{2}\right) = \frac{\mathring{q}l^2}{8}, \tag{2.24}$$

$$w(x) = \frac{\mathring{q}l^4}{2K}\left(1 - 2\frac{x^2}{l^2} + \frac{x^3}{l^3}\right)\frac{x}{l}, \tag{2.25}$$

$$\max_x w(x) = w\left(\frac{l}{2}\right) = \frac{5\mathring{q}l^4}{384K}. \tag{2.26}$$

Evaluating (2.13) with (2.25) we would get

$$\mathring{C} = \frac{\mathring{q}^2 l^5}{120K}. \tag{2.27}$$

This may be the value for our control problem. From (2.21) we have at once

$$\lambda_0 K\big(w''(x)\big)^2 = 1, \tag{2.28}$$

$$w''(x) = -1/\sqrt{\lambda_0 K}. \tag{2.29}$$

Integrating (2.6′) two times gives

$$Ks(x)\, w''(x) = \frac{-\mathring{q}l^2}{2}\frac{x}{l}\left(1 - \frac{x}{l}\right). \tag{2.30}$$

With respect of (2.29) follows

$$s(x) = \sqrt{\frac{\lambda_0}{K}}\,\frac{\mathring{q}l^2}{2}\frac{x}{l}\left(1 - \frac{x}{l}\right). \tag{2.31}$$

Integrating (2.29) we have

$$w(x) = \frac{1}{\sqrt{\lambda_0 K}}\frac{l^2}{2}\frac{x}{l}\left(1 - \frac{x}{l}\right). \tag{2.32}$$

Substituting (2.32) in (2.13) and considering (2.27) gives

$$\mathring{q}\,\frac{1}{\sqrt{\lambda_0 K}}\frac{l^2}{2}\frac{l}{6} = \frac{\mathring{q}^2 l^5}{120K}$$

or

$$\sqrt{\frac{K}{\lambda_0}} = \frac{\mathring{q}l^2}{10}. \tag{2.33}$$

From (2.31) and (2.33) results

$$s(x) = 5\,\frac{x}{l}\left(1 - \frac{x}{l}\right), \tag{2.34}$$

that is a parabolic distribution of bending stiffness along the slab. The deflection is given by

$$w(x) = \frac{\mathring{q}l^4}{20K} \frac{x}{l} \left(1 - \frac{x}{l}\right) \tag{2.35}$$

(note that the maximal deflection reaches only 96% of (2.26)). Compared with the slab with constant cross-section the load carrying capacity here can be increased to 125% while the material input is decreased to 83,3%.

Remarks

(i) The results are for example valid for a slab with double-T-cross-section with constant height and corresponding to $s(x)$ (2.34) varying thickness of profile as shown in fig. 4.

(ii) Of course from the practical point of view this solution is only of restricted worth (e.g. the shearing stresses near the bearing would exceed the admissible value).

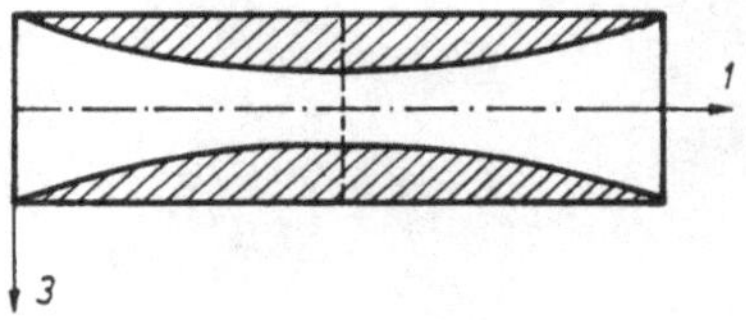

a)

Fig. 4a. Section alongside

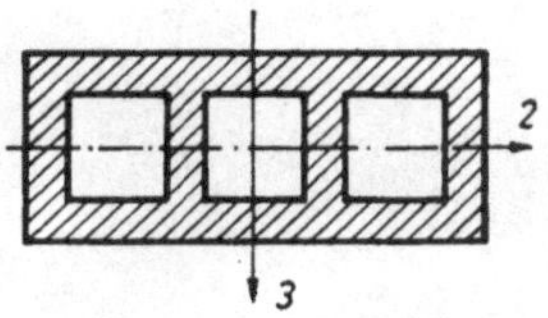

b)

Fig. 4b. Cross-section

Discussion

It is a matter of fact that the material efficiency of a structure can be considerably raised provided that the mathematical obstacles being inherent to the control problem can be overcome and the more difficult technological process can be governed. Clearly we can not call a structure optimal if we do not recognize the demands of production or usage in an appropriate manner. Beyond this, as in our cases with the intention to save material, aspects may come to the foreground that are otherwise of no importance. So for a slab as sketched in fig. 4 the transmission of sound (footfall) becomes essential.

This reveals a substantial feature of control problems in engineering. To avoid dissapointment one should not expect that the classical models are sufficient to tackle the arising questions in the first run. Besides it is necessary to study accessory circumstances to meet with additional demands and succesively to accomplish the modelization of control problems.

References

[1] W. Prager and J. E. Taylor, Problems of optimal structural design. J. Appl. Mech. **35** (1968), 102–106.

[2] N. Distefano, Dynamic Programming and a Max-Min-Problem in the Theory of Structures.

[3] B. Klosowicz and K. A. Lurie, On the optimal distribution of elastic moduli of a nonhomogeneous body. J. Optimization Theory Appl. 12, (1973), 32—42.
[4] J. L. P. Armand, Application of the theory of optimal control of distributed-parameter systems of structural optimization. Russ. Übers. MIR, Moskau 1977.
[5] N. W. Banitschuk et al., Über optimale Platten und eine Methode zur Lösung von Aufgaben der Optimierung von Konstruktionen. (russ.) Preprint No. 89, AdW SSSR, Moskau 1977.
[6] А. И. Егоров, Оптимальное управление тепловыми и диффузионными процессами. Наука, Москва, 1978.
[7] K. H. Schlüssler, Zur optimalen Steuerung mechanischer Eigenschaften in linear-elastischen Kontinua. Wiss. Zeitschrift IH Cottbus 2/79.

Address of the author:

Ingenieurhochschule Cottbus
Abteilung Mathematik/Naturwissenschaften
7500 Cottbus
Karl-Marx-Str. 17
DDR

Das asymptotische Verhalten von Toeplitzdeterminanten für einige Klassen von Erzeugerfunktionen

Bernd Silbermann, Karl-Marx-Stadt

Abstract. The lecture deals with the theory of Toeplitz determinants. For instance there are given some extensions of the Strong Szegö Limit Theorem.

I.

Mathematische, vor allem aber physikalische Probleme führen zu der Aufgabenstellung, die Determinanten von durch gewisse Erzeugerfunktionen $a(t) \in L^1(\Gamma)$, $\Gamma = \{|z| = 1\}$, erzeugten Toeplitzmatrizen zu berechnen:

$$D_n(a) = \det\, (a_{j-k})_{j,k=0}^{n},$$

$$a_n = \frac{1}{2} \int_{-\pi}^{\pi} a(e^{i\varphi})\, e^{-in\varphi}\, d\varphi \quad (n = 0, \pm 1, \ldots).$$

Von G. Szegö [4] wurde 1962 das folgende erstaunliche Resultat veröffentlicht.

Satz 1. *Sei $a(t) > 0$ und besitze eine hölderstetige Ableitung. Dann gilt*

$$\lim_{n\to\infty} \frac{D_n(a)}{G(a)^{n+1}} = E(a), \tag{1}$$

wobei

$$G(a) = \exp s_0, \quad E(a) = \exp \sum_{k=1}^{\infty} k s_k s_{-k}$$

mit

$$s_k = \frac{1}{2\pi} \int_{-\pi}^{\pi} \log a(e^{i\varphi})\, e^{-ik\varphi}\, d\varphi \quad (k = 0, \pm 1, \ldots).$$

In der Folgezeit wurden von verschiedenen Mathematikern große Anstrengungen unternommen, diesen Satz für allgemeinere Erzeugerfunktionen zu beweisen (Baxter, Hirschman, Devinatz, Krein, Widom). 1976 faßte H. Widom in seiner Arbeit [5] diese Bemühungen zusammen und bewies das sogenannte Strong Szegö Limit Theorem in folgender allgemeiner Form.

Satz 2. *Sei $a(t)$ ein invertierbares Element der Algebra $L^\infty(\Gamma) \cap l_2^{1/2}(\Gamma)$ vom Index Null. Dann gilt*

$$\lim_{n\to\infty} \frac{D_n(a)}{G(a)^{n+1}} = \det T_a T_{a^{-1}}. \tag{2}$$

Außerdem wird gezeigt, daß $\det T_a T_{a^{-1}} = E(a)$ gilt. Zur Erklärung sei gesagt, daß T_b der durch eine Funktion $b(t) \in L^\infty(\Gamma)$ erzeugte Toeplitzoperator im Raum $l_2^+ = \left\{(a_n)_{n=0}^\infty : \sum_{n=0}^\infty |a_n|^2 < \infty\right\}$ ist und $G(a)$, $E(a)$ wie in (1) erklärt sind, wobei $\log a$ auf geeignete Weise definiert ist.

In der Beweisführung für die Beziehung (2) ist bei H. Widom von entscheidender Bedeutung, daß

$$T_a T_{a^{-1}} = I - H_\varphi H_{\tilde{\varphi}^{-1}} \tag{3}$$

gilt, wobei H_b den durch die Funktion $b \in L^\infty$ erzeugten Hankeloperator darstellt. In der Beziehung (3) fand außerdem noch die Bezeichnung

$$\tilde{\varphi}(t) = \sum_{-\infty}^{\infty} \varphi_n \left(\frac{1}{t}\right)^n, \quad \varphi(t) = \sum_{-\infty}^{\infty} \varphi_n t^n$$

Berücksichtigung. Aus den Voraussetzungen von Satz 2 folgt nämlich, daß H_φ und $H_{\tilde{\varphi}^{-1}}$ Hilbert-Schmidt-Operatoren im Raum l_2^+ sind und somit $H_\varphi H_{\tilde{\varphi}^{-1}} \in S_1$, d. h. ein Spuroperator ist.

Andererseits folgt aus der genannten Arbeit von H. Widom, daß, falls $H_\varphi H_{\tilde{\varphi}^{-1}} \in S_1$, auch (2) gilt.

Interessant in diesem Zusammenhang ist ein Resultat von Devinatz [1]: Wenn für $a(t) \in L^\infty(\Gamma)$, $a(t) \geqq c > 0$ ($c = \text{const}$), der Grenzwert

$$\lim_{n\to\infty} \frac{D_n(a)}{G(a)^{n+1}}$$

existiert, dann gilt

$$a \in L^\infty \cap l_2^{1/2}.$$

Wir werden nun einige weitere Funktionenklassen angeben, für die

$$T_a T_{a^{-1}} - I \in S_1$$

und (2) gilt.

Sei

$$l_2^\alpha = \left\{\sum_{-\infty}^{\infty} a_k t^k \in L^2 : \sum_{0}^{\infty} |k|^{2\alpha} |a_k|^2 < \infty\right\} \quad (\alpha \geqq 0)$$

und

$$l_2^{\alpha,\beta} = \left\{\sum_{-\infty}^{\infty} a_k t^k \in L^2 : \sum_{0}^{\infty} a_k t^k \in l_2^\alpha, \sum_{-\infty}^{-1} a_k t^k \in l_2^\beta\right\}.$$

Wir setzen noch

$$F^{\alpha,\beta} = W \cap l_2^{\alpha,\beta},$$

wobei W die Wienersche Algebra ist. Es kann gezeigt werden, daß $F^{\alpha,\beta}$ eine Banach-Algebra ist, deren Raum der maximalen Ideale mit Γ übereinstimmt.

Satz 3. *Es gelte $a = a_- a_+$ mit $\bar{a}_-^{\pm 1} \in H^\infty \cap l_2^\beta$, $a_+^{\pm 1} \in H^\infty \cap l_2^\alpha$ und*

$$\alpha + \beta \geqq 1, \quad \max\{\alpha, \beta\} \neq \alpha + \beta = 1.$$

Dann ist $T_a T_{a^{-1}} - I \in S_1$, und es gilt (2).

Falls $\alpha + \beta = 1$, $\alpha \cdot \beta = 0$ gilt, ist ungeklärt, ob $T_a T_{a^{-1}} - I \in S_1$. Das folgende Resultat läßt die Vermutung zu, daß dem so ist.

Satz 4. *Sei a ein invertierbares Element der Algebra $F^{\alpha,\beta}$ vom Index Null und $\alpha + \beta = 1$ Dann gelten die Aussagen von Satz 2.*

Der Beweis dieser Aussage wird mittels Methoden geführt, die später in einem anderen Zusammenhang angedeutet werden.

Aus diesen Ausführungen ergibt sich eine Hypothese:

(i) Es gelte $a = a_- a_+$ mit $\bar{a}_-^{\pm 1} \in H^\infty \cap l_p^\beta$, $a_+^{\pm 1} \in H^\infty \cap l_q^\alpha$, $\dfrac{1}{p} + \dfrac{1}{q} = 1$ und

$\alpha + \beta = 1$.

Dann folgt $T_a T_{a^{-1}} - I \in S_1$.

(ii) $F_{p,q}^{\alpha,\beta} = W \cap l_{q,p}^{\alpha,\beta} \left(\dfrac{1}{q} + \dfrac{1}{p} = 1\right)$,

$$l_{p,q}^{\alpha,\beta} = \left\{\sum_{-\infty}^{\infty} a_k t^k \in L^1 \colon \sum_{0}^{\infty} |k|^{q\alpha} |a_k|^q < \infty, \sum_{-\infty}^{-1} |k|^{p\beta} |a_k|^p < \infty\right\},$$

ist eine Banach-Algebra, deren Raum der maixmalen Ideale mit Γ übereinstimmt.

(iii) $L^\infty \cap l_{q,p}^{1/p,1/q}$ ist eine Banach-Algebra und

$T_a T_{a^{-1}} - I \in S_1$,

falls a ein invertierbares Element dieser Algebra ist $\left(\dfrac{1}{p} + \dfrac{1}{q} = 1\right)$.

II.

Für die im Teil I erwähnten Klassen von Erzeugerfunktionen, für die das Szegö-Theorem gilt, kann — zumindest formal — der exakte Wert der Toeplitzdeterminante angegeben werden (n hinreichend groß). Um dies zu erläutern, führen wir in l_2^+ die Projektoren P_n ein ($n = 0, 1, 2, \ldots$):

$$P_n (a_k)_{k=0}^\infty = (a_0, \ldots, a_n, 0, \ldots).$$

Wir setzen noch $S_n = I - P_n$. Die an die Erzeugerfunktionen gestellten Voraussetzungen zeigen, daß bezüglich der Projektoren $\{P_n\}_{n=0}^\infty$ das sogenannte Reduktionsverfahren konvergiert. Als (nichttriviale) Konsequenz ergibt sich die wichtige Beziehung (n hinreichend groß)

$$\frac{D_{n-1}(a)}{D_n(a)} = \exp - s_0 - P_0 T_a^{-1} (S_n T_a^{-1} S_n)^{-1} S_n T_a^{-1} e_0, \quad l_0 = (1, 0, 0, \ldots). \tag{4}$$

Hieraus folgt nunmehr für hinreichend große n

$$D_n(a) = G(a)^{n+1} E(a) \prod_{k=n+1}^{\infty} \left(1 - \exp s_0 \{P_0 \dot{T}_a^{-1} (S_n T_a^{-1} S_n)^{-1} T_a^{-1} e_0\}\right). \tag{5}$$

Eine Auswertung dieser Formel kann nun für verschiedene Klassen von Erzeugerfunktionen vorgenommen werden, die in einem gewissen Sinne zusätzliche Glattheitsvoraussetzungen aufweisen. Aus Satz 3 und der eben angeführten Formel ergibt sich z. B. als Folgerung

Satz 5 ([3]). *Wenn der durch $a \in L^\infty$ erzeugte Toeplitz-Operator im Raum l_2^+ invertierbar ist und wenn für $b = a^{-1}$ gilt*

$$b_n = 0, \quad n \geqq n_0 \geqq 0$$

oder

$$b_{-n} = 0, \quad n \geqq n_0 \geqq 0,$$

dann folgt

$$D_n(a) = G(a)^{n+1} E(a), \quad n \geqq n_0.$$

Sei

$$W^{\alpha,\beta} = \left\{ f \in W : \sum_0^\infty k^\alpha |f_k| < \infty, \sum_{-\infty}^{-1} |k|^\beta |f_k| < \infty \right\} \quad (\alpha, \beta \geqq 0).$$

$W^{\alpha,\beta}$ ist eine Banach-Algebra, deren Raum der maximalen Ideale mit Γ übereinstimmt. Für Erzeugerfunktionen aus dieser Klasse ergibt eine weitere Auswertung der Formel (5) z. B. folgenden Satz.

Satz 6. *Sei $a \in W^{\alpha,\beta}$ ein invertierbares Element vom Index Null und $\alpha + \beta \geqq 1$. Dann gilt für hinreichend große n*

$$\log D_n(a) = (n+1)\, s_0 + \sum_{k=1}^\infty k s_k s_{-k} - \sum_{k=1}^\infty k \eta_{n+k} \xi_{-n-k} + o(n^{-2(\alpha+\beta)+1}). \tag{6}$$

mit

$$\sum_1^\infty k \eta_{n+k} \xi_{-n-k} = O(n^{-\alpha-\beta+1}).$$

Hierbei sind η_k, ξ_k entsprechend die Fourierkoeffizienten der Funktionen $a_- a_+^{-1}$ und $a_-^{-1} a_+$ ($a = a_+ a_-$ — die übliche kanonische Faktorisierung).

Dieser Satz kann auch für andere Klassen von Erzeugerfunktionen bewiesen werden. Als Beispiele seien die Algebren

$$L^\infty \cap l_2^{\alpha,\alpha} \left(\alpha > \frac{1}{2}\right), \quad H^\lambda \cap l_2^{1/2}$$

genannt. Dabei ist es möglich, in (6) weitere Glieder der Asymptotik anzugeben.

Satz 6 wurde erstmals für eine spezielle Klasse von Erzeugerfunktionen von R. E. Hartwig und M. E. Fisher mit anderen Methoden bewiesen (vgl. [2]).

III.

Wenn z. B. a ein invertierbares Element der Algebra $H^\lambda(\Gamma)$ (vom Index Null) ist, dann existiert für $\lambda > \frac{1}{2}$ der Grenzwert (1), für $\lambda \leqq \frac{1}{2}$ jedoch im allgemeinen nicht. Dennoch kann das asymptotische Verhalten der Toeplitzdeterminanten für $\lambda > \frac{1}{4}$ angegeben werden. Eine entsprechende Auswertung der Formel (4) liefert

$$\log D_n(a) = (n+1)\, s_0 + \sum_{i=1}^\infty \min\{i, n\}\, \eta_i \xi_{-i} + E'(a) + O(n^{-4\lambda+1}),$$

wobei $E'(a)$ eine gewisse Konstante ist. Ein Vergleich mit Satz 6 liefert folgende Hypothese

$$E'(a) = \sum_1^\infty k(s_k s_{-k} - \eta_k \xi_{-k}).$$

Wenn der Hölderexponent λ kleiner oder gleich $\frac{1}{4}$ ist, müssen weitere Korrekturglieder berücksichtigt werden, die auch angegeben werden können. Auf diese Weise entsteht eine abzählbare Menge von nicht klassischen Szegö-Theoremen. Diese Ausführungen gelten auch für andere Klassen von Erzeugerfunktionen, etwa für $W^{\alpha,\beta}$, $\alpha + \beta > \frac{1}{2}$, usw.

IV.

Satz 2 wurde in [5] nicht nur für den Skalarfall, sondern auch für Matrixerzeugerfunktionen bewiesen, deren Elemente zur Klasse $L^\infty \cap l_2^{1/2}$ gehören. Genauer gesagt, es gilt folgende Aussage.

Satz 7. *Sei* det $a \in L^\infty \cap l_2^{1/2}$ *ein invertierbares Element vom Index Null. Dann folgt*

$$\lim_{n\to\infty} \frac{D_n(a)}{G(a)^{n+1}} = \det T_a T_{a^{-1}},$$

wobei

$$G(a) = \lim_{\varrho\to 1} \exp\left\{\frac{1}{2\pi}\int_0^{2\pi} \log \det a_\varrho(e^{i\varphi})\, d\varphi\right\}$$

ist mit

$$a_\varrho(e^{i\varphi}) = \sum_{-\infty}^{\infty} a_k \varrho^{|k|} e^{ik\varphi}, \quad 0 < \varrho < 1.$$

Die Aussagen der Punkte I bis III können in entsprechender Form auch für Matrixerzeugerfunktionen bewiesen werden. Ausgangspunkt dafür ist eine Formel, die für hinreichend große n gilt, falls T_a und $T_{a^{-1}}$ invertierbare Toeplitzoperatoren sind:

$$\frac{D_{n-1}(a)}{D_n(a)} = \det\{\hat{P}_0 T_a^{-1} \hat{P}_0 - \hat{P}_0 T_a^{-1} (\hat{S}_n T_a^{-1} \hat{S}_n)^{-1} T_a^{-1} \hat{P}_0\},$$

wobei

$$\hat{P}_n = \begin{pmatrix} P_n & & & 0 \\ & \cdot & & \\ & & \cdot & \\ 0 & & & P_n \end{pmatrix}, \quad \hat{S}_n = I - \hat{P}_n.$$

Als Beispiel sei folgendes Resultat angeführt.

Satz 8. *Sei a eine Matrixfunktion mit Elementen aus der Klasse L^∞, für die der Toeplitzoperator T_a ein Fredholmoperator vom Index Null ist und für die $b = a^{-1}$ die Eigenschaft aufweist, daß*

$$b_n = 0, \quad n \geqq n_0 \geqq 0,$$

oder

$$b_{-n} = 0, \quad n \geqq n_0 \geqq 0$$

(b_k *sind hierbei die Blockfourierkoeffizienten von* b). *Dann gilt für hinreichend große* n

$$D_n(a) = G(a)^{n+1} \det T_a T_{a^{-1}}.$$

Literatur

[1] A. Devinatz, The strong Szegö limit theorem, Illinois J. Math. **11** (1967), 160–175.

[2] R. E. Hartwig and M. E. Fisher, Asymptotic behavior of Toeplitz matrices and determinants, Arch. Rat. Mech. Anal. **32** (1969), 190–225.

[3] B. Silbermann, Zur Berechnung von Toeplitz-Determinanten, die durch eine Klasse im wesentlichen beschränkter Funktionen erzeugt werden. Wiss. Z. Techn. Hochsch. Karl-Marx-Stadt **20** (1978), 683–687.

[4] G. Szegö, On certain Hermitian forms associated with the Fourier series of a positive function, Festschrift Marcel Riesz, Lund 1951.

[5] H. Widom, Asymptotic behavior of block Toeplitz matrices and determinants. II, Advances in Math. **21** (1976), 1–29.

Anschrift des Verfassers:

Technische Hochschule Karl-Marx-Stadt
Sektion Mathematik
9010 Karl-Marx-Stadt
Reichenhainer Str. 39
DDR

Optimization and best approximation

Ivan Singer, Bucharest

Abstract. In this expository paper we present some of our recent contributions to the interaction between optimization theory and the theory of best approximation.

§ 1. Introduction

At their beginnings, the theory of optimization (minimization or maximization of functionals, but the latter can be also expressed as minimization, on subsets of locally convex spaces) and the theory of best approximation (minimization or maximization of distance functionals and of some related functionals, on subsets of normed linear spaces) developed independently for some time. The fact that the problems of best approximation may be regarded as special problems of optimization, was observed in the 1960's (see e.g. [10, 11] and the references therein, [15], etc.). Some more systematic attempts to deduce known results of the theory of best approximation as particular cases of more general results of optimization theory, were made in the 1970's (see [14] and [13]). In [26, 28], we tried to obtain some new results on best approximation with this method.

Recently, starting with [24], we have suggested a program of work in the opposite direction. In the Introduction of [24] we wrote: "We want to show, in a series of papers, that much of the theory of best approximation in normed linear spaces can be generalized to convex optimization in locally convex spaces. There are only a few results known in this direction. Thus, E. W. Cheney and A. A. Goldstein [3] have proved some existence theorems, e.g., that if G is a closed convex set in a reflexive Banach space E, then every lower semi-continuous functional f on E having bounded convex level sets $\{y \in E \mid f(y) \leqq c\}$ attains its infimum on G. Also, R. B. Holmes [21] has introduced and studied an extension of a weaker version of the unique Hahn-Banach extension property (which plays some role in uniqueness problems of the theory of best approximation)". In some later papers we enlarged this goal, replacing "convex optimization" simply by "optimization" (not necessarily convex). Subsequently, the above point of view, on the usefulness of the generalizations of methods of best approximation to optimization, was adopted also by others, for example, V. I. Berdyshev [1] and M. Wriedt ([36, 37, 38]). In [38] M. Wriedt wrote: "Es hat sich in letzter Zeit gezeigt, daß die funktionalanalytischen Methoden, die in der Approximationstheorie zur Anwendung kommen, so weitreichend sind, daß sie für viele Probleme in der „allgemeineren" Optimierungstheorie das angemessene Mittel sind. Das hat dazu geführt, daß sich beide

Theorien zumindest in ihren abstrakten Teilen in einem Prozess der Annäherung befinden (z. B. [24]). Diese Arbeit will hierzu einen Beitrag leisten".

Thus, as shown by the above, there exists a fruitful interaction between optimization theory and the theory of best approximation, in both directions. In this brief expository paper we want to present some of our recent contributions along these lines. This will not be a survey of all interactions between the two theories; for example, we shall not consider here problems of existence, uniqueness, optimization operators (for some results in this direction see e.g. [36, 37, 38]) or numerical methods. We shall not give proofs, but we shall explain the underlying geometric ideas; we suggest to the reader to draw pictures, in order to visualize both the approximation theoretic and the optimization theoretic results. We shall assume that the reader is familiar with the elements of convex analysis. For simplicity, we shall consider here, without any special mention, only *real* spaces E; the extension to complex scalars can be obtained with the usual methods (see [22, 24]).

§ 2. Convex optimization: Minimization of convex functionals on convex sets

We recall that in a normed linear space E the *distance* from an element $x \in E$ to a subset G of E is the number

$$\operatorname{dist}(x, G) = \inf_{g \in G} \|x - g\| \tag{2.1}$$

and any $g_0 \in G$ for which this inf is attained, i.e., such that

$$\|x - g_0\| = \inf_{g \in G} \|x - g\|, \tag{2.2}$$

is called a *nearest point* to x in G, or a (an *element of*) *best approximation* of x in G. In the present section we shall consider only the case when G is a *convex* set.

Two of the main problems to be studied are those of finding convenient formulae for dist (x, G) and to give characterizations of elements of best approximation (i.e., necessary and sufficient conditions in order that g_0 satisfy (2.2)). For convex sets G there are some theorems which reduce these two main problems to the same problems for a suitable hyperplane in E. For example, concerning the first problem, M. Eidelheit [6] proved that if G is a linear manifold in a normed linear space E (i.e., $G = x_0 + S = \{x_0 + y \mid y \in S\}$, where S is a linear subspace of E and $x_0 \in E$), with $\bar{G} \neq E$, then we have

$$\operatorname{dist}(x, G) = \max_{\substack{H \in \mathscr{H} \\ H \supset G}} \operatorname{dist}(x, H), \tag{2.3}$$

where $\mathscr{H}$ denotes the collection of all hyperplanes in E (by "hyperplane" we shall mean, throughout the sequel, a closed hyperplane) and where the notation max means a supremum which is attained, so there exists a hyperplane H_0 with $H_0 \supset G$, such that

$$\operatorname{dist}(x, G) = \operatorname{dist}(x, H_0). \tag{2.4}$$

This theorem of M. Eidelheit was extended by A. L. Garkavi [9] (see also L. Nirenberg [17] and F. R. Deutsch and P. H. Maserick [4]) to an arbitrary convex set G, with $\bar{G} \neq E$, as follows: for each $x \in E \setminus \bar{G}$ we have

$$\operatorname{dist}(x, G) = \max_{H \in \mathscr{H}_{G,x}} \operatorname{dist}(x, H), \tag{2.5}$$

where $\mathscr{H}_{G,x}$ denotes the collection of all hyperplanes which support the set G and which separate G and x (and, again, there exists a hyperplane $H_0 \in \mathscr{H}_{G,x}$ satisfying (2.4)). We recall that in a locally convex space E, a hyperplane $H = \{y \in E \mid \Phi(y) = c\}$ (where $\Phi \in E^*$, the set of all continuous linear functionals on E, $\Phi \neq 0$, and $c \in R$, the real line) is said to *support* a set $G \subset E$, if G is contained in one of the two closed half-spaces $\{y \in E \mid \Phi(y) \leqq c\}$ or $\{y \in E \mid \Phi(y) \geqq c\}$ determined by H, but for each $\varepsilon > 0$, G is not contained in any one of the closed half-spaces $\{y \in E \mid \Phi(y) \leqq c - \varepsilon\}$, $\{y \in E \mid \Phi(y) \geqq c + \varepsilon\}$ (however, it is not required that $G \cap H \neq \emptyset$). Clearly, this happens if and only if either $c = \sup \Phi(G)$ or $c = \inf \Phi(G)$, so every hyperplane H supporting G can be written (replacing, if necessary, Φ by $-\Phi$) in the form

$$H = \{y \in E \mid \Phi(y) = \sup \Phi(G)\}, \tag{2.6}$$

where $\Phi \in E^*$, $\Phi \neq 0$, and conversely, every hyperplane of the form (2.6), with $\sup \Phi(G) < +\infty$, supports G; we use, for simplicity, the notation $\sup \Phi(G) = \sup_{g \in G} \Phi(g)$.

Such geometric theorems as the above ones seem fairly natural for the intuition and, in fact, their proofs are based on applying separation theorems. They are very useful, since the distance from an element $x \in E$ to a hyperplane $H = \{y \in E \mid \Phi(y) = c\}$, where $\Phi \in E^*$, $\Phi \neq 0$ and $c \in R$, can be expressed by the following particularly simple analytical formula of G. Ascoli (see e.g. [22], p. 24, lemma 1.2):

$$\operatorname{dist}(x, H) = \inf_{y \in H} \|x - y\| = \frac{|\Phi(x) - c|}{\|\Phi\|}; \tag{2.7}$$

indeed, using (2.7) and (2.6), Eidelheit's formula (2.3) and Garkavi's formula (2.5) can be expressed in the following analytical forms respectively (see [16] and [9]):

$$\operatorname{dist}(x, G) = \max_{\substack{\Phi \in S^\perp \\ \|\Phi\| = 1}} \{\Phi(x) - \Phi(x_0)\}, \tag{2.8}$$

$$\operatorname{dist}(x, G) = \max_{\substack{\Phi \in E^* \\ \|\Phi\| = 1}} \{\Phi(x) - \sup \Phi(G)\}, \tag{2.9}$$

where $S^\perp = \{\Psi \in E^* \mid \Psi(s) = 0 \ (s \in S)\}$. Note that, clearly, (2.8) is a particular case of (2.9), since for $G = x_0 + S$, where S is a linear subspace, we have $\sup \Phi(G) = \Phi(x_0)$ if $\Phi \in S^\perp$ and $= +\infty$ if $\Phi \in E^* \setminus S^\perp$. The usefulness of such "duality theorems" as (2.8), (2.9) for applications in various concrete spaces is due to the fact that for these spaces the general form of the continuous linear functionals $\Phi \in E^*$ is well known and simple.

Let us show the generalizations of the above duality theorems to convex optimization in locally convex spaces, i.e., to the problem of minimization of a convex functional f on a convex subset G of a locally convex space E; thus, we want to find formulae for $\inf f(G) = \inf_{g \in G} f(g)$. Since the particular case of best approximation can be obtained by taking E to be a normed linear space and f to be a convex functional on E of the form

$$f(y) = \|x - y\| \quad (y \in E), \tag{2.10}$$

the main idea is to replace in the above results on dist (x, G) involving hyperplanes H, the numbers dist $(x, H) = \inf_{y \in H} \|x - y\|$ by the numbers $\inf f(H)$. It turns out that such an approach works indeed, if we assume that f is a finite and continuous convex functional on E (this is satisfied by (2.10)). Thus, one can prove the following generalization of Eidelheit's formula (2.3):

Theorem 2.1 ([25], theorem 3.1). *Let E be a locally convex space, f a finite and continuous convex functional on E and $G = x_0 + S$, where $x_0 \in E$ and S is a linear subspace of E with $\bar{S} \neq E$. Then we have*

$$\inf f(G) = \max_{0 \neq \Phi \in S^{\perp}} \inf_{\substack{y \in E \\ \Phi(y) = \Phi(x_0)}} f(y), \tag{2.11}$$

hat is, geometrically,

$$\inf f(G) = \max_{\substack{H \in \mathscr{H} \\ H \supset G}} \inf f(H). \tag{2.12}$$

The duality formula (2.5) of Garkavi can be also generalized if we assume, in addition, that the constraint set G is "essential" i.e.,

$$\inf f(E) < \inf f(G), \tag{2.13}$$

or, equivalently, that there exist elements $x \in E$ satisfying

$$f(x) < \inf f(G) \tag{2.14}$$

(clearly, for f defined by (2.10), the assumption (2.14) is equivalent to $0 <$ dist (x, G), that is, to $x \in E \setminus \bar{G}$). Indeed, one can prove

Theorem 2.2 ([25], theorem 2.1). *Let E be a locally convex space, f a finite and continuous convex functional on E and G a convex subset of E, satisfying* (2.13) *and let x be any element of E satisfying* (2.14). *Then we have*

$$\inf f(G) = \max_{\substack{0 \neq \Phi \in E^* \\ \sup \Phi(G) \leqq \Phi(x)}} \inf_{\substack{y \in E \\ \Phi(y) = \sup \Phi(G)}} f(y), \tag{2.15}$$

that is, geometrically,

$$\inf f(G) = \max_{H \in \mathscr{H}_{G,x}} \inf f(H). \tag{2.16}$$

Note that in (2.15), (2.16) there occurs an $x \in E$ satisfying (2.14), but this also occurs in $\mathscr{H}_{G,x}$ of (2.5). One can show (see [25], corollary 2.1) that theorem 2.2 remains valid if the set $\{0 \neq \Phi \in E^* \mid \sup \Phi(G) \leqq \Phi(x)\}$ is replaced by the "smaller" set $\{0 \neq \Phi \in E^* \mid \sup \Phi(G) \leqq \inf \Phi(A)\}$, where

$$A = \{y \in E \mid f(y) < \inf f(G)\}. \tag{2.17}$$

Such duality theorems as theorems 2.1 and 2.2 above reduce the computation of $\inf f(G)$ to that of $\inf f(H)$ for suitable hyperplanes H in E, and therefore we have called them in [25] *hyperplane theorems*. The advantage of such results is that they permit to apply any formula known for $\inf f(H)$, where H is a hyperplane in E, to the computation of $\inf f(G)$. We have already seen that in the case of f defined by (2.10) this is indeed a great advantage, because of Ascoli's formula (2.7) for $\inf f(H)$ and another example will be given below (see the comment made after formula (2.21) for $\inf f(H)$).

Let us point out (see [25, 27, 30]) that the generalization of the theory of best approximation to optimization theory not only leads to new results, such as theorems 2.1 and 2.2 above, but has also an additional advantage. Namely, in the proofs of the results on best approximation one applies a separation theorem in order to separate a ball in E from a convex subset of E, which, in the case of optimization, generalizes naturally to the separation of a level set of a convex functional f on E from a convex subset of E. Thus, these proofs do not use the epigraph methods of convex analysis (which apply a separation theorem in order to separate a convex subset of $E \times R$ from the epigraph of a convex functional f on E, based on the fact that f is convex if and only if its epigraph is a convex subset of $E \times R$), so they do not "go up" to $E \times R$, but "remain down" in E; therefore, they can be extended also to functions which need not be convex, but have some (perhaps all) of their level sets convex. Actually, the assumptions of finiteness and continuity of f can be also relaxed, to some extent (see [25, 26]). Similar remarks are also valid for the results below, but we shall not mention them again in the sequel.

Let us give now a result in the opposite direction, showing how optimization theory leads to new results in the particular case of best approximation. In [26] we have deduced from the Fenchel-Rockafellar duality theorem of convex analysis (see e.g. [13], p. 68) the following Lagrangian duality theorem (called in [26] a "Lagrange multiplier theorem"):

Theorem 2.3 ([26], theorem 2.1). *Let E be a locally convex space, f a finite and continuous convex functional on E and G a convex subset of E. Then we have*

$$\inf f(G) = \max_{\Phi \in E^*} \inf_{y \in E} \{f(y) + \Phi(y) - \sup \Phi(G)\}. \tag{2.18}$$

The proof uses the convexity of the epigraph of f in $E \times R$, so the assumption of convexity of f is essential. The finiteness and continuity of f can be relaxed (it is enough to assume that $f: E \to \bar{R} = [-\infty, +\infty]$ is a proper convex functional, which is finite and continuous at one point of G), but we shall not be concerned here with this aspect. Any functional f for which the max in (2.18) is attained, may be called a *Lagrange multiplier* for the problem of minimizing f over G; indeed, this is justified, for example, by formula (2.20) below, applied to a finite-codimensional subspace of E.

In the particular case of best approximation, i.e., of the functional f defined by (2.10) on a normed linear space E, from theorem 2.3 we obtain

Theorem 2.4 ([26], theorem 2.2). *Let E be a normed linear space, G a convex subset of E, with $\bar{G} \neq E$, and $x \in E \setminus \bar{G}$. Then we have*

$$\operatorname{dist}(x, G) = \max_{\Phi \in E^*} \inf_{y \in E} \{\|x - y\| + \Phi(y) - \sup \Phi(G)\}. \tag{2.19}$$

Formula (2.19) is the first expression for $\operatorname{dist}(x, G)$ with the aid of a "penalty term" involving Lagrange multipliers, namely, $\Phi_0(y) - \sup \Phi_0(G)$, where $\Phi_0 \in E^*$, $\Phi_0 \neq 0$, is any Lagrange multiplier (i.e., for which the max in (2.19) is attained). In [26] we have shown how (2.19) implies again Garkavi's duality formula (2.9).

In the particular case when G is a linear manifold in E, say $G = x_0 + S$, where $x_0 \in E$ and S is a linear subspace of E, from theorem 2.3 we obtain the following formula of Lagrangian duality, which can be found in the book of S. Rolewicz ([18], theorem V. 2.1):

$$\inf f(G) = \max_{\Phi \in S^\perp} \inf_{y \in E} \{f(y) + \Phi(y) - \Phi(x_0)\}; \tag{2.20}$$

in particular, for a hyperplane $G = \{y \in E \mid \Phi(y) = c\}$, where $\Phi \in E^*$, $\Phi \neq 0$, $c \in R$, formula (2.20) yields

$$\inf f(H) = \inf_{\substack{y \in E \\ \Phi(y)=c}} f(y) = \max_{\alpha \in R} \inf_{y \in E} \{f(y) + \alpha\Phi(y) - \alpha c\}. \tag{2.21}$$

It turns out that the hyperplane theorem 2.2 and the Lagrange multiplier theorem 2.3, although obtained by different methods, are closely connected. Indeed, in [26] we have shown that theorem 2.3 and the inequality $\geqq$ in (2.15) (which is the simpler part of theorem 2.2) imply theorem 2.2 and, conversely, theorem 2.2, together with (2.21), imply theorem 2.3.

Let us turn now to the second main problem mentioned at the beginning of this section. In [20] (see also [22], p. 28) we have proved that if G is a linear subspace of a normed linear space E, with $\overline{G} \neq E$, and $x \in E \setminus \overline{G}$, an element $g_0 \in G$ satisfies (2.2) if and only if there exists a hyperplane $H_0 \supset G$ which supports the ball $B(x, \|x - g_0\|) = \{y \in E \mid \|x - y\| \leqq \|x - g_0\|\}$. By the translation method of [21], this result remains also valid for any linear manifold $G = x_0 + S$. V. N. Burov ([2], theorem 4; see also [22], Appendix I) has given the following characterization of best approximations by elements of a convex set G, with $\overline{G} \neq E$: An element $g_0 \in G$ satisfies (2.2) if and only if there exists a hyperplane H_0 containing g_0, which separates G and the ball $B(x, \|x - g_0\|)$; in [22], p. 262, it has been observed that any hyperplane H_0 which separates G and the ball $B(x, \|x - g_0\|)$ contains g_0 and hence supports the ball $B(x, \|x - g_0\|)$. For other characterizations of best approximations by elements of convex sets, in terms of continuous linear functionals, see [22], [23] and the references therein.

In order to show the generalization of Burov's result to convex optimization, we recall that if f is a functional on a locally convex space E and G a subset of E, an element $g_0 \in G$ is called a *solution of the program* $(G, f|_G)$ if we have

$$f(g_0) = \inf f(G); \tag{2.22}$$

we shall consider now the problem of finding necessary and sufficient conditions in order that g_0 satisfy (2.22) (i.e., the problem of characterization of the solutions of convex programs). Using theorem 2.2, one can prove

Theorem 2.5 ([25], theorem 2.2). *Let E be a locally convex space, f a finite and continuous convex functional on E and G a convex subset of E, satisfying* (2.13) *and let x be an element of E satisfying* (2.14). *For an element $g_0 \in G$ we have* (2.22) *if and only if there exists $\Phi_0 \in E^*$, $\Phi_0 \neq 0$, such that*

$$\Phi_0(g_0) = \sup \Phi_0(G) \leqq \Phi_0(x), \tag{2.23}$$

$$f(g_0) = \inf_{\substack{y \in E \\ \Phi_0(y)=\Phi_0(g_0)}} f(y), \tag{2.24}$$

that is, geometrically, if and only if there exists a hyperplane H_0 supporting H at g_0 and separating G and x, such that

$$f(g_0) = \inf f(H_0). \tag{2.25}$$

Another characterization theorem can be deduced from theorem 2.3, namely:

Theorem 2.6 ([26], theorem 2.3). *Let E be a locally convex space, f a finite continuous convex functional on E and G a convex subset of E. For an element $g_0 \in G$ we have* (2.22) *if and only if there exists $\Phi_0 \in E^*$ such that*

$$f(g_0) = \inf_{y \in E} \{f(y) + \Phi_0(y) - \sup \Phi_0(G)\}. \tag{2.26}$$

Since one can show ([26], proposition 2.1) that $\Phi_0 \in E^*$ satisfies (2.26) if and only if $\Psi_0 = -\Phi_0$ satisfies

$$\Psi_0 \in \partial f(g_0) \tag{2.27}$$

$$\Psi_0(g_0) = \inf \Psi_0(G), \tag{2.28}$$

it follows that theorem 2.6 is equivalent to the well known characterizations (2.27), (2.28) of the solutions of convex programs, due to Pshenichnyĭ and Rockafellar (see e.g. [13], pp. 30—31).

A number of other generalizations of some known characterizations of best approximations by convex sets (see [22, 23]) to characterizations of solutions of convex programs have been given in our paper [24]. The basic idea of [24] is to replace the set of all "maximal functionals" of $g_0 - x$, that is,

$$\{\Phi \in E^* \mid \Phi(g_0 - x) = \|x - g_0\|, \|\Phi\| = 1\}, \tag{2.29}$$

and the set of all extreme points of (2.29), which play the main role in the characterizations of best approximations mentioned above (such theorems are useful since in the usual concrete normed linear spaces E the general forms of the maximal functionals and of the extreme points of (2.29) are well known and simple) by the subdifferential $\partial f(g_0)$ and the set of all extreme points of $\partial f(g_0)$ respectively; in fact, one can show ([24], lemma 4.1) that for f defined by (2.10) $\partial f(g_0)$ coincides with (2.29). We refer the reader to [24] for more details.

§ 3. Maximization of convex functionals on bounded sets

We recall that if G is a bounded subset of a normed linear space E and $x \in E$, the deviation of G from x (see e.g. [22], p. 156, formula (6.61)) is the number

$$\delta(G, x) = \sup_{g \in G} \|g - x\|, \tag{3.1}$$

and any $g_0 \in G$ for which this sup is attained, i.e., such that

$$\|g_0 - x\| = \sup_{g \in G} \|g - x\|, \tag{3.2}$$

is called a *farthest point* to x in G; these may be also expressed as minimizations, writing $-\sup_{g \in G} \|g - x\| = \inf_{g \in G} (-\|g - x\|)$.

The usefulness of duality theorems for the distance and best approximation by convex sets, involving hyperplanes or continuous linear functionals (see § 2), suggests naturally to try to obtain such theorems also for the deviation and farthest points in bounded sets. Some results in this direction have been obtained in [7], for example, the following duality formula for the deviation ([7], theorem 2.1): If E is a normed linear space, G a bounded subset of E and $x \in E$, then we have

$$\sup_{g \in G} \|g - x\| = \sup_{H \in \mathscr{H}_G} \operatorname{dist}(x, H), \tag{3.3}$$

where $\mathscr{H}_G$ denotes the collection of all hyperplanes in E which support the set G, or, in equivalent analytic formulation,

$$\sup_{g \in G} \|g - x\| = \sup_{\substack{\Phi \in E^* \\ \|\Phi\|=1}} |\sup \Phi(G) - \Phi(x)| ; \tag{3.4}$$

however, in contrast with (2.5) and (2.9), the $\sup\limits_{H}$ and $\sup\limits_{\Phi}$ in (3.3) and (3.4) need not be attained, even when G is a convex weakly compact subset of the Hilbert space l^2 (see [7]). Let us also mention, for example, the following characterization of farthest points in a bounded subset G of a normed linear space E ([7], theorem 3.2): an element $g_0 \in G$ satisfies (3.2) if and only if there exists a hyperplane H_0 which supports G at g_0 and such that

$$\operatorname{dist}(x, H_0) \doteq \sup_{H \in \mathscr{H}_G} \operatorname{dist}(x, H) \tag{3.5}$$

(i.e., for which the sup in (3.3) is attained).

Looking at § 2 and, in particular, at formula (2.10), there arises naturally the idea of extending the results of [7] on deviation and farthest points to results on $\sup f(G)$, where f is a convex functional on a locally convex space E and G is any bounded subset of E and on maximum points of $f|_G$; these may be also expressed as minimizations, writing $-\sup f(G) = \inf(-f)(G)$, but since $-f$ and G need not be convex, this is no longer convex optimization. The main idea of this extension is, again, to replace in the results of [7] on deviation and farthest points in G, involving hyperplanes H related to G, the numbers $\operatorname{dist}(x, H)$ by the numbers $\inf f(H)$. Such an approach works indeed (however the methods of proofs, in the general case, are mostly different from those of [7]) and the duality results obtained in this way have the same advantages as those mentioned in § 2 for the case of convex optimization. Let us give, briefly, the generalizations of the above mentioned two results.

Theorem 3.1 ([27], theorem 2.1). *Let E be a locally convex space, $f: E \to [-\infty, +\infty)$ a lower semi-continuous convex functional and G a bounded subset of E. Then we have*

$$\sup f(G) = \sup_{0 \neq \Phi \in E^*} \inf_{\substack{y \in E \\ \Phi(y) = \sup \Phi(G)}} f(y), \tag{3.6}$$

that is, geometrically,

$$\sup f(G) = \sup_{H \in \mathscr{H}_G} \inf f(H). \tag{3.7}$$

In the particular case when $G = \{x_0\}$, a singleton (or, more generally, when G is weakly compact), one can prove more, namely:

Corollary 3.1 ([27], corollary 2.1). *Let E be a locally convex space and $f: E \to \bar{R} = [-\infty, +\infty]$ a lower semi-continuous convex functional. Then we have*

$$f(x_0) = \sup_{0 \neq \Phi \in E^*} \inf_{\substack{y \in E \\ \Phi(y) = \Phi(x_0)}} f(y) \quad (x_0 \in E), \tag{3.8}$$

that is, geometrically,

$$f(x_0) = \sup_{\substack{H \in \mathscr{H} \\ x_0 \in H}} \inf f(H) \; (x_0 \in E). \tag{3.9}$$

One can prove the following characterization of the *maximum points* of $f|_G$, i.e., of the elements $g_0 \in G$ satisfying

$$f(g_0) = \sup f(G): \tag{3.10}$$

Theorem 3.2 ([27], theorem 3.2). *Let E be a locally convex space, f a finite and continuous convex functional on E and G a bounded subset of E, satisfying*

$$\inf f(E) < \sup f(G). \tag{3.11}$$

For an element $g_0 \in G$ we have (3.10) *if and only if there exists $\Phi_0 \in E^*$ with $\Phi_0 \neq 0$, such that*

$$\Phi_0(g_0) = \sup \Phi_0(G), \tag{3.12}$$

$$\inf_{\substack{y \in E \\ \Phi_0(y) = \sup \Phi_0(G)}} f(y) = \sup_{0 \neq \Phi \in E^*} \inf_{\substack{y \in E \\ \Phi(y) = \sup \Phi(G)}} f(y), \tag{3.13}$$

that is, geometrically, if and only if there exists a hyperplane H_0 which supports G at g_0 and such that

$$\inf f(H_0) = \sup_{H \in \mathscr{H}_G} \inf f(H). \tag{3.14}$$

For the same reasons as in the case of theorems 2.2 and 2.5, we call the duality theorems 3.1 and 3.2 *hyperplane theorems.* One can also prove the following duality theorem of Lagrangian type, corresponding to theorem 2.3:

Theorem 3.3 ([28], theorem 2.1). *Let E be a locally convex space, f a proper lower semi-continuous convex functional on E and G a bounded subset of E. Then we have*

$$\sup f(G) = \sup_{\Phi \in E^*} \inf_{y \in E} \{f(y) - \Phi(y) + \sup \Phi(G)\}. \tag{3.15}$$

Similarly to the case of theorems 2.3 and 2.4, in the particular case of f defined by (2.10) theorem 3.3 yields a new formula for the deviation (3.1) ([28], theorem 2.2). Also, theorem 3.3 implies some necessary conditions for maximum points g_0 of $f|_G$ and it is related to the hyperplane theorem 3.1 in a similar way as theorem 2.3 to theorem 2.2 (see [28]). Finally, let us also mention that theorem 3.3 can be generalized to the following Fenchel-Rockafellar type duality theorem for maximization:

Theorem 3.4 ([29], theorem 1). *Let E be a locally convex space, f a proper lower semi-continuous convex functional on E and $h: E \to \bar{R} = [-\infty, +\infty]$ an arbitrary functional. Then we have*

$$\sup (f - h)(E) = \sup (h^* - f^*)(E^*). \tag{3.16}$$

In the particular case when $h = \delta_G$, the indicator of a bounded subset G of E, theorem 3.4 yields theorem 3.3.

§ 4. Minimization of convex functionals on complements of convex sets

The study of best approximation by complements of convex sets is motivated by its connection with the famous unsolved problem, whether in a Hilbert space E every Chebyshev set (i.e., such that each $x \in E$ has a unique element of best approximation in the set) is necessarily convex. Namely, it is known (see e.g. the references in [30]) that if a Hilbert space E contains a Chebyshev set which is not convex, then E also contains a Chebyshev set $E \setminus G$ which is the complement of an open bounded convex subset G of E. The study of the more general problem of minimization of convex functionals on complements of convex subsets of locally convex spaces (again, this is no longer convex optimization) is motivated by problems which may arise naturally in mathematical economics, for example, that of minimizing f on a subset of E of the form $\{x \in E \mid h(x) \geqq 0\}$, where h is a convex functional on E.

The usefulness of the duality theorems for approximation (see § 2 and § 3) suggests naturally to try to obtain such theorems also for best approximation by complements of convex sets. Some results in this direction have been obtained in [8], for example, the following duality formula for the distance ([8], theorem 2.1): If E is a normed linear space, G an open bounded convex subset of E and $g \in G$, then

$$\operatorname{dist}(g, E \setminus G) = \inf_{H \in \mathscr{H}_G} \operatorname{dist}(g, H), \tag{4.1}$$

where $\mathscr{H}_G$ denotes the collection of all hyperplanes in E which support the set G, or, in equivalent analytic formulation,

$$\operatorname{dist}(g, E \setminus G) = \inf_{\substack{\Phi \in E^* \\ \|\Phi\| = 1}} \{\sup \Phi(G) - \Phi(g)\}\,; \tag{4.2}$$

the $\inf\limits_{H}$ and $\inf\limits_{\Phi}$ in (4.1) and (4.2) need not be attained [8].

The main idea of the generalization is, again, to replace dist (g, H) by $\inf f(H)$, since the particular case of best approximation will be obtained by taking

$$f(y) = \|g - y\| \quad (y \in E). \tag{4.3}$$

Thus, one can prove

Theorem 4.1 ([30], theorem 2.1). *Let E be a locally convex space, f a finite and continuous convex functional on E and G a convex subset of E, with $G \neq E$, which is either closed or has non-empty interior, and assume that*

$$\inf f(G) \leqq \inf f(E \setminus G). \tag{4.4}$$

Then we have

$$\inf f(E \setminus G) = \inf_{\substack{0 \neq \Phi \in E^* \\ \sup \Phi(G) < +\infty}} \inf_{\substack{y \in E \\ \Phi(y) = \sup \Phi(G)}} f(y), \tag{4.5}$$

that is, geometrically,

$$\inf f(E \setminus G) = \inf_{H \in \mathscr{H}_G} \inf f(H). \tag{4.6}$$

In general, the assumption (4.4) in theorem 4.1 cannot be omitted ([30], example 2.1), but in the case when G is also bounded, the situation if different, as shown by

Theorem 4.2 ([30], theorem 2.2 and Addendum). *Let E be a locally convex space, f a finite and continuous convex functional on E and G a bounded convex subset of E, which is either closed or has non-empty interior (in the second case, E is necessarily normable). Then we have*

$$\inf f(E \setminus G) = \sup_{0 \neq \Phi \in E^*} \inf_{\substack{y \in E \\ \Phi(y) = \sup \Phi(G)}} f(y), \tag{4.7}$$

that is, geometrically, (4.6).

Note that, even in the particular case of best approximation, i.e., when E is a normed linear space and f is defined by (4.3), theorems 4.1 and 4.2 generalize the duality result (4.1) of [8] to the case when G is a convex subset of E, with $G \neq E$ (respectively, a bounded convex set), which is either closed or has non-empty interior. Let us mention that the necessity of proving theorem 4.2 has led to the discovery of some new separation theorems for bounded convex subsets of normed linear spaces (see [31]), which may have other applications as well.

In [8] and [30] there are also a number of characterization theorems, but we shall not give them here.

§ 5. Some applications

In the papers [32] and [33], applying some of the results of [27], [28] and [26], especially those on the particular case of a singleton $G = \{x_0\}$ (such as corollary 3.1 above), to suitable "primal functionals" f, we obtained some results concerning the following two problems:

1) Given two locally convex spaces F, X, a continuous linear mapping u of F into X (such a triple is called a *linear system*), a convex subset Ω of X with $u(F) \cap \Omega \neq \emptyset$ (called *target set*) and a convex functional $h: F \to \bar{R}$, find convenient formulae for

$$\inf_{\substack{y \in F \\ u(y) \in \Omega}} h(y), \tag{5.1}$$

under suitable assumptions on u, Ω and h.

2) Given two locally convex spaces F, X, a convex multifunction $\Gamma: X \to 2^F$ (called *constraints multifunction*), $x_0 \in X$ and a convex functional $h: F \to \bar{R}$, find convenient formulae for

$$\inf_{y \in \Gamma(x_0)} h(y), \tag{5.2}$$

under suitable assumptions on Γ and h.

For details we refer to [32] and [33]. For related results see also [19, 34, 35] and [5].

References

[1] В. И. Бердышев, Непрерывность оператора метрического проектирования и его обобщений. *In*: Internat. Confer. on Constructive Function Theory, Blagoevgrad 1977, Summaries, p. 5.

[2] В. Н. Буров, Аппроксимация со связами в линейных нормированных пространствах. Украйн. мат. ж. **15** (1963), 3—12, 135—144.

[3] E. W. Cheney and A. A. Goldstein, Tchebycheff approximation and related extremal problems. J. Math. Mech. **14** (1965), 87—98.

[4] F. R. Deutsch and P. H. Maserick, Applications of the Hahn-Banach theorem in approximation theory. SIAM Rev. **9** (1967), 516—530.

[5] S. Dolecki and S. Kurcyusz, On Φ-convexity in extremal problems. SIAM J. Contr. Optimiz. **16** (1978), 277—300.

[6] M. Eidelheit, Quelques remarques sur les fonctionnelles linéaires. Studia Math. **10** (1948), 140—147.

[7] C. Franchetti and I. Singer, Deviation and farthest points in normed linear spaces. Rev. roumaine math. pures et appl. **24** (1979), 373—381.

[8] C. Franchetti and I. Singer, Best approximation by elements of caverns in normed linear spaces. Boll. Un. Mat. Ital. (5) **17-B** (1980), 33—43.

[9] А. Л. Гаркави, Теоремы двойственности для приближений посредством элементов выпуклых множеств. Успехи мат. наук **16**, 4 (100) (1961), 141—145.

[10] Е. Г. Гольштейн, Об одном бесконечномерном аналоге задачи линейного программирования и его приложения к некоторым вопросам теории приближений. Доклады Акад. Наук СССР **140** (1961), 23—26.

[11] Е. Г. Гольштейн, Теория двойственности в математическом программировании и ее приложения. Наука, Москва 1971.

[12] R. B. Holmes, Smoothness indices for convex functions and the unique Hahn-Banach extension problem. Math. Zeitschr. **119** (1971), 95—110.

[13] R. B. Holmes, A course on optimization and best approximation. Lecture Notes in Math. 257, Springer-Verlag, Berlin—Heidelberg—New York 1972.

[14] P.-J. Laurent, Approximation et optimization. Hermann, Paris 1972.

[15] J. J. Moreau, Distance à un convexe d'un espace normé et caractérisation des points proximaux. Sémin. anal. unilatérale, Fac. Sci. Montpellier, 2 (1969), exposé no. 6.

[16] С. М. Никольский, Приближение функций тригонометрическими полиномами в среднем. Изв. Акад. Наук СССР, сер. мат. **10** (1946), 207—256.

[17] L. Nirenberg, Functional analysis. Dittoed notes, Courant Inst. Math. Sci. New York Univ., New York 1961.

[18] S. Rolewicz, Analiza funkcjonalna i teoria sterowania. PWN, Warszawa 1974; second ed. 1977.

[19] S. Rolewicz, On general theory of linear systems. Beiträge zur Analysis 8 (1976), 119—127.

[20] I. Singer, Caractérisation des éléments de meilleure approximation dans un espace de Banach quelconque. Acta Sci. Math **17** (1956), 181—189.

[21] I. Singer, Sur la meilleure approximation interpolative dans les espaces de Banach. Rev. math. pures et appl. **4** (1959), 95—113.

[22] I. Singer, Best approximation in normed linear spaces by elements of linear subspaces. Publ. House Acad., Bucharest and Springer-Verlag, Berlin—Heidelberg—New York, Grundl. Math. Wiss. 171, 1970.

[23] I. Singer, The theory of best approximation and functional analysis. CBMS Reg. Confer. Series in Applied Math. 13, SIAM, Philadelphia 1974.

24] I. Singer, Generalizations of methods of best approximation to convex optimization in locally convex spaces. I: Extension of continuous linear functionals and characterizations of solutions of continuous convex programs. Rev. roum. math. pures et appl. **19** (1974), 65–77.

[25] I. Singer, Generalizations of methods of best approximation to convex optimization in locally convex spaces. II: Hyperplane theorems. J. Math. Anal. Appl. **69** (1979) 571–584.

[26] I. Singer, Some new applications of the Fenchel-Rockafellar duality theorem: Lagrange multiplier theorems and hyperplane theorems for convex optimization and best approximation. Nonlinear Anal. Theory, Methods Appl. **3** (1979), 239–248.

[27] I. Singer, Maximization of lower semi-continuous convex functionals on bounded subsets of locally convex spaces. I: Hyperplane theorems. Appl. Math. Optimiz. **5** (1979), 349–362.

[28] I. Singer, Maximization of lower semi-continuous convex functionals on bounded subsets of locally convex spaces. II: Quasi-Lagrangian duality theorems. Result. Math. **3** (1980), 235–248.

[29] I. Singer, A. Fenchel-Rockafellar type duality theorem for maximization. Bull. Austral. Math. Soc. **20** (1979), 193–198.

[30] I. Singer, Minimization of continuous convex functionals on complements of convex subsets of locally convex spaces. Math. Operationsforsch. Stat., Ser. Optimiz., **11** (1980), 221–234.

[31] I. Singer, Extension with larger norm and separation with double support in normed linear spaces. Bull. Austral. Math. Soc. **21** (1980), 93–105.

[32] I. Singer, Duality theorems for linear systems and convex systems. J. Math. Anal. Appl. **76** (1980), 339–368.

[33] I. Singer, Duality theorems for constrained convex optimization Control and Cybernetics **9** (1980), 37–45.

[34] I. Singer, On the Pontryagin maximum principle for constant-time linear control systems in Banach spaces. J. Optimiz. Theory Appl. **27** (1979), 315–321.

[35] I. Singer, A. characterization of constant-time linear control systems satisfying the Pontryagin maximum principle. J. Optimiz. Theory Appl. **31** (1980) **32** (1980), 379–384.

[36] M. Wriedt, Stetigkeit von Optimierungsoperatoren. Arch. Math. **28** (1977), 652–656.

[37] M. Wriedt, Metrische Projektionen in lokalkonvexen Räumen. Rev. roum. math. pures et appl. **22** (1977), 701–728.

[38] M. Wriedt, Konvexe Optimierungsoperatoren (preprint).

Address of the author:

National Institute for Scientific and Technical Creation
Bd. Păcii 220
77 538 Bucureşti

and

Institute of Mathematics
Str. Academiei 14
70 109 Bucureşti
Romania

Control in coefficients for PDE

Jan Sokołowski, Warsaw

Introduction

In the paper some results concerning problems of optimal control in coefficients of linear partial differential equations are presented. Such optimization problems appear in different branches of technology and engineering.

The paper is divided into three sections. In the first section there are presented three examples of control problems for partial differential equations of elliptic, parabolic and hyperbolic types. Example 1 arises from the theory of optimal design of elastic plates [2]. Example 2 concerns the process of impurity diffusion in semiconductor [4]. Example 3 is a simplified version of optimal control of water quality [5].

In the second section an abstract parametric optimization problem for a weak variational problem in a Hilbert space is formulated and analyzed. Some sufficient conditions for existence of an optimal solution are given. The problem of differentiability of the cost functional is discussed.

In the last section of the paper a simple model problem for the second order linear elliptic equation on the plane is considered. An example of nonexistence [6] of a classical optimal solution is given, so-called generalized solutions to the optimal control problem are characterized.

1. Some examples of parametric optimization problems

In the section we give three examples of optimization problems for partial differential equations of elliptic, parabolic and hyperbolic types, which arise in real technological problems.

Example 1

Let us consider an elastic plate (Fig. 1) which is supported at the boundary $\Gamma = \partial\Omega$. We denote by $h(x_1, x_2)$, $(x_1, x_2) \in \Omega$, the variable thickness of the plate.

The plate is loaded by distributed transversal forces $q(x_1, x_2)$, $(x_1, x_2) \in \Omega$.

We assume that there are two parts Γ_1, Γ_2 of the boundary Γ, $\Gamma = \Gamma_1 \cup \Gamma_2$, such that the plate is fasten on Γ_1 and is hardly fasten on Γ_2.

Denote by $y(x_1, x_2)$ the flexure distribution in the state of the static equilibrium of the plate. Plate stiffness is evaluated by means of the maximal flexure:

$$J(h) = \max_{(x_1,x_2)\in\Omega} |y(x_1, x_2)| = \|y\|_{L_\infty(\Omega)}.$$

Under some assumptions the flexure distribution of the plate $y = y(h; x_1, x_2)$, $(x_1, x_2) \in \Omega$, in the static equilibrium state of the plate satisfies the following elliptic equation in the variational form:

$$\begin{cases} y \in V \\ \displaystyle\int_\Omega h^3 \left[\frac{\partial^2 y}{\partial x_1^2}\frac{\partial^2 \varphi}{\partial x_1^2} + \frac{\partial^2 y}{\partial x_2^2}\frac{\partial^2 \varphi}{\partial x_2^2} + \nu \frac{\partial^2 y}{\partial x_1^2}\frac{\partial^2 \varphi}{\partial x_2^2} + \nu \frac{\partial^2 y}{\partial x_2^2}\frac{\partial^2 \varphi}{\partial x_1^2} \right. \\ \displaystyle \left. + 2(1-\nu)\frac{\partial^2 y}{\partial x_1 \partial x_2}\frac{\partial^2 \varphi}{\partial x_1 \partial x_2}\right] dx_1\, dx_2 = \int_\Omega q\varphi\, dx_1\, dx_2, \quad \forall\, \varphi \in V \end{cases}$$

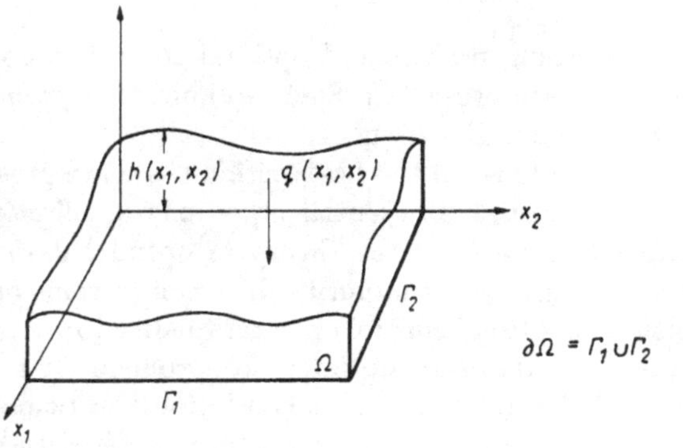

Fig. 1

where

$$V = \left\{\varphi \in H^2(\Omega) |\quad \varphi|_\Gamma = 0, \quad \frac{\partial \varphi}{\partial \nu}\bigg|_{\Gamma_2} = 0\right\}.$$

One can consider the following [2] optimal design problem:
minimize the performance index

$$J_p(h) = \|y(h)\|_{L_p(\Omega)}, \quad p \geqq 2,$$

subject to constraints

(i) $$\int_\Omega h(x_1, x_2)\, dx_1\, dx_2 = v,$$

where v is the volume of the plate,

(ii) $$0 < h_{\min} \leqq h(x_1, x_2) \leqq h_{\max}, \quad (x_1, x_2) \in \Omega$$

where v, $h_{\min}$, $h_{\max}$ are given constants.

Example 2

The second example arises from the technology of semiconductor devices. We consider a one-dimensional model of impurity diffusion in a semiconductor. In order to obtain the required impurity distribution in the semiconductor the so-called process of doping is performed.

The required impurity distribution is denoted by $N_d(x)$, $x \in (0, L)$. The process of doping can be divided into two steps.

An initial density distribution $\psi(x)$, $x \in (0, L)$, of impurity atoms in semiconductor is obtained after the first part of the doping process.

In the second part of the process the semiconductor heat treatment is applied. One can assume that the temperatur $u(t)$, $t \in (0, T)$ can be controlled during the treatment.

Impurity diffusion in semiconductor is described [4] by the linear parabolic equation of the form

$$\frac{\partial N}{\partial t}(x, t) = D(u)\frac{\partial^2 N}{\partial x^2}(x, t), \quad x \in (0, L), t \in (0, T)$$

with initial condition

$$N(x, 0) = \psi(x), \quad x \in (0, L)$$

and boundary conditions

$$\begin{cases} D(u)\dfrac{\partial N}{\partial x} + w(u)\, N = 0, & x = 0, \\ N = 0 & x = L, \end{cases}$$

where

$N = N(u; x, t)$ — distribution of the impurity density in semiconductor

$D(u) = D_0 \exp(-A/u)$ — diffusion coefficient

$w(u)$ — evaporation coefficient

The following optimization problem can be formulated:
minimize the performance index

$$J(u) = \frac{1}{2}\int_0^L [N(u; x, T) - N_d(x)]^2\, dx$$

over the set of admissible temperatures which satisfy the following constraints:

$$u_{\min} \leqq u(t) \leqq u_{\max}, \quad t \in (0, T).$$

Example 3

The example is a simplified version of an optimal water quality control problem [5] for the segment of a polluted river (fig. 2).

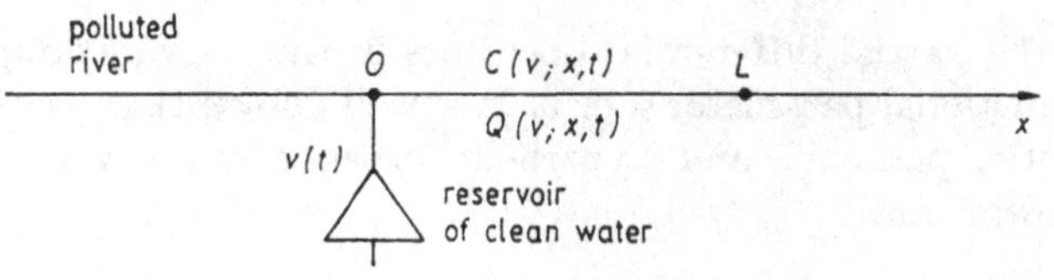

Fig. 2

Let us consider the segment $(0, L)$ of the polluted river. We assume for simplicity that the water quality in the segment depends only on the concentration $C(v; x, t)$ of dissolved oxygen (DO). On the other hand the river flow $Q(v; x, t)$ depends on the amount of water releases $v(t)$ from the reservoir of the clean water.

The simple mathematical model for DO has the form of one-dimensional linear hyperbolic equation

$$\frac{\partial C}{\partial t} = -\frac{1}{A}\,\frac{\partial (QC)}{\partial x} - kC + f$$

with boundary condition

$$C(0, t) = g(t)$$

and initial condition

$$C(x, 0) = C_0(x),$$

where $$Q(v; x, t) = Q_0(x, t) + \begin{cases} \int\limits_0^{at-x} e^{-\alpha(at-x-\tau)} v(\tau)\, d\tau, & x \leq at, \\ 0, & x > at. \end{cases}$$

The performance index to be minimized can be defined in the following way:

$$J(v) = \int\limits_0^T \int\limits_0^L P(x)\, \big(C_S - C(x, t)\big)^p\, dx\, dt + \varrho \int\limits_0^{T-T_0} \big(v(t) - v_d\big)^2\, dt, \quad p \geq 2.$$

The real control problem necessitates also some constraints on the functions $v(t)$ and $C(x, t)$, for example in the following form:

control:

$$v_{\min} \leq v(t) \leq v_{\max}, \quad t \in (0, T),$$

$$\int\limits_0^T v(t)\, dt \leq V_{\max};$$

state:

$$C(x, t) \geq C_{\min}, \quad x \in (0, L), \quad t \in (0, T).$$

2. Abstract parametric optimization problem

The coefficients of the partial differential operators in the above examples depend on control, say on a functional parameter u or v. It is well known that partial differential equations of elliptic, parabolic and hyperbolic types admit a weak variational formulation in the following form:

$$\left.\begin{aligned} & y_u \in W, \\ & E_u(y_u, \varphi) = L(\varphi), \quad \forall \in \varphi \in \Phi \subset W, \\ & L \in \Phi' \end{aligned}\right\} \tag{3.1}$$

with an appropriate choice of the bilinear form $E_u(\cdot, \cdot)$ and functional spaces Φ, W,

where $u \in U_{ad} \subset U$ is the functional parameter. The abstract results presented in the section can be applied in the case of a concrete equation. Consider the following parametric optimization problem:
(P) *minimize the cost functional*

$$J(u) = I(y_u) \tag{3.2}$$

over the given set of admissible parameters $U_{ad} \subset U$.

Remark. In order to assure the existence of an optimal solution to the problem (P) a method of regularization can be used. For example one can replace the cost functional (3.2) by the perturbed cost functional $J_\omega(u) = I(y_u) + \omega(u)$. In the sequel we indicate some possibilities of the suitable choice of the functional $\omega(u)$. We assume that W is a Hilbert space, $\Phi \subset W$ is a normed linear subspace, the set U_{ad} is a closed and bounded subset of a Banach space U, $E_u(\cdot, \cdot)$ is a linear form on $W \times \Phi$, $L \in \Phi'$ is a linear functional, where by Φ' we denote the conjugate space to Φ. The functional $I(\cdot)$ is assumed to be weakly lower semicontinuous on the space W.

For fixed parameter $u \in U_{ad}$ problem (3.1) has a unique solution if the following conditions are satisfied [12]:
(A1):

(i) $\forall \varphi \in \Phi$, *the linear form* $E_u(\cdot, \varphi)$ *is continuous on* W;

(ii) $\exists \beta > 0$ *such that* $\|M(u)\varphi\|_W \geqq \beta \|\varphi\|_\Phi$, $\forall \varphi \in \Phi$, *where the element* $M(u)\varphi \in W$ *is defined as follows:* $(w, M(u)\varphi)_W = E_u(w, \varphi)$, $\forall w \in W, \forall \varphi \in \Phi$;

(iii) *the set* $M(u)\Phi$ *is dense in* W;

furthermore we assume that the form $E_u(\cdot, \cdot)$ is continuous on the set U_{ad} with respect to the parameter $u \in U_{ad}$, namely:
(A 2):

$$\forall \{u_m\} \subset U_{ad} \text{ such that } u_m \xrightarrow{U} u_0, \text{ the following condition holds:}$$

$$\forall \varphi \in \Phi: \lim_{m\to\infty} \sup_{\|w\|_W \leqq 1} |E_{u_m}(w, \varphi) - E_{u_0}(w, \varphi)| = 0.$$

Lemma 1. *Assume that the conditions* (A_1), (A_2) *are satisfied. Then the cost functional* (3.2) *is lower semicontinuous on the set* U_{ad} *in the norm topology of the space* U.

Remark. If the set U_{ad} is compact in the norm topology of the space U then there exists a solution to the problem (P), otherwise a method of regularization should be used.

It can be shown that there exists a solution to the following regularized problem (P_ω)
(P_ω) *minimize the cost functional*

$$J_\omega(u) = I(y_u) + \omega(u)$$

over the given set $\tilde{U}_{ad} \subset U_{ad}$
if the functional $\omega(u)$ and the set $\tilde{U}_{ad}$ have the following form:

1) U is a reflexive Banach space. $\tilde{U}_{ad} = U_{ad}$; $\omega(u) = \varepsilon \|u - \bar{u}\|_U$, where $\varepsilon > 0$, $\bar{u} \in G$, and G is a G_δ subset [3] of the space U, the set G is dense in U.

2) U is a Hilbert space. $\tilde{U}_{ad} = U_{ad}$; $\omega(u) = \varepsilon \|u - \bar{u}\|_U^2$, where $\varepsilon > 0$, $\bar{u} \in G$ where G is a G_δ subset [1] of the space U, the set G is dense in U.

3) Assume that there exists a Hilbert space $U_1 \subset U$, U_1 dense in U, and the imbedding $U_1 \subset U$ is compact. In this case the set $\tilde{U}_{ad}$ is the intersection $\tilde{U}_{ad} = U_{ad} \cap U_1$ and $\omega(u) = \varepsilon \|u\|_{U_1}^2$, where $\varepsilon > 0$.

In practical applications one would like to use a numerical method of optimization, therefore it is important to know whenever the cost functional (3.2) is differentiable. We present conditions of directional differentiability of the cost functional (3.2) as well as its form. We need some additional assumptions:
let $u_0 \in U_{ad}$ be a fixed parameter;
(A3):
(i) *the functional $I(\cdot)$ is Fréchet differentiable on W*;
(ii) $\forall v \in U$, $\|v\|_U = 1$,

$$\lim_{t \to 0_+} \|y_{u_0+tv} - y_{u_0}\|_W = 0;$$

(iii) *there exists a linear operator $A(u_0; v) \in \mathscr{L}(\Phi; W)$, $v \in U$, such that*

$$\lim_{t \to 0_+} \big(E_{u_0+tv}(y, \varphi) - E_{u_0}(y, \varphi)\big)/t = \big(y, A(u_0; v)\, \varphi\big)_W,$$

$$\forall y \in W, \quad \forall \varphi \in \Phi.$$

Lemma 2. *Assume that conditions (A 1), (A 2), (A 3) are satisfied. Then there exists the directional derivative of the cost functional (3.2) at the point $u_0 \in U_{ad}$ in the following form:*

$$dJ(u_0; v) = \left(\frac{dI}{dy}(y_{u_0}), p(v)\right)_W,$$

where $p(v) \in W$,

$$E_{u_0}\big(p(v), \varphi\big) = \big(y_{u_0}, A(u_0; v)\, \varphi\big)_W, \quad \forall \varphi \in \Phi.$$

3. Model problem

Let $\Omega \subset R^2$ be a bounded domain with boundary $\Gamma = \partial\Omega$. We apply our abstract results to the following example:

$$W = \Phi = H_0^1(\Omega), \quad \Phi' = H^{-1}(\Omega), \quad U = L_2(\Omega), \tag{4.1}$$

$$U_{ad} = \{u \in L_2(\Omega) |\ 0 < a \leqq \alpha(x) \leqq u(x) \leqq \beta(x) \quad \text{a.e. in } \Omega\}, \tag{4.2}$$

$$E_u(y, \varphi) = \sum_{i=1}^{2} \int_\Omega u \frac{\partial y}{\partial x_i} \frac{\partial \varphi}{\partial x_i}\, dx, \quad \forall y, \varphi \in H_0^1(\Omega), \tag{4.3}$$

$$L(\varphi) = \int_\Omega f\varphi \, d\Omega, \tag{4.4}$$

$$I(y) = \frac{1}{2} \int_\Omega (y - z_d)^2 \, dx, \tag{4.5}$$

where $\alpha, \beta \in L_\infty(\Omega)$, $z_d, f \in L_2(\Omega)$ are given elements.

Problem (P) has the following from:

(P) *minimize the cost functional*

$$J(u) = \frac{1}{2} \int_{\Omega} (y(u; x) - z_d(x))^2 \, dx \tag{4.6}$$

over the set (4.2), *where* $y = y(u; x)$ *is the unique solution to the following elliptic equation:*

$$y \in H_0^1(\Omega),$$
$$\sum_{i=1}^{2} \int_{\Omega} u \frac{\partial y}{\partial x_i} \frac{\partial \varphi}{\partial x_i} \, dx = \int_{\Omega} f\varphi \, dx, \quad \forall \varphi \in H_0^1(\Omega). \tag{4.7}$$

It is easy to prove that in the case of the bilinear form (4.3) the assumptions (A1) (A2) are satisfied and the cost functional (4.6) is continuous in the norm topology of the space $L_2(\Omega)$, however the functional is not weakly lower semicontinuous.

The set (4.2) is weakly compact but it is not compact, hence we cannot use the Weierstrass theorem to obtain the existence of a solution to the problem (P). On the other hand it can be shown that for some suitably chosen elements z_d, $f \in L_2(\Omega)$ there is no solution to the problem (P). We give a counterexample [6]:

Counterexample 1

Consider the problem (P) with Ω in the following form

$$\Omega = \{(x_1, x_2) \in R_2 | -2 < x_1 < 2, -2 < x_2 < 2\}.$$

Let us define the elements z_d, f, α and β as follows:

$$z_d = \begin{cases} (x_1^2 + x_2^2 - 1)^2, & x_1^2 + x_2^2 \leq 1, \\ 0, & x_1^2 + x_2^2 > 1, \end{cases}$$

$$f = \frac{1}{2} \frac{\partial^2 z_d}{\partial x_1^2} + \frac{\partial^2 z_d}{\partial x_2^2},$$

$$\alpha = 1 - 1/\sqrt{2}, \quad \beta = 1 + 1/\sqrt{2}.$$

It can be shown that [6]

(i) there exists a sequence $\{u_n\}$,

$$u_n = u_n(x_1) = \begin{cases} 1 + 1/\sqrt{2}, & (x_1 + 2)/4 \in (2k/2n, (2k+1)/2n], \\ 1 - 1/\sqrt{2}, & (x_1 + 2)/4 \in ((2k+1)/2n, (2k+2)/2n], \end{cases}$$

such that the corresponding sequence $y_n = y(u_n)$ of solutions to (4.3) converges weakly,

$$y_n \rightharpoonup z_d \quad \text{weakly in} \quad H_0^1(\Omega),$$

hence one obtains:

$$\inf_{u \in U_{\mathrm{ad}}} J(u) = 0 ;$$

(ii) there is no element $\bar{u} \in U_{\mathrm{ad}}$ such that $y(\bar{u}) = z_d$, hence a *classical solution* to the problem (P) *does not exist.*

On the other hand it can be shown that for every $\varepsilon > 0$, there exists a G_δ set $G \subset L_2(\Omega)$ such that for every $g \in G$ there exists a solution to the perturbed problem

(P_ε) *minimize the cost functional*

$$J_\varepsilon(u) = J(u) + \varepsilon \, \|u - g\|^2_{L_2(\Omega)}$$

over the set U_{ad}.

We characterize so-called *generalized solutions* to the problem (P). Denote by $Z \subset H_0^1(\Omega)$ the following set

$$Z = \{y \in H_0^1(\Omega) \mid y = y(u) \quad \text{for some} \quad u \in U_{\mathrm{ad}}\}.$$

It can be shown that Z is a bounded and norm closed subset of the space $H_0^1(\Omega)$, however it is neither weakly closed nor convex. One can characterize [13] weak limits of the sequences $\{z_m\} \subset Z$ in the following way: given the sequence $\{y_m\} \subset Z$, corresponding to the sequence $\{u_m\} \subset U_{\mathrm{ad}}$, then for some subsequences

$$u_{m'} \rightharpoonup \bar{u} \quad \text{weakly in } L_2(\Omega),$$

$$y_{m'} \rightharpoonup z \quad \text{weakly in } H_0^1(\Omega),$$

where $z \in H_0^1(\Omega)$ is the unique solution to the following elliptic equation:

$$-\sum_{i=1}^{2} \frac{\partial}{\partial x_i} \left(a_{ij} \frac{\partial z}{\partial x_j} \right) = f \quad \text{in } \Omega$$

for some symmetric matrix a_{ij}. Furthermore, the eigenvalues λ_1, λ_2 of the matrix $[a_{ij}]$ have the following properties:

$$\lambda_2(x) = \bar{u}(x) \quad \text{a.e. in } \Omega,$$

$$\alpha(x) \leqq \lambda_1(x) \leqq \lambda_2(x) \leqq \beta(x),$$

$$\lambda_1(x)\,\lambda_2(x) - \lambda_1(x)\,\big(\alpha(x) + \beta(x)\big) + \alpha(x)\,\beta(x) \leqq 0.$$

References

[1] E. Asplund, Fréchet differentiability of convex functions, Acta Math. **121** (1968), 31–47.

[2] Н. В. Баничук, В. М. Картвелишвили, А. А. Миронов, Задачи оптимизации с локальными критериями качества в теории изгиба пластин, Механика Твёрдого Тела № 1 (1978), 124–131.

[3] J. Baranger, Quelques résultats en optimisation non convexe, Thesis, University of Grenoble, 1973.

[4] Б. В. Баталов, В. М. Петров, Е. В. Чеботаев, О возможности оптимального управления температурным режимом диффузионных процессов, Микроэлектроника, вып. 9 (1976), 78—90.

[5] A. Bogobowicz, J. Sokołowski, Optimal water quality control by treating the efluent at the pollution sources and flow regulation IAHS-AISH Publ. no. 129, 1980.

[6] Л. В. Корсакова, Пример несуществования решения задачи Лионса об оптимальном управлении, Проблемы Математического Анализа, вып. 6 (1977). Издательство Ленинградского Университета, 60—67.

[7] Л. С. Кучмент, Математическое моделирование речного стока, Гидрометеоиздат, Ленинград 1972.

[8] J. L. Lions, Sur le contrôle optimal de systèmes gouvernés par des équations aux derivées partielles, Dunod, Paris 1968.

[9] J. L. Lions, E. Magenes, Problèmes aux limites non homogènes et applications, Dunod, Paris 1968.

[10] F. Murat, Conterexamples pour divers problèmes où le contrôle intervient dans les coefficients, Ann. Math. Pura Appl. **112** (1977), 49—68.

[11] J. Sokołowski, Optimal control in coefficients for weak variational problems in Hilbert space Appl. Math. Optim. (to appear).

[12] T. Mazumdar, Generalized projection theorem and weak noncoercive evolution problems in Hilbert space, J. Math. Anal. Appl. **46** (1974), 143—168.

[13] У. Е. Райтум, К задачам оптимального управления для линейных эллиптических уравнений, ДАН СССР **244** (1979), 828—830.

Address of the author:

Systems Research Institute
Polish Academy of Sciences
Newelska 6
01-447 Warszawa
Poland

Towards a general-purpose difference scheme for the linear one-dimensional parabolic equation

Gisbert Stoyan, Berlin

1. Introduction

We consider here first and second boundary value problems for the partial differential equation

$$\begin{aligned} c\,\frac{\partial u}{\partial t} &= \frac{\partial}{\partial x}\left(a\,\frac{\partial u}{\partial x}\right) + b\,\frac{\partial u}{\partial x} - du + f, \quad (x, t) \in (0, 1) \times (0, T], \\ u(x, 0) &= u_0(x), \quad 0 \leqq x \leqq 1, \end{aligned} \tag{1}$$

with initial value $u_0(x)$.

This equation (and its nonlinear generalization) is important in many applications: The classical areas are diffusion and heat conduction, but (1) is used also in modelling air and water pollution [1], filtration (i.e. geohydrological [2], petroleum [3] and agricultural [4] problems) and problems of plasma physics [5].

Quite common in these applications is the occurence of small or great or rapidly changing parameters. For example, in groundwater problems the (storage) coefficient c is known to rise quickly by 3—4 magnitudes if the flow turns from confined to unconfined (see [2], p. 408). The linear term $-du$ may model a first order chemical reaction, then d is great for fast reactions. The (diffusion) coefficient a is mostly small as compared to the (convection) coefficient b, and this situation causes often discussed numerical diffuculties:

If (1) is approximated by a difference scheme, then the usual statement is that accuracy must be sacrificed for stability, e.g. by taking one-sided differences instead of the central difference for approximation of $b\,\frac{\partial u}{\partial x}$. But this "upstream" differencing method is rather unsatisfactory due to its artificial increasing (up to $a + |b|\,h/2$) of the diffusion coefficient which results, for example, into unphysically heavy smoothing of steep fronts.

To solve problems with small or great parameters successfully, methods are needed which continue to work in the degenerated case without loosing stability or basic physical properties of the solution of (1). Continuing research started in [6] and [7], we construct in this paper a difference scheme approximating (1) which comes near to such a general-purpose method:

This difference scheme satisfies the maximum principle and, hence, preserves monotonicity and positiveness, on any nonequidistant grid, for any coefficients satisfying only $a, c, d \geqq 0$ and $a + c > 0$. Degeneration of any coefficient is admitted and does not lead to loss of the announced properties or of local approximation. Especially, if $a \to 0$, the scheme turns to a two-parameter difference scheme for the degenerated equation containing the method of characteristics. This latter property renders possible the calculation of steep fronts with small numerical diffusion. The accuracy of the scheme is, in general, between first and second order on smooth solutions, but in some cases of constant and degenerating coefficients, exact difference schemes are approached.

The difference scheme is of the two-level, three-point type and therefore effectively solvable by the usual shortened Gauss elimination (see [8], p. 40)).

Paralleling the representation given in [7] for the case $d = 0$, the present paper describes the construction of the scheme and its boundary conditions, provides all formulae needed for programming (including automatic selection of the weighting parameters), and concludes with numerical examples.

For lack of time, neither local approximation, numerical diffusion, conservativity (see [8], p. 156, [9], p. 26, and [10]), nor problems of the implementation of the method are considered here, but we refer to [7].

2. Notations

Most notations used are the standard ones [8]. We introduce the spatial grids $\overline{\omega}_h = \{x_i, 0 = x_0 < x_1 < \cdots < x_N = 1\}$ and $\omega_h = \{x_i \in \overline{\omega}_h, 1 \leqq i \leqq N - 1\}$, the time grid $\overline{\omega}_\tau = \{t_j, 0 = t_0 < t_1 < \cdots < t_m = T\}$ and the discrete boundary $\gamma_{h\tau} = \{(x, 0), x \in \overline{\omega}_h\} \cup \{(x, t), x \in \overline{\omega}_h/\omega_h, t \in \overline{\omega}_\tau\}$. The discrete boundary consists of two parts: γ_2 is the set of points of $\gamma_{h\tau}$ in which there are given second kind boundary conditions, γ_1 is the remaining part, consisting at least of the points $\{(x, 0), x \in \overline{\omega}_h\}$.

The grid of all discretization points is $\overline{\omega}_{h\tau} = \overline{\omega}_h \times \overline{\omega}_\tau$. The spatial and time steplengths are denoted by $h_{i+1/2} = x_{i+1} - x_i$ and $\tau = \tau_{j+1/2} = t_{j+1} - t_j$. For functions y defined on $\overline{\omega}_h$ we write $y(x_i) = y_i = y$ and

$$y(x_i + rh_{i+1/2}) = y_{i+r}, \quad y(x_i - rh_{i-1/2}) = y_{i-r},$$

where r is a nonnegative number. The same notation is used for expressions containing grid functions. The forward and backward differences are y_x and $y_{\bar{x}}$,

$$y_x = y_{x,i} := (y_{i+1} - y_i)/h_{i+1/2}, \quad y_{\bar{x}} = y_{\bar{x},i} := (y_i - y_{i-1})/h_{i-1/2}.$$

For functions depending on time we write

$$y(x_i, t_j) = y_i^j = y_i = y, \quad y(x_i, t_{j+1}) = y_i^{j+1} = \hat{y}, \quad y_t := (\hat{y} - y)/\tau,$$

and $t_{j+s} = t_j + \tau_{j+1/2}s$, $y(x_i, t_{j+s}) = y_i^{j+s}$ for nonnegative numbers s.

3. Construction of the difference scheme for the case $d = 0$

The difference scheme is constructed by executing the following steps [7]:

1. Derivation of the exact difference scheme (for this term see [8], p. 203, or [9], p. 13) for the stationary equation (1) with constant coefficients a and b, $d = 0$, and for f linear in x, $f = \varphi + \psi x$.

2. Omission of terms not needed to have a second order scheme.

3. Transition to the time dependent case by substitution of $f - cy_t$ for f into the scheme obtained in Step 2.

4. For stability reasons, introduction of a weighting parameter σ into the spatial and time difference operator.

5. Transition to the case of variable coefficients by substitution of $a_{i+1/2}^{j+1/2}$ ($a_{i-1/2}^{j+1/2}$) for a where it appears in the difference scheme in connection with h_{i+1} ($h_{i-1/2}$), and by corresponding substitutions for b, c, f and σ.

The addition of the linear term $-du$ is carried out separately below. We remark that we don't start from the exact difference scheme [11] for the spatial differential operator (1) with constant coefficients a, b and d by two reasons: Firstly, to restrict complication of the formulae, and secondly, to be able to achieve a favorable behaviour of the scheme in the important case where the linear and the convection term are present but the diffusion coefficient is degenerated or nearly degenerated.

The decisive steps in the construction of the difference scheme are Step 1 and 4. Step 1 has been detailed in [7] for $x_i \in \dot{\omega}_h$.

Step 1 will now be executed for the boundary points x_0 and x_N in the case of second kind boundary conditions.

We consider the constant coefficient boundary value problem

$$Lu + \varphi + \psi x = 0, \quad 0 \leqq x \leqq x_1, \quad Lu := au'' + bu', \quad u' := \frac{du}{dx}, \qquad (2)$$
$$u'(0) = 0, \quad u(x_1) = u_1.$$

The solution of (2) is given by

$$u(x) = u_1 + \frac{1}{b}(x - x_1)\left[\frac{a}{b}\psi - \varphi - \frac{\psi}{2}(x + x_1)\right]$$
$$+ \left[\exp(-2q_{1/2}) - \exp\left(-\frac{bx}{a}\right)\right]\frac{a}{b^2}\left(\varphi - \frac{a}{b}\psi\right) \qquad (3)$$

as can be verified directly. Here $q_{1/2}$ denotes the "cell Reynolds number" of the interval $(0, x_1)$, $q = q_{1/2} := \frac{b}{2a} h_{1/2}$. We put $x = x_0 = 0$ in (3), observe that $x_1 = h_{1/2}$, and get

$$0 = u_1 - u_0 + \varphi\left[\frac{h}{b} + -\frac{a}{b^2}(\exp(-2q) - 1)\right]_{1/2}$$
$$+ \psi\left[\frac{h^2}{2b} - \frac{ah}{b^2} + \frac{a^2}{b^3}(1 - \exp(-2q))\right]_{1/2}. \qquad (4)$$

Now let be

$$\varrho = \varrho(q) := q \coth q, \quad \alpha = \alpha(q) := \frac{\varrho - 1}{2q}, \qquad (5)$$

then the factors multiplying φ and ψ can be expressed as

$$\left[\frac{h^2}{2a(\varrho + q)}(1 + 2\alpha)\right]_{1/2} \quad \text{and} \quad \left[\frac{h^2}{4a(\varrho + q)}\left(1 + 2\alpha - 2\frac{\alpha}{q}\right)\right]_{1/2}.$$

Exploiting the properties of hyperbolic functions, it can be shown that b, q and α are of the same sign, whereas ϱ is an even function of q. Moreover, we have the relations

$$\max(1, |q|) \leqq \varrho \leqq \min\left(1 + \frac{q^2}{3}, 1 + q^2/(1 + |q|)\right),$$
$$\frac{|q| - 1}{2|q|} \leqq |\alpha| \leqq \frac{|q|}{6}, \quad \alpha \in \left(-\frac{1}{2}, \frac{1}{2}\right), \quad \lim_{|q|\to\infty} \alpha(q) = \frac{1}{2} \operatorname{sign}(q). \tag{6}$$

We multiply (4) by $[a(\varrho + q)/h]_{1/2}$ and introduce $\bar{h}_0 := \left(h\left(\frac{1}{2} + \alpha\right)\right)_{1/2}$. Then (4) takes on the form

$$0 = a(\varrho + q)_{1/2}\, u_{x,0} + \bar{h}_0\left(\varphi + \frac{x_1}{2}\psi\right) - \frac{\psi}{2}\left(\frac{h^2\alpha}{q}\right)_{1/2}, \tag{7}$$

and Step 1 is finished. Introducing the notation

$$(\bar{h}M_\alpha f)_0 := \bar{h}_0 f_{1/2}$$

and remembering $f = \varphi + x\psi$, we rewrite (7) as

$$0 = a(\varrho + q)_{1/2}\, u_{x,0} + (\bar{h}M_\alpha f)_0 - \frac{\psi}{2}\left(\frac{h^2\alpha}{q}\right)_{1/2}. \tag{8}$$

For interior points x_i the difference scheme is

$$0 = a(\varrho + q)_{i+1/2}\, u_{x,i} - a(\varrho - q)_{i-1/2}\, u_{\bar{x},i} + (\bar{h}M_\alpha f)_i + \frac{\psi}{2}\left[\left(\frac{h^2\alpha}{q}\right)_{i-1/2} - \left(\frac{h^2\alpha}{q}\right)_{i+1/2}\right], \tag{9}$$

see [7], where

$$q_{i+1/2} := \frac{b}{2a} h_{i+1/2}, \tag{10}$$

and

$$(\bar{h}M_\alpha f)_i := \left(fh\left(\frac{1}{2} - \alpha\right)\right)_{i-1/2} + \left(fh\left(\frac{1}{2} + \alpha\right)\right)_{i+1/2}. \tag{11}$$

We remark that (8) follows from (9) formally by putting $i = 0$ and omitting all nondefined symbols, i.e. those indexed -1 and $-\frac{1}{2}$. Analogously, the exact difference scheme at $i = N$ is given by putting $i = N$ in (9) and omitting the symbols indexed $N + \frac{1}{2}$ and $N + 1$.

To perform Step 2, we observe that the factors multiplying $\frac{\psi}{2}$ in (9) are of order $O(h^2)$ by (6) and sum up to $O(h^2)$ over the whole grid. Hence, these terms will be neglected.

The difference scheme remaining is

$$0 = a(\varrho + q)_{i+1/2}\, y_{x,i} - a(\varrho - q)_{i-1/2}\, y_{\bar{x},i} + (\bar{h}M_\alpha f)_i, \quad x \in \omega_h, \tag{12}$$

with $\bar{h}M_\alpha f$ defined by (11) and with y approximating the exact solution $u(x)$. This scheme is, in general, at least of first (see [21], [13]) and at most of second order. In the case of equidistant spacing and first kind boundary conditions and for $f = \varphi + \psi x$ the scheme is exact. It is also exact for nonequidistant spacing and first or homogeneous second kind boundary conditions (the latter obtained from (12) by deleting all symbols not defined for $i = 0$ or $i = N$), if $f = \varphi = \text{const.}$

Step 3: The parabolic equation (1) is, obviously, approximated by

$$\bar{h}M_\alpha(cy_t) = \bar{h}\Lambda_\varrho y + \bar{h}M_\alpha f,$$

where

$$(\bar{h}\Lambda_\varrho y)_i := a(\varrho + q)_{i+1/2}\, y_{\bar{x},i} - a(\varrho - q)_{i-1/2}\, y_{\bar{x},i}.$$

An equivalent and, may be, more instructive form of Λ_ϱ is

$$(\bar{h}\Lambda_\varrho y)_i := \left(a + bh\left(\frac{1}{2} + \alpha\right)\right)_{i+1/2} y_{x,i} - \left(a - bh\left(\frac{1}{2} - \alpha\right)\right)_{i-1/2} y_{\bar{x},i},$$

cf. (6). Both forms are much more clear and usable than that given in [14]. To add here the approximation $-\bar{h}M_\alpha(dy)$ for the linear term $-du$ looks straightforward but proved to be unsatisfactory by reasons of flexibility and stability of the resulting scheme. More successful approaches will be considered below after finishing the construction of the scheme for (1), $d = 0$.

Step 4: A weighting parameter $\sigma \in [0, 1]$ is introduced now to assure stability. For the spatial operator this is done in the traditional way by taking $\bar{h}\Lambda_\varrho y^{(\sigma)}$ instead of $\bar{h}\Lambda_\varrho y$, where

$$y_x^{(\sigma)} := \sigma\hat{y}_x + (1 - \sigma)\, y_x.$$

The introduction of σ into the operator M_α acting on cy_t has been found by analyzing the stability properties of the scheme [6] and by comparison with the monotone approximation [15] of the pure transport equation (i.e. (1) in case $a = d = 0$).

We define the operator M_α^σ in the following way:

$$\begin{aligned}(\bar{h}M_\alpha^\sigma v)_i := {} & 2\sigma\big(h \max(0, -\alpha)\big)_{i-1/2}\, v_{i-1} \\ & + \left\{\left[h\left(\frac{1}{2} - \alpha - 2\sigma \max(0, -\alpha)\right)\right]_{i-1/2} \right. \\ & \left. + \left[h\left(\frac{1}{2} + \alpha - 2\sigma \max(0, \alpha)\right)\right]_{i+1/2}\right\} v_i + 2\sigma\big(h \max(0, \alpha)\big)_{i+1/2}\, v_{i+1}.\end{aligned}$$

We have, firstly, $M_\alpha^{1/2} = M_\alpha$. Further, if, say, $b < 0$, and $a \to 0$, then $\alpha \to -\dfrac{1}{2}$ by (6), and

$$\bar{h}M_\alpha^\sigma cy_t \to h_{i-1/2}\sigma cy_{t,i-1} + h_{i-1/2}(1 - \sigma)\, cy_{t,i},$$

$$\bar{h}\Lambda_\varrho y^{(\sigma)} \to bh_{i-1/2} y_{\bar{x},i}^{(\sigma)}.$$

We get a one-parameter difference scheme for the pure transport equation which is similar to [15]. Its maximum stability depends on σ and on the Courant number p,

$$p = p_{i+1/2}^{j+1/2} := b\tau_{j+1/2}/(ch_{i+1/2}), \tag{13}$$

in the following way [16]: If $|p| > 1$ ($|p| < 1$) then $\sigma = 1$ ($\sigma = 0$) ensures maximum norm stability, if $|p| = 1$ then the scheme is independent on σ and stable.

Step 5: We proceed now to the case of variable coefficients by taking all coefficients (including σ) at $t_{j+1/2}$ and in the middle of the spatial intervals (in the case of piecewise smooth coefficients other functionals of the coefficients must be used, see [8], p. 159, or [9]) as indicated by the grid step h. For example, q and p as given by (10) and (13) become then

$$q_{i+1/2}^{j+1/2} := \left(\frac{bh}{2a}\right)_{i+1/2}^{j+1/2}, \quad p_{i+1/2}^{j+1/2} := \left(\frac{b\tau}{ch}\right)_{i+1/2}^{j+1/2}. \tag{14}$$

All formulae of the resulting difference scheme will be presented in Point 5 after the introduction of the linear term below.

4. The linear term $-du$

To add this term we consider the initial value problem

$$c\dot{u} + du = f = \varphi + \psi t, \quad u(t_j) = u^j, \quad \dot{u} = \frac{du}{dt}, \tag{15}$$

where c, d, φ and ψ are constant. The exact difference scheme is constructed similarly as in the preceding point, the result being here

$$\left(\frac{c}{\tau} + \sigma_0 d\right) u^{j+1} = \left(\frac{c}{\tau} - (1 - \sigma_0)\, d\right) u^j + f^{(\sigma_0)}, \tag{16}$$

where

$$\sigma_0 = \sigma_0(r) := \frac{1}{2} + \alpha(r), \quad r = r^{j+1/2} := \frac{d}{2c}\, \tau_{j+1/2},$$

$$f^{(\sigma_0)} = \sigma_0 \hat{f} + (1 - \sigma_0)\, f = \varphi + \psi(\tau_{j+1/2}\sigma_0 + t_j),$$

and α is defined by (5). These formulae are identical to those in [17]. By (6) we know $\sigma_0 \in [0, 1]$ in general, and

$$\text{if} \quad d \geqq 0, \quad \text{then} \quad \sigma_0 \in \left[\frac{1}{2}, 1\right], \quad 1 - 2r(1 - \sigma_0) \geqq 0. \tag{17}$$

Now, the first possibility to exploit the scheme (16) in our context is the following: Rewrite the difference scheme of Point 3 as

$$\sum_{n=-1}^{1} B_{i+n}^{j+1} y_{i+n}^{j+1} = \sum_{n=-1}^{1} B_{i+n}^{j} y_{i+n}^{j} + \bar{h} M_\alpha f^{j+1/2}.$$

Then

$$\sum_{n=-1}^{1} (1 + 2r\sigma_0)\, B_{i+n}^{j+1} y_{i+n}^{j+1} = \sum_{n=-1}^{1} \big(1 - 2r(1 - \sigma_0)\big)\, B_{i+n}^{j} y_{i+n}^{j} + \bar{h} M_\alpha f^{j+\sigma_0}$$

will constitute an approximation to (1). This scheme works well as long as $c \geqq c_0 > 0$ and has the advantage that the automatic selection of the weighting parameter σ described in [7] and below for $d = 0$ may be used, with $\sigma_0(r)$ acting as the "mean" value of σ to be specified. This σ-selection is oriented on satisfaction of the conditions of the maximum principle requiring, essentially, that the factors of y_{i+n}^{j+1} be all nonnegative — which is not influenced by $1 + 2r\sigma_0$ or $\big(1 - 2r(1 - \sigma_0)\big)$ due to (17).

The drawbacks of this first way are that the multiplication by $1 + 2r\sigma_0$ and $1 - 2r \times (1 - \sigma_0)$ corresponds to an introduction of higher order differences presupposing more smoothness, and that in case $c \to 0$ the scheme does not approach an approximation of (1), $c = 0$.

The second way originates from considering u in (15) as depending on x and t, replacing c by $(b^2 + c^2)^{1/2}$, and interpreting u as directional derivative $\frac{du}{dn}$. The direction of interest is that of the characteristic of (1), $a = 0$, i.e. $n = (b^2 + c^2)^{-1/2} (-b, c)$. We have, identically,

$$n = \left((\min (|p|, 1)\, h)^2 + \left(\min \left(\frac{1}{|p|}, 1 \right) \tau \right)^2 \right)^{-1/2} \left(-\min (|p|, 1)\, h \operatorname{sgn} b, \min \left(\frac{1}{|p|}, 1 \right) \tau \right),$$

where p is the Courant number (13).

Requiring u^{j+1} and u^j in (16) to lie on grid lines, we insert u_i^{j+1} for u^{j+1} and $u_{i+\min(|p|,1)\operatorname{sgn} b}^{j+1+\min(1/|p|,1)}$ for u^j. The distance between the points is now $\min (\tau/c, h/|b|)\, (b^2 + c^2)^{1/2}$, which replaces τ in (16). Next, to get a difference scheme approximating the constant coefficient equation (1), $a = 0$, we substitute

$$y_p := \max \left(0, 1 - \frac{1}{|p|}\right) y_{i+sb}^{j+1} + \min \left(|p|, \frac{1}{|p|}\right) y_{i+sb}^{j} + \max (0, 1 - |p|)\, y_i^j,$$

$$sb := \operatorname{sgn} b, \quad \text{for} \quad u_{i+\min(|p|,1)sb}^{j+1+\min(1|p|,1)}, \quad \text{and} \quad \hat{y}_i \quad \text{for} \quad u_i^{j+1}.$$

The scheme

$$\left(\max \left(\frac{c}{\tau}, \frac{|b|}{h} \right) + \sigma_0 d \right) \hat{y}_i = \left(\max \left(\frac{c}{\tau}, \frac{|b|}{h} \right) - (1 - \sigma_0)\, d \right) y_p + f^{j+\sigma_0},$$
$$\sigma_0 := \frac{1}{2} + \alpha \left(\frac{d}{2} \Big/ \max \left(\frac{c}{\tau}, \frac{|b|}{h} \right) \right) \quad \text{(with } \alpha \text{ defined by (5))} \tag{18}$$

is maximum norm stable and of first order in general but exact in the constant coefficient case if either $c = 0$ or $b = 0$ or $|p| = 1$. In (18), the expression multiplied by d is

$$d \left[\sigma_0 \hat{y}_i + (1 - \sigma_0) \left(\max \left(0, 1 - \frac{1}{|p|}\right) \hat{y}_{i+sb} + \min \left(|p|, \frac{1}{|p|}\right) y_{i+sb} + \max (0, 1 - |p|)\, y_i \right) \right], \quad sb = \operatorname{sgn} b. \tag{19}$$

It is basically this expression we shall use for approximation of du. To have a continuous transition from the expression $dy^{(\sigma)}$ commonly used in difference approximations for (1), $b = 0$, $a > 0$, to the above one for $a = 0$, we replace now y_i^{j+l} by $\left[\left(\frac{1}{2} + \beta \right)_{i+1/2} + \left(\frac{1}{2} - \beta \right)_{i-1/2} \right] y_i^{j+l}$, and y_{i+sb}^{j+l} by $\left(\frac{1}{2} + \beta \right)_{i+1/2} y_{i+1}^{j+l} + \left(\frac{1}{2} - \beta \right)_{i-1/2} y_{i-1}^{j+l}$, $l = 0, 1$. Here $\beta = \beta_{i+1/2} = \beta(q_{i+1/2})$, $\operatorname{sgn} \beta = \operatorname{sgn} b$, $\beta \in \left[-\frac{1}{2}, \frac{1}{2} \right]$, $\lim_{q \to 0} \beta = 0$, $\lim_{|q| \to \infty} \beta = \frac{1}{2} \operatorname{sgn} b$.

The choice $\beta = \alpha(q)$ is at hand, see (6), but proved to be unsuitable in the stability analysis. Better suited occured to be

$$\beta = \beta_{i+1/2} := \left(\frac{q}{2\varrho}\right)_{i+1/2}. \tag{20}$$

Giving priority to the σ-selection mentioned above for the case $d = 0$, it may turn out to be impossible by stability reasons to take σ_0 in (19). Therefore σ_0 will be replaced by $\bar{\sigma}$ which will be selected (after selection of σ) according to stability requirements but nearest to σ_0.

Finally, to account for the nonequidistant grid $\overline{\omega}_h$, we multiply (19) by h and put the indices $i \pm 1/2$ in accordance with those of β. We are now on the level of Step 4 in Point 3, and Step 5 is performed by drawing $d\bar{\sigma}$ and $d(1 - \bar{\sigma})$ into the expressions and taking d, $\bar{\sigma}$ and p in the middle of the spatial intervals indicated by the indices of $h\left(\frac{1}{2} \pm \beta\right)$. Then we get the approximation of the linear term in the form $\bar{h}D^{\bar{\sigma}}_{\beta p}(dy)$ as written out in the next point below.

5. All formulae of the difference scheme

a) Approximation of equation (1)

$$\bar{h}M^{\sigma}_{\alpha}(cy_t) = \bar{h}\Lambda^{\sigma}_{\varrho}y - \bar{h}D^{\bar{\sigma}}_{\beta p}(dy) + \bar{h}M_{\alpha}f^{\sigma_0}, \quad x \in \omega_h. \tag{21}$$

Here ϱ, α, q, p, σ_0 and β are defined by (5), (14), (18) and (20). For the operators M, Λ, D we have the following definitions

$$\begin{aligned}
(\bar{h}M^{\sigma}_{\alpha}(cy_t))_i := {}& 2\big(\sigma ch \max(\alpha, 0)\big)_{i+1/2}\, y_{t,i+1} \\
&+ \left\{\left[ch\left(\frac{1}{2} + \alpha - 2\sigma \max(\alpha, 0)\right)\right]_{i+1/2}\right. \\
&\left.+ \left[ch\left(\frac{1}{2} - \alpha - 2\sigma \max(-\alpha, 0)\right)\right]_{i-1/2}\right\} y_{t,i} \\
&+ 2\big(\sigma ch \max(-\alpha, 0)\big)_{i-1/2}\, y_{t,i-1},
\end{aligned}$$

$$(\bar{h}\Lambda^{\sigma}_{\varrho}y)_i := \big(a(\varrho + q)\big)_{i+1/2}\, y^{(\sigma_{i+1/2})}_{x,i} - \big(a(\varrho - q)\big)_{i-1/2}\, y^{(\sigma_{i-1/2})}_{\bar{x},i},$$

$$y_x{}^{(\sigma)} := \sigma\hat{y}_x + (1 - \sigma)\, y_x,$$

$$\begin{aligned}
(\bar{h}D^{\bar{\sigma}}_{\beta p}(dy))_i := {}& \left\{\left[d\bar{\sigma}h\left(\frac{1}{2} + \beta\right)\right]_{i+1/2} + \left[d\bar{\sigma}h\left(\frac{1}{2} - \beta\right)\right]_{i-1/2}\right\}\hat{y}_i \\
&+ \left[d(1 - \bar{\sigma})\, h\left(\frac{1}{2} + \beta\right) \max\left(0, 1 - \frac{1}{|p|}\right)\right]_{i+1/2} \hat{y}_{i+1} \\
&+ \left[d(1 - \bar{\sigma})\, h\left(\frac{1}{2} - \beta\right) \max\left(0, 1 - \frac{1}{|p|}\right)\right]_{i-1/2} \hat{y}_{i-1} \\
&+ \left[d(1 - \bar{\sigma})\, h\left(\frac{1}{2} + \beta\right) \min\left(|p|, \frac{1}{|p|}\right)\right]_{i+1/2} y_{i+1}
\end{aligned}$$

$$+\left\{\left[d\,(1-\bar{\sigma})\,h\left(\frac{1}{2}+\beta\right)\max\,(0,1-|p|)\right]_{i+1/2}\right.$$

$$\left.+\left[d(1-\bar{\sigma})\,h\left(\frac{1}{2}-\beta\right)\max\,(0,1-|p|)\right]_{i-1/2}\right\}y_i$$

$$+\left[d(1-\bar{\sigma})\,h\left(\frac{1}{2}-\beta\right)\min\left(|p|,\frac{1}{|p|}\right)_{i-1/2}\right]y_{i-1},$$

$$(\bar{h}M_\alpha f^{\sigma_0})_i := \left(h\left(\frac{1}{2}+\alpha\right)f^{(\sigma_0)}\right)_{i+1/2}+\left(h\left(\frac{1}{2}-\alpha\right)f^{(\sigma_0)}\right)_{i-1/2}.$$

All coefficients (except f) are to be taken at $t = t_{j+1/2}$, whereas f may be taken at t_j and t_{j+1} (according to the quoted formulae), or at $t_{j+\sigma_0}$.

b) Boundary conditions

In the case of conditions of the first kind,

$$u(0,t) = \nu_1(t),\quad u(1,t) = \nu_2(t),\quad 0 < t \leqq T, \tag{22$_1$}$$

we use, as customary,

$$y_0^j = \nu_1^j,\quad y_N^j = \nu_2^j,\quad j = 0, 1, \ldots, m.$$

In the case of boundary conditions of the second kind,

$$-a\,\frac{\partial u}{\partial x} = \nu_1(t),\quad x = 0,\quad a\,\frac{\partial u}{\partial x} = \nu_2(t),\quad x = 1,\quad 0 < t \leqq T, \tag{22$_2$}$$

we take, inspired by the exact difference approximation of the second kind conditions found in Point 3, the "natural difference boundary conditions" defined quite simply as follows:

Put $i = 0$ and $i = N$ in (21), add ν_1 to $(\bar{h}M_\alpha f^{\sigma_0})_0$ and ν_2 to $(\bar{h}M_\alpha f^{\sigma_0})_N$ and delete all symbols indexed -1, $-\frac{1}{2}$, $N+\frac{1}{2}$ $N+1$, respectively.

Such boundary conditions, clearly, are advantageous if summing up all equations of the difference scheme to obtain the discrete conservation law, see [7].

6. Particular cases

The difference scheme looks rather complicated. Nevertheless, it is of the common two-level, three-point type, the summation identity obtained in [7] is simple, and for degenerating coefficients well-known difference schemes are contained in our formulae:

a) $a > 0$, $c > 0$, $b = 0$ (no convection)

$$\left(\bar{h}M_\alpha^\sigma(cy_t)\right)_i = \frac{1}{2}\left[(ch_{i+1/2} + (ch)_{i-1/2}\right]y_{t,i},\quad (\bar{h}\Lambda_\varrho^\sigma y)_i = a_{i+1/2}y_{x,i}^{(\sigma)} - a_{i-1/2}y_{\bar{x},i}^{(\sigma)},$$

$$\left(\bar{h}D_{\beta p}^{\bar{\sigma}}(dy)\right)_i = \left[(d\bar{\sigma}h)_{i+1/2} + (d\bar{\sigma}h)_{i-1/2}\right]\hat{y}_i + \left[\left(d(1-\bar{\sigma})\,h\right)_{i+1/2} + \left(d(1-\bar{\sigma})\,h\right)_{i-1/2}\right]y_i,$$

$$(\bar{h}M_\alpha f^{\sigma_0})_i = \frac{1}{2}\left[(hf^{(\sigma_0)})_{i+1/2} + (hf^{(\sigma_0)})_{i-1/2}\right].$$

This is the familar weighted difference scheme for (1) written on the nonequidistant grid ω_h in a special way which makes it easy to formulate sufficient conditions for the maximum principle to hold (see Point 7). Similarly, the "natural boundary conditions" turn out to be the usual second order approximations of (22_2), cf. [8], p. 294.

b) $a = 0$, $c > 0$ and, e.g., $b < 0$ (no diffusion)

All symbols are to be understood now as the limiting values corresponding to $a \to 0$. Then, by (6),

$$\alpha = -\frac{1}{2}, \quad a(\varrho + q) = 0, \quad a(\varrho - q) = |b|\, h, \quad \beta = -\frac{1}{2}.$$

Hence,

$$\left(\bar{h}M_\alpha^\sigma(cy_t)\right)_i = (ch\sigma)_{i-1/2}\, y_{t,i-1} + \left(ch(1-\sigma)\right)_{i-1/2}\, y_{t,i},$$

$$(\bar{h}\Lambda_\varrho^\sigma y)_i = (bh)_{i-1/2} y_{\bar{x},i}^{(\sigma)},$$

$$\left(\bar{h}D_{\beta p}^{\bar{\sigma}}(dy)\right)_i = (dh\bar{\sigma})_{i-1/2}\, \hat{y}_i + \left(dh(1-\bar{\sigma})\right)_{i-1/2}$$
$$\times \left[\max\left(0, 1 - \frac{1}{|p|}\right) \hat{y}_{i-1} + \min\left(|p|, \frac{1}{|p|}\right) y_{i-1} + \max\left(0, 1 - |p|\right) y_i\right],$$

$$(\bar{h}M_\alpha f)_i = (hf^{\sigma_0})_{i-1/2}.$$

In other notations and for $d = 0$ this is the scheme [16], p. 151. Especially, if $\sigma = \frac{1}{2}$, it is the Wendroff scheme [18]. In case $|p| = 1$ the scheme represents the method of characteristics.

It is important to remark that the same difference scheme results from the above "natural difference boundary conditions" in the case $\nu_2 = a = 0$, $b < 0$ $(i = N)$, $\nu_1 = a = 0$, $b > 0$ $(i = 0)$, i.e. for outflow boundaries:

Imagine a problem (1) in which $b > 0$ for all x and t (flow from the left), where there are first kind conditions (22_1) at $x = 0$ and homogeneous second kind conditions (22_2) at $x = 1$. If $a > 0$ for $t < t^*$ and $a = 0$ for $t \geqq t^*$ then the right-hand side boundary condition is superfluous for $t \geqq t^*$. The possibility for the boundary condition to "disappear" smoothly as $a \to 0$ is the above one: To turn from an approximation of (22_2), $a > 0$, to the same approximation of (1), $a = 0$, as is used in interior points. In this way boundary layers at $x = 1$ are avoided.

c) $c \equiv 0$, $a > 0$ (saturation of soil in agricultural problems)

$\left(\bar{h}M_\alpha^\sigma(cy_t)\right)_i = 0$, $(\bar{h}\Lambda_\varrho^\sigma y)_i$ as in the general case,

$$\left(hD_{\beta p}^{\bar{\sigma}}(dy)\right)_i = \left[\left(d\bar{\sigma}h\left(\frac{1}{2} + \beta\right)\right)_{i+1/2} + \left(d\bar{\sigma}h\left(\frac{1}{2} - \beta\right)\right)_{i-1/2}\right] \hat{y}_i$$
$$+ \left(d(1-\bar{\sigma})\, h\left(\frac{1}{2} + \beta\right)\right)_{i+1/2} \hat{y}_{i+1} + \left(d\,(1-\bar{\sigma})\, h\left(\frac{1}{2} - \beta\right)\right)_{i-1/2} \hat{y}_{i-1}$$

and $(\bar{h}M_\alpha f)$ as above. That is, we have used the limiting values for $|p| \to \infty$. Taking $\sigma = \bar{\sigma} = 1$ (and these values are automatically chosen by the σ-selection explained below), the scheme is an approximation of the elliptic equation (1), $c = 0$. Here the choice $\beta = \alpha$ would be possible and better suited to the operator Λ_ϱ.

Since the case $d = 0$ is trivial, we may say that degeneration of only one coefficient is settled directly by the considered difference scheme.

d) $a = b = 0$ (a being dispersion and modelled to be $a_0 |b|^k$, and $b = 0$)

In this case, the only gap is that q is not defined. It suffices to put $q = 0$ in the formulae of Point 5: We then get

$$\frac{1}{2}\left[(ch)_{i+1/2} + (ch)_{i-1/2}\right] y_{t,i} = -\frac{1}{2}\left[(d\bar{\sigma}h)_{i+1/2} + (d\bar{\sigma}h)_{i-1/2}\right]\hat{y}_i$$
$$-\frac{1}{2}\left[\left(d(1-\bar{\sigma})\,h\right)_{i+1/2} + \left(d(1-\bar{\sigma})\,h\right)_{i-1/2}\right] y_i$$
$$+\frac{1}{2}\left[(hf^{(\sigma_0)})_{i+1/2} + (hf^{(\sigma_0)})_{i-1/2}\right], \qquad (23)$$

which is just the difference scheme needed. We may take $\bar{\sigma} = \sigma_0$ as defined by (18).

e) $b = c = 0$

Now the Courant number p and σ_0 (see (18)) lack to be defined, and we put $|p| = \infty$ and $\sigma_0 = 1$. The σ-selection then chooses $\sigma = \bar{\sigma} = 1$ and the scheme takes on the form wanted:

$$0 = a_{i+1/2}\hat{y}_{x,i} - a_{i-1/2}\hat{y}_{\bar{x},i} - \frac{1}{2}\left[(dh)_{i+1/2} + (dh)_{i-1/2}\right]\hat{y}_i$$
$$+\frac{1}{2}\left[(h\hat{f})_{i+1/2} + (h\hat{f})_{i-1/2}\right].$$

f) $a_{i\pm 1/2} = 0$, $b_{i+1/2} < 0$, $b_{i-1/2} > 0$

In this one case only (no diffusion, turning point of velocity) the difference scheme (21) breaks down:
We have, by (6) and (20), $\alpha_{i+1/2} = \beta_{i+1/2} = -\frac{1}{2}$, $\alpha_{i-1/2} = \beta_{i-1/2} = 0$, and the difference scheme reduces to the identity $0 = 0$. The obvious exit from this situation is to assume $a = b = 0$ at x_i and to approximate (1) by (23).

Only in this case the requirement of homogeneity of the scheme ([9]) is violated.

Similarly, if the natural difference boundary conditions are considered and $a = 0$, $b < 0$ ($b > 0$) at the left (right) hand boundary (i.e. we have inflow boundaries), then this approximation becomes, in general, senseless (see [7] for an exception). But now we can't put $b = 0$ as for interior turning points. Rather, at inflow boundaries first kind boundary data must be available if $a = 0$.

It is clear now how to proceed in the case of three coefficients degenerating simultaneously:

If $c = 0$ is involved then we start from c) above, take $\sigma = \bar{\sigma} = 1$, and $q = 0$ if $a = b = 0$, or else, if $a = 0$, $b \neq 0$, $aq = bh/2$, $a\varrho = |b|\, h/2$, $\alpha = \beta = \frac{1}{2}\,\mathrm{sgn}\, b$. If $c > 0$ and $a = b = d = 0$ then we start from (23) where now $\sigma_0 = \frac{1}{2}$ by (18).

For shortness, the scheme considered in the Points 5 and 6 will be called the difference scheme (21), (23).

Moreover, boundary conditions (22) will be called *properly posed* if they are posed in agreement with the following requirements:

1. Second kind conditions are not to be posed in left (right) hand boundary points where $a = 0$, $b < 0$ ($b > 0$) at $x = x_{1/2}$ ($x = x_{N-1/2}$), $t = t_{j+1/2}$.
2. If $d_{i+1/2}^{j+1/2} = c_{i+1/2}^{j+1/2}$ for some j and for $0 \leqq i \leqq N-1$, then first kind boundary conditions are required at the left or the right hand boundary.

7. The automatic σ-selection

It will be shown now how to select σ and $\bar{\sigma}$ in order to satisfy the conditions of the maximum principle. For this aim, we rewrite the difference scheme (21), (23) as

$$\mathscr{L}(y)_i := B_i^{j+1} y_i^{j+1} - \left(B_{i+1}^{j+1} y_{i+1}^{j+1} + B_{i-1}^{j+1} y_{i-1}^{j+1} + \sum_{n=-1}^{1} B_{i+n}^{j} y_{i+n}^{j} \right) = (\bar{h} M_\alpha f^{\sigma_0})_i$$

for $(x, t) \in \omega_{h\tau} \cup \gamma_2$. This representation comprises, therefore, approximation of (1) and, at γ_2, of second kind boundary conditions. For points of γ_1 we have first kind conditions,

$$y_i^{j+1} = \nu_1^{j+1}, \quad x = 0, \quad y_i^{j+1} = \nu_2^{j+1}, \quad x = 1, \quad y_i^0 = u_0(x_i), \quad x_i \in \bar{\omega}_h.$$

For the maximum principle to be true we need the following conditions (see [8], p. 243—248):

$$\begin{aligned} & B_i^{j+1} > 0, \quad B_{i+n}^{j+l} \geqq 0, \quad l = 0, 1, \quad n = -1, 0, 1, \\ & B_i^{j+1} \geqq B_{i+1}^{j+1} + B_{i-1}^{j+1} + \sum_{n=-1}^{1} B_{i+n}^{j}. \end{aligned} \tag{24}$$

Moreover, in [8] there is the condition for the grid $\bar{\omega}_{h\tau}$ to be connected in the following sense: The point $(x_{i+n}, t_{j+l}) \in \bar{\omega}_{h\tau}$ is called neighbour of (x_i, t_{j+1}) if $B_{i+n}^{j+l} > 0$ ($l = 0, 1$, $n = -1, 0, 1$), and the grid is said to be connected, if any two points in the grid can be connected through a sequence of neighbours.

If looking at the proofs of the statements around the maximum principle then it becomes obvious that the strong requirement of connectedness is not necessary. Rather, we must have the property that every point may be connected through a sequence of neighbours with some point of a subset of $\bar{\omega}_{h\tau}$ for which the maximum principle is known to hold, e.g. the subset γ_1.

For example, consider the scheme (21), (23) in the following special case: Initial values are given at $t = 0$, first kind boundary conditions are specified at $x = 0$ and second kind conditions at $x = 1$. All coefficients are constant, especially $h = h_{i+1/2} = \text{const}$, $ch = b\tau > 0$ (i.e. $|p| = 1$). As mentioned in 6b), the scheme coincides now with the method of characteristics which is exact in that case. The grid $\overline{\omega}_{h\tau}$ occurs to be not connected, but every point of $\overline{\omega}_{h\tau}$ can be connected with γ_1, and the maximum principle, clearly, holds for both the equation (1), $a = 0$, and its exact discrete analogon.

If no such subset as γ_1 is known then we need the usual connectedness.

We shall now derive sufficient conditions on σ and $\bar{\sigma}$ which guarantee the inequalities (24) to hold. Firstly, we remark that $B_{i\pm1}^{j+1}$ contains only coefficients taken at $i \pm 1/2$, correspondingly, and B_i^{j+l} consists of the sum of two expressions taken at $i + 1/2$ and $i - 1/2$, respectively: $B_i^{j+l} = C_{i+1/2}^{j+l} + C_{i-1/2}^{j+l}$, $l = 0, 1$. Hence, for (24) it suffices to have

$$
\begin{aligned}
&C_{i-1/2}^{j+1} > 0, \quad B_{i-1}^{j+l} \geqq 0, \quad C_{i-1/2}^{j} \geqq 0, \quad C_{i-1/2}^{j+1} \geqq B_{i-1}^{j+1} + B_{i-1}^{j} + C_{i-1/2}^{j},\\
&C_{i+1/2}^{j+1} > 0, \quad B_{i+1}^{j+l} \geqq 0, \quad C_{i+1/2}^{j} \geqq 0, \quad C_{i+1/2}^{j+1} \geqq B_{i+1}^{j+1} + B_{i+1}^{j} + C_{i+1/2}^{j}.
\end{aligned}
\tag{25}
$$

In this way, pointwise conditions are obtained and there is no need for inequalities between expressions taken at $i + 1/2$ and at $i - 1/2$. In the case of second kind boundary conditions the inequalities (25) arise directly in such a separated form.

Checking the inequalities (25) for the difference scheme (21), we get the following eight conditions below in terms of the cell Reynolds number q, the Courant number p (14), of the weighting parameters σ and $\bar{\sigma}$, and of

$$
s := \left(\frac{dh}{2b}\right)_{i+1/2}^{j+1/2}. \tag{26}
$$

All these numbers are indexed $i + 1/2$, $j + 1/2$, and therefore these indices will be omitted.

1. $\sigma(\varrho - |q|)\,|p| \geqq (1 - \bar{\sigma})\, 4\,|qs| \left(\frac{1}{2} - |\beta|\right)(|p| - 1)$, if $|p| > 1$,

2. $\sigma\big(|p|\,(\varrho + |q|) - 2(\varrho - 1)\big) \geqq (1 - \bar{\sigma})\, 4\,|qs| \left(\frac{1}{2} + |\beta|\right)(|p| - 1)$, if $|p| > 1$,

3. $(1 - \sigma)\,(\varrho - |q|) \geqq (1 - \bar{\sigma})\, 4\,|qs| \left(\frac{1}{2} - |\beta|\right) \min\left(|p|, \frac{1}{|p|}\right)$,

4. $\big((1 - \sigma)\,|p|\,(\varrho + |q|) + 2\sigma(\varrho - 1)\big) \geqq (1 - \bar{\sigma})\, 4\,|pqs| \left(\frac{1}{2} + |\beta|\right) \min\left(|p|, \frac{1}{|p|}\right)$,

5. if $|p| < 1$, $2\,|q| \left(\frac{1}{2} - |\alpha|\right) + (1 - \sigma)\big(2(\varrho - 1) - |p|(\varrho + |q|)\big)$

   $\geqq (1 - \bar{\sigma})\, 4\,|pqs| \left(\frac{1}{2} + |\beta|\right)(1 - |p|)$,

6. $2\,|q| \left(\frac{1}{2} - |\alpha|\right) - (1 - \sigma)\,(\varrho - |q|)\,|p| \geqq (1 - \bar{\sigma})\, 4\,|pqs| \left(\frac{1}{2} - |\beta|\right)(1 - |p|)$, if $|p| < 1$,

7., if $0 \leqq |p| \leqq 1$, then $|p| \geqq 2(\varrho - 1)/(\varrho + |q|)$ or $\sigma = 0$,

8., if $|p| \geqq 1$, then $(1 - \sigma)\,|p| \leqq 2\,|q| \left(\frac{1}{2} - |\alpha| + 2\,|\alpha|\,(1 - \sigma)\right) \Big/ (\varrho + |q|)$ or $\sigma = 1$.

Only in the case considered in Point 6f, all coefficients B_{i+n}^{j+l} are zero, but then (23) is to be taken and this scheme is maximum norm stable for $1 \geqq \bar{\sigma} \geqq \sigma_0$ due to (17).

From conditions 1, 3 and 6 we see that it is desirable to have $\frac{1}{2} - |\beta| \to 0$ at least as quick as $\varrho - |q| \to 0$ if $|q| \to \infty$, which suggests to choose $\beta = q/2\varrho$ and excludes $\beta = \alpha$.

Fortunately, most of the conditions are redundant:

First of all, if $d = 0$, then only 7. and 8. remain.

Next, using the inequalities (6) and $|\beta| \geqq |\alpha|$, it follows that condition 2 implies 1, further, if $|p| > 1$, then 3. implies 2., condition 3 implies 4, and condition 5 implies 6.

Finally, if $|p| < 1$, then condition 3 implies 5 in case

$$\sigma(\varrho - |q|) \leqq 2\sigma - 1, \tag{27}$$

but for the inequality sign opposite to that in (27) condition 5 implies 3.

In summing up, if $|p| \geqq 1$ then conditions 3 and 8 are the decisive ones, and if $|p| < 1$ then conditions 3 and 7 (for (27) being true) and conditions 5 and 7 if (27) does not hold.

If $d = 0$ then there is a narrow domain of $|p|$-values for which 7. and 8. are satisfied independently on σ:

$$2(\varrho - 1)/(\varrho + |q|) \leqq |p| \leqq 2\,|q| \left(\frac{1}{2} + |\alpha|\right) \Big/ (\varrho + |q|). \tag{28}$$

Inside of this domain lies $p = 2\alpha$ (p and α are of the same sign) which ensures, firstly, second order accuracy of the difference scheme for constant coefficients (see [7]), and secondly, that the scheme turns to the method of characteristics if $a \to 0$: by (6) we have $|\alpha| \to \frac{1}{2}$ and hence $|p| \to 1$, compare now Point 6b. If $\sigma = 0$, then the stability domain (cf. condition 7) is $0 \leqq |p| \leqq 2\,|q| \left(\frac{1}{2} + |\alpha|\right) \Big/ (\varrho + |q|)$, if $\sigma = 1$, then it is $|p| \geqq 2(\varrho - 1)/(\varrho + |q|)$. Both stability domains overlap in (28), and so do the stability domains of all other σ. Hence, given any pair $|p|$, $|q|$ (that is, for any values of h, τ, a, b, c), there exists a σ which ensures the conditions (25) to hold. If some "mean" value σ^0 is specified (usually $\frac{1}{2}$ or (18)), then there is a unique σ having minimum distance from σ_0, and fulfilling the conditions 7 and 8, i.e. (25).

This value of σ is inserted now into condition 3 (if $|p| > 1$, or else if (27) is true) or else into condition 5. For $\bar{\sigma}$ we take that unique value which satisfies the corresponding inequality and is nearest to σ_0 as defined by (18).

This process of choosing uniquely σ and $\bar{\sigma}$ is called the *automatic σ-selection.*

Theorem. *The difference scheme* (21, (23) *with properly posed boundary conditions and with automatic σ-selection fulfils the condition of the maximum principle on any nonequidistant grid $\overline{\omega}_{h\tau}$ and for any coefficients a, b, c and d satisfying*

$$a \geqq 0, \quad c \geqq 0, \quad d \geqq 0, \quad a + c > 0.$$

Proof (sketch.) By the definition of the scheme and the automatic σ-selection, it remains to show that every point of $\overline{\omega}_{h\tau}$ can be connected with a subset for which the maximum principle is known to hold. Analyzing the several situations which may arise we find that every point not in γ_1 can be connected with points of γ_1 at the same time level or with points of the preceding level, if $c > 0$ in at least one point of the considered level t_{j+1}.

Hence, by induction, every point not in γ_1 can be connected with some point of γ_1 — or there is one time level, say t_{k+1}, at which $c_{i+1/2}^{k+1/2} = 0$, $0 \leqq i \leqq N - 1$. In the latter case the condition 2 of the definition of proper boundary conditions assures that the maximum principle holds for the level t_{k+1}.

In both bases, we have succeeded to find a subset of $\bar{\omega}_{h\tau}$ for which the maximum principle is true, and this proves the Theorem.

Remarks. 1. The maximum principle ensures unique solvability of the equations of the difference scheme. It provides also an estimate if a majorizing function Y (see [8], p. 247) is known. In the special case of first kind or homogeneous second kind conditions and

$$\min c(x, t) = c_0 > 0$$

we may take $Y = m_1 + m_2 t$. Then, for (21),

$$\bar{h}\left(M_\alpha^\sigma(cY_t) - \Lambda_\varrho^\sigma Y + D_{\beta p}^{\bar{\sigma}}(dY)\right)_i \geqq m_2 c_0 \left[\left(h\left(\frac{1}{2} + \alpha\right)\right)_{i+1/2} + \left(h\left(\frac{1}{2} - \alpha\right)\right)_{i-1/2}\right],$$

$$|\bar{h} M_\alpha f^{\sigma_0}| \leqq \max_{(x,t)\in\bar{\omega}_{h\tau}} |f(x, t)| \left[\left(h\left(\frac{1}{2} + \alpha\right)\right)_{i+1/2} + \left(h\left(\frac{1}{2} - \alpha\right)\right)_{i-1/2}\right],$$

and similarly, for (23),

$$\frac{1}{2}\left[(ch)_{i+1/2} + (ch)_{i-1/2}\right] Y_t + \frac{1}{2}\left[(d\bar{\sigma}h]_{i+1/2} + (d\bar{\sigma}h)_{i-1/2}\right] \hat{Y}_i$$
$$+ \frac{1}{2}\left[\left(d(1 - \bar{\sigma})\, h\right)_{i+1/2} + \left(d(1 - \bar{\sigma})\, h\right)_{i-1/2}\right] Y_i \geqq m_2 c_0 \frac{1}{2} (h_{i+1/2} + h_{i-1/2}),$$

$$\frac{1}{2}\left[(hf^{(\sigma_0)})_{i+1/2} + (hf^{(\sigma_0)})_{i-1/2}\right] \leqq \max_{(x,t)\in\bar{\omega}_{h\tau}} |f(x, t)| \frac{1}{2} (h_{i+1/2} + h_{i-1/2}).$$

Therefore we get the estimate

$$\max_{(x,t)\in\bar{\omega}_{h\tau}} |y(x, t)| \leqq \max_{(x,t)\in\gamma_1} \left(|u_0(x)|, |\nu_1(t)|, |\nu_2(t)|\right) + \frac{T}{c_0} \max_{(x,t)\in\bar{\omega}_{h\tau}} |f(x, t)|,$$

which shows the stability of the scheme in the considered case.

2. There may be doubts whether the automatic σ-selection will produce $\bar{\sigma}$-values other than $\bar{\sigma} = 1$.

Let be $\sigma \in [0, 1/(2 - \varrho + |q|)]$, e.g. $\sigma = \frac{1}{2}$, and $|p| \in [0, 2\,|\alpha|]$ if $\sigma = 0$, $|p| \in [2(\varrho - 1)/(\varrho + |q|), 2|\alpha|]$ if $\sigma > 0$.

Thus, conditions 7 and 8 are satisfied, and the σ-selection is directed by condition 5. In 5. the worst case occurs if $\sigma = 0$ and $|p| = 2\,|\alpha|$, that is,

$$\left(2|q|\left(\frac{1}{2} + |\alpha|\right)\Big/(\varrho + |q|) - 2|\alpha|\right) \varrho/4\,|q\alpha|\,(1 - 2|\alpha|) \geqq (1 - \bar{\sigma})\,|s|.$$

Using (6), the left hand side of this inequality can be shown to be $\geqq \frac{1}{4}$ with equality for $|q| = \infty$. Hence, $\bar{\sigma} \geqq 1 - 1/4\,|s|$, and we may take $\bar{\sigma} = \frac{1}{2} + \alpha(|s|)$ as long as $|s| \leqq \frac{1}{2}$, i.e., by (26), $d \leqq |b|/h$ (all inequalities to be understood pointwise).

8. Computational results

a) A series of computations was undertaken for (1), (22) with constant coefficients $c = 1$, $10^{-4} \leqq a \leqq 0.1$, $b = -5$, $d = 0$, $f = 0$, $T = 0.1$ and with discontinuous initial data

$$u(0, 0) = 1, \quad u(x, 0) = 0, \quad 0 < x \leqq 1.$$

Boundary conditions are of the first kind (22_1) with $\nu_1 = 1$ at $x = 0$ and of second kind (22_2) with $\nu_2 = 0$ at $x = 1$.

The solution y computed at $T = 0.1$ satisfies $0 \leqq y \leqq 1$ and is monotonously decreasing, exhibiting a steep front at $x = 0.5$, the right place. We check the accuracy by recalculating the coefficient a from y using the formula

$$\bar{a} \approx 1/4\pi T(y_x)^2$$

following from the well-known exact solution of the corresponding problem posed on $0 \leqq x < \infty$. Here y_x is the first difference of y at $x = 0.5$.

Some results for $\varrho_1 = q \coth q$ and its approximation $\varrho_2 = 1 + q^2(1 + 2\,|q|)/(3 + 3 \times |q| + 2q^2)$ are presented below for equidistant steps $h = 1/N$ and $\tau = 0.2/N$ (this choice being decisive here assuring $|p| = 1$).

ϱ_1, $a = 6.5 \cdot 10^{-3}$	N	20	40	80	160
	$\bar{a}$	4.929-3	5.483-3	6.302-3	6.445-3
	q	19.2308	9.6154	4.8077	2.4038

We remark that for $|q| > 20$ we have $\varrho_1 = |q|$ on the used computer with 16 decimal places. Then computing with ϱ_1 led to y equal to 1 (up to rounding errors in the last two decimal places) for $x \leqq 0.5$, and exactly zero for $x > 0.5$.

ϱ_2, $a = 10^{-4}$	N	20	40	80	160	320
	$\bar{a}$	2.021-3	5.300-4	1.593-4	7.608-5	8.403-5
	q	1250	625	312.5	156.25	78.125

b) We consider the following (academic) first kind boundary value problem (1), (22_1) with homogeneous initial data:

$T = 1, c = 1, a = 0$ for $z < 0$, $a = z^2$ for $z \geqq 0$ where

$$z = (x - 0.5)^2 + (t - 0.4)^2 - 0.09,$$

$b = 100\,v^3$ where $v = 0.4 - x$ if $x < 0.4$, $v = 0$ if $x \in [0.4, 0.6]$,
$v = 0.6 - x$ if $x > 0.1$, $d = 100(x + t - 1)^2$ if $x + t > 1$, else $\mathrm{d} = 0$.

The functions $\nu_1(t)$, $\nu_2(t)$ and $f(x, t)$ are chosen so as to fit to $u(x, t) = x(t + x)$.

The steps of the grid are equidistant, $h = 1/N$, and $\tau = 2/N$. The following maximum norm errors e of the difference solution have been found (using ϱ_2):

N	20	40	80	160
e	1.966-1	8.823-2	4.026-2	1.882-2

The accuracy turns out to be of first order. Note that in this example the difference scheme (23) has to be employed for $x \in [0.4, 0.6]$ and for t about 0.4.

c) An example not covered by the Theorem:

All data coinciding with those of b, but $c = 100z^2$, $z = 0$ for $1.1 \leqq x + t \leqq 1.3$, $z = x + t - 1.3$ for $x + t > 1.3$, $z = 1.1 - x - t$ for $x + t < 1.1$. Results:

N	20	40	80	160	320
e	1.337-1	4.493-2	1.800-2	2.952-2	1.632-2

Bibliography

[1] J. H. Seinfeld, M. Koda, Numerical implementation of distributed parameter filters with application to problems in air pollution. *In:* Lecture Notes in Control and Information Sciences 1, Springer-Verlag, Berlin—Heidelberg—New York 1978, 42—69.

[2] K.-F. Busch, L. Luckner, Geohydraulik. VEB Deutscher Verlag für Grundstoffindustrie, Leipzig 1973.

[3] G. Chavent, About the identification and modelling of miscible and immiscible displacements in porous media. *In:* Lecture Notes in Control and Information Sciences 1, Springer-Verlag Berlin—Heidelberg—New York 1978, 196—220.

[4] R. B. Guenther, Some mathematical problems in agriculture. Numerische Behandlung von Differentialgleichungen, Bd. 2. ISNM 31, Birkhäuser-Verlag, Basel und Stuttgart 1976, 65—78.

[5] А. А. Самарский, С. П. Курдюмов, П. П. Волосевич, Бегущие волны в среде с нелинейной теплопроводностью. Ж. вычисл. матем. и матем. физ., **5**, 2 (1965), 199—217.

[6] G. Stoyan, Monotone difference schemes for diffusion-convection problems. Z. Angew. Math. Mech. **59** (1979) 8, 361—372.

[7] G. Stoyan, On a monotone, maximum norm stable, and conservative difference approximation of the one-dimensional diffusion-convection equation. *In:* Proceedings of the Conference "Simulation der Migrationsprozesse im Boden- und Grundwasser" TU Dresden 1979, 139—160.

[8] А. А. Самарский, Теория разностных схем. Наука, Москва 1977.

[9] А. Н. Тихонов, А. А. Самарский, Об однородных разностных схемах. Ж. вычисл. матем. и матем. физ. **1**, 1 (1961), 5—63.

[10] R. Gorenflo, Conservative difference schemes for diffusion problems. Preprint N° 39, FU Berlin (West) 1977.

[11] Г. И. Шишов, В. А. Титов, Разностная схема для дифференциального уравнения с двумя малыми параметрами. Числ. мет. мех. спл. среды **7** (1976), 145—155.

[12] А. М. Ильин, Разностная схема для дифференциального уравнения с малым параметром при старшей производной. Мат. заметки **6** (1969) 2, 237—248.

[13] L. R. Abrahamsson, H. B. Keller, H. O., Kreiss, Difference approximations for singular perturbations of systems of ordinary differential equations. Numer. Math. **22** (1974), 367—391.

[14] K. E. Barrett, The numerical solution of singular-perturbation boundary value problems. Quart. J. Mech. Appl. Math. 27 (1974), 57–68.

[15] В. Я. Гольдин, Н. Н. Калиткин, Т. В. Шишова, Нелинейные разностные схемы для гиперболического уравнения. Ж. вычисл. матем. и матем. физ. 5 (1965) 5, 938–944.

[16] А. А. Самарский, А. В. Гулин, Устойчивость разностных схем. Наука, Москва 1973.

[17] W. Liniger, R. Willoughby, Efficient integration methods for stiff systems of ordinary differential equations. SIAM J. Num. Anal. 7 (1970), 47–66.

[18] B. Wendroff, On centered difference equations for hyperbolic equations. J. SIAM 8 (1960) 3, 549–555.

Address of the author:

Akademie der Wissenschaften der DDR
Institut für Mathematik
1080 Berlin
Mohrenstr. 39
DDR

Two-component flow round a solid sphere

Jurgis Szlaža, Berlin

The processes of mass and heat transfer are important in many practical situations. In order to solve these problems we have to know first of all the velocity distribution of appropriate flows. On the other hand the named processes take often place in the immediate neighbourhood of immersed particles like solid particles, drops, bubbles and so on. Therefore the problem of two-component flow round a solid sphere is of great importance.

Starting from the model idea for multicomponent flows as presented by L. D. Landau [1] and Ch. A. Rachmatulin [2] in this paper at first the solution of the hydrodynamic problem of this motion is given. In this case the formulas for velocities and pressures of the components and of the mixture are derived and a generalization of Stokes formula for the force on a small sphere is found.

In this connection it is also possible to solve the inverse problem, namely to find the mixture viscosity μ_G and the interaction coefficient K between the components.

In the second part of the paper the problem of convective mass transport in a two-component flow round a solid sphere is investigated and a formula for the total diffusion stream on a small sphere is given.

1. The formulation of the problem

We consider a sphere with radius a in a stationary laminar incompressible two-component flow. We assume that the real and reduced densities of both the components are constant, namely $\varrho_i^* = \text{const}$ and $\varrho_i = \text{const}$. Therefore the porosities $\varkappa_i = \frac{\varrho_i}{\varrho_i^*}$ are constant, too. The interaction coefficient K was also assumed to be constant.

Now let us consider a two-component flow with small Reynolds numbers $\text{Re}_i \ll 1$, i.e. the sphere radius a is assumed to be sufficiently small. Then in the equations of motion the inertia terms can be neglected in comparison with the friction and interaction terms. If in Rachmatulin's equations we eliminate the inertia terms and we assume that external forces do not exist the equation system will be given by

$$\text{grad}\, p = \mu_i\, \Delta \boldsymbol{v}_i + \frac{K}{\varkappa_i} \sum_{j=1}^{2} (\boldsymbol{v}_j - \boldsymbol{v}_i), \quad \text{div}\, \boldsymbol{v}_i = 0, \quad i = 1, 2, \tag{1.1}$$

where μ_i, $\boldsymbol{v}$, p are adequate the viscosities, velocities and pressure of both components.

For the complete formulation of the problem we need, besides the equations (1.1), the boundary conditions on the sphere surface and far from the sphere.

For the further investigations we choose the coordinate system so that the coordinate origin is placed at the centre of the sphere and the motion takes place parallel to the x-axis. Because of the form of the regarded body it is suitable to carry out the investigations in spherical coordinates (r, θ, φ). Therefore we have an axial symmetric flow for which

$$\frac{\partial \boldsymbol{v}_i}{\partial \varphi} = 0, \quad \boldsymbol{i}_\varphi \cdot \boldsymbol{v}_i = v_{i\varphi} = 0 \tag{1.2}$$

is valid and for which the x-axis is the axis of symmetry.

As is shown in [3] for such flows the stream functions ψ_i and ψ exist for the single components and for the mixture. With the help of these stream functions the different parameters of the two-component flow can be determined.

Therefore the problem is to solve the equations for the stream functions ψ_i derived from (1.1) considering the appropriate boundary conditions.

2. The velocities of the components and the mixture of the two-component medium

By the definition for the stream function the velocity components are given by

$$v_{ir} = -\frac{1}{r^2 \sin\theta} \frac{\partial \psi_i}{\partial \theta}, \quad v_{i\theta} = \frac{1}{r \sin\theta} \frac{\partial \psi_i}{\partial r}. \tag{2.1}$$

In [3] is shown that the equations of motion (1.1) can be decoupled and regarding the relations (2.1) the equations for the determination of the stream functions become the following:

$$E^4 \chi_1 = 0, \quad E^2[E^2 \chi_2 - \gamma^2 \chi_2] = 0. \tag{2.2}$$

The operator E^2 in (2.2) is given by

$$E^2 = \frac{\partial^2}{\partial r^2} + \frac{1-\eta^2}{r^2} \frac{\partial^2}{\partial \eta^2}, \quad \eta = \cos\theta. \tag{2.3}$$

There are

$$\chi_1 = \varkappa_1 \psi_1 + \varkappa_2 \mu^* \psi_2, \quad \chi_2 = \psi_1 - \psi_2, \tag{2.4}$$

and

$$\mu^* = \frac{\mu_2}{\mu_1}, \quad \gamma^2 = K_1 + K_2, \quad K_i = \frac{K}{\mu_i \varkappa_i}. \tag{2.5}$$

If we now determine the functions χ_i from (2.2), we shall, according to (2.3), find the stream functions ψ_i, namely

$$\psi_1 = \frac{\chi_1 + \varkappa_2 \mu^* \chi_2}{\varkappa_1 + \varkappa_2 \mu^*}, \quad \psi_2 = \frac{\chi_1 - \varkappa_1 \chi_2}{\varkappa_1 + \varkappa_2 \mu^*}. \tag{2.6}$$

With the use of the formula $\psi = \varkappa_1\psi_1 + \varkappa_2\psi_2$ we get the stream function of the mixture

$$\psi = \frac{1}{\varkappa_1 + \varkappa_2\mu^*}\left[\chi_1 + \varkappa_1\varkappa_2(\mu^* - 1)\,\chi_2\right]. \tag{2.7}$$

Considering the relations (2.6) we can compute the velocities of the components according to (2.1). Then the velocities of the mixture are given by the formulas

$$V_{q_k} = \varkappa_1 v_{1q_k} + \varkappa_2 v_{2q_k}, \quad k = 1, 2, \quad q_1 = r, \quad q_2 = \theta. \tag{2.8}$$

As in [3] is shown the general solution for the motion of a spherical particle in a two-component medium can be found by solving the equations (2.2). We then have

$$\chi_1 = \frac{1}{2}\sin^2\theta\,(A_1r^2 + B_1r^{-1} + C_1r^4 + D_1r),$$
$$\chi_2 = \frac{1}{2}\sin^2\theta\left[A_2\left(\gamma - \frac{1}{r}\right)e^{\gamma r} + B_2\left(\gamma + \frac{1}{r}\right)e^{-\gamma r} - \frac{1}{\gamma^2}(C_2r^2 + D_2r^{-1})\right]. \tag{2.9}$$

For the fixation of the boundary conditions we now consider the following problem: A motionless solid sphere is in the flow of a two-component medium the components of which have the velocities $\vec{U}_i = -\vec{\imath}_x U_i$ far from the sphere.

The boundary conditions on the sphere surface are the following

$$\psi_i|_{r=a} = 0, \quad \left.\frac{\partial\psi_i}{\partial r}\right|_{r=a} = 0. \tag{2.10}$$

Because the flow field far from the sphere is unperturbed the following relations must be satisfied there:

$$\psi_i \to \frac{1}{2}r^2U_i\sin^2\theta \quad \text{for} \quad r \to \infty. \tag{2.11}$$

In view of (2.4), (2.10) and (2.11) we obtain the boundary conditions for χ_1 and χ_2

$$\chi_1|_{r=a} = 0, \quad \left.\frac{\partial\chi_1}{\partial r}\right|_{r=a} = 0, \quad \chi_1 \to \frac{1}{2}r^2U^*\sin^2\theta \quad \text{for} \quad r \to \infty, \tag{2.12}$$

and

$$\chi_2|_{r=a} = 0, \quad \left.\frac{\partial\chi_2}{\partial r}\right|_{r=a} = 0, \quad \chi_2 \to \frac{1}{2}r^2\overline{U}\sin^2\theta \quad \text{for} \quad r \to \infty, \tag{2.13}$$

where

$$U^* = \varkappa_1U_1 + \varkappa_2\mu^*U_2, \quad \overline{U} = U_1 - U_2. \tag{2.14}$$

With the use of these boundary conditions we now determine χ_1 and χ_2 from (2.9). As according to (2.12) $\frac{\chi_1}{r_2} \to \frac{1}{2}U^*\sin^2\theta$ for $r \to \infty$ we obtain

$$C_1 = 0, \quad A_1 = U^*$$

immediately from (2.9), so that χ_1 is given by

$$\chi_1 = \frac{1}{2}\sin^2\theta\,(U^*r^2 + B_1r^{-1} + D_1r). \tag{2.15}$$

The conditions (2.12) for $r = a$ supply us with equations for the determination of B_1 and D_1, namely

$$\frac{B_1}{a} + D_1 a = -a^2 U^*, \quad -\frac{B_1}{a^2} + D_1 = -2aU^*,$$

from which we find

$$B_1 = \frac{1}{2} a^3 U^*, \quad D_1 = -\frac{3}{2} aU^*.$$

If we put the obtained coefficients into (2.15), we shall get to the relation

$$\chi_1 = \frac{1}{2} r^2 U^* \sin^2\theta \left[\frac{1}{2}\left(\frac{a}{r}\right)^3 - \frac{3}{2}\left(\frac{a}{r}\right) + 1\right]. \tag{2.16}$$

When we consider the boundary conditions (2.13) for χ_2 we see that $\frac{\chi_2}{r_2} \to \frac{1}{2} \overline{U} \sin^2\theta$ for $r \to \infty$. Because of (2.9) there are then

$$A_2 = 0, \quad C_2 = -\gamma^2 \overline{U},$$

so that we can write for χ_2

$$\chi_2 = \frac{1}{2} \sin^2\theta \left[B_2\left(\gamma + \frac{1}{r}\right) e^{-\gamma r} + \overline{U} r^2 - D_2 \frac{1}{\gamma^2 r}\right]. \tag{2.17}$$

Then the conditions (2.13) for $r = a$ supply us with the following equations for the determination of B_2 and D_2

$$B_2\left(\gamma + \frac{1}{a}\right) e^{-\gamma a} - \frac{1}{\gamma^2}\frac{D_2}{a} = -a_2 \overline{U},$$

$$-B_2\left[\frac{1}{a^2} + \gamma\left(\gamma + \frac{1}{a}\right)\right] e^{-\gamma a} + \frac{1}{\gamma^2}\frac{D_2}{a^2} = -2a\overline{U}.$$

After solving this system we find

$$B_2 = \frac{3a\overline{U}}{\gamma^2} e^{\gamma a}, \quad D_2 = a\overline{U}[\gamma^2 a^2 + 3(\gamma a + 1)].$$

If we put these coefficients into (2.17), we shall obtain, after some transformations, the formula

$$\chi_2 = \frac{1}{2} r^2 \overline{U} \sin^2\theta \left\{1 - \left(\frac{a}{r}\right)^3 \left[1 + \frac{3}{\gamma^2 a^2}\left((\gamma a + 1) - (\gamma r + 1) e^{-\gamma(r-a)}\right)\right]\right\}. \tag{2.18}$$

The χ_1 and χ_2 having been found, the stream functions ψ_1 and ψ_2 can be determined from (2.6) and the velocities of the components can be found from (2.1). We then have

$$\left.\begin{aligned} v_{1r} = -\frac{\cos\theta}{\varkappa_1 + \varkappa_2\mu^*} \Bigg\{&U^*\left[\frac{1}{2}\left(\frac{a}{r}\right)^3 - \frac{3}{2}\left(\frac{a}{r}\right) + 1\right] \\ &+ \varkappa_2\mu^*\overline{U}\left[1 - \left(\frac{a}{r}\right)^3\left[1 + \frac{3}{\gamma^2 a^2}\left((\gamma a + 1) - (\gamma r + 1) e^{-\gamma(r-a)}\right)\right]\right]\Bigg\}, \end{aligned}\right\} \tag{2.19}$$

$$\left.\begin{aligned}
v_{2r} &= -\frac{\cos\theta}{\varkappa_1+\varkappa_2\mu^*}\left\{U^*\left[\frac{1}{2}\left(\frac{a}{r}\right)^3-\frac{3}{2}\left(\frac{a}{r}\right)+1\right]\right.\\
&\quad\left.-\varkappa_1\bar{U}\left[1-\left(\frac{a}{r}\right)^3\left[1+\frac{3}{\gamma^2a^2}\left((\gamma a+1)-(\gamma r+1)\,\mathrm{e}^{-\gamma(r-a)}\right)\right]\right]\right\},\\
v_{1\theta} &= \frac{\sin\theta}{2(\varkappa_1+\varkappa_2\mu^*)}\left\{U^*\left[-\frac{1}{2}\left(\frac{a}{r}\right)^3-\frac{3}{2}\left(\frac{a}{r}\right)+2\right]\right.\\
&\quad\left.+\varkappa_2\mu^*\bar{U}\left[2+\left(\frac{a}{r}\right)^3\left[1+\frac{3}{\gamma^2a^2}\left((\gamma a+1)-(\gamma^2r^2+\gamma r+1)\,\mathrm{e}^{-\gamma(r-a)}\right)\right]\right]\right\},\\
v_{2\theta} &= \frac{\sin\theta}{2(\varkappa_1+\varkappa_2\mu^*)}\left\{U^*\left[-\frac{1}{2}\left(\frac{a}{r}\right)^3-\frac{3}{2}\left(\frac{a}{r}\right)+2\right]\right.\\
&\quad\left.-\varkappa_1\bar{U}\left[2+\left(\frac{a}{r}\right)^3\left[1+\frac{3}{\gamma^2a^2}\left((\gamma a+1)-(\gamma^2r^2+\gamma r+1)\,\mathrm{e}^{-\gamma(r-a)}\right)\right]\right]\right\}.
\end{aligned}\right\}\qquad(2.19)$$

In view of these expressions we can calculate the velocities of the mixture from (2.8), namely

$$\left.\begin{aligned}
v_r &= -\frac{\cos\theta}{\varkappa_1+\varkappa_2\mu^*}\left\{U^*\left[\frac{1}{2}\left(\frac{a}{r}\right)^3-\frac{3}{2}\left(\frac{a}{r}\right)+1\right]\right.\\
&\quad\left.+\varkappa_1\varkappa_2(\mu^*-1)\,\bar{U}\left[1-\left(\frac{a}{r}\right)^3\left[1+\frac{3}{\gamma^2a^2}\left((\gamma a+1)-(\gamma r+1)\,\mathrm{e}^{-\gamma(r-a)}\right)\right]\right]\right\},\\
v_\theta &= \frac{\sin\theta}{2(\varkappa_1+\varkappa_2\mu^*)}\left\{U^*\left[-\frac{1}{2}\left(\frac{a}{r}\right)^3-\frac{3}{2}\left(\frac{a}{r}\right)+2\right]\right.\\
&\quad+\varkappa_1\varkappa_2(\mu^*-1)\,\bar{U}\left[2+\left(\frac{a}{r}\right)^3\right.\\
&\quad\left.\left.\times\left[1+\frac{3}{\gamma^2a^2}\left((\gamma a+1)-(\gamma^2r^2+\gamma r+1)\,\mathrm{e}^{-\gamma(r-a)}\right)\right]\right]\right\}.
\end{aligned}\right\}\qquad(2.20)$$

If we put $U_1 = U_2$, $\mu_1 = \mu_2$ and $\varrho_1^* = \varrho_2^*$ into (2.19) and (2.20), i.e. if we have a quasi-homogeneous flow, the velocities shall be indepedent on the interaction between the components. In this case we get formulas which were obtained for a homogeneous medium [4].

3. The generalization of Stokes formula for the drag force of the small sphere and the determination of μ_G, K and the pressure of the two-component medium

For the case of an axial symmetric flow J. Happel and H. Brenner [4] have shown that the force acting on a small particle can be computed by

$$F_1 = \pi\mu_G\int\xi^3\frac{\partial}{\partial n}\left(\frac{E^2\psi}{\xi^2}\right)ds. \qquad (3.1)$$

For a small sphere this expression becomes

$$F_1 = \pi\mu_G \int_0^\pi a^3 \sin^3\theta \frac{\partial}{\partial r}\left(\frac{E^2\psi}{r^2\sin^2\theta}\right)\Bigg|_{r=a} a\, d\theta, \tag{3.2}$$

where μ_G is the mixture viscosity. On the other hand in [3] is explained that the force acting on a small particle for axial symmetric flows of a two-component medium can be found by the formula

$$F_2 = \pi\mu_1 \int \xi^3 \frac{\partial}{\partial n}\left(\frac{E^2\chi_1}{\xi^2}\right) ds \tag{3.3}$$

which for a small sphere is expressed by

$$F_2 = \pi\mu_1 \int_0^\pi a^3 \sin^3\theta \frac{\partial}{\partial r}\left(\frac{E^2\chi_1}{r^2\sin^2\theta}\right)\Bigg|_{r=a} a\, d\theta. \tag{3.4}$$

If we now consider our problem as a flow of a homogeneous medium round the sphere, we shall obtain, according to Stokes, the following formula of force on the sphere:

$$F_3 = -6\pi a\mu_G U_G, \tag{3.5}$$

where $U_G = \varkappa_1 U_1 + \varkappa_2 U_2$.

In view of (2.7), (2.16) and (2.18) we see that the expression for F_2 includes only the parameters of both the components, but F_1 additionally depends on μ_G and K and F_3 on μ_G. The mixture viscosity μ_G and the interaction coefficient K are not known. Since all three formulas (3.2), (3.4) and (3.5) represent the force on the sphere for the flow of the two-component medium, they have to be $F_1 = F_2 = F_3$. When we set $F_1 = F_3$ then we get the equation for the determination of K. The relation $F_2 = F_3$ gives us an equation for the compution of μ_G. If we put μ_G and K obtained in this way into (3.2), the relation $F_1 = F_2$ must be satisfied.

Now we compute the force F_2 according to (3.4). This force represents the generalization of the Stokes formula of the force on the sphere for the case of two-component flow.

Applying the operator E^2 according to (2.3) to (2.16) we obtain

$$E^2\chi_1 = \frac{3}{2} a U^* \frac{\sin^2\theta}{r}. \tag{3.6}$$

When we set this expression into (3.4) and we then integrate, the drag force of the sphere is given by

$$F = -6\pi a\mu_1(\varkappa_1 U_1 + \varkappa_2\mu^* U_2), \tag{3.7}$$

where it is assumed $F_2 = F$.

Now we compute F_1 from (3.2). According to (2.7) there is

$$\psi = \alpha\chi_1 + \beta\chi_2, \tag{3.8}$$

where

$$\alpha = \frac{1}{\varkappa_1 + \varkappa_2\mu^*}, \quad \beta = \frac{\varkappa_1\varkappa_2(\mu^* - 1)}{\varkappa_1 + \varkappa_2\mu^*}. \tag{3.9}$$

Applying the operator E^2 to (3.8) we obtain

$$E^2\psi = \alpha E^2\chi_1 + \beta E^2\chi_2. \tag{3.10}$$

With the use of (3.6) we can determine $E^2\chi_1$.

Applying (2.3) to (2.18) we find

$$E^2\chi_2 = \frac{3}{2}\, a\bar{U} \sin^2\theta \left(\gamma + \frac{1}{r}\right) e^{-\gamma(r-a)}. \tag{3.11}$$

The formulas (3.6) and (3.11) put into (3.10) give us

$$E^2\psi = \frac{3}{2}\, a \sin^2\theta \left[\alpha \frac{U^*}{r} + \beta\bar{U} \left(\gamma + \frac{1}{r}\right) e^{-\gamma(r-a)}\right]. \tag{3.12}$$

In view of this expression in (3.2) we obtain, after integration, F_1 as follows:

$$F_1 = -\frac{6\pi a\mu_G}{\varkappa_1 + \varkappa_2\mu^*}\left[(\varkappa_1 U_1 + \varkappa_2\mu^* U_2) + \frac{1}{3}\varkappa_1\varkappa_2(\mu^* - 1) \times (U_1 - U_2)(\gamma^2 a^2 + 3\gamma a + 3)\right], \tag{3.13}$$

where we have used (2.14) and (3.9).

The forces F_1 and F_2 having been found with the use of (3.13) and (3.7), we can now determine μ_G and K near the sphere. When we equate (3.5) and (3.7), i.e. $F_3 = F_2$, we obtain the equation for the determination of μ_G, namely

$$\mu_G U_G = \mu_1(\varkappa_1 U_1 + \varkappa_2\mu^* U_2),$$

thus is

$$\mu_G = \frac{\mu_1(\varkappa_1 U_1 + \varkappa_2\mu^* U_2)}{\varkappa_1 U_1 + \varkappa_2 U_2}. \tag{3.14}$$

When we set $F_1 = F_3$ according to (3.13) and (3.5), we get the relation

$$(\varkappa_1 U_1 + \varkappa_2 U_2)(\varkappa_1 + \varkappa_2\mu^*) - (\varkappa_1 U_1 + \varkappa_2\mu^* U_2) = \frac{1}{3}\varkappa_1\varkappa_2(\mu^* - 1)(U_1 - U_2)(\gamma^2 a^2 + 3\gamma a + 3)$$

from which we can find the interaction coefficient K. Solving this equation we obtain

$$\gamma a = -3. \tag{3.15}$$

According to (2.5) we see that

$$\gamma^2 = K\left(\frac{1}{\varkappa_1\mu_1} + \frac{1}{\varkappa_2\mu_2}\right)$$

and (3.15) then supplies us with a formula for K, namely

$$K = \frac{9}{a^2} \cdot \frac{\varkappa_1\mu_1\varkappa_2\mu_2}{\varkappa_1\mu_1 + \varkappa_2\mu_2}. \tag{3.16}$$

When we put the obtained expressions of μ_G and K, i.e. (3.14) and (3.16), into (3.13), we see that (3.7) and (3.13) are equal and thus is $F_1 = F_2$.

The found results show that μ_G and K can be computed with the use of the known parameters of both components of the medium. For the determination of the pressure p of the two-component medium we have according to [3] the formulas

$$\frac{\partial p}{\partial r} = -\frac{\mu_1}{r^2 \sin^2 \theta} \frac{\partial}{\partial \theta} (E^2 \chi_1),$$

$$\frac{\partial p}{\partial \theta} = \frac{\mu_1}{\sin \theta} \frac{\partial}{\partial r} (E^2 \chi_1).$$

We put in here $E^2\chi_1$ from (3.6) and get

$$\begin{aligned} dp &= \frac{\partial p}{\partial r} dr + \frac{\partial p}{\partial \theta} d\theta \\ &= -\frac{3}{2} aU^* \mu_1 \left(\frac{2 \cos \theta}{r^3} dr + \frac{\sin \theta}{r^2} d\theta \right) \\ &= \frac{3}{2} aU^* \mu_1 \, d \left(\frac{\cos \theta}{r^2} \right) \end{aligned}$$

from which follows

$$p = p_\infty + \frac{3}{2} \mu_1 a U^* \frac{\cos \theta}{r^2} \tag{3.17}$$

Here p_∞ is the pressure of the two-component medium far from the sphere.

4. The convective mass transport in a two-component flow round a solid sphere

Let us consider the diffusion process of a substance immersed in a two-component medium on a solid sphere. The appropriate hydrodynamic problem of the two-component flow round a solid sphere has been solved in the previous items of the paper. Starting from these results we can now deal with the solution of the diffusion problem.

We consider the following problem:

The two-component medium includes a little admixture which, in contact with the solid sphere, leads to chemical or physico-chemical changes. We search for the total diffusion stream to the solid spherical particle.

The concentration c of the admixture may be small so that the diffusion coefficient D can be assumed as constant. The Reynolds numbers Re_i have been supposed to be small. As we can see from the solution of the hydrodynamic problem the velocities v_θ and v_r decrease continuously with the distance from the particle surface and in the immediate neighbourhood of the solid sphere a hydrodynamic boundary layer does not exist. In spite of that however a diffusion boundary layer develops near the particle surface. It is connected with the fact that the appropriate Peclét numbers $\mathrm{Pe}_i = \dfrac{U_i a}{D}$

are many times greater than the Reynolds numbers Re_i. Thus there are $Re_i \ll 1$ and $Pe_i \gg 1$ simultaneously. As known for $Pe \gg 1$ the essential diffusion processes take place quite near the reaction surface. Here in the diffusion boundary layer we can observe a clear change of the concentration of the admixture.

The equation of the convective diffusion in the diffusion boundary layer in spherical coordinates is given by

$$v_r \frac{\partial c}{\partial r} + \frac{v_\theta}{r} \frac{\partial c}{\partial \theta} = D \left(\frac{\partial^2 c}{\partial r^2} + \frac{2}{r} \frac{\partial c}{\partial r} \right). \tag{4.1}$$

Here on the right side the term $\frac{1}{r^2 \sin \theta} \frac{\partial}{\partial \theta} \left(\sin \theta \frac{\partial c}{\partial \theta} \right)$ is omitted, because the derivatives along the sphere surface compared with the derivatives with respect to the radius vector are small. We can take the velocities v_r and v_θ according to (2.20).

For solving (4.1) we still need the boundary conditions. Since the diffusion process takes place in the immediate neighbourhood of the solid spherical particle, the condition far from the sphere is

$$c = c_0 \quad \text{for} \quad r \to \infty. \tag{4.2}$$

Now we consider such a diffusion process for which the condition

$$c = 0 \quad \text{for} \quad r = a \tag{4.3}$$

is satisfied on the sphere surface. That means that the admixture contacted with the particles reacts immediately with them and therefore the greatest possible diffusion stream is guaranteed. At last we assume that in the stagnation point singularities do not exist, i.e. it is

$$c = c_0 \quad \text{in the stagnation point.} \tag{4.4}$$

The problem now consists in solving the equation (4.1) with the boundary conditions (4.2)—(4.4) and in the determination of an expression of the total diffusion stream.

On account of the equation of continuity it is

$$v_r = -\frac{1}{r^2 \sin \theta} \frac{\partial \psi}{\partial \theta}, \quad v_\theta = \frac{1}{r \sin \theta} \frac{\partial \psi}{\partial r}, \tag{4.5}$$

where ψ is the stream function,

$$\psi = \frac{1}{\varkappa_1 + \varkappa_2 \mu^*} \left[\chi_1 + \varkappa_1 \varkappa_2 (\mu^* - 1) \chi_2 \right], \tag{4.6}$$

and χ_1, χ_2 can be found according to (2.16) and (2.18). Because the essential diffusion processes take place in the diffusion boundary layer, only solutions of the equation (4.1) for such values of r are interesting for us, which differ little from the sphere radius a. We set $x = r - a$. In view of (2.16) and (2.18) we then obtain for small x

$$\chi_1 = \frac{3}{4} U^* \sin^2 \theta \, x^2, \quad \chi_2 = \frac{3}{4} \bar{U} \sin^2 \theta \, (\gamma a + 1) \, x^2, \tag{4.7}$$

and according to (4.6) the stream function ψ is given by

$$\psi = \frac{3}{4} \sin^2 \theta\, x^2[\alpha U^* + \beta \overline{U}(\gamma a + 1)] \tag{4.8}$$

where

$$\alpha = \frac{1}{\varkappa_1 + \varkappa_2 \mu^*}, \quad \beta = \frac{\varkappa_1 \varkappa_2(\mu^* - 1)}{\varkappa_1 + \varkappa_2 \mu^*}.$$

When we change the variables θ, r of the equation (4.1) to the variables θ, ψ and we regard $\frac{\partial^2 c}{\partial x^2} \gg \frac{2}{a} \frac{\partial c}{\partial x}$ for $x \ll a$, we obtain the following equation at the diffusion boundary layer:

$$\frac{\partial c}{\partial \theta} = Da^2 \sin^2 \theta \frac{\partial}{\partial \psi} \left(a v_\theta \frac{\partial c}{\partial \psi} \right). \tag{4.9}$$

From (4.5) and (4.8) can be found

$$v_\theta = \frac{3}{2}\, x \frac{\sin \theta}{a} [\alpha U^* + \beta \overline{U}(\gamma a + 1)]. \tag{4.10}$$

Putting (4.10) into (4.9) and carrying out the substitution

$$\xi = Da^2 \sqrt{3[\alpha U^* + \beta \overline{U}(\gamma a + 1)]} \int \sin^2 \theta \, d\theta \tag{4.11}$$

we get the equation

$$\frac{\partial c}{\partial \xi} = \frac{\partial}{\partial \psi} \left(\sqrt{\psi} \frac{\partial c}{\partial \psi} \right). \tag{4.12}$$

Instead of (4.2)—(4.4) we now have the following boundary conditions:

On the particle surface

$$c = 0 \quad \text{for} \quad \psi = 0; \tag{4.13}$$

far from the particle

$$c = c_0 \quad \text{for} \quad \psi \to \infty; \tag{4.14}$$

at the stagnation point

$$c = c_0 \quad \text{for} \quad \theta = 0, \quad \psi = 0. \tag{4.15}$$

The solution of the equation (4.12) in view of the boundary conditions (4.13)—(4.15) was found in [5]. Therefore we can write at once

$$c = \frac{c_0}{1{,}15} \int_0^\eta \exp \left(-\frac{4}{9} \eta^3 \right) d\eta \tag{4.16}$$

where

$$\eta = \sqrt[3]{\frac{3[\alpha U^* + \beta \overline{U}(\gamma a + 1)}{4Da^2}} \cdot \frac{x \sin \theta}{\left(\theta - \frac{\sin 2\theta}{2} \right)^{1/3}}. \tag{4.17}$$

The diffusion stream on the particle surface can be determined by

$$j = D \frac{\partial c}{\partial x}\bigg|_{x=0}.$$

According to (4.16) and (4.17) we obtain

$$j = D \frac{c_0}{1{,}15} \sqrt[3]{\frac{3[\alpha U^* + \beta \overline{U}(\gamma a + 1)]}{4Da^2}} \frac{\sin\theta}{\left(\theta - \frac{\sin 2\theta}{2}\right)^{1/3}}. \tag{4.18}$$

The thickness of the diffusion boundary layer can be computed by

$$\delta = \frac{Dc_0}{j} = 1{,}15 \sqrt[3]{\frac{4Da^2}{3[\alpha U^* + \beta \overline{U}(\gamma a + 1)]}} \frac{\left(\theta - \frac{\sin 2\theta}{2}\right)^{1/3}}{\sin\theta}. \tag{4.19}$$

The total diffusion of the admixture on the solid sphere can be found by

$$J = \int j\, ds = 2\pi a^2 \int_0^\pi j \sin\theta\, d\theta.$$

In view of (4.18) we then obtain

$$J = 7{,}98 \cdot c_0 D^{2/3} a^{4/3} [\alpha U^* + \beta \overline{U}(\gamma a + 1)]^{1/3}. \tag{4.20}$$

This formula includes the interaction coefficient K. In the item 3 we have determined K in the immediate neighbourhood of the particle. When we set, according to (3.16), the obtained expression for K into (4.20), we then get

$$J = 7{,}98 \cdot c_0 D^{2/3} a^{4/3} \left\{\frac{1}{\varkappa_1 + \varkappa_2 \mu^*} [\varkappa_1 U_1 + \varkappa_2 \mu^* U_2 - 2\varkappa_1 \varkappa_2 (\mu^* - 1)(U_1 - U_2)]\right\}^{1/3} \tag{4.21}$$

in dependence on the parameters of the two-component medium. The function of θ from the formula (4.18) has, at the point $\theta = 0$, the value 1, at the point $\theta = \frac{\pi}{2}$ the value $\sqrt[3]{\frac{2}{\pi}}$ and for $\theta = \pi$ the value 0. Thus the diffusion stream has the greatest value at the stagnation point $\theta = 0$, it decreases with increasing θ.

The thickness δ of the diffusion boundary layer (4.19) becomes greater with increasing θ and goes to infinity for $\theta = \pi$. At the beginning we have assumed the thickness of the diffusion boundary layer to be much smaller than the particle radius. Thus we can conclude that beginning from some values θ placed near $\theta = \pi$, the considered theory is no longer applicable. Apart from that the domain $\theta \sim \pi$ however has little influence on the total stream J of the admixture.

In the end it may be remarked that the diffusion stream j, the thickness of the diffusion boundary layer δ and the total diffusion stream J of the admixture on the particle depend on the parameters of the components of the two-component medium. Already for a quasihomogeneous flow, i.e. when $U_1 = U_2$, the obtained formulas (4.18), (4.19) and (4.21) are converted into the formulas of a homogeneous flow [5].

In other words, when the velocities of both the components are equal far from the solid sphere, the interaction between the components is not important for solving the diffusion problem and we can immediately take the known formulas for a homogeneous medium.

References

[1] Л. Д. Ландау, Е. М. Лифшиц, Механика сплошных сред. ГИТТЛ, Москва 1953.
[2] Х. А. Рахматулин, Основы гидродинамики взаимопроникаюших движений сжимаемых сред. Приклад. мат. мех. **20** (1956), 2.
[3] Szlaža, J., Axialsymmetrische Strömungen von Zweikomponentenmedien bei kleinen Reynoldsschen Zahlen. Z. Angew. Math. Mech. (to appear).
[4] J. Happel, H. Brenner, Low Reynolds number hydroynamics. Prentice-Hall, New York 1965.
[5] В. Г. Левич, Физико-химическая гидродинамика. Физматгиз, Москва 1959.

Address of the author:

Akademie der Wissenschaften der DDR
Forschungsbereich Mathematik/Kybernetik
1199 Berlin
Rudower Chaussee 5
DDR

To the mathematical treatment of a technical heat control problem

Friedmar Unger, Freiberg

Abstract. In the paper first a technical heat control problem and the essential agreements of the corresponding mathematical model are formulated. Then some mathematical problems appearing at the treatment of the technical problem are discussed.

I. The technical problem

The object under consideration is a working roll in the stationary cold strip rolling process. The principle of the technical arrangement is to be seen in fig. 1. It includes especially a cooling system, which enables a thermal influence on the roll applying a cooling medium. The intensity of the cooling is assumed to be capable of being regulated for every $z \in [-a, a]$.

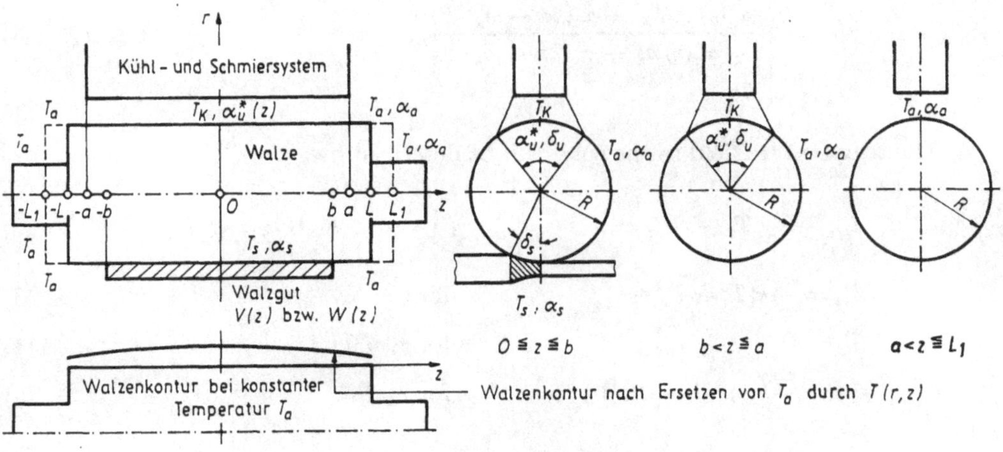

Fig. 1

To solve are now the following two problems:

Searched is a distribution of coolings intensity such that there will be realized

1. a given temperature distribution on the roll shell or
2. a certain wished contour of the roll shell.

II. To the mathematical model

The mathematical model consists essentially in the following agreements:

1. The roll is represented by a solid, which at room temperature T_a has the form of a straight circular cylinder with the modified length $2L_1 \geqq 2L$ and with the radius R.

2. The temperature in the solid always is symmetric with respect to the plane $z = 0$ (see fig. 1); there is only heat conduction; the solid is homogeneous, isotropic and includes no inner heat sources; considered is the part of the solid characterised by $z \geqq 0$.

3. The influence of the cooling medium to the solid is discribed in the following way: To the part of solids surface, which is characterised by a and δ_u (see fig. 1), a temperature T_K is joining; the coefficient of heat transmission there is a function of z: $\alpha_u^*(z)$ ($z \in [0, a]$).

4. To the part of solids surface fixed by b and δ_s (see fig. 1) is joining the roll gap temperature T_s; the coefficient of heat transmission between roll and the material being rolled is α_s. To all other parts of solids surface is joining the room temperature T_a (coefficient of heat transmission α_a).

5. The thermal influences called in 3. and 4. are replaced by the following mean influence to the whole circumferential surface of the solid:

$$\tilde{\tilde{u}}(z) = \begin{cases} \dfrac{\alpha_u^*(z)\,\delta_u T_K + \alpha_s\delta_s T_s + \alpha_a(2\pi - \delta_u - \delta_s)\,T_a}{\alpha_u^*(z)\,\delta_u + \alpha_s\delta_s + \alpha_a(2\pi - \delta_u - \delta_s)}, & 0 \leqq z \leqq b, \\ \dfrac{\alpha_u^*(z)\,\delta_u T_K + \alpha_a(2\pi - \delta_u)\,T_a}{\alpha_u^*(z)\,\delta_u + \alpha_a(2\pi - \delta_u)}, & b < z \leqq a, \\ T_a, & a < z \leqq L_1. \end{cases} \tag{1}$$

6. The temperature field in the solid will be described by

$$T_{rr} + \frac{1}{r}\,T_r + T_{zz} = 0 \qquad (0 < r < R,\ 0 < z < L_1), \tag{2}$$

$$T_z = -k[T - T_a] \qquad \text{for } z = L_1, \tag{3}$$

$$T_z = 0 \qquad \text{for } z = 0, \tag{4}$$

$$T_r = -\tilde{\tilde{h}}(z)\,[T - \tilde{\tilde{u}}(z)] \qquad \text{for } r = R \tag{5}$$

with

$$k = \frac{\alpha_a}{\lambda} \quad (\lambda\text{: coefficient of thermal conductivity}), \tag{6}$$

$$\tilde{\tilde{h}}(z) = \frac{1}{2\pi\lambda} \begin{cases} \delta_u\alpha_u^*(z) + \delta_s\alpha_s + (2\pi - \delta_u - \delta_s)\,\alpha_a, & 0 \leqq z \leqq b, \\ \delta_u\alpha_u^*(z) + (2\pi - \delta_u)\,\alpha_a, & b < z \leqq a, \\ 2\pi\alpha_a, & a < z \leqq L_1. \end{cases} \tag{7}$$

7. The displacement in r-direction of the points lying on the circumferential surface of the solid, which appears, if the temperature field $T \equiv T_a$ will be replaced by $T = T(r, z)$, is described approximately by

$$V_s(z) = \frac{2\beta}{R} \int_0^R r[T(r, z) - T_a]\, dr \tag{8}$$

(β: linear coefficient of thermal expansion).

Formula (8) we get by integration of the thermoelastic displacement equations in the stationary case assuming plane stress in the solid.

8. The technical problem can be said to be solved if

a. to a given temperature distribution $T(R, z) = T_0(z)$ on the circumferential surface has been found an $\alpha_u^*(z) \geqq 0$ such that this $\alpha_u^*(z)$ just generates the wished $T_0(z)$:

$$T[\alpha_u^*]\,(R, z) = T_0(z), \tag{9}$$

and if

b. to a given contour displacement $V_s(z) = g(z)$ ($z \in [0, L]$) has been found an $\alpha_u^*(z) \geqq 0$ such that under the influence of the cooling medium distribution, represented by $\alpha_u^*(z)$, just the given $g(z)$ will be realized:

$$V_s[\alpha_u^*]\,(z) = g(z), \tag{10}$$

respectively.

If $T_0(z)$ and $g(z)$, respectively, are not attainable, then we have to construct a control function $\alpha_u^*(z)$ such that the given functions will be approximated "optimally".

III. Mathematical problems

1. Determination of the heat transmission function $\alpha_u^*(z) \geq 0$ such that (9) is valid

a. The case $a = L = L_1$:

We assume, that $T_0(z)$ is given such that $T_K < T_0(z) < T_s$ and $\alpha_u^*(z)$, obtained by the following method, ist nonnegative. Denoting the right hand side in (5) by

$$f(z) = -\tilde{\tilde{h}}(z)\,[T(R, z) - \tilde{u}(z)] \tag{11}$$

we get, if $f(z) \in L_2[0, L_1]$, the solution of the problem (2), (3), (4) and

$$T_r = f(z) \qquad \text{for } r = R \tag{12}$$

in the form

$$T(r, z) = T_a + \sum_{m=1}^{\infty} \frac{1}{\varrho_m I_1(\varrho_m R)} f_m I_0(\varrho_m r)\, \varphi_m(z) \tag{13}$$

where is

$\{\varphi_m\}_{m=1,2,\dots}$ the eigenfunctions system of the Sturm-Liouville problem

$$\varphi'' + \varrho\varphi = 0, \qquad \varphi'(0) = 0, \qquad \varphi'(L_1) + k\varphi(L_1) = 0, \tag{14}$$

$\{\varrho_m\}_{m=1,2,\dots}$ the eigenvalues sequence of problem (14), arranged in ascending order of magnitude and

$\{f_m\}_{m=1,2,\ldots}$ the Fourier-coefficients sequence of $f(z)$ with respect to $\{\varphi_m\}_{m=1,2,\ldots}$ in $L_2[0, L_1]$.

Using (13) we get

$$T(R, z) - T_a = T_0(z) - T_a = \sum_{m=1}^{\infty} T_{0_m} \varphi_m(z) = \sum_{m=1}^{\infty} \frac{I_0(\varrho_m R)}{\varrho_m I_1(\varrho_m R)} f_m \varphi_m(z) \tag{15}$$

with

$$T_{0_m} = \int_0^{L_1} \left(T_0(z) - T_a\right) \varphi_m(z)\, dz \qquad (m = 1, 2, \ldots), \tag{16}$$

from which follows

$$f_m = \frac{\varrho_m I_1(\varrho_m R)}{I_0(\varrho_m R)} T_{0_m} \qquad (m = 1, 2, \ldots). \tag{17}$$

If $\{T_{0_m}\}_{m=1,2,\ldots}$ sufficiently strongly converges to 0, such that is valid

$$\sum_{m=1}^{\infty} f_m^2 < \infty, \tag{18}$$

$f(z)$ is also determined.

From (11), (7) and (1) the wanted $\alpha_u^*(z)$ can be calculated:

$$\begin{aligned}\alpha_u^*(z) = {} & \frac{1}{\delta_u[T_K - T_0(z)]} \{2\pi\lambda f(z) - \alpha_s\delta_s[T_s - T_0(z)] \\ & - \alpha_a(2\pi - \delta_u - \delta_s)\,[T_a - T_0(z)]\} \qquad \text{if } z \in [0, b],\end{aligned} \tag{19a}$$

$$\alpha_u^*(z) = \frac{1}{\delta_u[T_K - T_0(z)]} \{2\pi\lambda f(z) - \alpha_a(2\pi - \delta_u)\,[T_a - T_0(z)]\} \qquad \text{if } z \in (b, a]. \tag{19b}$$

The condition (18) expresses the incorrectness of the problem. In the numerical computation is worked with favourable truncated Fourier-series, what is corresponding to a regularisation of the problem.

b. The case $a = L < L_1$:

We continue $T_0(z)$, defined on $[0, L]$, to the intervall $(L, L_1]$:

$$T_{0f}(z) = \begin{cases} T_0(z) & z \in [0, L], \\ \dot{T}_0(z) & z \in (L, L_1]. \end{cases} \tag{20}$$

Here is $\dot{T}_0(z) = T(R, z)$ $(z \in (L, L_1])$, whereby $T(r, z)$ is determined by

$$T_{rr} + \frac{1}{r} T_r + T_{zz} = 0 \qquad (0 < r < R, 0 < z < L_1), \tag{21}$$

$$T_z = -k[T - T_a] \qquad \text{for } z = L_1, \tag{22}$$

$$T_z = 0 \qquad \text{for } z = 0, \tag{23}$$

$$T(R, z) = T_0(z) \qquad \text{for } 0 \leqq z \leqq L, \tag{24}$$

$$T_r = -\frac{\alpha_a}{\lambda} [T - T_a] \qquad \text{for } r = R \quad \text{and} \quad L < z \leqq L_1. \tag{25}$$

With this $T_{0f}(z)$ can be obtained as under 1a. a corresponding sequence $\{f_m\}_{m=1,2,\ldots}$. But now we can not find directly $\alpha_u^*(z)$ using $f(z)$, because the control only on $[0, a]$ is possible. But with

$$f_m = \int_0^{L_1} f(z)\, \varphi_m(z)\, dz \qquad (m = 1, 2, \ldots) \tag{26}$$

and (11) is valid

$$-\int_0^{L_1} \tilde{\tilde{h}}(z)\, [T_0(z) - \tilde{u}(z)]\, \varphi_m(z)\, dz = f_m \qquad (m = 1, 2, \ldots). \tag{27}$$

From (27) follows, if $\tilde{\tilde{h}}(z)$ and $\tilde{u}(z)$ are put according to (7) and (1), that the wanted $\varkappa_u^*(z) \geqq 0$ is solution of an infinite problem of moments

$$\int_0^{a} \varkappa_m(z)\, \alpha_u^*(z)\, dz = d_m \qquad (m = 1, 2, \ldots); \tag{28}$$

$\varkappa_m$ and d_m $(m = 1, 2, \ldots)$ are computable.

c. The case $a < L \leqq L_1$:

Here in general exists no solution, because the conditions (21), (22), (23) together with

$$T(R, z) = T_0(z) \qquad \text{for } 0 \leqq z \leqq L \tag{29}$$

and

$$T_r = -\frac{\alpha_a}{\lambda}\, [T - T_a] \qquad \text{for } r = R \text{ and } a \leqq z \leqq L_1 \tag{30}$$

with $a < L$ only can be fulfilled if just $T_0(z) = T^*(R, z)$ $(z \in [0, L])$, where $T^*(r, z)$ is solution of the problem (2)—(5) with certain $\alpha_u^*(z) \geqq 0$.

If $T_0(z)$ has not this property, arises (under additional assumptions) the following problem:

Wanted is $\alpha_u^*(z) \geqq 0$ $(z \in [0, a])$ such that

$$\|T[\alpha_u^*]\,(R, z) - T_0(z)\|^2_{L^2[0, L]} \to \min. \tag{31}$$

Remark. A problem of the same structure arises also in III 1a, if the there assumed properties of $T_0(z)$ are not fulfilled.

2. Determination of the heat transmission function $\alpha_u^*(z)$ such that (10) is valid

a. The case $a = L = L_1$:

Using $T(r, z)$ in the form (13) we have to find $\alpha_u^*(z)$ $(z \in [0, a])$ such that

$$V_s(z) = \frac{2\beta}{R} \int_0^{R} r \sum_{m=1}^{\infty} \frac{1}{\varrho_m I_1(\varrho_m R)}\, f_m I_0(\varrho_m r)\, \varphi_m(z)\, dr = g(z) \tag{32}$$

($z \in [0, L]$). If $g(z) \in L_2[0, L]$ is developed with respect to $\{\varphi_m\}_{m=1,2,...}$:

$$g(z) \sum_{m=1}^{\infty} g_m \varphi_m(z) \tag{33}$$

with

$$g_m = \int_0^{L_1} g(z)\, \varphi_m(z)\, dz \qquad (m = 1, 2, \ldots), \tag{34}$$

then follows by (allowed) link by link integration in (32) and comparison of coefficients

$$f_m = \frac{\varrho_m^2 g_m}{2\beta} \qquad (m = 1, 2, \ldots). \tag{35}$$

If $g(z)$ has the property

$$\sum_{m=1}^{\infty} f_m^2 < \infty \tag{36}$$

(this problem is also incorrect), then we can get $T(R, z)$ using $\{f_m\}_{m=1,2,...}$ and (13) and $f(z)$ by

$$f(z) = \sum_{m=1}^{\infty} f_m \varphi_m(z). \tag{37}$$

From (11), (7) and (1) follows again a formula for the wanted $\alpha_u^*(z)$, analoguously to (19).

Remark. We can write $V_s(z)$ in the form

$$V_s(z) = \varDelta R + V_s^{**}(z) \qquad (z \in [0, L]) \tag{38}$$

with $\varDelta R = V_s(L)$ and therefore $V_s^{**}(L) = 0$. Only $V_s^{**}(z)$ influences the flatness of the material being rolled. Therefore it is obvious to give $g(z)$ in the form

$$g(z) = \varDelta R + g^{**}(z) \qquad (z \in [0, L]) \tag{39}$$

with $g^{**}(L) = 0$ and $\varDelta R = \text{const.}$

If $\alpha_{u\varDelta R}^*(z)$ represents a distribution of the cooling medium which together with T_s and T_a induces the contour displacement $g(z) = \varDelta R + g^{**}(z)$ (g^{**} fixed), then we choose $\varDelta R$ such that $g(z)$ is realized with minimal cooling expense:

$$\int_0^a \left(\alpha_{u\varDelta R}^*(z)\, [T_K - T(R, z)]\right)^2 dz \to \min \tag{40}$$

under consideration of the restriction

$$0 \leqq \varDelta R_1 \leqq \varDelta R \leqq \varDelta R_2 \tag{41}$$

($\varDelta R_1$, $\varDelta R_2$ certain given constants).

b. The case $a \leqq L \leqq L_1$:

The treatment of the case $a = L < L_1$ takes place analoguously to section III 1b, certainly the condition (24) is to be replaced by

$$\frac{2\beta}{R}\int_0^R \left(T(r, z) - T_a\right) dr = g(z) \qquad \text{for} \quad 0 \leqq z \leqq L. \tag{42}$$

The case $a < L \leqq L_1$ leeds (under additional assumptions) to the optimization problem

$$\|V_s[\alpha_u^*](z) - g(z)\|_{L_2[0,L]}^2 \to \min \tag{43}$$

with

$$\alpha_u^*(z) \geqq 0 \quad \text{for} \quad z \in [0, a]. \tag{44}$$

3. Evaluation of $T(r, z)$ from (2)—(5)

The explicite evaluation of $T(r, z)$ is not necessary for solving the called inverse problems, but it can be used to check the results and can be realized for instance reducing it to an infinite linear equations system.

IV. To the instationary case

Starting from the notion, that the considered solid at the time $t = 0$ is a straight circular cylinder with the constant temperature $T \equiv T_a$, the following questions arise:

1. How to choose the function of heat transmission $\alpha_u^*(z, t)$ representing the intensity distribution of the cooling in time lapse such that a given temporal development of the circumferential surface temperature distribution of the solid will be realized?

2. Considering the instationary case the problem is important to choose $\alpha_u^*(z, t)$ such that the stress in φ-direction on the circumferential surface of the solid $\sigma_{\varphi\varphi}(R, z, t)$ in time lapse has a wanted progress.

3. As above we can looke again for a $\alpha_u^*(z, t)$ such that a given contour of the solid in time lapse will be generated.

In all three problems we assume $a = L = L_1$.

In order to treat the called problems we are obliged to modify the mathematical model, especially must be defined a mean temperature $T_M(z, t)$ (in analogy to $\tilde{\bar{u}}(z)$ in (1)) and a mean heat transmission function $h_M(z, t)$ (in analogy to $\tilde{\bar{h}}(z)$ in (7)), and the temperature of the considered solid will be described by

$$T_t = \bar{a}^2 \left(T_{rr} + \frac{1}{r} T_r + T_{zz}\right), \tag{45}$$

$$T(r, z, 0) = T_a, \tag{46}$$

$$T_z(r, 0, t) = 0, \tag{47}$$

$$T_z(r, L_1, t) = -k[T(r, L_1, t) - T_a], \tag{48}$$

$$T_r(R, z, t) = -h_M(z, t)\,[T(R, z, t) - T_M(z, t)], \tag{49}$$

$$T(0, z, t) \quad \text{finite.} \tag{50}$$

Further will be assumed a "quasistationary" development of the circumferential surface stresses in φ-direction and of the contour displacement in time lapse, i.e.:

The temperature $T = T(r, z, t)$ $\big(t \geqq 0,\ T(r, z, 0) = T_A(r, z)\big)$ induces

$$\sigma_{\varphi\varphi}(R, z, t) = \beta E \left(\frac{2}{R^2} \int_0^R r[T(r, z, t) - T_A(r, z)]\, dr - [T(R, z, t) - T_A(R, z)] \right) \tag{51}$$

(E: modulus of elasticity) (formula (51) results from the formula for $\sigma_{\varphi\varphi}(r, z)$ in the stationary case obtained by integration of the thermoelastic equations assuming plane stress in the solid) and (using (8)) the contour displacement

$$V(z, t) - V_A(z) = \frac{2\beta}{R} \int_0^R r[T(r, z, t) - T_A(r, z)]\, dr \tag{52}$$

$\big(V_A(z) = V(z, 0)\big)$.

The principle of solution for these problems is the following: Analoguously to (11) will be defined a function $F(z, t)$. Using the ansatzes

$$F(z, t) = \sum_{m=1}^{\infty} f_m(t)\, \varphi_m(z) \tag{53}$$

and

$$T(R, z, t) = \sum_{m=1}^{\infty} T_m(t)\, \varphi_m(z) \tag{54}$$

respectively ($\{\varphi_m\}_{m=1,2,\ldots}$ as in section III 1a) it is possible to construct conditions for the coefficient sequences $\{f_m\}_{m=1,2,\ldots}$ and $\{T_m\}_{m=1,2,\ldots}$ respectively such that are realized the properties of the solid, wanted in the formulated problems. These treatments still are in operation.

Numerical results to the sections III 2a are included in [1].

Bibliography

[1] Unger, F., Weber, K.-H., v. Wolfersdorf, L., Mathematische Untersuchungen zum stationären Kaltbandwalzprozess, Freiberger Forschungsheft B 205, VEB Deutscher Verlag für Grundstoffindustrie, Leipzig 1979.

Address of the author:

Bergakademie Freiberg
Sektion Mathematik
9200 Freiberg
Bernhard-v.-Cotta-Str. 2
DDR

Communications

On the radius of a set in a Hilbert space

Josef Daneš, Prague

In [1] we have proved the inequality $\chi(M) \leqq (1 - \delta(1))\, \alpha(M)$ between the Hausdorff and Kuratowski measures of noncompactness of any bounded set M in a normed linear space, where $\delta(\cdot)$ is the modulus of convexity of the space. This inequality is a direct consequence of the inequality $r(M) \leqq (1 - \delta(1))\, d(M)$ between the radius and diameter of M, derived there. In the case of a Hilbert space we have $1 - \delta(1) = \sqrt{3}/2$. The results presented here show that $\sqrt{3}/2$ can be replaced by $1/\sqrt{2}$ and that this last constant is the best possible in both inequalities provided the Hilbert space is infinite dimensional. Partial results were communicated at the ČSSR—GDR Summer School on Nonlinear Equations (Stará Lesná, High Tatras, October 1974).

Notation. In what follows, H is a real Hilbert space (all results remain true for a complex Hilbert space). For M a nonempty bounded subset of H, $B(M, r)$ is the closed r-ball centered at M (i.e. the set of all points x in H with $\operatorname{dist}(x, M) \leqq r$), $C(M, r) = \{x \in H : B(x, r) \supset M\}$, $d(M)$ is the diameter of M, $r(M) = \inf\{r > 0 : C(M, r) \neq \emptyset\}$ (we call this number the *radius of* M), $\chi(M) = \inf\{r > 0 : M$ has a finite r-net in $H\}$ (the Hausdorff measure of noncompactness of M), $\alpha(M) = \inf\{d > 0 : M$ can be covered by a finite number of set with diameters $\leqq d\}$ (the Kuratowski measure of noncompactness of M), $\operatorname{cocl}(M)$ denotes the closed convex hull of M and $\operatorname{sp}(M)$ the linear span of M. Furthermore, h denote the Hausdorff (pseudo-) metric in the space of all nonempty bounded subsets of H.

Lemma 1. *Let M be a non-empty bounded subset of H. Then:*

1) $d(C(M, r)) \leqq 2\sqrt{r^2 - r(M)^2}$ *for all* $r \geqq r(M)$;
2) $d(C(M, r)) \searrow 0$ *as* $r \searrow r(M)$;
3) $\bigcap \{C(M, r) : r > r(M)\}$ *consists of a unique point which we call the center of M and denote by $c(M)$; hence* $C(M, r(M)) = \{c(M)\}$;
4) *if* $r_n \to r(M)+$ *and* $x_n \in C(M, r_n)$ $(n \geqq 1)$, *then* $x_n \to c(M)$.

Theorem 1. *If M is a non-empty bounded subset of H, then there exists a unique smallest ball containing it, namely the ball* $B(c(M), r(M))$.

Lemma 2. *Let M and N be non-empty bounded subsets of H. Then:*

1) $|r(M) - r(N)| \leqq h(M, N)$;
2) $\|c(M) - c(N)\| \leqq \sqrt{h(M, N)\,(h(M, N) + r(M) + r(N))}$;
3) $r(\cdot)$ *and* $c(\cdot)$ *are continuous* ($r(\cdot)$ *is nonexpansive and* $c(\cdot)$ *is locally Hölder of order* $1/2$).

Lemma 3. *Let M be a non-empty bounded subset of H, $c = c(M)$ and $r = r(M)$. Then*

$$c \in \operatorname{cocl}\big(M \cap (B(c, r) \setminus B(c, r - \varepsilon))\big)$$

for each $\varepsilon \in (0, r)$.

Theorem 2. *Let M be a non-empty bounded subset of H. Then:*

1) $r(M) \leqq d(M)/\sqrt{2}$;
2) $\chi(M) \leqq \alpha(M)/\sqrt{2}$.

The constant $1/\sqrt{2}$ is exact provided H is infinite dimensional.

Remark 1. We are able to give three proofs of Theorem 2, part 1) (the simplest one is due to H. Steinlein [3]; the most natural one is based on a corresponding result for Euclidean spaces due to H. W. E. Jung [2]; the informative one is based on Lemma 3).

Remark 2. The whole paper will be published elsewhere.

References

[1] J. Daneš, On densifying and related mappings and their application in nonlinear functional analysis. *In:* Theory of Nonlinear Operators (Proc. Summer School Neuendorf, G.D.R. 1972), Akademie-Verlag, Berlin 1974, 15–56.

[2] H. W. E. Jung, Über die kleinste Kugel, die eine räumliche Figur einschließt. J. Reine Angew. Math. **123** (1901), 241–257.

[3] H. Steinlein, A private communication (1978).

Address of the author:

Mathematical Institute of Charles University
Sokolovská 83
18 600 Praha 8
Czechoslovakia

On the variational method without coerciveness

Zdzisław Denkowski, Kraków

The aim of this note is to state the theorem below. We admit the assumptions:

(A_1) *The topological vector spaces F and G form a dual system, the pairing will be denoted by* $\langle \cdot, \cdot \rangle$;

(A_2) *The function* $f\colon F \to \overline{\mathbb{R}}$ *is lower semi-continuous in the weak* $\sigma(F, G)$*-topology and satisfies the inequality* $f(v) \geqq \alpha(v)$, $v \in F$ *with some convex and proper function* α $(\alpha \in \Gamma_0(F))$;

(A_3) *The functional* $L\colon F \to \mathbb{R}$ *is linear and continuous, so we can write* $Lu = \langle u, l \rangle$ *with some* $l \in G$.

Under these assumptions we have

Theorem. *If the Young-Fenchel conjugate* α^* *of* α *is finite and continuous at the point* l *in the Mackey topology* $\tau(G, F)$ *and the function* f *is Gâteaux differentiable,* Au *being the gradient of* f *at* u, *then the equation (nonlinear in general)*

$$Au = l \tag{1}$$

has at least one solution. Moreover, the solution is unique provided f *is strictly convex and then it can be found as the* $\sigma(F, G)$*-limit of the minimizing sequence for the functional*

$$J(v) = f(v) - \langle v, l \rangle. \tag{2}$$

For the proof of the theorem it suffices to use the direct method of the calculus of variations when proving that the functional J attains its minimum. For the use of this method it is essential that the level sets

$$L_{l,\beta}(\alpha) = \{v \in F \mid \alpha(v) - \langle v, l \rangle \leqq \beta\} \quad (\beta \in \mathbb{R}) \tag{3}$$

should be compact in the $\sigma(F, G)$-topology (for we have with some $\beta \in \mathbb{R}$, $\{v_n\} \subset L_{l,\beta}(f) \subset L_{l,\beta}(\alpha)$, $\{v_n\}$ being the minimizing sequence for J). The last fact we get owing to the theorem obtained independently by Moreau [1] and Rockafellar [2] and saying that for $\alpha \in \Gamma_0(F)$ the $\sigma(F, G)$-compactness of level sets $L_{l,\beta}(\alpha)$ (for every $\beta \in \mathbb{R}$) is equivalent to the finiteness and $\tau(G, F)$-continuity at l of the conjugate α^*.

Remark. The above theorem when compared to the well known Lax-Milgram theorem (there $F = G = H$ is a Hilbert space and $f(v) = \frac{1}{2}\, a(v, v)$, where $a(u, v)$ is a continuous, symmetric and coercive bilinear form on H) relaxes the coercivity assumption. Indeed, the coercivity implies that all the level sets $L_{l,\beta}(f)$, $l \in H$, $\beta \in \mathbb{R}$, are bounded and in consequence compact in the weak topology. On the other hand assuming only "the coercivity in direction l" (i.e. we require the level sets $L_{l,\beta}(f)$, $\beta \in \mathbb{R}$, be $\sigma(F, G)$-compact) we get the existence of solution of equation (1) only for this l and not for all.

As a simple application of the theorem we give the following

Example. Let $F = H_0^1 = H_0^1(\Omega)$ be the Sobolev space over the open and bounded domain $\Omega \subset \mathbb{R}^n$, $G = H^{-1}$ denotes its topological dual and the pairing is defined by $\langle v, l \rangle = l(v)$. We put

$$f(v) = \left(c_1^2 + \sum_{i,j=1}^{n} \int_{\Omega} a_{ij} D_i v D_j v \, dx\right)^{1/2}, \quad v \in H_0^1, \tag{4}$$

where $D_i = \dfrac{\partial}{\partial x_i}$, $a_{ij} = a_{ji} \in L^\infty(\Omega)$ and are such that

$$c_1^2 |\xi|^2 \leqq \sum_{i,j=1}^{n} a_{ij} \xi_i \xi_j \leqq c_2^2 |\xi|^2, \quad \xi \in \mathbb{R}^n, \quad (0 < c_1 \leqq c_2). \tag{5}$$

Setting $\alpha(v) = c_1(1 + \|v\|_0^2)^{1/2}$, where the norm $\|\cdot\|_0$ is generated by the scalar product

$$(u, v)_0 = \sum_{i=1}^{n} \int_{\Omega} D_i u D_i v \, dx, \quad u, v \in H_0^1,$$

we find that

$$\alpha^*(v^*) = \begin{cases} -(c_1^2 - \|v\|_{H^{-1}}^2)^{1/2}, & \text{if } \|v\|_{H^{-1}} \leqq c_1 \\ +\infty, & \text{otherwise.} \end{cases}$$

Since in the considered case the Mackey topology $\tau(H^{-1}, H_0^1)$ coincides with the norm topology in H^{-1} we can apply the theorem and get the following

Corollary. *For each $l \in H^{-1}$ such that $\|l\|_{H^{-1}} < c_1$ there exists in H_0^1 the unique solution u of the integro-differential equation*

$$-\left(c_1^2 + \sum_{i,j=1}^{n} \int_{\Omega} a_{ij} D_i u D_j u \, dx\right)^{-1/2} \sum_{i,j=1}^{n} D_i(a_{ij} D_j u) = l. \tag{6}$$

This solution can be found as the weak limit of the minimizing sequence for the functional J.

It is worth noticing that the problem is not coercive for the functions $\alpha_1 = \alpha$ and $\alpha_2(v) = c_2(1 + \|v\|_0^2)^{1/2}$ have oblique asymptotes along each line passing through 0 in the space H_0^1, and $\alpha_1 \leqq f \leqq \alpha_2$.

Final Remark. It is possible (see [3]) to define the notion of G-convergence for a net of functions $\{f_\iota\}_I$ satisfying the boundedness condition $f_\iota \geqq \alpha$, $\iota \in I$, where α fulfils the growth condition only in some directions l (for instance α might be like in the example above), and then to prove the results analogous to those obtained by Marcellini [4], Boccardo and Marcellini [5], and others.

References

[1] J. J. Moreau, Sur la polaire d'une fonctionelle semi-continue supérieurement, C. R. Acad. Sci. Paris **258** (1964), 1128–1130.
[2] R. T. Rockafellar, Level sets and continuity of conjugate convex functions, Trans. Amer. Math. Soc. **123** (1966), 45–63.
[3] Z. Denkowski, On the Q-convergence of nets of functions in topological vector spaces, Zeszyty Naukowe UJ **22** (1981), 73–83.
[4] P. Marcellini, Su una convergenza di funzioni convesse, Boll. Un. Mat. Ital. **8** (1973), 137–158.
[5] L. Boccardo and P. Marcellini, Sulla convergenza delle soluzioni di disequazioni variazionali, Ann. Math. Pura Appl. **60** (1976), 137–159.

Address of the author:

Institute of Informatics
Jagiellonian University
ul. Kopernika 27
31-501 Kraków
Poland

Special control problems and monotone operators

Siegfried Dietze, Dresden

Abstract. In this information we will transform a constrained control-approximation problem into an unconstrained optimization problem.

In this information we consider the control-approximation problem

$$\inf \{\|Su - \hat{x}\|^2 : u \in \Omega\}. \tag{1}$$

Here are

H — a real Hilbert space with scalar product $(\cdot, \cdot)$ and norm $\|\cdot\|$,

$S \mid L_2^r[t_0, t_1] \to H$ — a continuous linear operator,

$\hat{x}$ — a fixed element of H,

$\Omega = \{u \in L_2^r[t_0, t_1] : |u_i(t)| \leqq 1$ for almost every $t \in [t_0, t_1]$ and for all $i = 1, \ldots r\}$.

The numbers r, t_0, t_1 are fixed.

There are many optimal control problems with linear state equation which have the form (1), for examples see for instance [7]. The cost functional in (1) is convex and continuous and therefore lower-semicontinuous with respect to the weak topology on H, and the set Ω is convex, closed, bounded and therefore weakly compact. Thus the convex optimization problem (1) has always a solution. There are different possibilities to solve problem (1). We can apply direct methods (for instance the conditional gradient method [7]) and indirect methods (use of optimality conditions). Furthermore, problem (1) can be transformed into several optimization problems in the space H, and it is possible that the solution of such a transformed problem is simpler than the solution of (1). The paper [4] gives a survey on transformations of (1) into optimization problems in the space H. In [4] the problem (1) is also considered in normed spaces and with more general sets Ω.

In the present information the problem (1) is transformed into an *unconstrained* convex optimization problem in the space H. The optimization problem in H is constructed with the aid of an optimality theorem for convex optimization problems and a theorem of Rockafellar on the existence of a convex function which has a subdifferential equal to a given point-set operator (see theorem 3). This information generalizes investigations on a special *normal* problem (1) which are contained in [2], [3] (definition of "normal" see below). The considerations in [2], [3] require the normality of (1), only the use of the above mentioned theorem of Rockafellar enables the generalization. The present information is the summary of a part of [5]. The detailed paper [5] includes the proofs of the theorems given below and extensions, among other things the regularized problem corresponding to (1)

$$\inf \{\|Su - \hat{x}\|^2 + c\,\|u - \hat{u}\|^2 : u \in \Omega\}$$

($c > 0$, $\hat{u} \in L_2^r[t_0, t_1]$ given) and the differentiability of the transformed cost functionals are investigated.

We begin with a characterization theorem for the solutions of (1). It can be proved via optimality theorems for convex optimization problems.

Theorem 1. *Let the equation*

$$0 = \lambda + Su(\cdot; \lambda, v) - \hat{x} \tag{2}$$

for the pair $(\lambda, v) \in H \times \Omega$ *be given with*

$$u_i(t; \lambda, v) = \begin{cases} \operatorname{sgn} (S^*\lambda)_i (t) & \text{for almost every } t \in [t_0, t_1] \\ & \text{with } (S^*\lambda)_i (t) \neq 0, \\ v_i(t) & \text{for almost every } t \in [t_0, t_1] \\ & \text{with } (S^*\lambda)_i (t) = 0 \end{cases} \tag{3}$$

($i = 1, \ldots, r$; $\lambda \in H$, $v \in \Omega$). *The operator* $S^* \mid H \to L_2^r[t_0, t_1]$ *is the adjoint of* S. *Then we have:*

a) $\hat{u}$ *solution of* (1), $\hat{\lambda} := \hat{x} - S\hat{u}$, $\hat{v} := \hat{u} \Rightarrow (\hat{\lambda}, \hat{v})$ *solution of* (2)
b) $(\hat{\lambda}, \hat{v})$ *solution of* (2) $\Rightarrow \hat{u} := u(\cdot\,; \hat{\lambda}, \hat{v})$ *solution of* (1) *and* $\hat{\lambda} = \hat{x} - S\hat{u}$.

Discussion. Always the λ-part $\hat{\lambda}$ of a solution (λ, v) of equation (2) determines the minimal distance

$$\min \{\|Su - \hat{x}\|^2 : u \in \Omega\} = \|\hat{\lambda}\|^2$$

of problem (1). The problem (1) is said to be *normal* if the sets

$$\{t \in [t_0, t_1] : (S^*\lambda)_i (t) = 0\}, \quad (i = 1, \ldots, r)$$

have the measure zero for all $\lambda \neq 0$. If the problem (1) is normal and $\min \{\|Su - \hat{x}\|^2 : u \in \Omega\} > 0$, the quantity v in theorem 1 is unimportant and with the solutions $\hat{\lambda}$ of (2) the solutions $\hat{u}$ of (1) are given. For this reasons it is justified that we concentrate on the investigation of the λ-part in the solutions of equation (2).

The investigation of the λ-part in the solutions of equation (2) is equivalent to the computation of all $\lambda = \hat{\lambda}$ with the property

$$0 \in \{\lambda + Su(\cdot; \lambda, v) - \hat{x} : v \in \Omega\} \tag{4}$$

respectively

$$0 \in f(\lambda), \tag{5}$$

where f is the point-set operator

$$f \mid H \to 2^H, \quad f(\lambda) = \{\lambda + Su(\cdot; \lambda, v) - \hat{x} : v \in \Omega\}. \tag{6}$$

Furthermore, let $F \mid H \to R^1$ be a convex function with the subdifferential

$$\partial F = f. \tag{7}$$

Then problem (5) is equivalent to

$$\min \{F(\lambda) : \lambda \in H\} \tag{8}$$

(optimality condition for the minimum points of a convex function). In other words, if a function F with the mentioned properties exists, $\hat{\lambda}$ solves (5) if and only if $\hat{\lambda}$ is a minimum point of F. With regard to this we ask the

Question. Exists a convex function $F \mid H \to R^1$ with (7)?

In the following we will answer this question and calculate F in the case of existence. We begin with basic results of convex analysis and the theory of monotone operators. The theorems 2, 3 can be found for instance in [1, 6].

Definition. The operator $f \mid H \to 2^H$ is said to be

- *cyclically monotone* if

$$(y_0, \lambda_0 - \lambda_1) + \cdots + (y_{n-1}, \lambda_{n-1} - \lambda_n) + (y_n, \lambda_n - \lambda_0) \geqq 0$$

  for all $n = 1, 2, \ldots, \lambda_i \in H, y_i \in f(\lambda_i)$ $(i = 0, 1, \ldots, n)$,
- *maximal cyclically monotone* if f is cyclically monotone and $[g \mid H \to 2^H$ cyclically monotone, $g(\lambda) \supset f(\lambda)$ for all $\lambda \in H \Rightarrow g = f]$.

Theorem 2 (Minty). *Let $f \mid H \to 2^H$ be cyclically monotone and*

$$\{y + \lambda : \lambda \in H, y \in f(\lambda)\} = H.$$

Then f is maximal cyclically monotone.

Theorem 3 (Rockafellar). *Let $f \mid H \to 2^H$. For the existence of a lower-semicontinuous proper convex function $F \mid H \to \bar{R}^1$ such that $\partial F = f$, it is necessary and sufficient that f is a maximal cyclically monotone operator. Moreover, in the case of existence f determines F uniquely apart from an additive constant.*

Theorem 4. *Let $F \mid H \to \bar{R}^1$ be a lower-semicontinuous proper convex function and $\partial F(\lambda) \neq \emptyset$ for all $\lambda \in H$. Then*

a) *F is an operator from H even to R^1.*
b) *F is continuous.*
c) $F'(\lambda; h) = \max \{(y, h) : y \in \partial F(\lambda)\}$

for all $\lambda, h \in H$, where $F'(\lambda; h)$ is the directional derivative of F at λ in the direction h (theorem of Moreau/Pshenichniy).

d) $F(\lambda) = F(0) + \int_0^1 F'(t\lambda; \lambda)\, dt$

for all $\lambda \in H$. Here $F'(t\lambda; \lambda)$ is monotonically increasing with respect to $t \in R^1$, which implies that the integral is a R-integral.

It can be shown that the operator (6) is cyclically monotone, and moreover with theorem 2 that it is maximal cyclically monotone. This yields by theorem 3 and theorem 4a, b the existence of a continuous convex function $F \mid H \to R^1$ with the subdifferential

(6). The representation for F (see (9)) follows from theorem 4c, d. This and the above considerations imply the following result.

Theorem 5. *The function* $F \mid H \to R^1$,

$$F(\lambda) = \frac{\|\lambda\|^2}{2} - (\lambda, \hat{x}) + \sum_{i=1}^{r} \int_{t_0}^{t_1} |(S^*\lambda)_i(t)| \, dt \tag{9}$$

is continuous, convex and has the subdifferential

$$\partial F(\lambda) = \{\lambda + Su(\cdot; \lambda, v) - \hat{x} : v \in \Omega\}$$

($u(\cdot\,; \lambda, v)$ *see* (3)). *The function* F *is even strongly convex. Furthermore,* F *has an unique (global) minimum point* $\hat{\lambda}$.

The minimum point $\hat{\lambda}$ *corresponds to the minimal distance of problem* (1), *it is*

$$\hat{\lambda} = \hat{x} - S\hat{u}$$

for all solutions $\hat{u}$ *of* (1) *and thus*

$$\min \{\|Su - \hat{x}\|^2 : u \in \Omega\} = \|\hat{\lambda}\|^2.$$

If the problem (1) *is "normal" (definition of "normal" see to theorem* 1) *and* $\hat{\lambda} \neq 0$, (1) *is uniquely solvable and the solution* $\hat{u}$ *is given by*

$$\hat{u}_i(t) = \operatorname{sgn} (S^*\hat{\lambda})_i (t), \quad (t \in [t_0, \dot{t_1}], \ i = 1, \ldots, r).$$

References

[1] Barbu, V., and Precupanu, Th., Convexity and optimization in Banach spaces, Editura Aacdemiei, Bucuresti, Sijthoff & Noordhoff, Int. Publ. 1978.

[2] Dietze, S., Zur Lösung gewisser linearer optimaler Prozesse mit dem Maximumprinzip, Z. Angew. Math. Mech. **48** (1968), 57–64.

[3] Dietze, S., Zur Lösung zweier optimaler Steuerprobleme mit dem Maximumprinzip, Dissertation, TU Dresden 1971.

[4] Dietze, S., Steuer-Approximationsprobleme und Fenchel-Rockafellar-Dualität, Math. Operationsforsch. Statist. Ser. Optimization **12** (1981) No. 3 (to appear).

[5] Dietze, S., Spezielle Steuer-Approximationsprobleme und monotone Operatoren, Beiträge zur Analysis (to appear).

[6] Kluge, R., Nichtlineare Variationsungleichungen und Extremalaufgaben, VEB Deutscher Verlag der Wissenschaften, Berlin 1979.

[7] Krabs, W., Einführung in die Kontrolltheorie, Wissenschaftliche Buchgesellschaft, Darmstadt 1978.

Address of the author:

Technische Universität Dresden
Sektion Mathematik
8027 Dresden
Mommsenstr. 13
DDR

A generalization of an optimality theorem

Siegfried Gähler, Berlin

1. Introduction. Let X be an open set of $\mathbb{R}^n$ and $f = (f_1, \ldots, f_m)$ a C^2-mapping of X into $\mathbb{R}^m$. f is said to have a *local* (*Pareto*) *optimum at* x if there exists a neighbourhood U of x in $\mathbb{R}^n$ such that $f_i(y) \geqq f_i(x)$, $i = 1, \ldots, m$, and $y \in X \cap U$ imply $f(y) = f(x)$. f is said to have a *local strict* (*Pareto*) *optimum at* x if there exists a neighbourhood U of x in $\mathbb{R}^n$ such that $f_i(y) \geqq f_i(x)$, $i = 1, \ldots, m$, and $y \in X \cap U$ imply $y = x$.

In [4] Smale proved the following optimality theorem: *If f has a local optimum at x, then there exist real numbers $\psi_1, \ldots, \psi_m \geqq 0$, not all 0 such that $\sum_{i=1}^{m} \psi_i df_i(x) = 0$. If there exist such numbers and if, furthermore, the symmetric bilinear form $\sum_{i=1}^{m} \psi_i d^2 f_i(x)$ is negative definit on thel inear space $\{y \in \mathbb{R}^n;\ \psi_i df_i(x).y = 0$ for all $i\}$, then f has a strict local optimum at x.*

It is the aim of the underlying contribution to give a generalization of this theorem.

2. Generalized notions of differentiability. Let X be a subset of $\mathbb{R}^n$ and x a point of X. $C_x = \{y \in \mathbb{R}^n;\ \exists\, \varepsilon > 0 : x + [0, \varepsilon]\, y \subseteqq X\}$ is a cone in $\mathbb{R}^n$ with 0 as vertex. Assume C_x is closed and $\neq \{0\}$ and there exists a neighbourhood V of the point 0 in $\mathbb{R}^n$ such that $(X - x) \cap V = C_x \cap V$. These assumptions are fulfilled if X is open or if X is a euclidean polyhedron.

Let Z be an ordered T_2-topological vector space, i.e. a T_2-topological vector space which is equipped with a partial ordering $\leqq$ by $z \leqq z' \Leftrightarrow z' - z \in K$ using a closed proper cone K in Z with 0 as vertex, the so-called *positive cone*. K to be *proper* means that K is convex and $K \cap (-K) = \{0\}$.

A mapping $f : X \to Z$ is said to be *l-differentiable* (*lower differentiable*) *at* x if there exists a mapping $\Phi : C_x \to Z$ with the following properties:

i. $\Phi(\alpha y) - \alpha \Phi(y) \in K$ for every $\alpha \in (0, 1)$ and $y \in C_x$.

ii. For every $y \in C_x$ and every neighbourhood W of the point 0 in Z, there exists an $\varepsilon > 0$ and a neighbourhood V of the point 0 in $\mathbb{R}^n$ such that

$$\frac{f\big(x + \varepsilon'(y + y')\big) - f(x)}{\varepsilon'} \in \Phi(y) + W + K$$

whenever $\varepsilon' \in (0, \varepsilon]$, $y' \in V$, and $x + \varepsilon'(y + y') \in X$.

Every such mapping Φ is called (*first*) *l-derivative* ((*first*) *lower derivative*) *of f at x*. From properties i and ii, there follows $\Phi(0) = 0$. The set $D_l f(x)$ of all l-derivatives of f at x

is convex. Analogously as the notion of an l-derivative of f at x, the notion of a (*first*) *u-derivative* ((*first*) *upper derivative*) *of* f *at* x is to be introduced. $D_u f(x) = -D_l(-f)(x)$ is the set of all u-derivatives of f at x.

Let $\mathscr{Z}$ denote the vector space of all positive homogeneous continuous mappings of C_x into Z which is equipped with the T_2-vector topology of uniform convergence on bounded sets of C_x and which is (partially) ordered by use of the closed proper cone $\mathscr{K} = \{\Phi \in \mathscr{Z}; \Phi[C_x] \subseteq K\}$. f is said to be *twice l-differentiable at* x if in $\mathbb{R}^n$, there exists a neighbourhood U of x with the following properties:

i. $C_x \subseteq C_{x'}$ for every $x' \in X \cap U$.
ii. f admits a positive homogeneous continuous l-derivative $\Phi_{x'}$ at every point $x' \in X \cap U$.
iii. The mapping $\Phi: X \cap U \to \mathscr{Z}$ given by $\Phi(x') = \Phi_{x'} \mid C_x$ is l-differentiable at x.

Every l-derivative Φ^2 of every such mapping Φ is called *second l-derivative* (*second lower derivative*) *of* f *at* x and Φ_x is called an *associated* (*first*) *l-derivative*. The set $D_l^2 f(x)$ of all second l-derivatives of f at x is convex. Analogously as the notion of a second l-derivative of f at x, the notion of a *second u-derivative* (*second upper derivative*) *of* f *at* x is to be introduced. $D_u^2 f(x) = -D_l^2(-f)(x)$ is the set of all second u-derivatives of f at x.

In [1] slight modifications of the given notions of first and second lower and upper derivatives are studied in detail.

3. Application to optimality problems. In what follows let $\underline{K} \neq \emptyset$. A mapping f as in the preceding section is said to have a *local* (*Pareto*) *optimum at* x if there exists a neighbourhood U of x in $\mathbb{R}^n$ such that $f[X \cap U] - f(x) \subseteq Z \setminus (K \setminus \{0\})$. f is said to have a *local strict* (*Pareto*) *optimum at* x if there exists a neighbourhood U of x in $\mathbb{R}^n$ such that $\big(f[(X \cap U) \setminus \{x\}] - f(x)\big) \cap K = \emptyset$.

Theorem 1 ([1], Theorem 12). *If* f *has a local optimum at* x, *then* $D_l f(x)\,[C_x] \subseteq Z \setminus \underline{K}$.

By K^* we denote the algebraically dual cone of K, i.e. the set of all linear functionals ψ on Z such that $\psi[K] \subseteq [0, \infty)$. As $\underline{K} \neq \emptyset$, every $\psi \in K^*$ is continuous. A mapping Φ of a convex set $C \subseteq \mathbb{R}^n$ into Z is said to be *concave* if $\alpha \in (0, 1)$ and $y, y' \in C$ imply

$$\Phi\big(\alpha y + (1 - \alpha)\, y'\big) - \alpha \Phi(y) - (1 - \alpha)\, \Phi(y') \in K.$$

Corollary (See [1], Theorem 12). *Let* $\Phi: C_x \to Z$. *If there exists a* $\psi (\neq 0) \in K^*$ *such that* $\psi \circ \Phi[C_x] \subseteq (-\infty, 0]$, *then* $\Phi[C_x] \subseteq Z \setminus \underline{K}$. *In the case of* C_x *being convex and* Φ *being concave*, $\Phi[C_x] \subseteq Z \setminus \underline{K}$ *conversely implies the existence of a* $\psi(\neq 0) \in K^*$ *such that* $\psi \circ \Phi[C_x] \subseteq (-\infty, 0]$.

A mapping g of a set $X' \subseteq \mathbb{R}^n$ into Z is said to be *lower semicontinuous at a point* $x' \in X'$ if for every neighbourhood W of the point 0 in Z, there exists a neighbourhood V of the point 0 in $\mathbb{R}^n$ such that $g[(x' + V) \cap X'] \subseteq g(x') + W + K$.

Theorem 2 (Modification of [1], Theorem 15). *Let there exist a neighbourhood* V *of the point* x *in* $\mathbb{R}^n$ *such that for every* $y(\neq 0) \in \mathbb{R}^n$ *for which* $x + [0, 1]\, y \subseteq X \cap V$, $f(x + [0, 1]\, y)$ *is lower semicontinuous at* $x + y$. *Let* f *be twice u-differentiable at* x, Φ^2 *denote a second u-derivative of* f *at* x, *and* Φ^1 *be an associate d u-derivative. For every* $y(\neq 0) \in C_x$, *let there exist a* $y' = \varepsilon y$, $\varepsilon > 0$, *such that* $\Phi^1(y') \in Z \setminus \underline{K}$ *and in the case* $\Phi^1(y') \in K \setminus \underline{K}'$ *let there exist a* $\psi \in K^*$ *and a neighbourhood* V *of the point* 0 *in* $\mathbb{R}^n$ *such that* $\psi \circ \Phi^1[(y' + V) \cap C_x] \subseteqq (-\infty, 0]$ *and* $\psi \circ \Phi^2(y', y') < 0$. *Then* f *has a local strict optimum at* x.

4. Concluding remarks. To show that Theorem 1 and Theorem 2 and the Corollary generalize the theorem of Smale, we now specialize our assumptions.

i. Let X be open. Then $C_x = \mathbb{R}^n$.

ii. Let Z be the space $\mathbb{R}^m$ equipped with the natural vector topology and (partially) ordered by use of the closed proper cone $K = \{(\alpha_1, \ldots, \alpha_m) \in \mathbb{R}^m; \alpha_i \geqq 0 \text{ for } i = 1, \ldots, m\}$. Under this restriction on Z, the notions of a local (strict) optimum given in section 1 and in section 3 are identical. A functional ψ on $\mathbb{R}^m$ belongs to K^* if and only if there exist real numbers $\psi_1, \ldots, \psi_m \geqq 0$ such that $\psi(z) = \sum_{i=1}^{m} \psi_i \zeta_i$ where $z = (\zeta_1, \ldots, \zeta_m)$.

iii. Let $f = (f_1, \ldots, f_m)$ be a C^2-mapping. Then $df(x) = (df_1(x), \ldots, df_m(x)) \in D_l f(x) \cap D_u f(x)$ and $d^2 f(x) = (d^2 f_1(x), \ldots, d^2 f_m(x)) \in D_l^2 f(x) \cap D_u^2 f(x)$ and $df(x)$ is associated to $d^2 f(x)$.

It is easy to be seen now that the necessary condition for a local optimum given in the theorem of Smale is an immediate consequence of Theorem 1 and the Corollary. In what follows let the assumptions of the sufficient condition for a local strict optimum given in the theorem of Smale be fulfilled and let ψ be the element of K^* for which $\psi(z) = \sum_{i=1}^{m} \psi_i \zeta_i$, $z = (\zeta_1, \ldots, \zeta_m)$. For every $y \in \mathbb{R}^m$, by the Corollary, $df(x).y \in Z \setminus \underline{K}$. If $df(x).y \in K \setminus \underline{K}$, then $df_i(x).y \geqq 0$, $i = 1, \ldots, m$, and because of $\psi_i \geqq 0$, $i = 1, \ldots, m$, and $\psi \circ df(x) = 0$, we get $\psi_i df_i(x).y = 0$, $i = 1, \ldots, m$, hence $\psi \circ d^2 f(x).(y, y) < 0$. By Theorem 2, consequently f has a local strict optimum at x. Thus the sufficient condition for a local strict optimum given in the theorem of Smale follows from Theorem 2 and the Corollary. This we wanted to prove.

References

[1] S. Gähler, Generalized notions of differentiability and their application in optimality theory. *In:* Convergence Structures and Applications to Analysis (Proc. Internat. Summer School Frankfurt/Oder 1978), Abh. Akad. Wiss. DDR, Jahrgang 1979, Nr. 4 N, Akademie-Verlag, Berlin 1979, 67–88.

[2] —, Optimality conditions in polypotimization. *In:* Theory of Nonlinear Operators (Proc. Internat. Summer School Berlin 1977), Abh. Akad. Wiss. DDR, Jahrg. 1978, Nr. 6N, Akademie-Verlag, Berlin 1978, 353–355.

[3] —, Generalized derivatives and optimality conditions. *In:* Variationsrechnung und optimale Prozesse. Preprint-Reihe Mathematik, Preprint 1, Ernst-Moritz-Arndt-Universität Greifswald, 1979.

[4] S. Smale, Sufficient conditions for an optimum. *In:* Dynamical Systems — Warwick 1974, Lecture Notes in Mathematics, vol. 468, Springer-Verlag, Berlin—Heidelberg—New York 1975, 287–292.

Address of the author:

Akademie der Wissenschaften der DDR
Institut für Mathematik
1080 Berlin
Mohrenstr. 39
DDR

Mixed variational formulation of unilateral problems

Jaroslav Haslinger, Praha, and Ján Lovíšek, Bratislava

Let Ω be a bounded domain with Lipschitz boundary. We shall use the Sobolev spaces $H^k(\Omega)$, $k = 1, 2, \ldots$, of functions the generalized derivatives of which up to the order k exist and are square-integrable in Ω.

We consider the following model problem

$$-\Delta u + u = f \quad \text{in } \Omega \tag{1}$$

with unilateral boundary conditions

$$u \geqq g, \frac{\partial u}{\partial n} \geqq 0, \quad (u - g)\frac{\partial u}{\partial n} = 0, \text{ a.e. on } \Omega \tag{2}$$

where $\frac{\partial u}{\partial n}$ denotes the normal derivative with respect to the outward normal n, $f \in L_2(\Omega)$, $g \in H^{1/2}(\partial\Omega)$.

Variational formulation of the problem

In order to give a variational formulation to problem (1), (2) we introduce the convex set

$$K(\Omega) = \{v \in H^1(\Omega) \mid \gamma_0 v \geqq g \quad \text{a.e. on } \partial\Omega\} \tag{3}$$

(where γ_0 is the trace operator) and the functional (potential energy)

$$J(v) = \|v\|^2_{H^1(\Omega)} - 2(f, v)_{L_2(\Omega)}. \tag{4}$$

Then the problem ($\mathscr{P}$) to find u such that

$$\left.\begin{array}{l} u \in K(\Omega), \\ J(u) \leqq J(v) \quad \text{for any} \quad v \in K(\Omega), \end{array}\right\} \tag{$\mathscr{P}$}$$

represents a variational formulation of problem (1), (2).

Theorem 1. *There exists a unique solution of the problem ($\mathscr{P}$) and moreover u can be characterized as the solution of*

$$\left.\begin{array}{l} u \in K(\Omega), \\ (\operatorname{grad} J(u), v - u)_{H^1(\Omega)} = (u, v - u)_{H^1(\Omega)} \geqq (f, v - u)_{L_2(\Omega)} \end{array}\right\} \tag{5}$$

for any $v \in K(\Omega)$.

Mixed formulation for ($\mathscr{P}$)

Let

$$\left.\begin{aligned}\Lambda(\partial\Omega) = \{\mu \in H^{-1/2}(\partial\Omega) \mid \langle\mu, \gamma_0 v\rangle_{H^{1/2}(\partial\Omega)} \geqq 0, \\ \text{for any } \gamma_0 v \in H^{1/2}(\partial\Omega), \gamma_0 v \geqq 0\}.\end{aligned}\right\} \tag{6}$$

Then it is possible to check that

$$v \in K(\Omega) \Leftrightarrow v \in H^1(\Omega) \quad \text{and} \quad \langle\mu, \gamma_0 v - g\rangle_{H^{1/2}(\partial\Omega)} \geqq 0 \tag{7}$$

for any $\mu \in \Lambda(\partial\Omega)$.

Variational mixed formulation for $\mathscr{P}$

Find $[u^*, \lambda^*] \in H^1(\Omega) \times \Lambda(\partial\Omega)$ satisfying

$$\left.\begin{aligned}(u^*, v)_{H^1(\Omega)} - \langle\lambda^*, \gamma_0 v\rangle_{H^{1/2}(\partial\Omega)} = (f, v)_{L_2(\Omega)} \quad \text{for any} \quad v \in H^1(\Omega), \\ \langle\lambda - \lambda^*, \gamma_0 u^*\rangle_{H^{1/2}(\partial\Omega)} \geqq \langle\lambda - \lambda^*, g\rangle_{H^{1/2}(\partial\Omega)} \text{ for any } \lambda \in \Lambda(\partial\Omega).\end{aligned}\right\} \tag{$\mathscr{P}^*$}$$

Theorem 2. There exists a unique solution of the problem ($\mathscr{P}^*$) and $[u^*, \lambda^*] = \left[u, \frac{\partial u}{\partial n}\right]$, *where u is a solution of the problem* ($\mathscr{P}$).

Remark 1. Note that $\left[u, \frac{\partial u}{\partial n}\right]$ is characterized as the saddle point of the Lagrangian $\mathscr{L}(v, \lambda)$ over $H^1(\Omega) \times \Lambda(\partial\Omega)$ where

$$\mathscr{L}(v, \lambda) = J(v) - \langle\lambda, \gamma_0 v - g\rangle_{H^{1/2}(\partial\Omega)}; \tag{8}$$

i.e.

$$\begin{aligned}\inf J(v) &= \inf_{v\in H^1(\Omega)} \sup_{\lambda\in\Lambda(\partial\Omega)} \mathscr{L}(v, \lambda) = \sup_{\lambda\in\Lambda(\partial\Omega)} \inf_{v\in H^1(\Omega)} \mathscr{L}(v, \lambda) \\ &= \mathscr{L}\left(u, \frac{\partial u}{\partial n}\right).\end{aligned} \tag{9}$$

Hence λ is the Lagrange multiplier associated with the constraint $u \in K(\Omega)$.

Finite element approximations of the mixed problems ($\mathscr{P}^*$)

Let $\Omega \subset R^2$ be a polygonal bounded domain. We consider a regular family of triangulations $\{\mathscr{T}_{h_\Omega}\}$ of $\bar{\Omega}$, where $h_\Omega = \max_{T\in\mathscr{T}_{h_\Omega}} \operatorname{diam} T$.

Let $\{\mathscr{T}_{h_{\partial\Omega}}\}$ be a family of partitions of $\partial\Omega$, independent of $\{\mathscr{T}_{h_\Omega}\}$, i.e.

$$\partial\Omega = \bigcup_{i=1}^{m} a_i a_{i+1}, \quad a_1 \equiv a_{m+1}.$$

We denote by h_i = length $a_i a_{i+1}$ and $h_{\partial\Omega} = \max_i h_i$.

Next we shall consider a regular family $\{\mathcal{T}_{h_{\partial\Omega}}\}$ in the following sense: there exists a constant $\alpha_0 > 0$ such that $h_{\partial\Omega|h_i} \leqq \alpha_0$, for any $i = 1, 2, \dots, m$. Let us define the space

$$\left.\begin{aligned} V_{h_\Omega}(\Omega) &= \{v \in C(\bar{\Omega}) |\ \ v|_T \in P_1(T) \ \text{ for any } \ T \in \mathcal{T}_{h_\Omega}\}, \\ \tilde{\Lambda}_{h_{\partial\Omega}}(\partial\Omega) &= \{\mu \in L_2(\partial\Omega) \,|\, \mu_{a_i a_{i+1}} \in P_0(a_i, a_{i+1});\ i = 1, 2, \dots, m\} \end{aligned}\right\} \tag{10}$$

and

$$\Lambda_{h_{\partial\Omega}}(\partial\Omega) = \{\mu_h \in \tilde{\Lambda}_{h_{\partial\Omega}}(\partial\Omega) \mid \mu_h \geqq 0 \ \text{ on } \ \partial\Omega\}$$

where $P_1(T)$, $P_0(a_i a_{i+1})$ denotes the space of all polynomials of degree $k = 1$, and the constant functions on $a_i a_{i+1}$, respectively.

• Approximation of $\mathcal{P}^*$

Let us define the following problem: find an element $[u_h^*, \lambda_h^*] \in V_{h_\Omega}(\Omega) \times \Lambda_{h_{\partial\Omega}}(\partial\Omega)$ which satisfies

$$\left.\begin{aligned} &(u_h^*, v_h)_{H^1(\Omega)} - \langle\lambda_h^*, \gamma_0 v_h\rangle_{H^{1/2}(\partial\Omega)} = (f, v_h)_{L_2(\Omega)} \\ &\text{for any } v_h \in V_{h_\Omega}(\Omega), \\ &\langle\mu_h - \lambda_h^*, \mu_h^*\rangle_{H^{1/2}(\partial\Omega)} \geqq \langle\mu_h - \lambda_h^*, g\rangle_{H^{1/2}(\partial\Omega)} \\ &\text{for any } \mu_h \in \Lambda_{h_{\partial\Omega}}(\partial\Omega) \end{aligned}\right\} \tag{$\mathcal{P}_h^*$}$$

with

$$\langle\mu_h, \gamma_0 v_h\rangle_{H^{1/2}(\partial\Omega)} = \int_{\partial\Omega} \mu_h \gamma_0 v_h \, ds \quad \text{for any} \quad v_h \in V_h(\Omega), \quad \mu_h \in \Lambda_{h_{\partial\Omega}}(\partial\Omega).$$

The existence and uniqueness of the solution of $\mathcal{P}^*$

In what follows we shall study the existence and the uniqueness of the solution of $(\mathcal{P}_h^*)$.

Remark 2. Let us set

$$\begin{aligned} K(h_\Omega, h_{\partial\Omega}) = \{v_h \in V_{h_\Omega}(\Omega) \mid &\langle\mu_h, \gamma_0 v_h\rangle_{\Lambda(\partial\Omega)} \\ &\geqq \langle\mu_h, g\rangle_{\Lambda(\partial\Omega)} \quad \text{for any} \quad \mu_h \in \Lambda_{h_{\partial\Omega}}(\partial\Omega).\} \end{aligned} \tag{11}$$

Then $K(h_\Omega, h_{\partial\Omega})$ is the finite dimensional approximation of $K(\Omega)$.

Next, let the pair $[u_h^*, \lambda_h^*] \in V_{h_\Omega}(\Omega) \times \Lambda_{h_{\partial\Omega}}(\partial\Omega)$ denote the solution of the problem $(\mathcal{P}_h^*)$.

Then u_h is the unique solution of the problem: Find an element $u_h^* \in K(h_\Omega, h_{\partial\Omega})$ such that

$$(u_h^*, v_h - u_h^*)_{H^1(\Omega)} \geqq (f, v_h - u_h^*)_{L_2(\Omega)} \quad \text{for any} \quad v_h \in K(h_\Omega, h_{\partial\Omega}). \tag{12}$$

Furthermore, on the other hand a pair $[u_h^*, \lambda_h^*] \in V_{h_\Omega}(\Omega) \times \Lambda_h(\partial\Omega)$ is characterized as the saddle-point of the Lagrangian $\mathcal{L}$ over the space $V_{h_\Omega}(\Omega) \times \Lambda_{h_{\partial\Omega}}(\partial\Omega)$, where $\mathcal{L}$ is defined by

$$\mathcal{L}(v_h, \mu_h) = J(v_h) - \langle\mu_h, \gamma_0 v_h - q\rangle_{\Lambda(\partial\Omega)}. \tag{13}$$

Lemma 1. *Let $h_\Omega/h_{\partial\Omega}$ be sufficiently small. Then the following estimates hold.*

$$\left.\begin{aligned} &\sup_{v_h \in V_{h_\Omega}(\Omega)} \frac{\langle \mu, \gamma_0 v_h \rangle_{\Lambda(\partial\Omega)}}{\|v_h\|_{H^1(\Omega)}} \geqq \beta \|\mu\|_{H^{-1/2}(\partial\Omega)}, \\ &\sup_{v_h \in V_{h_\Omega}(\Omega)} \frac{\langle \mu, \gamma_0 v_h \rangle_{\Lambda(\partial\Omega)}}{\|v_h\|_{H^1(\Omega)}} \geqq \beta h_{\partial\Omega}^{1/2} \|\mu\|_{L_2(\partial\Omega)} \quad \text{for any} \quad \mu \in \tilde{\Lambda}_{h_{\partial\Omega}}(\partial\Omega), \end{aligned}\right\} \tag{14}$$

where the constant $\beta > 0$ is independent of h.

Theorem 3. *There exists a unique solution $[u_h^*, \lambda_h^*] \in V_{h_\Omega}(\Omega) \times \Lambda_{h_{\partial\Omega}}(\partial\Omega)$ of the problem $(\mathscr{P}_h^*)$.* (Clearly exists a saddle-point of the Lagrangian $\mathscr{L}$ over finite-dimensional spaces $V_{h_\Omega}(\Omega) \times \Lambda_{h_{\partial\Omega}}(\partial\Omega)$. We note that the weak lower semi-continuity and coerciveness of J and (14) (imply the uniqueness of the saddle-point of the Lagrangian $\mathscr{L}$.)

Error estimates for variational inequalities

Let $\mathscr{H}(\Omega) = H^1(\Omega) \times H^{-1/2}(\partial\Omega)$ be a Hilbert space with norm

$$|||V||| = (\|v\|^2_{H^1(\Omega)} + \|\mu\|^2_{H^{-1/2}(\partial\Omega)})^{1/2} \tag{15}$$

where $V = (v, \mu) \in \mathscr{H}(\Omega)$. Let

$$\mathscr{A}(U, V) = (u, v)_{H^1(\Omega)} - \langle \lambda, \gamma_0 v \rangle_{\Lambda(\partial\Omega)} + \langle \mu, \gamma_0 u \rangle_{\Lambda(\partial\Omega)}, \tag{16}$$

be a continuous symmetric bilinear form on $\mathscr{H}(\Omega) \times \mathscr{H}(\Omega)$ and let $\mathbb{K}(\Omega) = H^1(\Omega) \times \Lambda(\partial\Omega)$ be a non-empty closed convex subset of $\mathscr{H}(\Omega)$. We denote by M a positive constant such that

$$|\mathscr{A}(U, U)| \leqq M\, |||U|||\, |||V||| \quad \text{for any} \quad U, V \in \mathscr{H}(\Omega).$$

On the other hand we see that

$$\mathscr{A}(U, U) = (u, u)_{H^1(\Omega)}. \tag{17}$$

Then we consider the variational problem $(\tilde{\mathscr{P}})$ which corresponds to the following data, find

$$\left.\begin{aligned} &U^* = [u^*, \lambda] \in \mathbb{K}(\Omega) \quad \text{such that} \\ &\mathscr{A}(U^*, V - U^*) \geqq \mathscr{F}(V - U^*) \quad \text{for any} \quad V \in \mathbb{K}(\Omega), \end{aligned}\right\} \tag{18}$$

where

$$\mathscr{F}(V) = \int_\Omega fv \, d\Omega + \langle \mu, g \rangle_{\Lambda(\partial\Omega)}.$$

Lemma 2. *The problem $(\tilde{\mathscr{P}})$ is equivalent to the problem $(\mathscr{P}^*)$.*

Internal approximation of a problem $\tilde{\mathscr{P}}$

We set $K(h_\Omega, h_{\partial\Omega}) = V_{h_\Omega}(\Omega) \times \Lambda_{h_{\partial\Omega}}(\partial\Omega)$. Then $K(h_\Omega, h_{\partial\Omega})$ is the internal approximation of the convex set $K(\Omega)$. We look for the element

$$\left.\begin{array}{l} U_h^* = [u_h^*, \lambda_h^*] \in K(h_\Omega, h_{\partial\Omega}) \quad \text{such that} \\ \mathscr{A}(U_h^*, \; - V_h - U_h^*) \geqq \mathscr{F}(V_h - U_h^*) \quad \text{for any} \quad V_h \in K(h_\Omega, h_{\partial\Omega}). \end{array}\right\} \tag{19}$$

It is easy to see that (19) is equivalent to $(\mathscr{P}_h^*)$.

Lemma 3. *We have*

$$\|u^* - u_h^*\|_{H^1(\Omega)}^2 \leqq \{\mathscr{A}(U^* - U_h^*, U^* - V_h^*) + \mathscr{A}(U^*, V_h - U^*) + \mathscr{F}(U^* - V_h)\} \quad \textit{for any} \quad V_h \in K(h_\Omega, h_{\partial\Omega}). \tag{20}$$

Lemma 4. *We have*

$$\left.\begin{array}{l} \|\lambda^* - \lambda_h^*\|_{H^{-1/2}(\partial\Omega)} \leqq \left\{C\, \|u^* - u_h^*\|_{H^1(\Omega)} + \inf\limits_{\mu_h \in \Lambda_{h\partial\Omega}(\partial\Omega)} \|\lambda^* - \mu_h\|_{H^{-1/2}(\partial\Omega)}\right\}, \\ \|\lambda^* - \lambda_h^*\|_{L_2(\partial\Omega)} \leqq C h_{\partial\Omega}^{-1/2}\, \|\lambda^* - \lambda_h^*\|_{H^{-1/2}(\partial\Omega)} \quad \textit{if} \quad \lambda^* \in L_2(\partial\Omega). \end{array}\right\} \tag{21}$$

Theorem 4. *Let u^* denote the solution of the minimization problem $(\mathscr{P})$ such that $u^* \in K(\Omega) \cap H^2(\Omega), \gamma_0 u^*, g \in H^1_\infty(a_i, a_{i+1}), i = 1, 2, \ldots, m$. Let the set of points where u^* changes from $u^* > g$ to $u^* = g$ be finite. Then, we have*

$$\left.\begin{array}{l} \|u^* - u_h^*\|_{H_1(\Omega)} \leqq C(u^*, g, \lambda^*)\, [h_\Omega + h_{\partial\Omega}], \\ \|\lambda^* - \lambda_h^*\|_{H^{-1/2}(\partial\Omega)} \leqq C(u^*, g, \lambda^*)\, [h_\Omega + h_{\partial\Omega}], \\ \|\lambda^* - \lambda_h^*\|_{L_2(\partial\Omega)} \leqq C(u^*, g, \lambda^*) h_{\partial\Omega}^{-1/2}\, [h_\Omega + h_{\partial\Omega}]. \end{array}\right\} \tag{22}$$

Numerical algorithms (based on duality) of problems $(\mathscr{P}_h^*)$

Our aim is to find a saddle-point of the Lagrangian

$$\mathscr{L}(v_h, \mu_h) = 1/2(v_h, v_h)_{H^1(\Omega)} - (f, v_h)_{L_2(\Omega)} - \langle \mu_h, \gamma_0 v_h - g_h \rangle_{\Lambda(\partial\Omega)} \tag{23}$$

over the space $V_{h_\Omega}(\Omega) \times \Lambda_{h_{\partial\Omega}}(\partial\Omega)$ by duality techniques — Uzawa's algorithm.

References

[1] Cea, J., Optimization, theorie et algorithmes. Dunod, Paris 1971.

[2] Glowinski, R., Lions, J. L., Trémolières, R., Analyse numérique des inéquations variationnelles. T. 1, 2. Dunod—Bordas, Paris 1976.

[3] Haslinger J., Lovíšek, J., Mixed variational formulation of unilateral problems. Comment. Math. Univ. Carolin. 21 (1980) 2.

[4] Lions, J. L., Magenes E., Problèmes aux limites non homogènes. T. 1, Dunod, Paris 1968.

Address of the authors:

J. Haslinger
Faculty of Mathematices and
Physics of Charles University
Malostranské nám. 2
118 00 Praha 1
Czechoslovakia

J. Lovíšek
Stavebná fakulta, SVŠT
Radlinského 11
Bratislava
Czechoslovakia

Sufficient conditions for optimality of a control problem with discrete time

Armin Hoffmann, Ilmenau

Abstract. The given sufficient conditions for optimality of a control problem with discrete time are similar to the so-called Bellman equations of dynamic programming. The occuring unknown functions are approximately computed by a set of first order and second order conditions. By using these approximations of the functions a local sufficient optimality condition is given for the problem.

Problem and preliminary assumptions

Let $\hat{X}_i \subseteq \mathfrak{R}^n$, $\hat{U}_j \subseteq \mathfrak{R}^{m_j}$ be non-empty open sets and let $(x_j, u_j) \to g_j(x_j, u_j) \in \mathfrak{R}^n$, $(x_j, u_j) \to f_j(x_j, u_j) \in \mathfrak{R}^1$, $x_{N+1} \to f_{N+1}(x_{N+1}) \in \mathfrak{R}^1$ be twice continuously differentiable functions defined on $\hat{X}_j \times \hat{U}_j$ and on X_{N+1}, respectively, for all $i = 1, 2, \ldots, N+1$; $j = 1, 2, \ldots, N$. The *gradients* $\nabla f_j(x_j, u_j)$ and $\nabla f_{N+1}(x_{N+1})$ are row-vectors and $\nabla g_j(x_j, u_j)$ is a matrix with n rows and $n + m_j$ columns. $\nabla_1 f_j(x_j, u_j)$, $\nabla_2 f_j(x_j, u_j)$ (analogous for g_j) denote the gradients with respect to the first or second argument, respectively. $\nabla^2 f_j(x_j, u_j)$ (analogous for f_{N+1}, g_j) denotes the *Hessian* of f_j at (x_j, u_j). We write also

$$\nabla^2 f_j(x_j, u_j) = \begin{pmatrix} \nabla^2_{11} f_j(x_j, u_j) & \nabla^2_{12} f_j(x_j, u_j) \\ \nabla^2_{21} f_j(x_j, u_j) & \nabla^2_{22} f_j(x_j, u_j) \end{pmatrix}$$

using *partial gradients*.

Consider the following *problem* **P**:

Minimize the functional (cf. [2], [3])

$$\sum_{j=1}^{N} f_j(x_j, u_j) + f_{N+1}(x_{N+1}) \tag{1}$$

subject to the constraints

$$x_{j+1} = g_j(x_j, u_j), \qquad j = 1, 2, \ldots, N, \tag{2}$$

$$\begin{aligned} x_i &\in X_i \subseteq \hat{X}_i, \qquad i = 1, 2, \ldots, N+1, \\ u_j &\in U_j \subseteq \hat{U}_j, \qquad j = 1, 2, \ldots, N. \end{aligned} \tag{3}$$

We assume that the Problem **P** has at laest one solution. A point $(x, u) := (x_1, \ldots, x_{N+1}, u_1, \ldots, u_N)$ satisfying (2) and (3) is called an *admissible point* of the problem **P**. A point $(\hat{x}, \hat{u})$ solving the problem **P** with respect to a neighbourhood W of $(\hat{x}, \hat{u})$ is called a *local solution* of the Problem **P**.

We use the following abbreviations: $(x_i^1, \ldots, x_i^n) =: x_i^T$ $(i = 1, 2, \ldots, N+1)$, u_j^T analogous; x_i and u_j are the related column-vectors. $f_j(x_j, u_j) =: f_j$, $f_j(\bar{x}_j, \bar{u}_j) =: \bar{f}_j$, analogous for $g_j, \bar{g}_j, \ldots, \nabla\bar{f}_j, \nabla^2\bar{f}_j, \nabla\bar{g}_j, \nabla^2\bar{g}_j, \ldots$ Let $h_j \in \mathfrak{R}^n$ be a vector, then $h_j^T\bar{g}_j$ means the real scalar product and $h_j^T \nabla\bar{g}_j$ means the usual matrix product between a row-vector and a matrix, analogous for similar products.

$\forall\, i$ means: i runs from 1 step 1 to $N+1$,
$\forall\, j$ means: j runs from 1 step 1 to N.

Results

The following Theorem 1 gives a sufficient condition for an admissible point to be a solution of the Problem **P** (cf. [1] pp. 73—84).

Theorem 1. *Let* $(\bar{x}, \bar{u})$ *be admissible and* $\forall\, i\, \forall j\ (x_j, u_j) \to p_j(x_j, u_j) \in \mathfrak{R}^n$, $x_i \to q_{i-1}(x_i) \in \mathfrak{R}^n$ *be functions defined on* $\hat{X}_j \times \hat{U}_j$ *and* $\hat{X}_i$, *respectively.*

If the following conditions (4), (5) *and* (6) *can be fulfilled then* $(\bar{x}, \bar{u})$ *is a solution of the problem* **P**.

$$\forall\, i, j\ \forall\, x_i \in X_i\ \forall\, u_j \in U_j: \tag{4}$$
$$-q_0(x_1)^T (x_1 - \bar{x}_1) \geqq 0,$$

$$\big(p_j(x_j, u_j) - q_j(g_j(x_j, u_j))\big)^T (g_j - \bar{g}_j) \geqq 0, \tag{5}$$

$$f_j - \bar{f}_j - p_j^T(g_j - \bar{g}_j) + q_{j-1}^T(x_j - \bar{x}_j) \geqq 0,$$
$$f_{N+1} - \bar{f}_{N+1} + q_N^T(x_{N+1} - \bar{x}_{N+1}) \geqq 0. \tag{6}$$

The relations contain the unknown functions q_{i-1} and p_j which can be compared with the so-called Bellman functions F_{i+1} (cf. [3] p. 297) of dynamic programming. However, be aware of the fact by dynamic programming is minimized over u_j and the conditions like (6) become an equality and must hold over all $x_j \in X_j$. In Theorem 1 is minimized over $x_j \in X_j$, $u_j \in U_j$, $x_{N+1} \in X_{N+1}$, i.e. it holds an inequality for all $x_i \in X_i$, $u_j \in U_j$. If X_i, U_j are not finite sets the determination of F_{i+1} is very difficult. Our conditions allow to determine q_{i-1}, p_j approximately (Theorem 2, 3) and set up a set of local conditions giving a local sufficient optimality condition for the problem **P** (Theorem 4). Finally the fitted conditions seem to be useful by using penalty methods, modified Lagrangian methods or other related methods basing on sequentially unconstrained minimization (see formulas (6), (9)—(12)). The investigations on this topic are running and have not been closed yet.

Remarks. The condition (5) is satisfied if we choose

$$p_j(x_j, u_j) := q_j\big(g_j(x_j, u_j)\big) \qquad \forall\, j\, \forall\, x_j \in \hat{X}_j\, \forall\, u_j \in \hat{U}_j. \tag{7}$$

We now assume generally that relation (7) *is fulfilled.*

Let $T_r(X, x)$ be the *tangent cone of the order r of the set X at the point* $x \in \mathrm{cl}\,(X)$, i.e.

$$T_r(X, x) := \left\{z \in \mathfrak{R}^n \mid o_z, \{t_n\} : x + t_n z + o_z(t_n^r) \in X,\ t_n \downarrow 0,\ \lim_{t \downarrow 0} \frac{o_z(t)}{t} = 0\right\}.$$

It holds $r' \geqq r \geqq 1 \Rightarrow T_r(X, x) \supseteq T_{r'}(X, x) \supseteq \mathrm{int}\, T_1(X, x)$.

Theorem 2. *If* (7) *holds and* $(\bar{x}, \bar{u})$ *is an admissible point then holds:*

in the case $q_{i-1}(x_i) := h_{i-1} + R_{0,i-1}(x_i, \bar{x}_i) \quad \forall\, i \,\forall\, x_i \in X_i$

$$\text{with} \qquad h_0 = 0 \quad \text{and} \quad \lim_{x_i \to \bar{x}_i} \|R_{0,i-1}(x_i, \bar{x}_i)\| = 0 \tag{8a}$$

the conditions

$$\left.\begin{aligned} &\forall\, i, j \,\forall\, z_i \in T_1(X_i, \bar{x}_i), \qquad v_j \in T_j(U_j, \bar{u}_j): \\ &\varphi_{1j}(z_j, v_j) := (\nabla_1 \bar{f}_j + h_{j-1}^T - h_j^T \nabla_1 \bar{g}_j, \nabla_2 \bar{f}_j - h_j^T \nabla_2 \bar{g}_j) \begin{pmatrix} z_j \\ v_j \end{pmatrix} \geqq 0, \\ &\varphi_{1N+1}(z_{N+1}) := (\nabla_1 \bar{f}_{N+1} + h_N^T)\, z_{N+1} \geqq 0 \end{aligned}\right\} \tag{9}$$

are necessary for the validity of the inequalities (4) *and* (6); *in the case*

$$q_{i-1}(x_i) := h_{i-1} + D_{i-1}(x_i - \bar{x}_i) + R_{1,i-1}(x_i, \bar{x}_i) \;\forall\, i \,\forall\, x_i \in X_i$$

$$\text{with} \qquad h_0 = 0 \quad \text{and} \quad \lim_{x_i \to \bar{x}_i} \|R_{1,i-1}(x_i, \bar{x}_i)\| / \|x_i - \bar{x}_i\| = 0 \tag{8b}$$

the conditions (9),

$$\forall\, z_1 \in T_1(X_1, x_1) : z_1^T(-D_0)\, z_1 \geqq 0, \tag{10}$$

$$\left.\begin{aligned} &\forall\, i, j \,\forall\, z_i \in T_2(X_i, x_i), \quad v_j \in T_2(U_j, u_j): \text{ (9) becomes an equality:} \\ &\varphi_{2j}(z_j, v_j) := (z_j^T, v_j^T) \left[\nabla^2 \bar{f}_j - \nabla^2 h_j^T \bar{g}_j + 2 \begin{pmatrix} D_{j-1}^T & 0 \\ 0 & 0 \end{pmatrix} \right. \\ &\qquad \left. - 2\nabla \bar{g}_j^T D_j^T \nabla \bar{g}_j \right] \begin{pmatrix} z_j \\ v_j \end{pmatrix} \geqq 0, \\ &\varphi_{2N+1}(z_{N+1}) := z_{N+1}^T [\nabla^2 \bar{f}_{N+1} + 2D_N^T]\, z_{N+1} \geqq 0 \end{aligned}\right\} \tag{11}$$

are necessary for the validity of the inequalities (4) *and* (6).

Remark. From (11) follows for $v_j = 0$ or $z_j = 0 : \forall\, j$

$$z_j^T(\nabla_1^2 \bar{f}_j - \nabla_1^2 h_j^T \bar{g}_j + 2D_{j-1}^T - 2\nabla_1 \bar{g}_j^T D_j^T \nabla_1 \bar{g}_j^T)\, z_j \geqq 0, \tag{11a}$$

$$u_j^T(\nabla_2^2 \bar{f}_j - \nabla_2^2 h_j^T \bar{g}_j - 2\nabla_2 \bar{g}_j^T D_j^T \nabla_2 g_j^T)\, u_j \geqq 0. \tag{11b}$$

We denote by $D_{i-1}(\forall\, i)$ *square matrices of the order* n.

Theorem 3. *If* $\bar{x}_i \in \operatorname{int} X_i$, $\bar{u}_j \in \operatorname{int} U_j(\forall\, i;\, j)$ *and* $(\bar{x}, \bar{u})$ *is a unique local solution of the problem* **P**, *then there are positive constants* $C_1, C_2, \ldots, C_{N+1}$ *such that the relations* (10), (11a), (11b) *and the condition with the number* $N + 1$ *from* (11) *are fulfilled whenever the matrices* $D_{i-1}(\forall\, i)$ *are chosen in the following way:*

$$\left.\begin{aligned} &D_N^T := -\frac{1}{2} \nabla_1^2 \bar{f}_{N+1} + \varepsilon_N E \;(0 \leqq \varepsilon_N \leqq C_{N+1}) \left(E := (\delta_{kl})_{nn}\right), \\ &D_{j-1}^T := -\frac{1}{2} \nabla_1^2 \bar{f}_j + \frac{1}{2} \nabla_1^2 h_j^T \bar{g}_j + \nabla_1 \bar{g}_j^T D_j^T \nabla_1 \bar{g}_j + \varepsilon_{j-1} E \\ &(0 \leqq \varepsilon_{j-1} \leqq C_j). \end{aligned}\right\} \tag{12}$$

$D_{i-1}(\forall\, i)$ *are negative semidefinite and in the case* $\varepsilon_{i-1} < C_i(\forall\, i)$ *negative definite.*

Theorem 4. (cf. [1] p. 29). *Let* $(\bar{x}, \bar{u})$ *be admissible and let* $(\forall\ i)$ C_i *and* d_i *be positive constants with*

$$C_j(\|x_j - \bar{x}_j\|^2 + \|u_j - \bar{u}_j\|^2) \leqq \varphi_{1j}(x_j - \bar{x}_j, u_j - \bar{u}_j) + \frac{1}{2}\,\varphi_{2j}(x_j - \bar{x}_j, u_j - \bar{u}_j),$$

$$C_{N+1}\,\|x_{N+1} - \bar{x}_{N+1}\|^2 \leqq \varphi_{1N+1}(x_{N+1} - \bar{x}_{N+1}) + \frac{1}{2}\,\varphi_{2N+1}(x_{N+1} - \bar{x}_{N+1})$$

for all (x, u) *satisfying* (3) *with respect to a certain neighbourhood of* $(\bar{x}, \bar{u})$.

Then $(\bar{x}, \bar{u})$ *is a local solution of the problem* **P**.

References

[1] Hoffmann, A., Ein hinreichendes Optimalitätskriterium bei Optimierungsproblemen in Banachräumen und seine Anwendungsmöglichkeiten. Dissertation, TU Dresden, 1975.

[2] Йоффе, А. Д., Тихомиров, В. М., Теория экстремальных задач. Наука, Москва 1974.

[3] Boltianski, V; Commande optimale de systèmes discrets (traduit du russe). Mir, Moscou 1976.

Address of the author:

Technische Hochschule Ilmenau
Sektion Mathematik, Rechentechnik und ökonomische Kybernetik
6300 Ilmenau
Am Ehrenberg, Block G
DDR

A controllability problem of nonlinear dynamical systems

T. B. Kopeikina, Minsk

A controllability problem of dynamical systems takes an important place in the general theory of optimal control. While solving many applied problems, before seeking the control, transfering the initial state x_0 of the system

$$\dot{x}(t) = f(x(t), u(t), t), \qquad x(t_0) = x_0, \qquad x \in \mathbb{R}^n, \qquad u \in \mathbb{R}^r, \tag{1}$$

to the terminal state x_1 and minimizing some functional on the trajectories of (1), it is necessary to clear the matter up: does there exist even one admissible control transfering the trajectory $x(t)$ from the initial state x_0 to the terminal state x_1? This basic question is the problem of controllability. A controllability problem is closely connected with many other properties of the system (1), for example, with such as stabilization, reconstructability, invertibility, modal controllability, etc.

The theory of controllability has been formed as independent division of the theory of optimal processes following the report of R. E. Kalman at the First IFAC Congress in 1960 [1]. According to R. E. Kalman a dynamical system

$$\dot{x}(t) = Ax(t) + Bu(t), \qquad x(t_0) = x_0, \qquad x \in \mathbb{R}^n, \qquad u \in \mathbb{R}^r, \tag{2}$$

is called completely controllable if for any state x_0 there is a number t_1 and a piecewise continuous control $u(t)$, $t_0 \leqq t \leqq t_1$, such that the trajectory $x(t)$ of the system (2) satisfies the condition $x(t_1) = 0$. In order that the system (2) be completelyc ontrollable it is necessary and sufficient that

$$\operatorname{rank} \{B, AB, \ldots, A^{n-1}B\} = n.$$

The controllability notion introduced by R. E. Kalman had been extended then to nonlinear systems and studied by many authors [2—10].

Controllability of the nonlinear system $\dot{x}(t) = f(x(t), u(t))$, $f(0, 0) = 0$, $x(0) = x_0$ in a direction in the critical case, i.e. in the case when the corresponding linear model $\dot{x}(t) = Ax(t) + Bu(t)$ is not controllable in this direction, is considered in this paper.

1. A formula for variations relative to the controls

Let us consider a dynamical system whose behaviour may be described by the equation

$$\dot{x}(t) = f(x(t), u(t)), \qquad x(0) = 0, \qquad f(0, 0) = 0, \qquad t \in T = [0, t_1], \tag{3}$$

where x is an n-dimensional state vector, $u(t)$ is a scalar control belonging to the set U of piecewise continuous functions, $f(x, u)$ is a vector-function which is assumed to be continuous together with its partial derivatives with respect to x, u up to the second order inclusively. Suppose that to each admissible control $u(t)$, $t \in T$, there corresponds a unique trajectory $x(t)$ of the system (3) starting from the origin $x(0) = 0$.

Let us find the conditions under which for any state $x_1 \in \mathbb{R}^n$ there exists an admissible control $u(t) \in U$, $\|u(t)\| \leqq L$, transfering the trajectory $x(t)$ of the system (3) from $x(0) = 0$ to $x(t_1) = x_1$.

Definitions [11]. 1. The origin of coordinates is said to be *TL-controllable in the direction* p, $p \in \mathbb{R}^n$, $\|p\| = 1$, $\|p\| = \left(\sum_{i=1}^{n} p_i^2\right)^{1/2}$, if there exists such number α_0, $\alpha_0 = \alpha_0(t_1, L, p) > 0$, that for each α, $|\alpha| \leqq \alpha_0$ there is an admissible control $u(t) \in U$, $|u| \leqq L$, such that $p'x(t_1) = \alpha$.

2. The dynamical system (3) is called *TL-controllable* if the origin of coordinates is TL-controllable in any direction $\left(\inf_{\|p\|=1} \alpha_0(t_1, L, p) > 0\right)$.

3. The origin of coordinates is said to be *controllable in the direction* p, $\|p\| = 1$, if it is TL-controllable in this direction for some $t_1 = t_1(p)$, $L = L(p) < \infty$.

4. The system (3) is *controllable* if the origin of coordinates is controllable in any direction $\left(\inf_{\|p\|=1} \alpha_0(t_1(p), L(p), p) > 0\right)$.

5. The origin of coordinate is *completely controllable in the direction* p if it is TL-controllable in this direction with any $t_1 > 0$, $L > 0$ ($\alpha_0(t_1, L, p) > 0$ for any $t_1 > 0$, $L > 0$).

6. The system (3) is *completely controllable* if the origin of coordinates is completely controllable in any direction ($\inf_{\|p\|=1} \alpha_0(t_1, L, p) > 0$ for any $t_1 > 0$, $L > 0$).

The method of investigation of controllability in a direction had been described in [11], where it was shown that with the aid of introducing the functional

$$I(u) = p'x(t_1) \tag{4}$$

on the trajectories of the system (3) and a formula for the increment of $I(u)$ with respect to control it is not difficult to get controllability conditions for the linear system

$$\dot{x}(t) = Ax(t) + bu(t) \tag{5}$$

or for the nonlinear system (3) under the assumption that its linear model (5) $\left(A = \frac{\partial f(0, 0)}{\partial x(t)},\ b = \frac{\partial f(0, 0)}{\partial u(t)}\right)$ is controllable (completely controllable). Namely, from the controllability of the system (5) we make the conclusion about the controllability of the nonlinear system (1) at some restraints of its right part.

Let us suppose however that the linear model (5) for the system (3) is not controllable (completely controllable).

Definition 7. A direction p, $\|p\| = 1$, is called *critical* for the system (3) if its linear model (5) is not TL-controllable (controllable, completely controllable) in this direction.

If the system (3) possesses at least one critical direction then we shall say that it is a *critical case* of the controllability problem of the system (3).

To investigate a critical case of controllability let us consider the functional (4) on the trajectories of the system (3) given in the form

$$I(u) = p'x(t_1) = \int_0^{t_1} [\dot{\psi}'(t)\, x(t) + \psi'(t)\, \dot{x}(t) + \dot{x}'(t)\, \Psi(t)\, x(t) + x'(t)\, \dot{\Psi}(t)\, x(t) + x'(t)\, \Psi(t)\, \dot{x}(t)]\, dt, \tag{6}$$

and let us substitute for $\dot{x}(t)$ in (6) expression $\dot{x}(t)$ written as

$$\dot{x}(t) = Ax(t) + bu(t) + \frac{1}{2}\, x'(t)\, Cx(t) + Dx(t)\, u(t) + \frac{1}{2}\, gu^2(t) + o(\|x\|^2, |u|\, \|x\|, |u|^2), \tag{7}$$

where

$$A = \frac{\partial f_i(0,0)}{\partial x_j(t)},\ b = \frac{\partial f_i(0,0)}{\partial u(t)},\ C = \frac{\partial^2 f_i(0,0)}{\partial x_j(t)\, \partial x_k(t)},\ D = \frac{\partial^2 f_i(0,0)}{\partial x_j(t)\, \partial u(t)},\ g = \frac{\partial^2 f_i(0,0)}{\partial u^2(t)},$$

and the n-vector-function $\psi(t)$ and $n \times n$-matrix function $\Psi(t)$ satisfy the differential equations

$$\begin{aligned} &\dot{\psi}(t) = -A'\psi(t), \qquad \psi(t_1) = p, \\ &\dot{\Psi}(t) = -A'\Psi(t) - \Psi(t)\, A - \frac{1}{2}\, \psi'(t)\, C, \qquad \Psi(t_1) = 0_n. \end{aligned} \tag{8}$$

Taking into account (7), (8), formula (6) takes the form

$$I(u) = \int_0^{t_1} \psi'(t)\, bu(t)\, dt + \frac{1}{2} \int_0^{t_1} \psi'(t)\, gu^2(t)\, dt + \int_0^{t_1} \psi'(t)\, Dx(t)\, u(t)\, dt + \int_0^{t_1} \big(b'\Psi(t)\, x(t) + x'(t)\, \Psi(t)\, b\big)\, u(t)\, dt + o_1(\|x\|^2, |u|\, \|x\|, |u|^2). \tag{9}$$

If p is a critical direction then it is clear [11] from (9) that $\psi'(t)\, b \equiv 0$, $t \in [0, t_1]$. Consequently having in mind that the leading term in $u(t)$ of the trajectory $x(t)$ may be described as $x(t) = \int_0^t F(t, \tau)\, bu(\tau)\, d\tau$, where $F(t, \tau) = F(t)\, F^{-1}(\tau)$ is a fundamental matrix of the solutions of (5) $\big(\dot{F}(t, \tau) = AF(t, \tau),\ F(t, t) = E_n\big)$ we have from (9)

$$I(u) = \frac{1}{2} \int_0^{t_1} \psi'(t)\, gu^2(t)\, dt + \int_0^{t_1}\int_0^{t} K(t, \tau)\, u(\tau)\, d\tau\, u(t)\, dt + o_1. \tag{10}$$

Here $K(t, \tau) = \psi'(t)\, DF(t, \tau) + b'\big(\Psi(t)\, F(t, \tau) + F'(t, \tau)\, \Psi(t)\big)\, b$.

2. Controllability of nonlinear systems

Let $u(t)$ be a solution of a homogeneous integral Volterra equation of the second kind:

$$u(t) = \lambda \int_0^t K(t, \tau)\, u(\tau)\, d\tau, \qquad \lambda \neq 0. \tag{11}$$

Then it follows from (10) that

$$I(u) = \int_0^{t_1} \left(\frac{1}{2}\, \psi'(t)\, g + \frac{1}{\lambda}\right) u^2(t)\, dt,$$

and the following results are true.

Theorem 1. *If $u(t)$ is a solution of* (11), $|u(t)| \leqq L$, $\det A \neq 0$, *all characteristic roots of A have nonzero imaginary parts, then the system* (3) *is TL-controllable* (*controllable, completely controllable*).

Theorem 2. *If the origin of coordinates of the system* (3) *is controllable in the critical direction p for sufficiently small $t_1 > 0$, $u(t)$ is a solution of the equation* (11), $|u| \leqq L$, *then*

$$\frac{1}{2}\, p'g + \frac{1}{\lambda} = 0, \qquad p'Ag = 0, \ldots, \qquad p'A^{n-1}g = 0.$$

Corollary 1. *If the system* (3) *is controllable in the critical case then for each critical direction p the condition*

$$\operatorname{rank}\left\{\frac{1}{2}\, g + \frac{1}{\lambda}\, p,\ Ag, \ldots, A^{n-1}g\right\} < n$$

is satisfied.

In the case when the set of admissible controls is a set of constant functions and t_1 is a sufficiently small value, controllability conditions of the system (1) transform to the following ones:

Lemma 1. *If the system* (3) *is controllable in the critical case then*

$$\frac{1}{2}\, \psi'(t)\, g + \int_0^t \left[\psi'(t)\, DF(t, \tau)\, b + b'\big(\Psi(t)\, F(t, \tau) + F'(t, \tau)\, \Psi(t)\big)\, b\right] d\tau \equiv 0. \tag{12}$$

From (12) for some partial cases of the system (3) it is not difficult to obtain the controllability conditions expressed directly through the parameters of the system (7).

Suppose that the quadratic model (7) of the system (3) is not essentially nonlinear, i.e. $C = 0_n$ (see, for example, [12]).

Theorem 3. *If the origin of coordinates of the system* (3) *is controllable in the critical direction p then it satisfies the following equalities:*

$$p'\, e^{At_1} Q_k = 0, \qquad k = 0, 1, 2, \ldots,$$

where

$$Q_k = (-1)^k \cdot \frac{1}{2} A^k g + \sum_{i=0}^{k-1} (-1)^{k-i-1} C_k^{i+1} A^{k-i-1} D A^i b, \qquad k = 0, 1, 2, \ldots$$

For the controllability of the system (3) *in the critical case it is necessary that*

$$\operatorname{rank} \{Q_k, k = 0, 1, 2, \ldots\} < n.$$

References

[1] Калман Р. Е., Об общей теории систем управления. Труды I конгресса ИФАК, т. 2, Изд-во АН СССР, Москва 1960.

[2] Петров Н. Н., Об управляемости автономных систем. Дифференц. уравнения 4 (1968) 4, 606—617.

[3] Rink, R. E., Mohler, R. R., Completely controllable bilinear systems. SIAM J. Control 6 (1968) 3, 477—486.

[4] Haynes, G. W., Hermes, H., Nonlinear controllability via Lie theory. SIAM J. Control 8 (1970) 4, 450—460.

[5] Davison, E. J., Kunze, E. G., Some sufficient conditions for the global and local controllability of nonlinear time-varying systems. SIAM J. Control 8 (1970) 4, 489—497.

[6] Ли Э. Б., Маркус Л., Основы теории оптимального управления. Наука, Москва 1971.

[7] Семенов В. Н., Об управляемости нелинейных динамических систем. Кибернет. и вычислит. техника, Респ. межвед. сб., вып. 8, 1971, 34—40.

[8] Sussman, H. J., Jurdjevic, V., Controllability of nonlinear systems. J. Diff. Equations 12 (1972) 1, 95—116.

[9] Коробов В. И., Управляемость, устойчивость некоторых нелинейных систем. Дифференц. уравнения 9 (1973) 4, 614—619.

[10] Yamamoto, Y., Controllability of nonlinear systems. J. Optimization Theory Appl. 22 (1977) 1, 41—50.

[11] Габасов Р., Кириллова Ф., Качественная теория оптимальных процессов. Наука, Москва 1971.

[12] Bruni, C., Di Pillo, G., Koch, G., Bilinear systems: an appealing class of "nearly linear" systems in theory and applications. IEEE Trans. Automat. Control 19 (1974) 4, 334—348.

Address of the author:

Институт математики АН БССР
Типографская, 11
Минск, ГСП
СССР

On the convergence of the finite elements method for some initial problem

Katarzyna Litewska and Jerzy Muszyński, Warsaw

The problem of the convergence of the succesive approximations given by the method of finite elements to the solution of the evolution problem was considered by many authors. The main difficulty in these considerations is the necessity to omit the troubles connected with the problem that there are given two initial conditions and none of them at the end of the time interval (see for example [3]). These conditions are natural for the hyperbolic problems (with such conditions the problems are well posed). Some modification of the time-space method of finite elements, given in [2] and considered among others by the authors of this paper, leads in the convergence problems to the separation of the convergence in space variables (this is known) and convergence for the system of linear differential equations of the second order with initial conditions. For this purpose the authors consider the simplest problem in this region as a starting point. Next will be given the method of finite elements for the given problem (it is in some sense near to the method given in [1]), and observations concerning its convergence.

Let $a, b \in R$, $x_0, v_0 \in R$ and $f: [0, 1] \to R$ be an integrable function. Consider the initial problem: equation

$$x'' + ax' + bx = f(t) \tag{1}$$

and initial conditions

$$x(0) = x_0, \qquad x'(0) = v_0. \tag{2}$$

This problem in this paper will be solved by the method of finite elements and afterwards it will be explained in what manner the approximate solutions tend to the known solution of the problem (1), (2).

Divide the interval $I = [0, 1]$ into n parts by the points $x_k = \frac{k}{n}$, $k = 0, 1, \ldots, n$, and consider $n + 1$ functions $\overset{n}{\varphi}_k: [0, 1] \to [0, 1]$, linear on every $\overset{n}{I}_k = \left[\frac{k}{n}, \frac{k+1}{n}\right]$ and such that $\overset{n}{\varphi}_k\left(\frac{r}{n}\right) = \delta_{rk}$, $k, r = 0, 1, \ldots, n$ (see fig. 1).

The approximate solutions $\overset{n}{x} = \overset{n}{x}(t)$ will be looked for in the form

$$\overset{n}{x} = \sum_{k=0}^{n} \overset{n}{\alpha}_k \overset{n}{\varphi}_k(t), \tag{3}$$

where $\overset{n}{\alpha}_k \in R$ are the unknown coefficients which will be found by the typical procedure in the method of finite elements, i.e. they will be found from the initial conditions (2) and from the "Galerkin rule" of the form

$$\int_0^1 \left[y'(t)\, \overset{n}{x}'(t) - y(t) \left(a\overset{n}{x}'(t) + b\overset{n}{x}(t) - f(t) \right) \right] dt = 0. \tag{4}$$

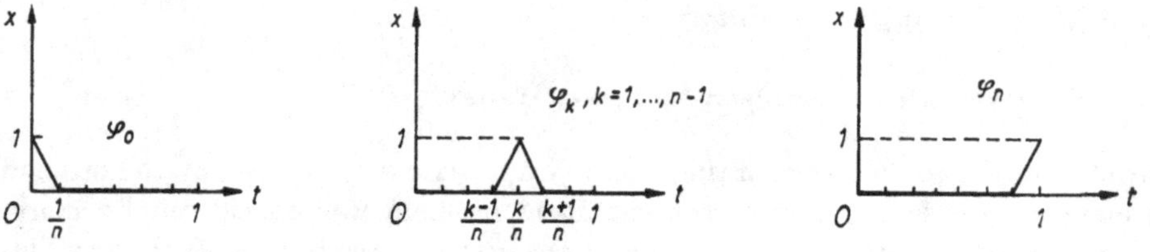

Fig. 1

For $y = y(t)$ will be taken the functions $\overset{n}{\varphi}_p$ for $p = 1, 2, \ldots, n-1$ (not all of the "basic" functions, only $n-1$ of them). More precisely the condition (4) will be considered in the form

$$\sum_{r=0}^{n-1} \int_{r/n}^{(r+1)/n} \left[\overset{n}{\varphi}_p'(t)\, \overset{n}{x}'(t) - \overset{n}{\varphi}_p(t) \left(a\overset{n}{x}'(t) + b\overset{n}{x}(t) - f(t) \right) \right] dt = 0,$$

$$p = 1, 2, \ldots, n-1. \tag{5}$$

The initial conditions will be considered in the form

$$\overset{n}{x}(0) = x_0, \qquad \overset{n}{x}'(0) = v_0, \tag{6}$$

where x_0, v_0 are the numbers given above. From the formulae (3) and conditions (6) we have

$$x_0 = \sum_{k=0}^{n} \overset{n}{\alpha}_k \overset{n}{\varphi}_k(0) = \alpha_0, \qquad v_0 = \sum_{k=0}^{n} \overset{n}{\alpha}_k \overset{n}{\varphi}_k'(0) = \alpha_0(-n) + \alpha_1 n$$

and thus

$$\alpha_0 = x_0, \qquad \alpha_1 = \frac{1}{n}(v_0 + nx_0). \tag{7}$$

The conditions (5) for $\overset{n}{x}$ given by (3) lead to

$$\sum_{k=0}^{n} \overset{n}{\alpha}_k \sum_{r=0}^{n-1} \int_{r/n}^{(r+1)/n} \left[\overset{n}{\varphi}_p'(t)\, \overset{n}{\varphi}_k'(t) - a\overset{n}{\varphi}_p(t)\, \overset{n}{\varphi}_k'(t) - b\overset{n}{\varphi}_p(t)\, \overset{n}{\varphi}_k(t) \right] dt = -\overset{n}{f}_p, \; p = 1, 2, \ldots, n-1, \tag{8}$$

where

$$\overset{n}{f}_p = \int_0^1 f(t)\, \overset{n}{\varphi}_p(t)\, dt. \tag{9}$$

Taking in account the form of the functions $\overset{n}{\varphi}_k$ we obtain that

$$\overset{n}{\alpha}_{k-1}\left(-n+\frac{a}{2}-\frac{b}{6n}\right)+\overset{n}{\alpha}_k\left(2n-\frac{2b}{3n}\right)+\overset{n}{\alpha}_{k+1}\left(-n-\frac{a}{2}-\frac{b}{6n}\right)=-\overset{n}{f}_k,$$

$$k = 1, 2, \ldots, n-1$$

or

$$\overset{n}{\alpha}_{k+1}+\overset{n}{\mathfrak{R}}_1\overset{n}{\alpha}_k+\overset{n}{\mathfrak{R}}_2\overset{n}{\alpha}_{k-1}=\overset{n}{p}_k, \qquad k = 1, 2, \ldots, n-1, \tag{10}$$

where

$$\overset{n}{\mathfrak{R}}_1=\frac{-12n^2+4b}{6n^2+3an+b}, \quad \overset{n}{\mathfrak{R}}_2=\frac{6n^2-3an+b}{6n^2+3an+b}, \quad \overset{n}{p}_k=\frac{6n\overset{n}{f}_k}{6n^2+3an+b}. \tag{11}$$

Solving the reccurence equation (10) with given (by (7)) initial conditions $\overset{n}{\alpha}_0$, $\overset{n}{\alpha}_1$ we obtain that

$$\overset{n}{\alpha}_k=\frac{1}{\sqrt{\overset{n}{\Delta}}}\left\{\frac{ny_0\left(\overset{n}{\lambda}_2-1\right)-y_1}{n}\left(\overset{n}{\lambda}_1\right)^k+\frac{ny_0\left(1-\overset{n}{\lambda}_1\right)+y_1}{n}\left(\overset{n}{\lambda}_2\right)^k\right.$$
$$\left.+\sum_{s=0}^{k-1}\overset{n}{p}_{s+1}\left[\left(\overset{n}{\lambda}_2\right)^{k-s-1}-\left(\overset{n}{\lambda}_1\right)^{k-s-1}\right]\right\}, \tag{12}$$

where $k = 2, 3, \ldots, n$ and

$$\overset{n}{\Delta}=\frac{36n^2(a^2-4b)+12b^2}{(6n^2+3an+b)^2}, \quad \overset{n}{\lambda}_1=\frac{-\overset{n}{\mathfrak{R}}_1-\sqrt{\overset{n}{\Delta}}}{2}, \quad \overset{n}{\lambda}_2=\frac{-\overset{n}{\mathfrak{R}}_1+\sqrt{\overset{n}{\Delta}}}{2}. \tag{13}$$

In the case where $a^2 - 4b + \dfrac{b^2}{3n^2} < 0$ the numbers $\overset{n}{\lambda}_1, \overset{n}{\lambda}_2, \sqrt{\overset{n}{\Delta}} \in C$, but obtained as a result of the formulae (12) the approximate solution $\overset{n}{x}$ will be real.

We obtain all the coefficients $\overset{n}{\alpha}_k$, $k = 0, 1, \ldots, n$ and thus from the equation (3) we have the approximate solution $\overset{n}{x}$. On the interval $\overset{n}{I}_k = \left[\dfrac{k}{n}, \dfrac{k+1}{n}\right]$ all the functions $\overset{n}{\varphi}_p$, except $\overset{n}{\varphi}_k$ and $\overset{n}{\varphi}_{k+1}$, are equal to zero, and $\overset{n}{\varphi}_k\big|_{\overset{n}{I}_k} = -nx + k + 1$, $\overset{n}{\varphi}_{k+1}\big|_{\overset{n}{I}_k} = nx - k$. Then from the formula (3) we have

$$\overset{n}{x}\big|_{\overset{n}{I}_k}=\overset{n}{\alpha}_k(-nx+k+1)+\overset{n}{\alpha}_{k+1}(nx-k), \tag{14}$$

where $\overset{n}{\alpha}_k$, $\overset{n}{\alpha}_{k+1}$ are given by (12).

The investigation of the formulae (14) shows that:

1. for any $t \in I_k$ we have $\lim_{n\to\infty} \overset{n}{x}(t) = x(t)$, where $x = x(t)$ is the exact solution of the problem (1), (2), in any case of the sign of $a^2 - 4b$,

2. taking in I_k

$$\overset{n}{x}'\Big|_{\overset{n}{I_k}} = -n\overset{n}{\alpha}_k + n\overset{n}{\alpha}_{k+1}$$

we have $\lim_{n\to\infty} \overset{n}{x}'(t) = x'(t)$ for any $t \in I_k$.

The proofs of these facts are very hard and complicated. We have to take in account that for any fixed t the interval $\overset{n}{I_k}$, containing this t, depends on n, we have then $k = k(n)$, but $\frac{k(n)}{n}$ and $\frac{k(n)+1}{n}$ tend to t when $n \to \infty$. In the case when $f \neq 0$ in the considerations concerning the convergence we used the definition of the Stieltjes integral. We omit these proofs, which are simple from the principal point of view but analytically very hard.

In forthcoming papers the authors will generalize these results to systems of second order linear equations and to nonlinear equations.

Bibliography

[1] Argyris, J. H., Scharpf, D. D., Finite elements in tim e and space, Aeronaut. J. Roy. Aeronaut Soc. **73** (1969).

[2] Kączkowski, Z., The method of finite space-time elements in dynamics of structures, J. Tech. Phys. **16** (1975), 1.

[3] Mitchell, A. R., Wait, R., The finite element method in partial differential equations, Chesterfield and others, 1977.

Address of the authors:

Mathematical Institute
Technical University of Warsaw
Pl. Jednosci Robotniczej 1
00-661 Warszawa
Poland

On the normality of Lagrange multiplier rules for optimal control problems

Rainer Lorentz, Ilmenau

In recent years many authors published papers concerning necessary conditions of optimality for abstract optimization problems, see e.g. Girsanov [2], Ioffe/Tikhomirov [4], Warga [8], Kurcyusz [5], Dolecki [1], and Goebel [3]. Most of them obtained necessary conditions of optimality of John type. Now it is important to find an answer to the following question: Are there multipliers with a vanishing multiplier of the objective functional or not? In the first case we will say that the Lagrange multiplier rule (L-m-r) is abnormal, in the second case we will say that L-m-r is normal. Our aim is to give a necessary and sufficient condition of normality of a L-m-r for a special problem of optimal control formulated in Banach spaces.

We consider the following problem:

$$(P) \qquad \begin{cases} \text{Minimize } J(x, u) \\ \text{subject to } T(x, u) := x - g(u)\, KF(x) = \Theta, \qquad u \in U. \end{cases}$$

We assume that the following statements are fulfilled: Let B_1, B_2, and B_3 be real Banach spaces, J a real-valued Fréchet differentiable functional defined on $B_1 \times B_2$, g a real-valued continuously Fréchet differentiable functional defined on B_2, $F: B_1 \to B_3$ a continuously Fréchet differentiable operator, and $K: B_3 \to B_1$ a continuous linear operator. Further let $U \subseteqq B_2$ be a convex set with non-empty interior, $T_x(x, u) = I - g(u)\, KF_x(x)$ a Fredholm operator, and the pair $(x^0, u^0) \in B_1 \times B_2$ a solution of (P).

By *nontrivial Lagrange multipliers* for (P) at $(x^0, u^0) \in B_1 \times B_2$ we mean a pair $(a_0, v^*) \in R^1 \times B_1^*$ satisfying

$$(a_0, v^*) \neq \Theta, \qquad a_0 \geqq 0; \tag{1}$$

$$\big(I - g(u^0)\, F_x^*(x^0)\, K^*\big)\, v^* = a_0 J_x(x^0, u^0); \tag{2}$$

$$\big\langle u - u^0, \langle KF(x^0), v^*\rangle\, g_u(u^0) + a_0 J_u(x^0, u^0)\big\rangle \geqq 0 \quad \text{for all} \quad u \in U. \tag{3}$$

By *normal Lagrange multipliers* for (P) at $(x^0, u^0) \in B_1 \times B_2$ we mean a pair $(a_0, v^*) \in R^1 \times B_1^*$ satisfying (2), (3) with $a_0 = 1$.

If there are only normal Lagrange multipliers for (P) at $(x^0, u^0) \in B_1 \times B_2$, then we say that the L-m-r is *normal* at $(x^0, u^0) \in B_1 \times B_2$, otherwise we say that the L-m-r is *abnormal* at $(x^0, u^0) \in B_1 \times B_2$.

For holding a L-m-r for (P) at $(x^0, u^0) \in B_1 \times B_2$ it is sufficient that the above-mentioned assumptions are fulfilled (Lorentz [6]).

Let be

$$Q_1 := \{\hat{u} \in U \mid \langle u - \hat{u}, g_u(\hat{u})\rangle \geqq 0 \quad \text{for all} \quad u \in U\},$$

$$Q_2 := \{\hat{u} \in U \mid \langle u - \hat{u}, g_u(\hat{u})\rangle \leqq 0 \quad \text{for all} \quad u \in U\},$$

$$Q := Q_1 \cup Q_2.$$

The set $Q \subseteqq B_2$ is the set of all points $u \in U$ satisfying the necessary condition of optimality of the problems:

$$\begin{cases} \text{minimize } g(u) \\ \text{subject to } u \in U \end{cases} \quad \text{or} \quad \begin{cases} \text{maximize } g(u) \\ \text{subject to } u \in U. \end{cases}$$

Further, we denote by n the dimension of the Fredholm operator $T_x(x^0, u^0)$, and in case $n = 1$ let be $q^* \in N(T_x^*(x^0, u^0))$, $q^* \neq \Theta$.

Theorem 1. $n = 0$: *The L-m-r is normal.*
$n = 1$: *The L-m-r is normal iff* $\langle x^0, q^*\rangle \neq 0$ *and* $u^0 \notin Q$.
$n \geqq 2$: *The L-m-r is abnormal.*

Proof. If $n = 0$ the result is obvious, because the operator $(T_x(x^0, u^0))^{-1}$ exists. If $n = 1$ we distinguish the following cases:

(a) $\langle x^0, q^*\rangle = 0$,

(b) $\langle x^0, q^*\rangle \neq 0, \quad u^0 \in Q$,

(c) $\langle x^0, q^*\rangle \neq 0, \quad u^0 \notin Q$.

In case (a) it is easy to see that the pair $(0, q^*) \in R^1 \times B_1^*$ fulfils (1)—(3), because $g(u^0)\langle KF(x^0), v^*\rangle = \langle x^0, v^*\rangle$. Hence, the L-m-r is abnormal at $(x^0, u^0) \in B_1 \times B_2$.

In case (b) the pair $(0, aq^*) \in R^1 \times B_1^*$ with

$$a = \begin{cases} \operatorname{sign}\left(g(u^0)\langle x^0, q^*\rangle\right), & u^0 \in Q_1 \setminus Q_2 \\ -\operatorname{sign}\left(g(u^0)\langle x^0, q^*\rangle\right), & u^0 \in Q_2 \end{cases}$$

satisfies (1)—(3), therefore the L-m-r is abnormal.

In case (c) we assume that the L-m-r is abnormal. Then we get a $v^* \in B_1^*$, $v^* \neq \Theta$; with $T_x^*(x^0, u^0)\, v^* = \Theta$, that is $v^* = aq^*$ with $a \neq 0$. Consequently, we conclude

$$a \frac{1}{g(u^0)} \langle u - u^0, g_u(u^0)\rangle \langle x^0, q^*\rangle \geqq 0 \quad \text{for all} \quad u \in U.$$

Because $\langle x^0, q^*\rangle \neq 0$ and $u^0 \notin Q$ we find $a = 0$ and $v^* = \Theta$, respectively, in contradiction to (1). Hence, the L-m-r is normal at $(x^0, u^0) \in B_1 \times B_2$.

Finally let be $n \geqq 2$, and let $q_1^*, q_2^* \in N(T_x^*(x^0, u^0))$ be linearly independent elements. If $\langle x^0, q_1^*\rangle = 0$, then the pair $(0, q_1^*) \in R^1 \times B_1^*$ satisfies (1)—(3).

If $\langle x^0, q_1^*\rangle \neq 0$, then the pair

$$(0, \langle x^0, q_2^*\rangle\, q_1^* - \langle x^0, q_1^*\rangle\, q_2^*) \in R^1 \times B_1^*$$

fulfils (1)—(3).

Now the proof is complete.

The following condition is well-known in literature (see e.g. Ioffe/Tikhomirov [4]):

(*) *There is a real number* $r > 0$ *such that for all* $q \in S(\Theta, r) \subseteqq B_1$ *there exists a pair* $(\bar{x}, \bar{u}) \in B_1 \times U$ *satisfying*

$$\big(I - g(u^0)\, KF_x(x^0)\big)\, \bar{x} + \langle \bar{u} - u^0, g_u(u^0)\rangle\, KF(x^0) = q.$$

It is easy to show that condition (*) is sufficient for normality of the above-mentioned L-m-r of problem (P). Furthermore condition (*) is necessary for normality of the L-m-r. This can be shown as follows.

In case $n = 0$ the result is trivial.

In case $n = 1$ we have to show that there is a real number $r > 0$ such that for all $q \in S(\Theta, r) \subseteqq B_1$ there is a $\bar{u} \in U$ with

$$\big\langle q - \langle \bar{u} - u^0, g_u(u^0)\rangle\, KF(x^0), q^*\big\rangle = 0.$$

Now using theorem 1 we get $u^0 \notin Q$ and $\langle x^0, q^*\rangle \neq 0$. Consequently, if r is sufficiently small, there is for all $q \in S(\Theta, r)$ a $\bar{u} \in U$ with the mentioned property, because the set U is a convex set.

In case $n \geqq 2$ normality is impossible.

These results we summarize:

Theorem 2. *The L-m-r* (1)—(3) *is normal iff condition* (*) *is fulfilled.*

References

[1] Dolecki, S., A general theory of necessary optimality conditions. Polish Academy of Sciences, Institute of Mathematics, preprint 146 (1978).

[2] Гирсанов, И. В., Лекции по математической теории экстремальных задач. Изд. МГУ, Москва 1970.

[3] Goebel, M., Zur Theorie optimaler Steuerprobleme im Banach-Raum und deren Anwendung auf elliptische Systeme. Dissertation B, Bergakademie Freiberg 1979.

[4] Иоффе, А. Д., Тихомиров, В. М., Теория экстремальных задач. Наука, Москва 1974.

[5] Kurcyusz, S., On the existence and nonexistence of Lagrange multipliers in Banach spaces. Politechnika Warszawska, Technical Report No. 10 (1974).

[6] Lorentz, R., Notwendiges Optimalitätskriterium für eine Steueraufgabe mit nicht eindeutig lösbarer Zustandsgleichung. *In:* Theory of Nonlinear Operators (Proc. Summer School Berlin 1975). Abh. Akad. Wiss. DDR, Jahrg. 1977, Nr. 1N, Akademie-Verlag, Berlin 1977, 337—340.

[7] Вайнберг, М. М., Треногин, В. А., Теория ветвления решений нелинейных уравнений. Наука, Москва 1969.

[8] Warga, J., Optimal control of differential and functional equations. Academic Press, New York 1972.

Address of the author:

Technische Hochschule Ilmenau
Sektion Mathematik, Rechentechnik und ökonomische Kybernetik
6300 Ilmenau
Am Ehrenberg
DDR

Some methods of solving optimal control problems for free boundary processes

Marek Niezgódka, Warsaw

Abstract. In the communication an optimal control problem for a process governed by a parabolic free boundary problem is formulated. Two approaches to solving such a problem are presented.

1. Introduction

Problems of Stefan type belong to the class of parabolic problems with free boundary They are essentially nonlinear due to existence of the free boundary, by conditions given over it and defining its evolution [11].

Stefan type problems arise in mathematical modelling of numerous real processes involving phase-change phenomena such as melting or solidification [2, 6, 9, 11]. In particular they can describe the physical fields of temperature and concentration of a component in mathematical models of multi-component crystal-growth processes [2, 6, 9]. In applications it is often of great importance to have the possibility of controlling the crystal-growth dynamics in an optimal way.

Our aim is to present some approaches to the analysis of the problems of such a type. We do not discuss here properties of the models having form of the Stefan problems, referring only to the paper of I. Pawłow [10] in these proceedings as well as to the papers [1, 4, 7] for details.

2. Formulation of the Stefan problem

We shall consider a process governed by a two-phase Stefan problem involving mixed boundary conditions.

We introduce the following notations:

$Q^T \triangleq (0, L) \times (0, T)$ where L, T are some given finite constants,

S — free boundary, treated as graph of a function $x = s(t)$, $t \in [0, T]$ which is unknown a priori and should be determined in course of solving the problem,

$$Q_1^T \triangleq \{(x, t) \in Q^T \mid 0 < x < s(t), 0 < t < T\},$$

$$Q_2^T \triangleq \{(x, t) \in Q^T \mid s(t) < x < L, 0 < t < T\},$$

y_i, $i = 1, 2$ — functions defining the state of the process, defined over Q_i^T,

a_i, λ_i, $i = 1, 2$ — given positive finite constants, $a_1 \neq a_2$, $\lambda_1 \neq \lambda_2$.

The two-phase Stefan problem can formulated as follows: process equations

$$\frac{\partial y_i}{\partial t} - a_i^2 \frac{\partial^2 y_i}{\partial x^2} = 0 \quad \text{in} \quad Q_i^T, \quad i = 1, 2, \tag{1}$$

conditions along the free boundary S

$$y_i\big(s(t), t\big) = 0, \qquad t \in (0, T], \tag{2}$$

$$\frac{ds}{dt}(t) = \lambda_1 \frac{\partial y_1}{\partial x}\big(s(t)-, t\big) - \lambda_2 \frac{\partial y_2}{\partial x}\big(s(t)+, t\big), \quad t \in (0, T]; \tag{3}$$

initial conditions

$$s(0) = s_0, \quad \text{where} \quad 0 < s_0 < L, \tag{4}$$

$$y_i(x, 0) = y_{i0}(x), \qquad \begin{array}{l} x \in (0, s_0) \ \text{ if } \ i = 1, \\ x \in (s_0, L) \ \text{ if } \ i = 2; \end{array} \tag{5}$$

boundary conditions

$$y_1(0, t) = u(t), \qquad t \in (0, T], \tag{6}$$

$$\frac{\partial y_2}{\partial x}(L, t) = v(t), \quad t \in (0, T]. \tag{7}$$

The process equations and the conditions (2)—(5), (7) are to be understood in the strong sense, whereas the boundary condition (6) has the following weak sense:

$$\lim_{x \to 0+} \int_0^T [y_1(x, t) - u(t)]\, p(t)\, dt = 0 \quad \forall\, p \in C_0^\infty(0, T). \tag{8}$$

Existence and uniqueness of the classical solution $\{y_1, y_2, s\}$ to the Stefan problem (1)—(7) have been proved in [7]. Moreover the solution is continuously dependent upon the data, in particular

$$\|s - \bar{s}\|_{L^2(0,T)} \leqq k\, \|u - \bar{u}\|_{H^1(0,T)}, \tag{9}$$

where s, $\bar{s}$ correspond to u, $\bar{u}$, respectively.

3. Control of the process

The basic way of controlling processes governed by parabolic free boundary problems consists in using boundary controls. In the Stefan problem (1)—(7) either the function u or v can be taken as the boundary control.

We take $u \in H^1(0, T)$ satisfying the conditions

$$u_m \leqq u(t) \leqq u_M < 0, \quad t \in [0, T], \tag{10}$$

$$\left|\frac{du}{dt}(t)\right| \leqq u_M, \quad \text{a.e. over } (0, T), \tag{11}$$

$$u(0) = y_{10}(0) \tag{12}$$

as the control variable, putting $v(t) = 0$ at the same time.

By U_{ad} we shall denote the set of all admissible controls. The set U_{ad} is a bounded, closed convex subset of the space $H^1(0, T)$.

For the process (1)—(7) we formulate the following optimal control problem, whose purpose consists in determining a control u that assures the desired evolution of the free boundary S:

Find $\hat{u} \in U_{ad}$ *such that*

$$J(\hat{u}) = \inf_{u \in U_{ad}} J(u) \tag{13}$$

where

$$J(u) = \|s(u) - s_d\|_{L^2(0,T)} \tag{14}$$

with a given function $s_d \in C[0, T]$.

Existence of optimal controls $\hat{u}$ follows by Weierstrass theorem [8].

Remark. The above problem is of importance in several technical applications, in particular for processes of steel or glass solidification where flexibility and purity of the final product depend upon the movement of the phase interface.

4. The proposed methods of solving the optimal control problem

I. A smoothing technique

The method consists in introducing a sequence of auxiliary "regularized" control problems, involving processes with smoothed coefficients:

$$\frac{\partial y^n}{\partial t} - a^n(y^n) \frac{\partial^2 y^n}{\partial x^2} = f^n(x, t, y^n), \quad (x, t) \in Q^T, \tag{15}$$

$$y^n(x, 0) = y_0(x), \quad x \in (0, L), \tag{16}$$

$$x^n(0,t) = u(t), \quad t \in (0, T], \tag{17}$$

$$\frac{\partial y^n}{\partial x}(L, t) = 0, \quad t \in (0, T]. \tag{18}$$

Here $f^n(x, t, y^n)$, $a^n(y^n)$ are some smooth functions [8, 10].

One can prove the convergence of optimal controls for the regularized problems to a certain control, optimal for problem (1)—(7), (13), (14) both in the sense of values of functional (14) and in the norm of $H^1(0, T)$. These results were established in [8].

The above approach seems to be profitable from the computational point of view [8], since it enables us to omit difficulties associated with an immediate numerical treatment of the Stefan problem. This approach can be also applied in case of several space dimensions [8, 10].

II. Introduction of some inverse formulations

We propose also another approach to solving control problems for the Stefan process (1)—(7). The approach consists in exploiting an integral representation of problem (1)—(7), obtained via thermal potentials techniques [3, 6, 7, 11] and introducing an inverse formulation to the problem, involving

$$s(t) = s_d(t), \qquad t \in [0, T]. \tag{19}$$

The control problem treated as such an inverse parabolic problem has the advantage of the linearity with respect to y_1, y_2 and u. Its integral representation has the form of the following system of linear Volterra equations [6, 7]:

$$z(t) - \int_0^t k_1(t, \tau; s_d|_{[0,t)})\, z(\tau)\, d\tau = b_1(t; s_d|_{[0,t)}), \quad t \in (0, T], \tag{20}$$

$$\int_0^t k_2(t, \tau; s_d|_{[0,t)})\, \varphi(\tau)\, d\tau = b_2(t; s_d|_{[0,t)}, z|_{[0,t)}), \quad t \in (0, T] \tag{21}$$

with respect to $z \triangleq \left.\frac{\partial y_2}{\partial x}\right|_S$, $\varphi \triangleq \frac{du}{dt}$.

Equation (20) has a unique solution $z \in C[0, T]$ as a Volterra equation of the second kind with the continuous kernel k_1 [7], where asfor solving (21), being a Volterra equation of the first kind, one has to apply a regularization method [5, 12]. In particular using Tichonov regularization procedure one can obtain a sequence convergent to some optimal control $\hat{u}$ [5].

References

[1] J. R. Cannon, D. B. Henry, D. B. Kotlow, Classical solutions of the one-dimensional two-phase Stefan problem. Ann. Mat. pura appl. (IV) **57** (1976).

[2] M. C. Flemings, Solidification processing. Mc Graw-Hill, New York 1974.

[3] A. Friedman, Partial differential equations of parabolic type. Prentice-Hall, Englewood Cliffs, N. J., 1964.

[4] A. Friedman, The Stefan problem in several space dimensions. Trans. Amer. Math. Soc. **132** (1968).

[5] V. K. Ivanov, V. V. Vasin, V. P. Tanana, Linear non-correct problems — theory and applications. Nauka, Moskva 1978 (in Russian).

[6] M. Niezgódka, Control of parabolic systems with free boundaries — application of inverse formulations. Control Cybernet. **8** (1979), 3.

[7] M. Niezgódka, On some properties of two-phase parabolic free boundary value control problems. Ibid. **8** (1979), 1.

[8] M. Niezgódka, I. Pawłow, Optimal control problems for parabolic systems with free boundaries — existence of optimal controls, approximation results. Proc. of 9th IFIP Conference on Optimization Techniques, Part 1, 412—420. Lecture Notes in Control and Information Sciences Vol. 22, Springer-Verlag, Berlin—Heidelberg—New York 1980.

[9] R. L. Parker, *in:* R. A. Laudise, R. L. Parker, Single crystals growth. Mir, Moskva 1974 (in Russian).

[10] I. Pawłow, Parabolic problems with free boundaries — existence and properties of solutions, optimal control problems, These Proceedings, 221—231.

[11] L. I. Rubinstein, The Stefan problem. Zvajgzne, Riga 1967.

[12] A. N. Tichonov, V. Ya. Arsenin, Methods of solving non-correct problems. Nauka, Moskva 1974 (in Russian).

Address of the author:

Systems Research Institute
Polish Academy of Sciences
Newelska 6
01-447 Warszawa
Poland

Ein Approximations-Iterationsverfahren für nichtlineare Volterrasche Integralgleichungen zweiter Art

Günter Porath, Güstrow

Abstract. This treatment is an iteration method for the solution of the nonlinear Volterra integral equation of the second kind of Hammerstein type. In every iteration step the nonlinear part is approximated by a function with a foregiven accuracy.

1.

In der vorliegenden Arbeit werden für die nichtlineare Volterrasche Integralgleichung zweiter Art vom Hammersteinschen Typ

$$y(x) = g(x) + \int_a^x k(x, s)\, F\big(y(s)\big)\, ds \tag{1.1}$$

spezielle Iterationsverfahren untersucht. Das klassische Iterationsverfahren besitzt die Form

$$y_{n+1}(x) = g(x) + \int_a^x k(x, s)\, F\big(y_n(s)\big)\, ds \tag{1.2}$$

mit einem Startelement $y_0(x)$. Bei der praktischen Anwendung dieses Verfahrens ist die Berechnung des Bestandteils $F\big(y_n(s)\big)$ mit Schwierigkeiten verbunden. Bei dem folgenden Iterationsverfahren wird bei jedem Iterationsschritt $F\big(y_n(s)\big)$ durch eine Funktion $z_n(s)$ approximiert. Es werden Konvergenz- und Fehlerbetrachtungen durchgeführt. C bezeichnet den Banach-Raum der stetigen Funktionen auf dem Intervall $[a, b]$ mit der Norm

$$\|y\| = \max_{a \leq x \leq b} |y(x)|. \tag{1.3}$$

Voraussetzung 1. Die reelle Funktion $k(x, s)$ ist auf dem Dreieck $D = \{(x, s): a \leqq s \leqq x \leqq b\}$ stetig. Es gelte

$$|k(x, s)| \leqq p. \tag{1.4}$$

Der zugehörige Volterrasche Integraloperator K im Raum C besitzt die Gestalt

$$Ky = \int_a^x k(x, s)\, y(s)\, ds. \tag{1.5}$$

Die reelle Funktion $F(u)$ ist für $u \in \mathbb{R}$ definiert und stetig. F erfülle die Lipschitz-Bedingung

$$|F(u_2) - F(u_1)| \leqq m\,|u_2 - u_1|. \tag{1.6}$$

Satz 1. *Die Gleichung* (1) *besitzt eine eindeutig bestimmte stetige Lösung* $y(x)$. *Es sei* $y_0(x)$ *ein Startelement mit der Bedingung*

$$\|y_0 - g - KF(y_0)\| \leqq c. \tag{1.7}$$

Für die Folge (y_n) *des Iterationsverfahrens* (2)

$$y_{n+1} = g + KF(y_n), \qquad n = 0, 1, \ldots \tag{1.8}$$

gilt

$$\lim_{n\to\infty} y_n = y. \tag{1.9}$$

Es besteht die Abschätzung

$$\|y_n - y\| \leqq c\left[\exp pm(b-a) - \sum_{\nu=0}^{n-1} \frac{(pm)^\nu}{\nu!}(b-a)^\nu\right] \tag{1.10}$$

für $n \geqq 1$.

Das Approximations-Iterationsverfahren mit dem Startelement $\bar{y}_0 = y_0$ *ist durch die folgenden Beziehungen definiert:*

$$\|F(\bar{y}_n) - z_n\| \leqq \varepsilon q^n, \qquad 0 < q \leqq 1, \qquad 0 < \varepsilon, \tag{1.11}$$

$$\bar{y}_{n+1} = g + Kz_n, \qquad n = 0, 1, \ldots \tag{1.12}$$

Dann gilt die Abschätzung

$$\|y_n - \bar{y}_n\| \leqq \frac{\varepsilon q^n}{m}\left[\exp\frac{pm}{q}(b-a) - 1\right] \tag{1.13}$$

für $n \geqq 1$. *Im Fall* $0 < q < 1$ *ist*

$$\lim_{n\to\infty} \bar{y}_n = y. \tag{1.14}$$

2.

Der folgende Abschätzungssatz verwendet im Raum C die äquivalente Norm

$$\|y\|_1 = \max_{a \leqq x \leqq b} \{\exp[-\alpha(x-a)]\,|y(x)|\}, \quad \alpha > 0 \tag{2.1}$$

mit der Bedingung

$$0 < \frac{pm}{\alpha} < 1. \tag{2.2}$$

Es ist

$$\exp[-\alpha(b-a)]\,\|y\| \leqq \|y\|_1 \leqq \|y\|. \tag{2.3}$$

Voraussetzung 2. Die reelle Funktion $h(x, s)$ ist auf dem Dreieck D (vgl. Voraussetzung 1) stetig, und es gelte

$$|h(x, s)| \leqq p.$$

Der zugehörige Volterrasche Integraloperator sei H. Der Kern $k(x, s)$ sei ein Näherungskern für den Kern $h(x, s)$ mit der Abschätzung

$$|k(x, s) - h(x, s)| \leqq \delta \quad \text{für} \quad (x, s) \in D. \tag{2.4}$$

Neben der Gleichung (1.1) wird jetzt die Gleichung

$$w(x) = f(x) + \int_a^x h(x, s)\, F\bigl(w(s)\bigr)\, ds \tag{2.5}$$

betrachtet. Diese Gleichung besitzt eine eindeutig bestimmte stetige Lösung $w(x)$. Die Gleichung (1.1), für welche das Approximations-Iterationsverfahren durchgeführt wird, sei eine Näherungsgleichung für die Gleichung (5). Der Fehler $\|w - \bar{y}_{n+1}\|_1$ kann unter Verwendung von $\|\bar{y}_{n+1} - \bar{y}_n\|_1$ abgeschätzt werden.

Satz 2. *Es sei* $w(x)$ *die eindeutig bestimmte Lösung der Gleichung* (5). *Für die Folge* $(\bar{y}_n)$ *des Iterationsverfahrens* (1.11) *und* (1.12) *besteht die Abschätzung*

$$\|w - \bar{y}_{n+1}\|_1 \leqq \left(1 - \frac{pm}{\alpha}\right)^{-1} \left[\frac{pm}{\alpha} \|\bar{y}_{n+1} - \bar{y}_n\|_1 + \frac{p}{\alpha}\, \varepsilon q^n\right] + \left(1 - \frac{pm}{\alpha}\right)^{-1} \left[\|f - g\|_1 + \frac{\delta}{\alpha} \|F(\bar{y}_{n+1})\|_1\right] \tag{2.6}$$

für $n \geqq 0$.

3.

Das Iterationsverfahren (1.11) und (1.12) für die Näherungsgleichung (1.1) der Gleichung (2.5) läßt sich auf der Grundlage der Algebra und Analysis der Tschebyscheffschen Polynome rechentechnisch realisieren. Als Integrationsintervall wird dann das Intervall $[-1, +1]$ benutzt. Ein beliebiges Intervall $[a, b]$ läßt sich durch eine lineare Substitution auf das Intervall $[-1, +1]$ zurückführen. Im folgenden sei $[a, b]$ das Intervall $[-1, +1]$. Die numerische Verwendung von T-Polynomen der Form

$$\frac{1}{2} c_0 + \sum_{\nu=1}^{n} c_\nu T_\nu(x)$$

bei Iterationsverfahren für lineare Volterrasche Integralgleichungen ist in der Arbeit [2] dargestellt. Die entwickelten numerischen Algorithmen werden für das vorliegende Iterationsverfahren mit Approximation genutzt.

Approximation durch T-Polynome: Es sei $y(x)$ eine Funktion aus $C_{[-1,+1]}$. Das Interpolationspolynom zu $y(x)$ mit den Interpolationsknoten

$$x_k = \cos k \frac{\pi}{N}, \quad k = 0, \ldots, N,$$

ist dann von der Form

$$p_N(x) = \sum_{\nu=0}^{N} \lambda_\nu c_\nu T_\nu(x),$$

vgl. [2]. Es gelten die Beziehungen

$$\lambda_\nu = \begin{cases} \frac{1}{2} & \text{für} \quad \nu = 0, N, \\ 1 & \text{für} \quad \nu = 1, 2, \ldots, N-1, \end{cases}$$

$$c_\nu = \frac{2}{N} \sum_{k=0}^{N} \lambda_k f\left(\cos k \frac{\pi}{N}\right) \cos k \frac{\pi}{N}, \quad \nu = 0, 1, \ldots, N.$$

Der Fehler dieser Approximation ist näherungsweise gegeben durch die Größe

$$\varepsilon_{1N} = |c_N| + |c_{N-1}|.$$

Ist in der Gleichung (2.5) die Funktion f ein Polynom, dann wird für f die T-Polynomdarstellung bestimmt. Ist f eine analytische bzw. stetige Funktion, dann wird f durch ein T-Polynom g approximiert.

Die folgenden Näherungskerne können verwendet werden.

a. Zweidimensionale Interpolation des stetigen Kerns $h(x, s)$, $-1 \leqq x, s \leqq +1$, durch den Kern

$$k_1(x, s) = \sum_{\nu=0}^{N} \sum_{\mu=0}^{N} \lambda_\nu \lambda_\mu c_{\nu\mu} T_\nu(x)\, T_\mu(s)$$

(vgl. dazu [2]).

b. Approximation des analytischen Kerns $h(x, s)$, $-1 \leqq x, s \leqq +1$, durch einen Kern

$$k_2(x, s) = \sum_{\nu=0}^{N} \sum_{\mu=0}^{N} b_{\nu\mu} x^\nu s^\mu.$$

c. Approximation des Differenzenkerns

$$h(x - s) = \sum_{\nu=1}^{\infty} a_\nu \frac{(x - s)^{\nu-1}}{(\nu - 1)!}$$

durch den Differenzenkern

$$k_3(x - s) = \sum_{\nu=1}^{N} a_\nu \frac{(x - s)^{\nu-1}}{(\nu - 1)!}.$$

Die Funktion $\bar{y}_n(x)$, $n \geqq 0$, aus Satz 1 sei ein T-Polynom. Dann wird $F(\bar{y}_n(x))$ nach dem angegebenen Approximationsverfahren durch ein G-Polynom $z_n(x)$ mit der vorgegebenen Genauigkeit approximiert. Die Funktion $\bar{y}_{n+1}(x) = g + Kz_n$ ist dann ein T-Polynom. Die Koeffizienten von $\bar{y}_{n+1}(x)$ entstehen aus den Koeffizienten von $z_n(x)$ nach bestimmten Algorithmen, vgl. dazu [2]. Die Näherungsfunktionen des Verfahrens (1.11) und (1.12) sind also T-Polynome. Die Fehlerabschätzung (2.6) aus Satz 2 liefert ein Abbruchkriterium für das Iterationsverfahren.

Das dargestellte Iterationsverfahren mit T-Polynomen wurde für ESER-Computer implementiert. Die Beweise für die Sätze 1 und 2 und numerische Ergebnisse sind in einer weiteren Arbeit [4] zu finden.

Literatur

[1] Fox, L., and Parker, I. B., Chebyshev polynomials in numerical analysis, Oxford University Press, London 1968.

[2] Porath, G., und Tabbert, E., Das Tschebyscheffsche Iterationsverfahren für lineare Volterrasche Integralgleichungen zweiter Art, Beitr. Numer. Math. 7 (1979), 83–96.

[3] Sag, T. W., Chebyshev iteration methods for integral equations of the second kind, Math. Comp. 24 (1970), 341–355.

[4] Porath, G., Approximations-Iterationsverfahren für nichtlineare Volterrasche Integralgleichungen zweiter Art, Wiss. Z. PH Güstrow (1979).

Anschrift des Verfasser:

Pädagogische Hochschule Güstrow
Sektion Mathematik
2600 Güstrow
Goldberger Straße 12
DDR

Secondary axisymmetric buckled states of thin elastic spherical shells

Lutz Recke, Berlin

Let H be the closure of $C^1[0, \pi]$ in the norm $\|\cdot\|$ defined by the inner product $(u, v) := \int_0^\pi (u'v' + (1 - \nu)\, uv) \sin^3 \theta \, d\theta$. Poisson's ratio ν $(0 < \nu < 1)$ is a physical constant.

We consider the operator equation (cf. [1])

$$F(\alpha, \lambda, f) := f - \lambda A f + \alpha^2 A^2 f + \alpha \left(B(f, Af) + \frac{1}{2} AB(f, f) \right) + \frac{1}{2} B(f, B(f, f)) = 0, \quad (1)$$

where $A : H \to H$ is a bounded linear, self-adjoint, positive, compact operator defined by $(Au, v) := \int_0^\pi uv \sin^3 \theta \, d\theta$, $v \in H$. The bounded bilinear operator $B : H \times H \to H$ is defined by $\big(B(u, v), w\big) := \int_0^\pi uvw \cos \theta \sin^3 \theta \, d\theta$.

Equation (1) describes the axisymmetric buckled states of an uniformly compressed thin elastic spherical shell. The function f is a measure of the angle of rotation of the meridional tangent, θ is the polar angle. The parameter λ is a measure of the load, $\alpha > 0$ is defined by $\alpha^2 := 3(1 - \nu^2) \left(\frac{R}{h}\right)^2$, where R and h are the radius and wall thickness of the shell. The trivial solution $f = 0$, which exists for any parameter values α and λ, corresponds to the unbuckled uniformly contracted state. Solutions f, which are not equal to zero, correspond to nonspherical buckled states of the shell.

The characteristic values of A are all simple and uniquely determined by $\mu_n := n(n + 1) - 1 - \nu$ $(n = 1, 2, \ldots)$ (cf. [1]). Let f_n be the associated eigenfunctions, normed by $(Af_n, f_m) = \delta_{nm}$. For fixed $\alpha > 0$ the eigenvalues of the eigenvalue problem $f - \lambda Af + \alpha^2 A^2 f = 0$ are

$$\lambda_n(\alpha) := \mu_n + \frac{\alpha^2}{\mu_n} \quad (n = 1, 2, \ldots). \quad (2)$$

They are all simple unless $\alpha^2 = \mu_n \mu_m$ for $n \neq m$, in which case $\lambda_n(\alpha) = \lambda_m(\alpha)$ has multiplicity two.

Suppose $\lambda_0 := \lambda_n(\alpha_0) = \lambda_m(\alpha_0)$ $(\alpha_0{}^2 = \mu_n \mu_m)$ is the smallest of the eigenvalues (2). Then it follows (see [1]) that $m = n \pm 1$. It suffices to consider the case when $n = 2k$ and $m = 2k + 1$, the case when $n = 2k$ and $m = 2k - 1$ then follows in an analogous way.

In order to apply the Liapunov-Schmidt procedure (see, for instance, [2]) we denote by $P: H \to \ker \frac{\partial F}{\partial f}(\alpha_0, \lambda_0, 0)$ the orthogonal projection and we write $f_0 := (I - P) f$, $f = u_1 f_{2k} + u_2 f_{2k+1} + f_0$, $(u_1, u_2) \in \mathbb{R}^2$.

The auxiliary equation

$$(I - P) F(\alpha, \lambda, u_1 f_{2k} + u_2 f_{2k+1} + f_0) = 0 \tag{3}$$

can be uniquely solved (by the implicit function theorem) for small f_0 in terms of $(\alpha, \lambda, u_1, u_2) \in \mathbb{R}^4$. Thus $f_0 = f_0(\alpha, \lambda, u_1, u_2)$ in a neighborhood of $(\alpha_0, \lambda_0, 0, 0)$.

The problem of finding all small solutions of (1) for small $|\lambda - \lambda_0| + |\alpha - \alpha_0|$ reduces to the equivalent problem of solving the two-dimensional analytic bifurcation equations (see [1])

$$G_1(\alpha, \lambda, u_1, u_2) := \big(\lambda_{2k}(\alpha) - \lambda\big) u_1 + a u_1^2 + b u_2^2 + O\big((u_1^2 + u_2^2)(|u_1| + |u_2| + |\alpha - \alpha_0|)\big) = 0, \tag{4}$$

$$G_2(\alpha, \lambda, u_1, u_2) := \big(\lambda_{2k+1}(\alpha) - \lambda\big) u_2 + 2b u_1 u_2 + O\big((u_1^2 + u_2^2)(|u_1| + |u_2| + |\alpha - \alpha_0|)\big) = 0 \tag{5}$$

The coefficients a and b are determined by $a := \left(f_{2k+1}, B(f_{2k}, Af_{2k}) + \frac{1}{2} AB(f_{2k}, f_{2k})\right)$ and $b := \left(B(f_{2k+1}, Af_{2k+1}) + \frac{1}{2} AB(f_{2k+1}, f_{2k+1}), f_{2k+1}\right)$ and we have (see [1])

$$2b > a > 0. \tag{6}$$

In order to get more information about the bifurcation equations (4) (5) we apply a known property of bifurcation problems which are invariant under symmeries (see, for instance, [3, 4]):

Lemma. *The function G_1 is an even function of the variable u_2, G_2 is an odd function of u_2.*

Proof. We consider the operator $T: H \to H$ defined by $(Tf)(\theta) := -f(\pi - \theta)$. It is easy to see that $TA = AT$ and $TB(\cdot, \cdot) = B(T\cdot, T\cdot)$. That means, that for given parameter values α and λ any nonspherical buckled state is a solution of (1) if and only if the symmetric with respect to the equator of the sphere buckled state is also a solution of (1). The eigenfunctions satisfy (see [1]) $Tf_{2k} = f_{2k}$ and $Tf_{2k+1} = -f_{2k+1}$. From the uniqueness of the solution of (3) it follows $Tf_0(\alpha, \lambda, u_1, u_2) = f_0(\alpha, \lambda, u_1, -u_2)$. From $T = T^*$ it follows $G_1(\alpha, \lambda, u_1, u_2) = \big(Tf_{2k}, F(\alpha, \lambda, u_1 f_{2k} + u_2 f_{2k+1} + f_0(\alpha, \lambda, u_1, u_2))\big)$ $= \big(f_{2k}, F(\alpha, \lambda, u_1 f_{2k} - u_2 f_{2k+1} + Tf_0(\alpha, \lambda, u_1, u_2))\big) = G_1(\alpha, \lambda, u_1, -u_2)$ and analogous $G_2(\alpha, \lambda, u_1, u_2) = -G_2(\alpha, \lambda, u_1, -u_2)$.

The Weierstrass preparation theorem implies that there are analytic functions H_i $(i = 1, \ldots, 5)$ such that

$$G_1(\alpha, \lambda, u_1, u_2) = \big(u_1^2 + u_1(\lambda_{2k}(\alpha) - \lambda) H_1(\alpha, \lambda, u_2^2) + u_2^2 H_2(\alpha, \lambda, u_2^2)\big) H_3(\alpha, \lambda, u_1, u_2^2),$$

$$G_2(\alpha, \lambda, u_1, u_2) = u_2\big(u_1 + (\lambda_{2k+1}(\alpha) - \lambda) H_4(\alpha, \lambda, u_2^2)\big) H_5(\alpha, \lambda, u_1, u_2^2)$$

and $H_i \neq 0$ $(i = 1, \ldots, 5)$ for $\alpha = \alpha_0$, $\lambda = \lambda_0$, $u_1 = u_2 = 0$. For that reason there are exactly two solutions of (4) (5) with $u_2 = 0$:

(i) $u_1 = u_2 = 0$

(ii) $u_1 = \big(\lambda - \lambda_{2k}(\alpha)\big) H_1(\alpha, \lambda, 0)$, $\quad u_2 = 0$

We once more apply the Weierstrass preparation theorem which implies that there are analytic functions H_i $(i = 6, 7, 8)$ such that

$$G_1(\alpha, \lambda, (\lambda - \lambda_{2k+1}(\alpha)) H_4(\alpha, \lambda, u_2^2), u_2)$$
$$= \big(u_2^2 - (\lambda - \lambda_{2k+1}(\alpha)) (\lambda - H_6(\alpha)) H_7(\alpha, \lambda)\big) H_8(\alpha, \lambda, u_2^2)$$

and $H_6(\alpha) = \dfrac{1}{2b - a} \big(2b\lambda_{2k}(\alpha) - a\lambda_{2k+1}(\alpha)\big) + O\big((\alpha - \alpha_0)^2\big)$, $\quad H_i \neq 0 \quad (i = 7, 8)$ for $\alpha = \alpha_0$, $\lambda = \lambda_0$, $u_2 = 0$. Besides the solutions (i) and (ii) there are exactly the following two solutions of (4) (5):

$$\text{(iii)}\ u_1 = \big(\lambda - \lambda_{2k+1}b(\alpha)\big) H_4\big(\alpha, \lambda, (\lambda - \lambda_{2k+1}(\alpha)) (\lambda - H_6(\alpha)) H_7(\alpha, \lambda)\big),$$
$$u_2 = \pm\big(\lambda - \lambda_{2k+1}(\alpha)\big) \big((\lambda - H_6(\alpha)) H_7(\alpha, \lambda)\big)^{1/2}.$$

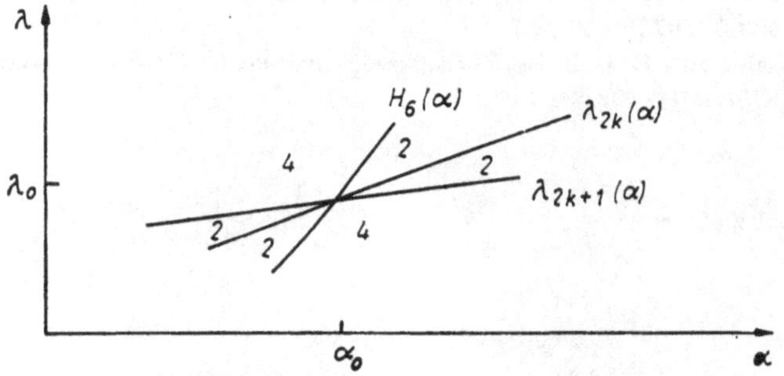

Fig. 1. Number of solutions

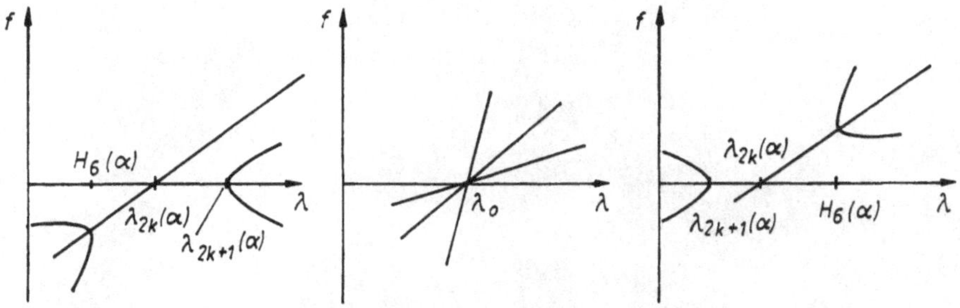

Fig. 2. $\alpha < \alpha_0$ Fig. 3. $\alpha = \alpha_0$ Fig. 4. $\alpha > \alpha_0$

The solutions (iii) are real for $\big(\lambda - \lambda_{2k+1}(\alpha)\big) \big(\lambda - H_6(\alpha)\big) \geqq 0$ (cf. (6) and Fig. 1) and intersect solution (i) for $\lambda = \lambda_{2k+1}(\alpha)$ and solution (ii) for $\lambda = H_6(\alpha)$ (cf. Fig. 2, 3 and 4).

The corresponding to (ii) solution $f(\alpha, \lambda) := \big(\lambda - \lambda_{2k}(\alpha)\big) H_1(\alpha, \lambda, 0) f_{2k} + f_0\big(\alpha, \lambda, (\lambda - \lambda_{2k}(\alpha)\big) H_1(\alpha, \lambda, 0)\big)$ of (1) describes symmetric buckled states of the shell, because $Tf(\alpha, \lambda) = f(\alpha, \lambda)$. The solutions (iii) correspond to nonsymmetric buckled states.

It is known (see [5, 6]), that in general secondary bifurcation fails to take place, if the symmetry property of the bifurcation equations is destroyed even in terms of arbitrary high order. For that reason it is not possible to determine the intersection property of solutions (ii) and (iii) by means of the method of formal asymptotic expansions which is applied to equation (1) in [7].

References

[1] Knightly, G. H., and D. Sather, Existence and stability of axisymmetric buckled states of spherical shells, Arch. Rat. Mech. Anal. **63** (1977), 305–319.

[2] Wainberg, M. M., und W. A. Trenogin, Theorie der Lösungsverzweigung bei nichtlinearen Gleichungen, Akademie-Verlag, Berlin 1973.

[3] Б. В. Логинов, В. А. Треногин, Об использовании групповых свойств для определения многопараметрических семейств решений нелинейных уравнений, Мат. сб. **85** (1971), 440–454.

[4] Sattinger, D. H., Group representation theory and branch points of nonlinear functional equations, SIAM J. Math. Anal. **8** (1977), 179–201.

[5] Mallet-Paret, J., Buckling of cylindrical shells with small curvature, Quart. Appl. Math. **35** (1977), 383–400.

[6] Vanderbauwhede, A., Generic and nongeneric bifurcation for the von Kármán equations, J. Math. Anal. Appl. **66** (1978), 550–573.

[7] Bauer, L., H. B. Keller and E. L. Reiss, Multiple eigenvalues lead to secondary bifurcation, SIAM Review **17** (1975), 101–122.

Address of the author:

Humboldt-Universität zu Berlin
Sektion Mathematik
1080 Berlin
Unter den Linden 6
DDR

Perturbation and branching of the solutions of nonlinear eigenvalue problems

Thomas Riedrich, Dresden

Introduction

In this note we consider the classical bifurcation problem (small solutions of the equation $Ax = \lambda x$) for a completely continuous nonlinear operator $A : E \to E$ in a real Banach space E with $A(o) = o$ (the zero element) which is of the form

$$A = T + F,$$

where $T = (DA)\ (o)$ (the Fréchet derivative of A at zero) is a linear completely continuous operator with positive spectral radius $\lambda_0 = \text{spr}\ T$. We assume that λ_0 is an eigenvalue of T. The question, whether this λ_0 is a bifur cation point of the operator A is considered under special assumptions by K. Stathakopoulos, [7] but by avoiding the assumption of odd algebraic multiplicity of the eigenvalue λ_0, for which the existence of branching solutions is a classical result. Because it seems to be worthwhile from the viewpoint of applications to discuss bifurcation effects at the spectral radius of the linear part of the operator A in more detail — independent on the algebraic multiplicity of λ_0 — we present some results in this direction, which especially are related to perturbation theorems for nonlinear mappings in cones (see [5]). A main tool is a generalization of a cone construction, which goes back to M. A. Krasnoselskij [3].

In the following, a cone in a normed vector space denotes a closed convex nonvoid set $K \neq \{o\}$, which contains with $x \in K \setminus \{o\}$ also tx for all $t \geqq 0$, but not the element $(-x)$. A subset of a vector space is called radially bounded, if the intersection of this set with any line through the zero element is a bounded subset of this line. For topological notations we refer to be $\bar{A}$; int A; ∂A the closure, the interior, the boundary of A respectively. A continuous mapping is called compact, if the closure of its image set is compact. A continuous mapping from a normed space into a topological space is called completely continuous, if its restriction to any bounded set of its domain of definition is a compact mapping.

1. A perturbation theorem

Theorem 1 (see [5], p. 132; Satz 4.27). *Let (E, τ) be a complete metrisable locally convex space, K_0 a (closed) cone in E and U a closed starshaped neighborhood of zero in E for which $U \cap K_0$ is radially bounded. Let*

$$T : K_0 \cap \partial U \to E$$

be a compact mapping with $T[K_0 \cap \partial U] \subseteq K_0$ *and* $0 \notin \overline{T[K_0 \cap \partial U]}$. *The eigenvalue problem*

$$Tx = \lambda x \qquad (x \in K_0 \cap \partial U;\ \lambda > 0)$$

has after assumption exactly one solution $\lambda = \lambda_0 > 0$ *with a corresponding nonvoid set of eigenvectors* X_0. *Let furthermore exist a closed neighborhood of zero* V_0 *in* E *with*

$$(X_0 + V_0) \cap \partial U \subseteq K_0 \cap \partial U.$$

Then there is for every $\varepsilon > 0$ *und for every neighborhood of zero* $W \subseteq V_0$ *a neighborhood* $V = V(\varepsilon, W)$ *so, that every compact map*

$$\tilde{T} : (X_0 + V_0) \cap \partial U \to E$$

with

$$\tilde{T}[(X_0 + V_0) \cap \partial U] \subseteq K_0$$

and with

$$\tilde{T}x - Tx \in V \qquad \big(x \in (X_0 + W) \cap \partial U\big)$$

has an eigenvalue $\tilde{\lambda}_0 > 0$ *and a corresponding eigenvector* $\tilde{x}_0 \cap K_0(\partial U)$, *so that*

$$|\tilde{\lambda}_0 - \lambda_0| \leqq \varepsilon \quad \text{and} \quad x_0 \in X_0 + W.$$

(Proof of th. 1 by modification of the proof of the theorem quoted above.)

Example 1. Let $E = C[0, 1]$ be the space of realvalued continuous functions on $[0, 1]$ with sup-norm $\|x\| = \sup_{0 \leqq t \leqq 1} |x(t)|$ $(x \in E)$. The set $U = \left\{x \in E \,\middle|\, \int_0^1 |x(t)|\, dt = 1\right\}$ is a closed, starshaped, radially bounded (but not bounded) neighborhood of zero in E and $K_0 = \{x \in E \mid 0 \leqq x(t) \text{ for } t \in [0, 1]\}$ is a closed cone in E. The functions $K(s, t)$ and $R(s, t)$ are assumed to be continuous on the square $0 \leqq s, t \leqq 1$ and fulfil the inequalities $K(s, t) > 0$ resp. $R(s, t) \geqq 0$ $(0 \leqq s, t \leqq 1)$. For a fixed real $p > 1$ and for $0 \leqq \mu < +\infty$ we introduce the mappings

$$T_\mu : K_0 \cap \partial U \to E$$

defined by

$$T_\mu(x)\,(s) = \int_0^1 K(s, t)\, x(t)\, dt + \mu^{p-1} \int_0^1 R(s, t)\, |x(t)|^p\, dt \qquad (0 \leqq s \leqq 1).$$

We denote the mapping T_0 by the symbol T (the unperturbed map). After the classical theorem of P. Jentzsch [1, 6], T possesses in $K_0 \cap \partial U$ exactly one eigenvector x_0 with (a unique) eigenvalue $\lambda_0 > 0$ (see [5], p. 134). The relations $T[K_0 \cap \partial U] \subseteq K_0$ and $T_\mu[K_0 \cap \partial U] \subseteq K_0$ $(0 < \mu < \infty)$ are obvious. The mapping T is compact and fulfils the relation $0 \notin \overline{T[K_0 \cap \partial U]}$ (see [5], p. 134); the mappings T_μ $(\mu > 0)$ are not compact. There is ([5], p. 134) a neighborhood of zero V_0 in E such that $(x_0 + V_0) \cap \partial U \subseteq K_0 \cap \partial U$ which can be choosen as a ball $V_0 = \{x \in E \mid \|x\| \leqq r\}$ $(r > 0)$. The restriction of T_μ to $(x_0 + V_0) \cap \partial U$ is a compact map $(0 < \mu < \infty)$. If μ is small enough $(0 < \mu \leqq \mu_0)$, the pair (T, T_μ) fulfils the assumptions of theorem 1 with $\tilde{T} = T_\mu$ and with $X_0 = \{x_0\}$. In consequence of theorem 1, the nonlinear integral operator T_μ possesses eigenvectors and eigenvalues arbitrary close to x_0 and λ_0 if μ is (positive and) small enough. From this result it can be easily shown that λ_0 is a bifurcation value of the operator

$$x(\cdot) \to \int_0^1 \left(K(\cdot, t)\, x(t) + R(\cdot, t)\, |x(t)|^p\right) dt \qquad \left(x(\cdot) \in C[0, 1]\right).$$

2. Bifurcation at the spectral radius of the linear part of a nonlinear mapping

2.1.

In the following, we consider completely continuous mappings $A : E \to E$ from a real Banach space $(E, \|\cdot\|)$ into itself of the form

$$A = T + F, \tag{1}$$

where T is linear and completely continuous and maps E into itself, $F : E \to E$ is a (continuous) mapping with

$$\lim_{\|x\| \to 0} \frac{\|F(x)\|}{\|x\|} = 0. \tag{2}$$

As a consequence we have the relations $A(o) = o$ and $T = (DA)\,(o)$ (the Fréchet derivative).

We assume that the spectral radius spr T of the operator T is positive and that spr T is an eigenvalue of T. We denote this eigenvalue by λ_0,

$$\lambda_0 = \operatorname{spr} T > 0. \tag{3}$$

Of course, λ_0 is an eigenvalue of finite algebraic multiplicity. We assume that all other real numbers of the spectrum of T are of smaller module than λ_0:

$$\lambda \in \sigma(T) \Rightarrow |\lambda| < \lambda_0. \tag{4}$$

2.2.

With the assumptions of 2.1. we decompose the space E into the direct (topological) sum of two factor spaces E_0 and E_1, which are invariant with respect to the operator T:

$$E = E_0 + E_1 \tag{5}$$

where E_0 is the space

$$E_0 = \operatorname{Ker} (T - \lambda_0 I)^\nu$$

and

$$E_1 = \operatorname{Ran} (T - \lambda_0 I)^\nu$$

(The natural number $\nu \geqq 1$ denotes the index of the operator $T - \lambda_0 I$, I the identity). The (continuous) projection from E onto the space E_1 we denote by $P : E \to E_1$. In the space E_0 we can introduce a basis of eigenvectors (corresponding to λ_0) and principal vectors for which the finite-dimensional operator $T \mid E_0$ is a Jordan block matrix with respect to the eigenvalue λ_0 (which is the only eigenvalue of $T \mid E_0$). This basis is denoted by $(g_j)_{1 \leqq j \leqq N}$, where N is the algebraic multiplicity of λ_0 (= dimension of E_0). We can choose the (g_j) in such a way that we have the relations (alternatively)

$$\begin{aligned} &Tg_j = \lambda_0 g_j + g_{j+1} \quad (g_j \text{ a principal vector}) \\ \text{or} \quad &Tg_j = \lambda_0 g_j \quad (g_j \text{ an eigenvector}). \end{aligned} \tag{6}$$

One can construct a dual basis $(g_j^*)_{1 \leqq j \leqq N}$ for the space Ker $(T^* - \lambda_0 I^*)^\nu$ in an analogous manner (T^* denotes the dual operator for T) such that the g_j^* are continuous linear functionals on E with

$$g_j^*(g_k) = \begin{cases} 0 & j \neq k \\ 1 & j = k \end{cases}$$

and $g_j^*(x) = 0$ ($x \in E_1$) and with

$$\begin{aligned} &T^* g_j^* = \lambda_0 g_j^* + g_{j-1}^* \\ \text{or} \quad &T^* g_j^* = \lambda_0 g_j^* \end{aligned} \tag{7}$$

(for details and proof see [8], p. 330—344).

2.3.

In the space E_1 we can introduce a new, equivalent, norm $\|\cdot\|_0$ for which we have

$$\|Tx\| \leqq k\lambda_0 \|x_0\| \qquad (x \in E_1) \tag{8}$$

with a fixed positive $k < 1$ (see [3], p. 91). We extend this norm to an equivalent norm on the whole space E by putting

$$\|x\|_0 = \|Px\|_0 + \sum_{j+1}^{N} |g_j^*(x)| \qquad (x \in E). \tag{9}$$

Generalizing an idea of Krasnoselskij ([3], p. 231) we introduce the following cone K_0 in E:

$$K_0 = \left\{x \in E \mid 0 \leqq g_j^*(x) \quad (j = 1, \ldots, N) \quad \text{and} \quad \|Px\|_0 \leqq \sum_{j=1}^{N} g_j^*(x)\right\}. \tag{10}$$

K_0 is a closed, convex cone in E with $K_0 \cap E_1 = \{o\}$ and int $K_0 \neq \emptyset$ (e.g. the element $\sum_{j=1}^{N} g_j$ belongs to int K_0). Therefore, we have $K_0 - K_0 = E$.

2.4.

In the following we denote the set $\{x \in E \mid \|x\|_0 \leqq r\}$ $(r > 0)$ by $B(o, r)$.

Theorem 2. *Let $A : E \to E$ be an operator which satisfies the assumptions formulated in 2.1. Additionally we assume that there is a $r_0 > 0$ such that*

$$0 \leqq g_j^*(Ax) \qquad \big(x \in K_0 \cap B(o, r_0);\ j = 1, \ldots, N\big). \tag{11}$$

Then

$$\lambda_0 = \operatorname{spr}(DA)(o)$$

is a bifurcation value of A.

(The proof of th. 2 will be published elsewhere.)

Remark 1. Without the assumption (11) the conclusion of theorem 2 generally does not hold. This can easily be seen by considering the example of K. Stathakopoulos [7] (which goes back to Krasnoselskij [2]), where $E = R^2$ with the usual euclidean norm $\|x\| = \sqrt{\xi_1^2 + \xi_2^2}$ $(x = (\xi_1, \xi_2))$ and the mapping $A : E \to E$ defined by

$$Ax = \begin{pmatrix} \xi_1 - \xi_2(\xi_1^2 + \xi_2^2) \\ \xi_2 + \xi_1(\xi_1^2 + \xi_2^2) \end{pmatrix} = \begin{pmatrix} \xi_1 \\ \xi_2 \end{pmatrix} + (\xi_1^2 + \xi_2^2) \begin{pmatrix} -\xi_2 \\ \xi_1 \end{pmatrix} = Tx + Fx \; (x \in E).$$

Here $\lambda_0 = \operatorname{spr} T = 1$ is of algebraic multiplicity 2. We have (see 2.1.—2.4.) $E_0 = E$; $E_1 = \{o\}$, $K_0 = R_+^2 = \{x \in R^2 \mid \xi_1 \geqq 0, \xi_2 \geqq 0\}$, $g_1^* = pr_1$, $g_2^* = pr_2$ and $g_1^*(Ax) = \xi_1 - \xi_2 \times (\xi_1^2 + \xi_2^2)$. There is no $\delta_0 > 0$ (resp. $r_0 > 0$) such that $g_1^*(Ax) \geqq 0$ for all $x \in K_0 \cap B(o; \delta_0)$. The other assumptions of theorem 2 are fulfilled. But A has no eigenvalues.

Example 2. Replacing the minus-sign in the first coordinate of $A(x)$ in remark 1 by a plus-sign, then we get an operator for which all the assumptions of theorem 2 are fulfilled and then $\lambda_0 = 1$ is a bifurcation value for A (though λ_0 has algebraic multiplicity 2 (!)).

References

[1] Jentzsch, P., Über Integralgleichungen mit positivem Kern. J. reine angew. Math. **141** (1912), 235—244.

[2] Красносельский, М. А., Топологические методы в теории нелинейных интегальных уравнений. Москва 1956.

[3] Красносельский, М. А., Положительные решения операторных уравнений. Москва 1962.

[4] Riedrich, T., Über Störungen nichtlinearer Operatorengleichungen. *In:* Theory of Nonlinear Operators (Proc. Summer School Berlin 1975) Abh. Akad. Wiss. DDR, Jahrg. 1977, Nr. 1 N, Akademie-Verlag, Berlin 1977, 229—237.

[5] Riedrich, T., Vorlesungen über nichtlineare Operatorengleichungen. Teubner-Texte zur Mathematik, Leipzig 1976.

[6] Schmeidler, W., Integralgleichungen mit Anwendungen in Physik und Technik. 2. Auflage. Akademische Verlagsgesellschaft, Leipzig 1955.

[7] Stathakópoulos, K., Ein Beitrag zur Verzweigungstheorie. Manuscripta math. **15** (1975), 211—218.

[8] Zaanen, A. C., Linear Analysis. North-Holland Publishing Co. Amsterdam 1953.

Address of the author:

Technische Universität Dresden
Sektion Mathematik
8027 Dresden
Mommsenstr. 13
DDR

Zur Existenz der optimalen Steuerung bei Volterraschen Integralgleichungen

Hartmut Schmeling, Greifswald

Abstract. The paper deals with the existence problem of optimal controls with nonlinear integral equations of Volterra type.

1962 führte E. Roxin [1] die Existenz einer optimalen Steuerung bei gewöhnlichen Differentialgleichungen auf die Abgeschlossenheit der Erreichbarkeitsmenge zurück. Wir untersuchen in der vorliegenden Arbeit die Struktur der Erreichbarkeitsmenge für nichtlineare Volterrasche Integralgleichungen mit verallgemeinerten Steuerungen (nach J. Warga [2]). Derartige Steuerungen untersuchte auch R. W. Gamkrelidze [3] für den Fall gewöhnlicher Differentialgleichungen. Für den vorliegenden Fall leiten wir Bedingungen her, unter denen die verallgemeinerte optimale Steuerung eine gewöhnliche ist.

Definition. Seien s_e, $r_0 \in R$ positiv, m, $r \in N$; $g:[0, s_e] \to R^m$, $K:[0, s_e] \times [0, s_e] \times R^m \times R^r \to R^m$ seien gegebene stetige Funktionen mit den Argumenten s bzw. (s,t,x,u) und g', K'_s, K'_x, K'_u deren stetige (partielle) Ableitungen nach den entsprechenden Argumenten; $|\cdot|$ bezeichne die Maximumnorm in endlichdimensionalen Räumen bzw. für Matritzen die von der Maximumnorm abgeleitete Supremumsnorm; $U \subset R^r$ sei kompakt; frm ([2], S. 65) bezeichne die Menge der endlichen, regulären Borel-Maße über U; $\mathcal{S} \subset ([0, s_e] \to \text{frm})$ bezeichne die Menge der Lebesgue-meßbaren Familien $(\sigma_t) \in \mathcal{S}$ nichtnegativer Maße σ_t, $t \in [0, s_e]$, mit der Eigenschaft $\sigma_t(U) = 1$ für f. a. $t \in [0, s_e]$, und (σ_t) heiße *verallgemeinerte Steuerung* ([2], S. 294). Sei $u(\cdot)$ meßbar mit Werten $u(t) \in U$ und $\sigma_t = \delta_{u(t)}$ (Dirac-Maß in $u(t)$) für f. a. $t \in [0, s_e]$, dann heiße diese spezielle verallgemeinerte Steuerung *gewöhnliche Steuerung.*

Setzen wir für $(\sigma_t) \in \mathcal{S}$

$$K(s, t, x, \sigma_t) := \int_U K(s, t, x, u)\, \sigma_t(du), \tag{1}$$

dann folgt $K(s, t, x, \delta_{u(t)}) = K\big(s, t, x, u(t)\big)$. Deswegen identifizieren wir gelegentlich die Maßfamilie $\delta_{u(\cdot)}$, erzeugt durch die Funktion $u(\cdot)$, mit der letzteren.

Für meßbare und im wesentlichen beschränkte Maßfamilien erhalten wir in (1) eine bez. t integrierbare und bez. s, x stetige Funktion ([2], S. 297).

Als *Phasenfunktion* bezeichnen wir jede stetige Lösung der Gleichung

$$x(s) = \int_0^s K\big(s, t, x(t), \sigma_t\big)\, dt + g(s), \qquad s \in [0, s_e], \tag{2}$$

und $(\sigma_t) \in \mathcal{S}$ als *dazugehörige verallgemeinerte Steuerung.* Eine Phasenfunktion $x(\cdot)$ heiße *zulässig*, wenn sie der Ungleichung $|x(t)| \leqq r_0$, $t \in [0, s_e]$, genügt.

Sei

$$\Omega_s := \{z \in R^m : z = x(s),\ x(\cdot) \text{ zulässige Phasenfunktion}\},\ s \in [0, s_e] \tag{3}$$

und $\Omega := \bigcup\limits_{s\in[0,s_e]} \Omega_s$, dann heiße Ω Erreichbarkeitsmenge und Ω_s heiße *Schnitt von Ω in s.*

Satz 1. *Sei die Erreichbarkeitsmenge $\Omega \neq \emptyset$. Dann sind die Menge Ω und ihre Schnitte Ω_s, $s \in [0, s_e]$, abgeschlossen* (s. analogen Beweis in [2], Satz VII. 1.1.).

Wir geben nun ein **Beispiel** für eine nichtlineare Volterrasche Integralgleichung an, welches zeigt, daß die Erreichbarkeitsmenge, erzeugt durch gewöhnliche Steuerungen, nicht abgeschlossen zu sein braucht, obwohl in unserem Falle die Mengen $K(s, t, x, U) := \{K(s, t, x, u) : u \in U\}$ konvex sind. Sei

$$x_1(s) = \int\limits_0^s \{(1 - t)\, u(t) + (s - t) u^2(t)\}\, dt,$$

$$x_2(s) = \int\limits_0^s \left(x_1(t) - \frac{1}{2} t^2\right)^2 dt,$$

$s \in [0, 1]$, $u(t) \in [-1, 1] =: U$ für f. a. $t \in [0, 1]$. Es läßt sich zeigen, daß $\left(\frac{1}{2}, 0\right)^T \notin \Omega$, jedoch enthalten ist in ihrer Abschließung $\bar{\Omega}$ (s. [4]).

Definition. Sei $\Omega \subset R^m$. $\delta z \in R^m$ heiße *Richtungsgrenzwert von Ω in $z_0 \in R^m$,* wenn es eine Nullfolge positiver Zahlen (γ_k) und eine Folge von Resten (r_k), $r_k \in R^m$, gibt, so daß

(i) $z_k := z_0 + \gamma_k(\delta z + r_k) \in \Omega$

für alle k,

(ii) $r_k \to 0$ bei $k \to \infty$ gilt.

Satz 2. *Sei $z_0 \in \Omega$* (Definition entsprechend (2), (3)) *mit dazugehörigen s_0, $x_0(\cdot)$, (σ_t^0) und $\varkappa \in R^m$ ein beliebiger Häufungspunkt der beschränkten Folge*

$$\left(\frac{1}{l_k}(s - s_0)\right) \int\limits_{s_0}^{s_0 + l_k(s - s_0)} K\big(s_0, t, x_0(t), \sigma_t^0\big)\, dt,\ s \in [0, s_e] \setminus \{s_0\},$$

wobei (l_k) eine beliebige Nullfolge positiver Zahlen sei. Für beliebige $(\sigma_t) \in \mathcal{S}$ ist dann

$$\delta z := \left(\int\limits_0^{s_0} K_s'\big(s_0, t, x_0(t), \sigma_t^0\big)\, dt + \varkappa + g'(s_0)\right)(s - s_0)$$

$$+ \int\limits_0^{s_0} \int\limits_0^{\tau} Y(\tau; s_0)\, K\big(\tau, t, x_0(t), \sigma_t - \sigma_t^0\big)\, dt\, d\tau$$

$$+ \int\limits_0^{s_0} K\big(s_0, t, x_0(t), \sigma_t - \sigma_t^0\big)\, dt$$

ein Richtungsgrenzwert von Ω in z_0 und im Falle $s := s_0$ ein Richtungsgrenzwert des Schnittes Ω_{s_0} von Ω in s_0. Dabei genügt die Matrixfunktion $Y(\cdot; s_0)$ der (bis auf Äquivalenzen) eindeutig lösbaren linearen Volterraschen Integralgleichung

$$Y(t; s_0) - \int_0^{s_0} Y(\tau; s_0)\, K'_x\big(\tau, t, x_0(t), \sigma_t^0\big)\, d\tau = K'_x\big(s_0, t, x_0(t), \sigma_t^0\big)$$

für f. a. $t \in [0, s_0]$ *und* $\operatorname{ess\,sup} |Y(t; s_0)| < \infty$.

Beim Beweis benutzen wir Satz 1 aus [5].

Ausgehend von Satz 1 und Satz 2 erhalten wir für ein spezielles Steuerproblem eine Aussage über die Existenz einer gewöhnlichen optimalen Steuerung, die A. Langenbach [6] in analoger Form durch die Untersuchung von Minimalfolgen erhielt.

Sei $m := n + 1$, $\bar{x} := (x_1, \ldots, x_n)^T$. $K(\cdot)$ hänge nicht von der $(n + 1)$-ten Komponente von x ab, so daß wir $K(s, t, \bar{x}, u) := K(s, t, x, u)$ setzen können. Weiterhin setzen wir $f_0 := K_{n+1}$ und $\bar{K} := (K_1, \ldots, K_n)^T$ sowie $g_{n+1} := 0$ und $\bar{g} := (g_1, \ldots, g_n)^T$. Dann betrachten wir die Aufgabe

$$\int_0^{s_e} f_0\big(s_e, t, \bar{x}(t), \sigma_t\big)\, dt = \min! \tag{4}$$

bei $|\bar{x}(s)| \leqq r_0$ und

$$\bar{x}(s) = \int_0^{s} \bar{K}\big(s, t, \bar{x}(t), \sigma_t\big)\, dt + \bar{g}(s), \tag{5}$$

$s \in [0, s_e]$, $(\sigma_t) \in \mathcal{S}$.

Aufgabe (4), (5) entspricht der Minimierung der $(n + 1)$-ten Koordinate aller Punkte von Ω_{s_e}, wobei wir die ursprüngliche Formulierung der Integralgleichung in (2) zugrunde legen. Dann erhalten wir

Satz 3. *Sei $\Omega \neq \emptyset$. Zusätzlich fordern wir:*

(i) *U sei konvex;*

(ii) $K'_{ix_j}(s, t, x, u) \geqq 0$ *für* $i, j = 1, \ldots, m$, $s, t \in [0, s_e]$, $|x| \leqq r_0$, $u \in U$;

(iii) *$\bar{K}_i(s, t, x, \cdot)$ sei konvex für $i = 1, \ldots, n$ und $f_0(s_e, t, x, \cdot)$ streng konvex auf U für beliebige $s, t \in [0, s_e]$ und $|x| \leqq r_0$.*

Unter den Voraussetzungen (i) *bis* (iii) *von Satz 3 existiert dann für Aufgabe* (4), (5) *eine verallgemeinerte optimale Steuerung, und jede derartige Maßfamilie ist gleich einer gewöhnlichen Steuerung.*

Literatur

[1] Roxin, E., The existence of optimal controls. Mich. Math. J. **9** (1962), 109–119.

[2] Warga, J., Optimal control of differential and functional equations (russ. Übers.), Nauka, Moskau 1977.

[3] Гамкрелидце, Р. В., Основы оптимального управления, Изд. Тбилиского университета, Тбилиси 1975.
[4] Schmeling, H., Zur Existenz der optimalen Steuerung bei Aufgaben mit nichtlinearen Volterraschen Integralgleichungen. Preprint-Reihe Mathematik, Preprint 4, Ernst-Moritz-Arndt-Universität Greifswald, 1979.
[5] Bittner, L., On optimal control of processes governed by abstract functional, integral and hyperbolic differential equations, Math. Operationsforsch. Statist. 6, 107–134 (1975).
[6] Langenbach, A., Stark konvexe Steuerprobleme, Math. Nachr. 77, 353–360 (1977).

Anschrift des Verfassers:

Ernst-Moritz-Arndt-Universität Greifswald
Sektion Mathematik
2200 Greifswald
Friedrich-Ludwig-Jahn-Str. 15a
DDR

The finite difference method for quasilinear elliptic equations with rapidly increasing coefficients

Rainer Schumann, Leipzig

Abstract. In this paper it is indicated how to use a general approximation-sovability result to prove the convergence of the finite difference method for quasilinear elliptic partial differential equations with rapidly increasing coefficients.

1. Introduction

We are going to consider the following equation

$$G\colon \sum_{|\alpha|\leqq m} (-1)^{|\alpha|} D^\alpha A_\alpha(x, Du(x)) = f(x), \tag{1}$$

$$\partial G\colon D^\beta u(x) = 0 \quad \text{for} \quad \beta\colon 0 \leqq |\beta| \leqq m-1,$$

where the coefficients A_α have "rapid growth". ($G \subset R^N$, G open and bounded, $Du = (D^\beta u)_{|\beta|\leqq m}$, $m \geqq 1$.)

Example. $G\colon \sum_{i=1}^{N} - D_i(D_i u(x) \exp |D_i u(x)|) = f(x),$

$\partial G\colon u(x) = 0.$

Using complementary systems of Sobolev-Orlicz spaces the existence theory for generalized solutions of equation (1) was studied by Gossez [1, 2]. It is our aim to prove the convergence of the finite difference approximation to equation (1). The convergence will follow from a result concerning the approximation-solvability of functional equations. This result is of the following type (cf. Petryshyn [4]):

The solvability of the equation $Au = b$ implies

(i) $Au = b$ is uniquely solvable,
(ii) the approximate equations $A_n u_n = b_n$ are uniquely solvable,
(iii) convergence $u_n \to u$ in some sense,

provided that some consistency and stability hypotheses are satisfied. The considerations of this paper follow the lines of Schumann, Zeidler [6] (see Zeidler [9], too) where the convergence of the finite difference method for equation (1) was proved under the hypothesis that the coefficients A_α have polynomial growth. Allowing rapidly increasing coefficients, however, gives rise to some complications since, in general, Sobolev-Orlicz spaces are neither separable nor reflexive and the domains of the operators involved fail to be linear subspaces of the Sobolev-Orlicz spaces.

2. An external approximation scheme

We define an external approximation scheme (in the sense of Temam [8]) with respect to the Sobolev-Orlicz space $Y = \mathring{W}^m L_H(G)$ (see fig. 1), where

(i) $\mathscr{G}_h$ denotes the set of grid points in G (points $\circ$ and $\bullet$ in fig. 2 if $m = 1$).

(ii) $\mathring{\mathscr{W}}^m \mathscr{L}_H(\mathscr{G}_h)$ is a discrete Sobolev-Orlicz space consisting of grid functions u_h defined on $\mathscr{G}_h$ and vanishing at boundary grid points (points $\circ$ in fig. 2 if $m = 1$).

(iii) $p_h u_h = (\mathring{p}_h \nabla^\alpha u_h)_{|\alpha| \leq m}$ with $\nabla^\alpha u_h$ being difference quotients and

$$\mathring{p}_h u_h(x) = \sum_{M \in \mathscr{G}_h} u_h(M)\, \chi_{c_h(M)}(x),$$

where $\chi_{c_h(M)}$ denotes the characteristic function of the cube $c_h(M) \subset R^N$, i.e. $\mathring{p}_h u_h(x) = u_h(M)$ if $x \in c_h(M)$.

(iv) $(r_h u)(M) = h^{-N} \int_{c_h(M)} u(x)\, dx$ if $M \in \mathscr{G}_{h,m}$ (set of inner grid points; grid points $\bullet$ in Fig. 2 if $m = 1$), $(r_h u)(M) = 0$ if $M \notin \mathscr{G}_{h,m}$.

(v) $\omega u = (D^\alpha u)_{|\alpha| \leq m}$.

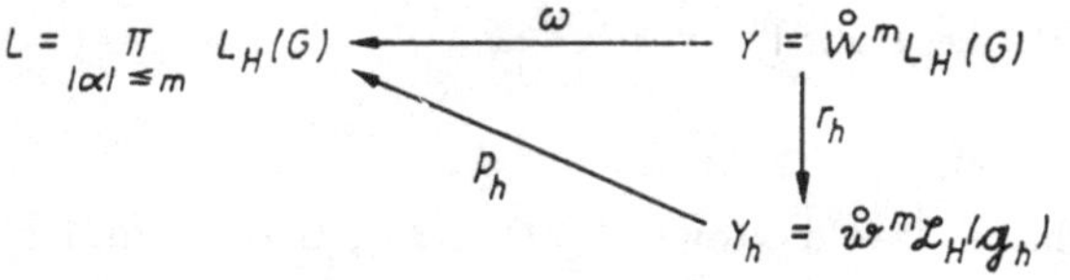

Fig. 1

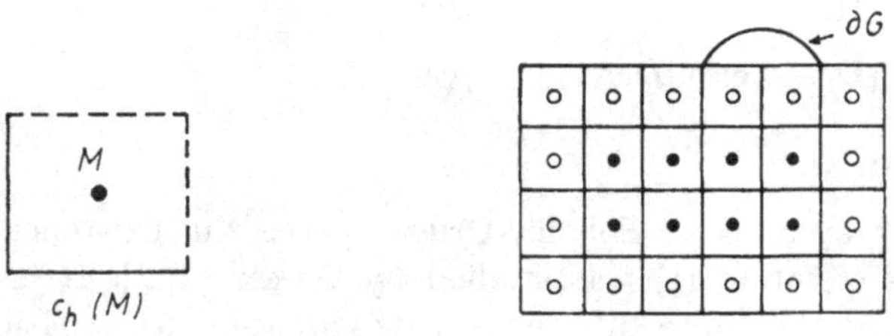

Fig. 2

The following definition plays a crucial role in the proofs of convergence in [6], [7]:

A sequence $(u^*_{h_n})$ of functionals $u^*_{h_n} \in Y^*_{h_n}$ is said to *converge discretely** to $u^* \in Z \cong Y^*_0$ if $\langle u_{h_n}, u^*_{h_n} \rangle \to \langle u, u^* \rangle$ for all bounded sequences $(u_{h_n} \in Y_{h_n})$ and elements $u \in Y$ with $p_{h_n} u_{h_n} \to \omega u$ in $\sigma(L, E)$ ($\sigma(L, E)$ denotes the weak* topology on L with respect to $E = \prod_{|\alpha| \leq m} E_{H^*}(G)$). If $u^*_{h_n}$ converges discretely* to u^* we write $u^*_{h_n} \xrightarrow{d^*} u^*$.

3. Operator equations

We are going to define generalized solutions of (1) using the following complementary system (see Gossez [1], [2]):

$$(Y, Y_0; Z, Z_0) = \big(\mathring{W}^m L_H(G), \mathring{W}^m E_H(G); W^{-m} L_{H^*}(G), W^{-m} E_{H^*}(G)\big).$$

Problem (P). Given $f \in E_{H^*}(G)$ find $u \in D(A) \subset Y = \mathring{W}^m L_H(G)$ with

$$(1')\ a(u, v) = b(v) \text{ for all } v \in Y_0 = \mathring{W}^m E_H(G), \tag{1'}$$

where

$$a(u, v) = \int_G \sum_{|\alpha| \leq m} A_\alpha\big(x, Du(x)\big)\, D^\alpha v(v)\, dx, \quad b(v) = \int_G f(x)\, v(x)\, dx,$$

$$D(A) = \{u \in Y : A_\alpha(\cdot, Du) \in L_{H^*}(G) \quad \text{for} \quad \alpha : |\alpha| \leq m\}.$$

If $u \in D(A)$ we define $Au \in Z$ by $\langle v, Au \rangle = a(u, v)$ for all $v \in Y_0$. Then (1') is equivalent to the equation $Au = b$.

To solve Problem (P) approximatively by a finite difference method we replace

- the functions $u, v \in \mathring{W}^m L_H(G)$ by grid functions $u_h, v_h \in \mathring{\mathscr{W}}^m \mathscr{L}_H(\mathscr{G}_h)$,
- the derivatives $D^\alpha u$ by discrete derivatives $\nabla^\alpha u_h$ (difference quotients),
- the integrals by sums.

Problem (P_h). Find a grid function $u_h \in Y_h = \mathring{\mathscr{W}}^m \mathscr{L}_H(\mathscr{G}_h)$ with

$$a_h(u_h, v_h) = b_h(v_h) \quad \text{for all} \quad v_h \in Y_h, \tag{2}$$

where $a_h(u_h, v_h)$ (resp. $b_h(v_h)$) are discrete forms arising from $a(u, v)$ (resp. $b(v)$) by means of the replacements mentioned above. If we define $A_h u_h \in Y_h^*$ by

$$\langle v_h, A_h u_h \rangle = a_h(u_h, v_h) \quad \text{for all} \quad v_h \in Y_h,$$

then (2) is equivalent to the equation $A_h u_h = b_h$.

4. Convergence Theorem

Let H be a Young function generated by the function φ:

$$H(t) = \int_0^t \varphi(s)\, ds, \qquad H^* \in \Delta_2.$$

Suppose Ψ to be another Young function with $\Psi \ll H$ (see e.g. Kufner, John, Fučik [3]). Assume that the following conditions are satisfied:

(a) The functions A_α are continuous for $|\alpha| \leq m$.

(b) Growth condition: there exist numbers $c, c_\beta > 0$ with

$$|A_\alpha(x, D)| \leq \text{const} \left(1 + \sum_{|\beta| \leq m-1} (H^*)^{-1} H(cD^\beta) + \sum_{|\beta| = m} \varphi(c_\beta\, |D^\beta|)\right)$$

for all points $x \in G$ and all real vectors $D = (D^\beta)_{|\beta| \leq m}$. Here, $u \in D(A)$ implies $\varphi(c_\beta D^\beta u) \in L_{H^*}(G)$ for all $\beta : |\beta| = m$.

(c) Monotonicity: there exist numbers $c_0 > 0$, $a > 0$ with

$$\sum_{|\alpha| \leq m} \big(A_\alpha(x, D) - A_\alpha(x, D')\big)\, (D^\alpha - D'^\alpha) \geq c_0 \sum_{|\beta| = m} H\big(a(D^\beta - D'^\beta)\big)$$

for all $x \in G$ and all real vectors $D = (D^\beta)$, $D' = (D'^\beta)$.

Then the following conclusions can be drawn:

(i) Problem (P) has a unique solution $u \in D(A)$.

(ii) Problem (P_h) has a unique solution $u_h \in Y_h$ if $h \in [0, h_0[$.

(iii)
$$\mathring{p}_h \nabla^\alpha u_h \to D^\alpha u \quad \text{in} \quad E_H(G) \quad \text{for} \quad |\alpha| \leqq m - 1 (h \to +0),$$
$$\mathring{p}_h \nabla^\alpha u_h \to D^\alpha u \quad \text{in} \quad E_\Psi(G) \quad \text{for} \quad |\alpha| = m \ (h \to +0)$$

(i.e. the derivatives of order m converge in a coarser norm only).

(iv) There exists $\gamma > 0$ with
$$\int_G H\big(\gamma(\mathring{p}_h \nabla^\alpha u_h(x) - D^\alpha u(x))\big)\, dx \to 0 \quad \text{for} \quad |\alpha| = m \ (h \to +0).$$

Sketch of the proof. One has to verify some consistency and stability hypotheses. Then, as mentioned in the Introduction, the solvability of $Au = b$ implies the assertion. The solvability of $Au = b$ is guaranteed by a theorem of Gossez [1, 2]. The most important step is to prove
$$A_{h_n} r_{h_n} u \xrightarrow{d^*} Au \ (h_n \to +0 \quad \text{if} \quad n \to \infty). \tag{3}$$
As contrasted with the case of polynomial growth (where $Y = \mathring{W}^m_p(G)$) (3) is not valid for all $u \in Y = \mathring{W}^m L_H(G)$ but only for $u \in \mathring{W}^m E_H(G)$. Thus an additional approximation result which is proved by means of convex analysis (cf. Rockafellar [5]) is needed. The complete proof is given in Schumann [7].

References

[1] J. P. Gossez, Nonlinear elliptic boundary value problems for equations with rapidly or slowly increasing coefficients, Trans. Amer. Math. Soc. **190** (1974), 163–205.

[2] J. P. Gossez, Orlicz-Sobolev spaces and nonlinear elliptic boundary value problems: *in:* S. Fučik, A. Kufner (eds.), Nonlinear analysis, function spaces and applications. Proc. Spring School Horni Bradlo 1978, Teubner-Texte zur Mathematik, Teubner Verlag, Leipzig 1979, 59–94.

[3] A. Kufner, O. John. S. Fučik, Function spaces, Academia, Prague 1977.

[4] W. V. Petryshyn, Nonlinear equations involving noncompact operators. *In:* Nonlinear Functional Analysis, Proc. Symp. Pure. Math. **18** (1970) 1, 206–233, Amer. Math. Soc., Providence, R. I.

[5] R. T. Rockafellar, On the maximal monotonicity of subdifferential mappings, Pacific J. Math. **33** (1970), 209–216.

[6] R. Schumann, E. Zeidler, The finite difference method for quasilinear elliptic equations of order $2m$, J. Numer. Funct. Anal. Optim. **1** (1979) 2, 161–195.

[7] R. Schumann, The finite difference method for quasilinear elliptic equations with rapidly increasing coefficients. Math. Nachr. (toappear).

[8] R. Temam, Analyse numérique, Presses Univ. de France, 1970.

[9] E. Zeidler, Vorlesungen über nichtlineare Funktionalanalysis I–IV, Teubner-Texte zur Mathematik, Teubner Verlag Leipzig, 1976, 1977, 1978, 1980. (Engl. transl. in preparation, Springer-Verlag, Berlin).

Address of the author:

Karl-Marx-Universität Leipzig
Sektion Mathematik
7010 Leipzig
Karl-Marx-Platz
DDR

A constructive proof of the implicit function theorem for analytic operators

Hans-Dieter Sparing, Ilmenau

It is an old problem to determine a sphere for a given operator $F(x, y)$ of two variables, a sphere in which the solution of the implicit equation $F(x, y) = 0$ is unique and regular.

Although Banach's fixed point theorem, taken together with some regularity conditions, in the case of the well-known suppositions

$$F(x_0, y_0) = 0,$$

there exists a bounded inverse to the linear operator $F_y(x_0, y_0)$

guarantees the existence of a unique solution in a neighborhood of the point (x_0, y_0), there is nothing known about the size of this sphere.

In 1950 T. Ważewski [1] developed some conditions by which it was possible to compute a uniqueness sphere of the solution. To obtain this, he needed the existence of the inverse operators to the partial differentials for y in a whole neighborhood of (x_0, y_0) and the uniform boundedness of these operators and of the partial differentials for x.

Under similar conditions it is possible to derive theorems about the global unique solution of an implicit operator from the first chapter of J. Schwartz' "Nonlinear functional analysis" (1969). The advantage of these theorems lies in the fact, that the class of regularity of the implicit operator is nearly unrestricted. However, it is normally not easy to examine all the differentials in a whole sphere.

Another possibility seemed to be the use of higher classes of regularity in exchange to a restriction to the point (x_0, y_0) itself.

Beginnings for this were to be found in the method of E. Goursat [3] used in the classical case. The disadvantage of this method is that it provides a sphere of regularity for the solution, but it guarantees the uniqueness of the solution only in the class of analytic mappings.

Nevertheless, first investigations went in this direction. In 1960 A. E. Gelman [4] published a generalization of the classical theorem without proof. The statement of his theorem was repeated by M. Wainberg and W. Trenogin in [5], but the proof of the theorem seemed doubtful.

So let me once more return to this theorem and demonstrate the difficulty in the proof:

Theorem. *Consider the Banach spaces* E_1, E_2, D *over the same field of scalars* K. *Suppose* $F : E_1 \times E_2 \to D$, *defined on the open set* $K(x_0, r) \times K(y_0, \varrho)$, *is an operator satisfying* *the following conditions:*

$$F(x_0, y_0) = 0,$$

F *analytic,*

$$\exists\, B = (''d_{(x_0,y_0)}\, F)^{-1} \in L(D, E_2),$$

then:

$$\exists\, 0 < \sigma < r, \quad 0 < \varepsilon < \varrho, \quad \exists\, f : K(x_0, \sigma) \to K(y_0, \varepsilon) \quad \textit{so that}$$

$$f(x_0) = y_0,$$

$$F(x, f(x)) = 0 \quad \forall\, x \in K(x_0, \sigma),$$

f *analytic.*

Proof. The equality $F(x, y) = 0$ is equivalent to

$$h = R_{01}(g) + \sum_{i+j \geqq 2} R_{ij}(h^i, g^j)$$

where

$$R_{ij} = \frac{1}{i!j!} B \circ \frac{\partial^{i+j} F(x_0, y_0)}{\partial y^i\, \partial x^j}, \qquad g = x - x_0, \quad h = y - y_0.$$

To obtain some necessary conditions for the existence of an analytic solution f, we use the power series representation of the unknown function:

$$h = \sum_{k=1}^{\infty} \Psi_k(g^k) = R_{01}(g) + \sum_{i+j \geqq 2} R_{ij}\left(\left(\sum_{k=1}^{\infty} \Psi_k(g^k)\right)^i, g^j\right). \tag{+}$$

Comparing the coefficients standing with equal powers of g, we obtain:

$$\Psi_k(g^k) = \sum_{i+j=2} \sum_{S_{k-1}} \frac{i!}{\prod\limits_{l}^{k-1} i_l!} \tilde{R}_{\tilde{i}j}\left(\Psi_1(g)^{i_1}, \ldots, \Psi_{k-1}(g^{k-1})^{i_{k-1}}, g^j\right) + \sum_{i+j=k} R_{ij}\left(\Psi_1(g)^i, g^j\right) \tag{++}$$

where S_k means the summations $\sum\limits_1^k i_l = i, \quad \sum\limits_1^k l_{i_l} = k + 1 - j$.

Here it is necessary to observe that $\tilde{R}_{\tilde{i}j}$ does not any more represent the i-homogeneous polynomial with respect to the first index, but the corresponding polar form, that is the i-linear, symmetric mapping which gives the polynomial when applied to equal arguments. This is the heart of the matter because in the case of abstract Banach spaces the norms of the two mappings mentioned are not the same.

For instance, take $P : R^2 \in (x, y) \to 2xy \in R$, which has the norm $\|P\| = \frac{1}{2}$. On the other hand the norm of the corresponding polar form $\tilde{P} : R^2 \times R^2 \ni ((u, v), (x, y)) \to uy + xv \in R$ is $\|\tilde{P}\| = 1$.

The above mentioned formula (++) guarantees the uniqueness of the analytic solution if its existence, that is the convergence of the constructed series has been shown. For this aim one can use the method of majorants developed by Cauchy and Goursat. Two lemmas are needed before:

Lemma 1.

$$\sum_{r=0}^{n-1} \left(e^{\frac{2\pi i r}{n}}\right)^k = \begin{cases} n & \exists\, m \in Z : k = nm, \\ 0 & \nexists\, m \in Z : k = nm. \end{cases}$$

Lemma 2.

$$\varphi \in W_n^F(E, D);\quad \text{a) } K = C \Rightarrow \|\tilde{\varphi}\| \leqq \frac{n^n}{n!} \|\varphi\|,$$

$$\text{b) } K = R \Rightarrow \|\tilde{\varphi}\| \leqq \frac{(2n)^n}{n!} \|\varphi\|.$$

The first part of lemma 2 was proved by Gawurin in 1939 and the proof was based on lemma 1. The second part of lemma 2 is a rough estimation which is easy to obtain using the isomorphism between homogeneous polynomials and polar forms. In spite of all efforts it could not be improved yet.

For further computation we need

$$M = \sup \{\|R_{ij}(h^i, g^j)\| : \|h\| < \varrho, \quad \|g\| < r, \quad i, j \in N\}.$$

This demand seems rather pretentious, in the case of complex spaces, however, it is possible to lower it so far as to take for M simply the supremum of the function in the mentioned sphere, because Cauchy's inequalities are valid in the abstract case, too.

Denoting $\overline{M} = M \|B\|$, we obtain the estimations:

$$\|R_{ij}\| \leqq \overline{M} r^{-j} \varrho^{-i} \quad \text{and} \quad \|\tilde{R}_{ij}\| \leqq \frac{(ti)^i}{i!} \overline{M} r^{-j} \varrho^{-i}$$

with $t = 1, 2$ depending on the real or complex field of scalars. Now it is possible to find a majorant for the right-hand side of $(+)$:

$$\overline{M} \left(\frac{\xi}{r} + \sum_{i+j \geqq 2} \left(\frac{\xi}{r}\right)^j \left(\frac{\eta}{\varrho}\right)^i \frac{(ti)^i}{i!} \right) \leqq \overline{M}\, \mathrm{e}^{-1} \left[\left(\sum_{j=0}^{\infty} \left(\frac{\xi}{r}\right)^j \right) \left(\sum_{i=0}^{\infty} \left(\frac{t\, \mathrm{e}\, \eta}{\varrho}\right)^i \right) - 1 - \frac{t\, \mathrm{e}\, \eta}{\varrho} \right].$$

This allows us to adopt the following majorant equation:

$$\eta = \overline{M}\, \mathrm{e}^{-1} \left[\left(1 - \frac{\xi}{r}\right)^{-1} \left(1 - \frac{t\, \mathrm{e} \eta}{\varrho}\right)^{-1} - 1 - \frac{t\, \mathrm{e} \eta}{\varrho} \right].$$

The equation has exactly one solution $\eta = \eta(\xi)$, satisfying the condition $\eta(0) = 0$, namely:

$$\eta = \frac{\varrho^2}{2t\, \mathrm{e}(\varrho + t\overline{M})} \left[1 - \left(1 - \frac{\xi}{r}\right)^{-1/2} \left(1 - \frac{\xi}{\alpha}\right)^{1/2} \right], \quad \alpha = r \left(\frac{\varrho}{\varrho + 2t\overline{M}} \right)^2.$$

Both parts of the product are absolutely convergent as $\xi < \alpha$, so the solution is convergent for all x with:

$$\|x - x_0\| < r \left(\frac{\varrho}{\varrho + 2tM \|B\|} \right)^2.$$

At the end of this note it should be remarked, that the problem of the norm change for polynomials and polar forms does not exist in the case of Hilbert spaces because Banach's lemma $\|\varphi\| = \|\tilde{\varphi}\|$ is valid there, as was shown by J. Siciak and J. Bochnak [7] in 1971 for real scalars, a statement which is easy to extend to the complex situation.

References

[1] Ważewski, T., Sur l'évaluation du domaine d'existence des fonctions implicites dans les cas des espaces abstraits Fund. Math. **37** (1950).
[2] Schwartz, J., Nonlinear functional analysis, New York 1969.
[3] Goursat, E., Course d'analyse mathématique, Paris 1929.
[4] Gelman, A. E., Theorems on an implicit abstract function (Russian), Dokl. Akad. Nauk SSSR, **132** (1960), 3.
[5] Wainberg, M., Trenogin, W., Theorie der Lösungsverzweigung bei nichtlinearen Gleichungen, Berlin 1973.
[6] Gawurin, M. K., On the k-linear operators in Banach spaces (Russian), Dokl. Akad. Nauk SSSR **22** (1939).
[7] Bochnak, J., Siciak, J., Polynomials and multilinear mappings in topological vector spaces, Stud. Math. **39** (1971).

Address of the author:

Technische Hochschule Ilmenau
Sektion Mathematik, Rechentechnik und ökonomische Kybernetik
6300 Ilmenau
Am Ehrenberg
DDR

Some remarks to existence problems of optimal control for parabolic equations

Wolfram Sperber, Cottbus

1. Introduction

The problem which is investigated here, is a question of existence of an optimal control for processes governed by differential equations of parabolic type. Especially we consider boundary control, that means that the control functions enters only into the boundary conditions of the parabolic initial-boundary-value problem.

The optimal control problem is now formulated as follows. It may be given the initial-boundary-value problem:

$$Lw(x, t) = f(x, t) \quad \text{in } D, \tag{1}$$

$$w(x, 0) = e(x) \quad \text{in } B, \tag{2}$$

$$\frac{\partial w}{\partial r}(x, t) = g_0(x, t, w) - g_1(x, t, u) \quad \text{in } S \tag{3}$$

with B — a bounded domain in R^n, $n \geqq 2$, with boundary ∂B,

$$D = B \times (0, T), \qquad S = \partial B \times (0, T).$$

L denotes the general linear parabolic differential operator:

$$Lw(x, t) = \sum_{j=1}^{n} a_{ij}(x, t) \frac{\partial^2 w}{\partial x_i \, \partial x_j}(x, t) + \sum_{i=1}^{n} b_i(x, t) \frac{\partial w}{\partial x_j}(x, t)$$
$$+ c(x, t)\, w(x, t) - \frac{\partial w}{\partial t}(x, t),$$

$\dfrac{\partial w(x, t)}{\partial r(\xi, t)} = \sum_{i=1}^{n} a_{ij}(\xi, t) \cos(N_\xi, x_j) \dfrac{\partial w}{\partial x_i}(x, t)$ denotes the directional derivative of $w(x, t)$ in the inward conormal direction in the point $(\xi, t) \in \partial B$.

The control function is a function

$$u \in U = \{u \in L_\infty(0, T) \mid a \leqq u(t) \leqq b \quad \text{almost everywhere}\}.$$

u is to be chosen so that the aim of control is fulfilled:

$$\int_B F(x, w(x, T))\, dx \to \min_{u \in U}! \tag{4}$$

At first the existence of a solution of the parabolic initial-boundary-value problem is investigated for every admissible control u. It is usual to define generalized solutions and to show the existence of these, because the control functions need not be smooth.

Here they are defined in the sense of A. Friedman [2] and L. v. Wolfersdorf [3, 4]; the considered nonlinear initial-boundary-value problem is reduced to a system of integral equations. By a substitution the system of integral equations can be reduced to a parabolic initial-boundary-value problem which has a classical solution. The existence of a unique classical solution can be shown generalizing some theorems of A. Friedman [1]. To prove the existence of an optimal control a sequence is constructed which converges to the minimum of the functional in (4) for all $u \in U$, here we mention a work of E. Sachs [5].

2. The existence of a solution for the initial-boundary-value problem

At first we give the generalizations of the theorems of A. Friedman used later.

The following problem is considered:

(5) $$Lw(x, t) = f(x, t) \quad \text{in } D,$$

(6) $$w(x, 0) = e(x) \quad \text{on } B,$$

(7) $$\frac{\partial w}{\partial r}(x, t) = g\big(x, t, w(x, t), \varphi(x, t)\big) \quad \text{on } S;$$

B, D, S, L and $\frac{\partial w}{\partial r}$ have the same meaning as in Problem (1)—(3).

Now the theorems can be formulated:

Theorem 1. *It is assumed that*

(8) $$\partial B \in C^{1+\lambda} \quad (0 < \lambda < 1),$$

(9) *L is parabolic in $\bar{D}$,*

the coefficients a_{ij}, b_i, c of L are continuous in $\bar{D}$ and $c \leqq 0$,

(10) *$g(x, t, w, v)$ is strictly increasing in w.*

Then there exists at most one solution of (5)—(7) *in D.*

Theorem 2. *It is assumed that* (8) *and* (10) *are valid,*

(11) *L is parabolic in $\bar{D}$ and $a_{11}(x, t) \geqq K_1$, $b_1(x, t) \geqq -K_2$ for some positive constants K_1, K_2;*

the coefficients a_{ij}, b_i, c of L are continuous in $\bar{D}$ and $c \leqq 0$,

(12) *$g(x, t, w, v) \to \pm\infty$ as $w \to \pm\infty$, uniformly with respect to (x, t) in $\bar{S}$ and in bounded sets.*

If w denotes a solution of (5)—(7) *in D, then*

$$|w(x, t)| \leqq H \quad \text{in } D,$$

where H is a constant depending only on g, K_1, K_2, the diameter of B and on upper bounds on $|f|$, $|e|$, $|\varphi|$.

Theorem 3. *It is assumed that* (8), (10), (12) *are valid,*

(13) *L is parabolic in $\bar{D}$ and $a_{11}(x, t) \geqq K_1$, $b_1(x, t) \geqq -K_2$ for some positive constants K_1, K_2; the coefficients a_{ij}, b_i, c are Hölder continuous and in D and $c \leqq 0$,*

(14) *$g(x, t, w, v)$ is a continuous function for $(x, t) \in \bar{S}$,*
$-\infty < w < \infty, \quad -\infty < v < \infty$,

(15) *$f(x, t)$ is Hölder continuous in x, uniformly in bounded subsets of $\bar{D}$,*

(16) *$\varphi(x, t)$ is a continuous function in S,*

(17) *$e(x)$ is a continuous function in B with compact support.*

Then there exists a unique solution of (5)—(7).

Remark. The theorems 1—3 can be proved in the same way as the theorems 11, 12, 13 in [1], chapter 7, section 5.

Now the existence of a solution for the initial-boundary-value problem (1)—(3) for every admissible control u is investigated.

To this aim we define generalized solutions for (1)—(3).

Definition. Every continuous function w in $\bar{D}$ is a *generalized solution* of (1)—(3), if w is described by

(18)
$$\begin{aligned} w(x, t) = &-\int_0^t \int_B G(x, t, \xi, \tau)\, f(\xi, \tau)\, d\xi\, d\tau + \int_B G(x, t, \xi, 0)\, e(\xi)\, d\xi \\ &+ \int_0^t \int_{\partial B} G(x, t, \xi, \tau)\, g_0\big(\xi, \tau, w_1(\xi, \tau)\big)\, dB_\xi\, d\tau \\ &- \int_0^t \int_{\partial B} G(x, t, \xi, \tau)\, g_1\big(\xi, \tau, u(\tau)\big)\, dB_\xi\, d\tau \quad \text{in } D \end{aligned}$$

and if there exists a continuous function w_1 in $\bar{S}$ solving the integral equation

(19)
$$\begin{aligned} w_1(x, t) = &-\int_0^t \int_B G(x, t, \xi, \tau)\, f(\xi, \tau)\, d\xi\, d\tau + \int_B G(x, t, \xi, 0)\, e(\xi)\, d\xi \\ &+ \int_0^t \int_{\partial B} G(x, t, \xi, \tau)\, g_0\big(\xi, \tau, w_1(\xi, \tau)\big)\, dB_\xi\, d\tau \\ &- \int_0^t \int_{\partial B} G(x, t, \xi, \tau)\, g_1\big(\xi, \tau, u(\tau)\big)\, dB_\xi\, d\tau \quad \text{in } S. \end{aligned}$$

Remark. $G(x, t, \xi, \tau)$ denotes the Green (or Neumann) function for the problem (with sufficiently smooth data) $Lw(x, t) = f(x, t)$ in D, $w(x, 0) = e(x)$ on B and $\frac{\partial w}{\partial r}(x, t) + \beta(x, t)\, w(x, t) = g(x, t)$ on S.

Theorem 4. *It is assumed that* (8), (13), (15), (17) *are valid,*

(20) $g_0(x, t, w_1)$ *is a continuous function for* $(x, t) \in \bar{S}$, $-\infty < w_1 < \infty$, $g_0(x, t, w_1) \to \pm\infty$ *for* $w_1 \to \pm\infty$, *uniformly with respect to* (x, t) *in* $\bar{S}$; $g_0(x, t, w_1)$ *is strictly increasing in* w_1,

(21) $g_1(x, t, u)$ *is a continuous function for* $(x, t) \in \bar{S}$, $-\infty < u < \infty$.

Then there exists a generalized solution of (1)—(3) *for every* $u \in U$.

Idea of proof: The function φ defined by the formula

$$\varphi(x, t) := \int_0^t \int_{\partial B} G(x, t, \xi, \tau)\, g_1\big(\xi, \tau, u(\tau)\big)\, dB_\xi\, d\tau \quad \text{on } S \tag{22}$$

is a continuous function in $\bar{S}$ for a given measurable and bounded function u, because it can be shown that the operator R defined by

$$Rv(\xi, \tau) := \int_0^t \int_{\partial B} G(x, t, \xi, \tau)\, v(\xi, \tau)\, dB_\xi\, d\tau \quad \text{in } S$$

has the properties

$R\colon L_p(S) \to C(\bar{S})$ for p sufficiently large and R is linear and completely continuous.

Substituting

$$\overline{w}_1(x, t) := w_1(x, t) + \varphi(x, t) \quad \text{on } W,$$

$$\overline{w}(x, t) := w(x, t) + \int_0^t \int_{\partial B} G(x, t, \xi, \tau)\, g_1\big(\xi, \tau, u(\tau)\big)\, dB_\xi\, d\tau \quad \text{on } D$$

to (18) and (19) we now get the form:

$$\begin{aligned}\overline{w}(x, t) = &-\int_0^t \int_B G(x, t, \xi, \tau)\, f(\xi, \tau)\, d\xi\, d\tau + \int_B G(x, t, \xi, 0)\, e(\xi)\, d\xi \\ &+ \int_0^t \int_{\partial B} G(x, t, \xi, \tau)\, g_0\big(\xi, \tau, \overline{w}_1(\xi, \tau) - \varphi(\xi, \tau)\big)\, dB_\xi\, d\tau \quad \text{in } D,\end{aligned} \tag{23}$$

$$\begin{aligned}\overline{w}_1(x, t) = &-\int_0^t \int_B G(x, t, \xi, \tau)\, f(\xi, \tau)\, d\xi\, d\tau + \int_B G(x, t, \xi, 0)\, e(\xi)\, d\xi \\ &+ \int_0^t \int_{\partial B} G(x, t, \xi, \tau)\, g_0\big(\xi, \tau, \overline{w}_1(\xi, \tau) - \varphi(\xi, \tau)\big)\, dB_\xi\, d\tau \quad \text{in } S.\end{aligned} \tag{24}$$

Now we study the initial-boundary-value problem corresponding to (23)—(24)

$$Lw(x, t) = f(x, t) \quad \text{in } D, \tag{25}$$

$$w(x, 0) = e(x) \quad \text{on } B, \tag{26}$$

$$\frac{\partial w}{\partial r}(x, t) = g_0\big(x, t, w(x, t) - \varphi(x, t)\big) \quad \text{on } S. \tag{27}$$

It is easy to show that it is possible to use the theorems 1—3 (by substituting $g(x, t, w, \varphi) := g_0(x, t, w - \varphi)$) which guarantees a unique classical solution (that means by definition, that $\overline{w} \in C(\overline{D}) \cap C^{2,1}(D)$ and fulfils (25)—(27)).

Remark. The solution of the problem (25)—(27) is a solution of the system of integral equations (23)—(24), too. It can be shown that the problem (23)—(24) cannot have more than one solution.

3. The existence of a solution for the optimal control problem

The uniquely defined solution $w(x, t)$ of (23)—(24) will be called in the further context "*the solution* $w(u, x, t)$ *of* (18)—(19)".

Theorem 5. *Let all assumptions of theorem 4 be valid and assume that*

(28) *$g_1(x, t, u)$ is of the form $g_1(x, t, u) = g_2(x, t)\, u$ and $g_2(x, t)$ is a continuous function for $(x, t) \in \overline{S}$;*

(29) *F is continuously depending on x, w.*

Then there exists a solution for the optimal control problem (1)—(4).

Idea of proof: We construct a minimizing sequence $\{u_n\}$ with $u_n \in U$:

(31) $$\lim_{n\to\infty} \int_B F\big(x, w(u_n, x, T)\big)\, dx = \inf_{u \in U} \int_B F\big(x, w(u, x, T)\big)\, dx.$$

For this sequence $\{u_n\}$ there exists a subsequence, also called $\{u_n\}$, which is weakly convergent in $L_p(Q, T)$, because U is a weakly (sequentially) compact subset of L_p. The limit of $\{u_n\}$ is denoted by $\overline{u}$:

(32) $$\lim_{n\to\infty} u_n = \overline{u} \quad \text{in the weak } L_p\text{-sense.}$$

The idea is now to show that $w(u_n, \cdot, T)$ converges to $w(\overline{u}, \cdot, T)$ in the maximum norm.

It can be shown that $w(u, \cdot, \cdot)$ and $w_1(u, \cdot, \cdot)$ are uniformly bounded for $u \in U$. It follows the existence of a subsequence $w_1(u_n, \cdot, \cdot)$ and of an element $\overline{w}_1 \in L_p(S)$ such that

(33) $$\lim w_1(u_n, \cdot, \cdot) = \overline{w}_1 \quad \text{in weak } L_p\text{-sense}$$

and also

(34) $$g_0\big(\cdot, \cdot, w_1(u_n, \cdot, \cdot)\big) \quad \text{is weakly convergent in } L_p.$$

We write the corresponding sequence of integral equations (19) as operator equations:

(35) $$w_1(u_n, \cdot, \cdot) = Rg_0\big(\cdot, \cdot, w_1(u_n, \cdot, \cdot)\big) - Rg_2(\cdot, \cdot)\, u_n + F(\cdot, \cdot)$$

in S, where $F(x, t)$ is the symbol for

$$F(x, t) := -\int_0^t \int_B G(x, t, \xi, \tau)\, f(\xi, \tau)\, d\xi\, d\tau + \int_B G(x, t, \xi, 0)\, e(\xi)\, d\xi.$$

Taking the limit on both sides of (35) we get the weak convergence of $w_1(u_n, \cdot, \cdot)$ by (32), (33), (34), the complete continuity of R:

$$\overline{w}_1 = Rg_0(\cdot, \cdot, \overline{w}_1) - Rg_2(\cdot, \cdot)\, \overline{u} + F(\cdot, \cdot). \tag{36}$$

This means that $\overline{w}_1 = w_1(\overline{u}, \cdot, \cdot)$. It is possible to show that $w_1(u_n, \cdot, \cdot)$ converges to $w_1(\overline{u}, \cdot, \cdot)$ not only in the weak L_p-sense but in the maximum norm, too.

By investigating (23) we can show that

$$\lim_{n\to\infty} w(u_n, \cdot, \cdot) = w(\overline{u}, \cdot, \cdot) \text{ in the maximum norm and}$$

$$\lim_{n\to\infty} w(u_n, \cdot, T) = w(\overline{u}, \cdot, T) \text{ in the maximum norm, too.}$$

Hence by continuity of F

$$\lim_{n\to\infty} \int_B F\big(x, w(u_n, x, T)\big)\, dx = \int_B F\big(x, w(\overline{u}, x, T)\big)\, dx$$

results. This completes the proof.

References

[1] A. Friedmann, Partial differential equations of parabolic type, Prentice-Hall, Englewood Cliffs 1964.

[2] A. Friedmann, Optimal control for parabolic equations, J. Math. Anal. Appl. **18** (1967), 479–491.

[3] L. v. Wolfersdorf, Optimal control for processes governed by mildly nonlinear differential equations of parabolic type. I, Z. Angew. Math. Mech. **56** (1976), 531–538.

[4] L. v. Wolfersdorf, Optimale Steuerung einer Klasse nichtlinearer Aufheizungsprozesse. Z. Angew. Math. Mech. **55** (1975), 353–362.

[5] E. Sachs, A parabolic control problem with a boundary condition of the Stefan-Boltzmann type, Z. Angew. Math. Mech. **58** (1978), 443–449.

Address of the author:

Ingenieurhochschule Cottbus
Abteilung Mathematik/Naturwissenschaften
7500 Cottbus
Karl-Marx-Str. 17
DDR

Construction of local solutions for Maxwell equations

Wolfgang Sprößig, Karl-Marx-Stadt

As is well-known, Maxwell equations have in isotropic media the following form

$$\left.\begin{aligned} &\operatorname{rot} E = -\mu \frac{\partial}{\partial t} H, \\ &\operatorname{rot} H = \varepsilon \frac{\partial}{\partial t} E + \varkappa E, \\ &\operatorname{div} \varepsilon E = \varrho, \\ &\operatorname{div} \mu H = 0. \end{aligned}\right\} \qquad (1)$$

Here denote $\mu = \mu(p, H)$ permeability, $\varepsilon = \varepsilon(p, E)$ dielectric constant, $\varrho = \varrho(p)$ density of loading, $\varkappa = \varkappa(p, \varepsilon, E)$ conductivity, E electric field, H magnetic field, and $p = (p_1, p_2, p_3)$ is the vector of independent variables.

Applying of the multidimensional function theory developed in papers by R. Fueter [1], A. W. Bizadse [2], W. S. Winogradow [3], E. Stein/G. Weiß [4], R. Delanghe [5], G. Hile [6] and by the author [7]—[9] the system (1) can be reduced to the system

$$\begin{aligned} &\nabla \circ u = A_+^{(+)} u + A_-^{(+)} v - \frac{\varrho}{\varepsilon} \\ &(J\nabla) \circ v = A_-^{(-)} u - A_+^{(-)} v + \frac{\varrho}{\varepsilon}. \end{aligned} \qquad (2)$$

Here we write $u = E^* + H^*$, $v = E^* - H^*$, $E^*(\hat{p}) = E_0^*(p_0)\, E_1^*(p)$ $H^*(\hat{p}) = H(t, p)$ with $\hat{p} = (p_0, p) = \left(\frac{1}{\mu} t, p\right)$ $J\hat{u} = (u_0, -u)$, $\hat{u} = (u_0, u)$, $\nabla = \left(\frac{\partial}{\partial p_0}, \frac{\partial}{\partial p_1}, \frac{\partial}{\partial p_2}, \frac{\partial}{\partial p_3}\right)$,

$$A_\mp^{(\pm)} = \left\| \begin{matrix} \ddots & & 0 \\ & \mu_i \mp \varepsilon_i^\pm & \\ 0 & & \ddots \end{matrix} \right\|, \quad \mu_i = \frac{1}{\mu}\left(\frac{\partial}{\partial p_i} \mu\right), \quad \varepsilon_i^{(\pm)} = \left(\frac{\partial}{\partial p_i} \varepsilon\right) \frac{1}{\varepsilon} \pm \left((\mu + \varepsilon) \frac{E_0^*}{E_0} + \varkappa\right) \qquad i = 0, 1, 2, 3,$$

and "$\circ$" is the quaternionic multiplication.

Now, it is possible to write the system of differential equations (2) in the following form

$$V \circ \hat{w} = A\hat{w} + f + Vw_0, \tag{3}$$

where

$$V = \begin{Vmatrix} V & 0 \\ 0 & JV \end{Vmatrix}, \quad A = \begin{Vmatrix} A_-^{(+)} & A_+^{(+)} \\ A_-^{(-)} & A_+^{(-)} \end{Vmatrix}, \quad w_0 \begin{Vmatrix} u_0 \\ 0 \\ v_0 \\ 0 \end{Vmatrix}, \quad f = \begin{Vmatrix} -\frac{\varrho}{\varepsilon} \\ 0 \\ \frac{\varrho}{\varepsilon} \\ 0 \end{Vmatrix}$$

and $u = (u_1, u_2, u_3)$, $v = (v_1, v_2, v_3)$, $\hat{v} = (v_0, v)$, $\hat{u} = (u_0, u)$.

Instead of the equation (3) we consider the integral equation

$$\hat{w} = \boldsymbol{T}A\hat{w} + \boldsymbol{T}f + \boldsymbol{T}Vw_0 + \Phi, \tag{4}$$

where $V \circ \Phi = 0$,

$$\boldsymbol{T} = \begin{Vmatrix} T & 0 \\ 0 & T^* \end{Vmatrix}, \quad (T\cdot)(p) = \frac{1}{\pi^2} \int_G \frac{\hat{p} - \hat{q}}{|\hat{p} - \hat{q}|^4} \circ \cdot \, dG\hat{q} \tag{5}$$

and $\hat{p} = (p_0, p)$, $\hat{q} = (q_0, q)$. By using the formula of Borel-Pompeiu in [7] we receive the equation

$$w = P_1\boldsymbol{T}Aw + \boldsymbol{T}f + P_1[\Phi_1 - \boldsymbol{T}Aw_0], \tag{6}$$

$$V \circ \Phi_1 = 0 \quad \text{and} \quad P_1\hat{u} = (0, u).$$

The sign "$*$" means, that in definition (5) the vector $\hat{p} - \hat{q}$ must be replaced by $J(\hat{p} - \hat{q})$. If the function w^* is a solution of equation (4) then it is also a solution of equation (3). It is easy to see that $\|T\|_{L(C(G))} \leqq 2 \operatorname{diam} G$ holds.

We assume that the functions ε_i, μ_i and $\varkappa$ satisfy Lipschitz conditions for the variables u and v and the inequality

$$\left[-\frac{t_1}{\varepsilon_0} + t_0, \frac{t_2}{\mu_0} + t_0\right] + 2 \operatorname{diam} G_0 < \min\left[\frac{1}{2M}, \frac{1}{2K}\right]$$

holds. Here we use the notation $M = \|A\|_{L(C(\bar{G}))}$, K is the maximal Lipschitz constant, t_0, t_1 and t_2 are fixed time instants, and if $\varepsilon_0 = \min \varepsilon$, $\mu_0 = \min \mu$ the inclusion

$$G \subset \left[-\frac{t_1}{\varepsilon_0} + t_0, \frac{t_2}{\mu_0} + t_0\right] \times G_0$$

holds, where G_0 is a domain from the space R^3. Under these conditions the sequence of successive approximations

$$w^{(n+1)} = P_1\boldsymbol{T}Aw^{(n)} + \boldsymbol{T}f + P_1[\Phi_1 - \boldsymbol{T}Aw_0] \tag{7}$$

is uniformly convergent, see [9]. Furthermore we have the estimate

$$\|w^{(n)} - w^*\|_{C(\bar{G})} \leqq \frac{1}{2^n},$$

where w^* is the exact solution.

References

[1] Fueter, R., Die Theorie der regulären Funktionen einer Quaternionenveränderlichen. *In:* Proceedings of the International Congress of Mathematicians, Oslo 1936, 75—91.

[2] Bizadse, A. W., Grundlagen der Theorie der analytischen Funktionen, Akademie-Verlag, Berlin 1973.

[3] Виноградов, В. С., Об аналоге системы Коши-Римана в четырмерном иространстве. ДАН СССР **154** (1964), 16—19.

[4] Stein, E., Weiß, G., Generalization of the Cauchy-Riemann equations and representations of the rotation group. Amer. J. Math. **90** (1968), 163—196.

[5] Delanghe, R., On the regular points and Liouville's theorem for functions with values in a Clifford algebra. Simon Stevin **44** (1970/71), 55—56.

[6] Hile, G., Representations of solutions of a special class of first order systems. J. Differential Equations **25** (1977), 410—424.

[7] Sprößig, W., Analoga zu funktionentheoretischen Sätzen im R^n. Beiträge zur Analysis **12** (1978), 113—126.

[8] Sprößig, W., Räumliches Analogon zum komplexen T-Operator. Beiträge zur Analysis **12** (1978), 127—137.

[9] Sprößig, W., Methods of function theory for the solution of non-linear partial differential equations. *In:* Theory of Nonlinear Operators (Proc. Internat. Summer School Berlin 1977), Abh. Akad. Wiss. DDR, Jahrg. 1978 Nr. 6N, Akademie-Verlag, Berlin 1978, 399—402.

Address of the author:

Technische Hochschule Karl-Marx-Stadt
Sektion Mathematik
9000 Karl-Marx-Stadt
Reichenhainer Str. 39
DDR

An abstract equation modelling diffusion in solids with bidisperse pore structure

Reinhard Steudel, Magdeburg

The diffusion in porous pellets formed from zeolite crystals usually involves two main mechanisms: diffusion through the micropore structure within the crystals and diffusion through macropore structure between crystals.

In dimensionless form we consider in this paper a nonlinear Ruckenstein model (see [2]) with nonlinear diffusion coefficients and a nonlinear isotherm:

$$\frac{1}{r^2}\frac{\partial}{\partial r}\left(r^2\frac{\partial F(u)}{\partial r}\right)=\frac{\partial u}{\partial t}+\left.\frac{\partial G(\tilde{u})}{\partial \varrho}\right|_{\varrho=1}, \tag{1}$$

$$\frac{1}{\varrho^2}\frac{\partial}{\partial \varrho}\left(\varrho^2\frac{\partial G(\tilde{u})}{\partial \varrho}\right)=\frac{\partial \tilde{u}}{\partial t}, \tag{2}$$

$$u|_{r=1}=1,\quad \left.\frac{\partial u}{\partial r}\right|_{r=0}=0,\quad \left.\frac{\partial \tilde{u}}{\partial \varrho}\right|_{\varrho=0}=0,\quad \tilde{u}|_{\varrho=1}=h(u), \tag{3}$$

$$u|_{t=0}=u_0,\qquad \tilde{u}|_{t=0}=\tilde{u}_0. \tag{4}$$

(1)—(4) discribes the diffusion in a solid with a bipore distribution.

1. Notation

Let $S=[0,T]$ be a bounded (time) interval, $I=\,]0,1[$. We set $H:=\{u\mid ru\in L^2(I)\}$, $\tilde{H}:=\{\tilde{u}\mid r\varrho\tilde{u}\in L^2(I\times I)\}$, $\|u\|_H:=\int_I r^2u(r)^2\,dr$, $\|\tilde{u}\|_{\tilde{H}}=\int_{I\times I} r^2\varrho^2\tilde{u}(r,\varrho)^2\,dr\,d\varrho$, $H^1:=\left\{u\in H\,\middle|\,\frac{du}{dr}\in H\right\}$, $H_0^1:=\{u\in H^1\mid u(1)=0\}$, $\|u\|_{H_0^1}:=\left\|r\frac{du}{dr}\right\|_{L^2(I)}$. The duality mapping K of H_0^1 onto the dual space $H^{-1}:=(H_0^1)^*$ is defined by $\langle Ku,f\rangle:=\int_I r^2\frac{du}{dr}\frac{df}{dr}\,dr$, u, $f\in H_0^1$. We set $V:=\{u\in H_0^1\mid Ku\in H\}$, $(u,v)_V:=(Ku,Kv)_H$, $\tilde{H}^1:=\left\{\tilde{u}\in\tilde{H}\,\middle|\,\frac{\partial\tilde{u}}{\partial\varrho}\in\tilde{H}\right\}$, $\tilde{H}_0^1:=\{\tilde{u}\in\tilde{H}^1\mid\tilde{u}(\cdot,1)=0\}$, $\|\tilde{u}\|_{\tilde{H}_0^1}:=\left\|r\varrho\frac{\partial\tilde{u}}{\partial\varrho}\right\|_{L^2(I\times I)}$.

The duality mapping $\tilde{K}$ of $\tilde{H}_0^1$ onto the dual space $\tilde{H}^{-1}:=(\tilde{H}_0^1)^*$ is defined by

$\langle\langle \tilde{K}\tilde{u}, \tilde{f}\rangle\rangle := \int\limits_{I\times I} r^2\varrho^2 \frac{\partial \tilde{u}}{\partial \varrho} \frac{\partial \tilde{f}}{\partial \varrho}\, d\varrho\, dr, \tilde{u}, \tilde{f} \in \tilde{H}_0^1$. We set $\tilde{V} := \{\tilde{u} \in \tilde{H}_0^1 \mid \tilde{K}\tilde{u} \in \tilde{H}\}$, $(\tilde{u}, \tilde{v})_{\tilde{V}} := (\tilde{K}\tilde{u}, \tilde{K}\tilde{v})_{\tilde{H}}$.

Let E be a Hilbert space. We denote by $L_\lambda^2(S; E)$ the space of all square integrable functions on S with values in E for which $\int\limits_S e^{-2\lambda t} \|u(t)\|_E^2\, dt < \infty$. $(u, v)_{\mathscr{H}} := \int\limits_S e^{-2\lambda t}\big(u(t), v(t)\big)_H\, dt$. The parameter $\lambda > 0$ will be determined appropriately (see 3.) We set $\mathscr{H} := L_\lambda^2(S; H)$, $\mathscr{H}_0^1 := L_\lambda^2(S; H_0^1)$, etc.

It is well known that $\mathscr{V} \subset \mathscr{H}_0^1 \subset \mathscr{H} \subset \mathscr{H}^{-1}$, $\tilde{\mathscr{V}} \subset \tilde{\mathscr{H}}_0^1 \subset \tilde{\mathscr{H}} \subset \tilde{\mathscr{H}}^{-1}$ with dense and continuous imbeddings.

We write v' for the derivative of v (taken in the sense of the space $\mathscr{D}^*(S; E)$ of distributions on $]0, T[$ with values in E) with respect to $t \in S$ and set $\mathscr{W} := \{[u, \tilde{u}] \in \mathscr{V} \times \tilde{\mathscr{V}} \mid [u', \tilde{u}'] \in \mathscr{H} \times \tilde{\mathscr{H}}\}$. We define the operator $R_1 \in (\tilde{V} \to H)$ by $R_1\tilde{u} := \frac{\partial \tilde{u}}{\partial \varrho}(\cdot, 1)$. R_1 has the property (see [3])

$$\|R_1\tilde{u}\|_H^2 \leqq 2\, \|\tilde{u}\|_{\tilde{H}_0^1} \|\tilde{u}\|_{\tilde{V}}, \quad \forall \tilde{u} \in \tilde{V}. \tag{5}$$

2. Assumptions. Formulation of the problem

We suppose that the following conditions are satisfied:

a) $F(s) = \int\limits_0^s f(p)\, dp$, $G(s) = \int\limits_0^s g(p)\, dp$, $s \in R$,

$0 < f_1 \leqq f(s) \leqq f_2 < \infty$, $0 < g_1 \leqq g(s) \leqq g_2 < \infty$, f, g measurable, $s \in R$,

b) $|h(s) - h(t)| \leqq h_1\, |s - t|$, $s, t \in R$, $h_1 = \text{const}$,

c) $[u_0, \tilde{u}_0] \in H \times \tilde{H}$,

and seek functions $[u, \tilde{u}] \in \mathscr{H}^1 \times \tilde{\mathscr{H}}^1$ such that

$$\begin{gathered}[u - 1, \tilde{u} - h(u)] \in \mathscr{H}_0^1 \times \tilde{\mathscr{H}}_0^1, \quad [u', \tilde{u}'] \in \mathscr{H} \times \tilde{\mathscr{H}}, \\ u' + KF(u) + R_1\big(G(\tilde{u}) - G(h(u))\big) = 0,\end{gathered} \tag{6}$$

$$\tilde{u}' + \tilde{K}G(\tilde{u}) = 0, \tag{7}$$

$$u(0) = u_0, \quad \tilde{u}(0) = \tilde{u}_0, \tag{8}$$

where we consider K as a mapping belonging to $\mathscr{L}(\mathscr{H}_1^0, \mathscr{H}^{-1})$. In the same way we deal with the operators $\tilde{K}$, R_1, G and F.

Besides the problem (6)—(8) we now consider the problem

$$Aw' + (L + R)\, w = 0, \quad w(0) = 0, \quad w = [v, \tilde{v}] \in \mathscr{W}, \tag{9}$$

where the operators A, L and R are defined by

$$Az := \begin{pmatrix} F^{-1}(y + F(1)) - u_0 \\ \sigma G^{-1}(\tilde{y} + G(h(F^{-1}(y + F(1))))) - \sigma\tilde{u}_0 \end{pmatrix} \in (\mathscr{H} \times \tilde{\mathscr{H}} \to \mathscr{H} \times \tilde{\mathscr{H}}),$$
$$z = [y, \tilde{y}] \in \mathscr{H} \times \tilde{\mathscr{H}},$$

$$Lz := \begin{pmatrix} Ky \\ \sigma\tilde{K}\tilde{y} \end{pmatrix} \in (\mathscr{V} \times \tilde{\mathscr{V}} \to \mathscr{H} \times \tilde{\mathscr{H}}),\ z = [y, \tilde{y}] \in \mathscr{V} \times \tilde{\mathscr{V}},$$

$$Rz := \begin{pmatrix} R_1 y \\ 0 \end{pmatrix} \in (\mathscr{V} \times \tilde{\mathscr{V}} \to \mathscr{H} \times \tilde{\mathscr{H}}),\ z = [y, \tilde{y}] \in \mathscr{V} \times \tilde{\mathscr{V}}.$$

The parameter $\sigma > 0$ will be determined appropriately later on. The problems (6)—(8) and (9) are equivalent in the following sense:

Theorem 1. *If* $[u, \tilde{u}]$ *is a solution of the problem* (6)—(8), *then* $w(t) := [v(t), \tilde{v}(t)]$, $v(t) := \int_0^t F(u(s)) - F(1))\, ds$, $\tilde{v}(t) := \int_0^t (G(\tilde{u}(s)) - G(h(u(s))))\, ds$ *is a solution of the problem* (9).

Conversely, let w *be a solution of* (9) *and let, in addition,* $[u, \tilde{u}] \in \mathscr{H}^1 \times \tilde{\mathscr{H}}^1$, $[u', \tilde{u}'] \in \mathscr{H} \times \tilde{\mathscr{H}}$. *Then* $[u, \tilde{u}]$, *defined by*

$$u := F^{-1}(v' + F(1)), \quad \tilde{u} := G^{-1}(\tilde{v}' + G(h(F^{-1}(v' + F(1))))) \tag{10}$$

is a solution of (6)—(8).

Proof. [1], Theorem 2.2 and [3], Lemma 3.1.

With regard to Theorem 1 it is useful to introduce the following

Definition. Let w be a solution of (9). Then we say that (10) is a *generalized solution of* (1)—(4).

3. Iteration method

From the assumptions a)—c) we easily conclude that the operator A satisfies the following basic conditions:

$$(Az_1 - Az_2, z_1 - z_2)_{\mathscr{H}\times\tilde{\mathscr{H}}} \geqq m \|z_1 - z_2\|^2_{\mathscr{H}\times\tilde{\mathscr{H}}}, \quad \forall z_1, z_2 \in \mathscr{H} \times \tilde{\mathscr{H}},$$

$$\|Az_1 - Az_2\|_{\mathscr{H}\times\tilde{\mathscr{H}}} \leqq M \|z_1 - z_2\|_{\mathscr{H}\times\tilde{\mathscr{H}}}, \quad \forall z_1, z_2 \in \mathscr{H} \times \tilde{\mathscr{H}},$$

where m and M are defined by

$$m = \frac{3f_1^2 g_1^2}{g_2^4 + 3g_1^2 f_1^2}, \quad M = \left(\max\left\{\frac{1}{f_1^2} + \frac{\sigma^2 g_2^2}{3g_1^2 f_1^2}, \frac{\sigma^2}{f_1^2}\right\}\right)^{1/2},$$

if we choose $\sigma = \dfrac{2m}{g_2}$. The operator $L + R + \dfrac{1}{2\sigma^2} I \in (\mathscr{V} \times \tilde{\mathscr{V}} \to \mathscr{H} \times \tilde{\mathscr{H}})$ is strongly positive and continuous (see (5); I denotes the imbedding operator from $\mathscr{V} \times \tilde{\mathscr{V}}$ into $\mathscr{H} \times \tilde{\mathscr{H}}$).

Theorem 2. *There exists a unique generalized solution* w *of the problem* (1)—(4). *Let* $w_0 \in \mathscr{W}$ *and* $\alpha \in \left]0, \dfrac{2m}{M^2}\right[$ *be arbitrary.*

Then the sequence $(w_i)_{i=1,2,\dots}$ *defined by*

$$\alpha(R + L)\, w_i + w_i' = w_{i-1}' - \alpha A w_{i-1}', \quad w_i \in \mathscr{W}, \quad w_i(0) = 0$$

converges to w strongly in $\mathscr{W}$, if we introduce the norm

$$\|z\|_{\mathscr{W}}^2 = \frac{\alpha^2\beta}{2(\beta-1)} \|z\|_{\mathscr{V}\times\tilde{\mathscr{V}}}^2 + \|z'\|_{\mathscr{H}\times\mathscr{H}}^2 \quad \textit{with} \quad \lambda = \frac{\alpha^2(1+\beta)^2}{4\sigma^3} \quad \textit{in} \quad \mathscr{W}.$$

Proof. [3], Lemma 2.9 and Satz 2.7.

Remark. Combining the iteration method and the Galerkin method one obtains a so-called projection-iteration method. The convergence of this method we proved in [3]. In the same paper we also proved a regularity theorem for the solution of (9).

References

[1] H. Gajewski, On the iterative solution of nonlinear heat conduction and diffusion problems. Apl. Mat. **22** (1977), 77–91.

[2] E. Ruckenstein, A. S. Waidyansthan, G. R. Yongquist, Sorption by solids with bidisperse pore structure. Chem. Eng. Science **26** (1971), 1305.

[3] R. Steudel, Iterative Lösung eines nichtlinearen Operatordifferentialgleichungssystems und ihre Anwendung auf Modelle biporöser Systeme. Thesis, Magdeburg 1979.

Address of the author:

Technische Hochschule „Otto von Guericke“ Magdeburg
Sektion Mathematik und Physik
3010 Magdeburg
Boleslaw-Bierut-Platz 5
PSF 124
DDR

Bifurcation of axially compressed circular cylindrical shells with free edges

Dietmar Wegner, Berlin

Let us regard the buckling problem for a thin-walled circular cylindrical shell, the edges of which are free of all stresses except those caused by a given axial pressure λ. The same problem was considered by Hünlich [1]. We choose the same mathematical formulation, but we use methods of Keller/Langford [2] to show the bifurcation and moreover to construct the solutions near the bifurcation points.

The deformation of the shell is described by the displacement vector $u(x) = \big(u_1(x), u_2(x), u_3(x)\big)$, $x = (x_1, x_2) \in [-l, l] \times [0, 2\pi R] = \bar{\Omega}$, where $2l$ is the length and R is the radius of the shell; u is $2\pi R$-periodic in the variable x_2. Then the buckling can be statet as generalized operator equation [see 1]

$$u + T_2 u + T_3 u = \lambda f, \tag{1}$$

where $f \in H$ is given and $T_i : H \to H$ are i-power operators ($i = 2, 3$); H is a suitable Hilbert space.

First we search for rotational-symmetric solutions $u^0(x_1)$ of (1). From (1) we get

$$\frac{h^2}{12} u^0_{3,1111} + \frac{1 - \nu^2}{R^2} u^0_3 + \lambda u^0_{3,11} + \lambda \frac{\nu}{R} = 0, \tag{2}$$

$$u^0_{1,1} - \frac{\nu}{R} u^0_3 + \frac{1}{2} (u^0_{3,1})^2 + \lambda = 0, \tag{3}$$

$$u^0_{2,11} = 0 \tag{4}$$

and the boundary conditions

$$u^0_{3,11} = 0 \quad \text{for} \quad x_1 = \pm l, \tag{5}$$

$$\frac{h^2}{12} u^0_{3,11} + \lambda u^0_{3,1} = 0 \quad \text{for} \quad x_1 = \pm l, \tag{6}$$

$$u^0_{1,1} + \frac{1}{2} (u^0_{3,1})^2 - \frac{\nu}{R} u^0_3 = -\lambda \quad \text{for} \quad x_1 = \pm l, \tag{7}$$

$$u^0_{2,1} = 0 \quad \text{for} \quad x_1 = \pm l, \tag{8}$$

where h is the thickness of the shell.

The solutions of (2)—(8) are for all λ

$$u^0(x_1) = \left(-\frac{\lambda}{1-\nu^2}\,x_1 + a_1,\; a_2,\; -\frac{\lambda\nu R}{1-\nu^2}\right)$$

with arbitrary a_1, a_2.

The constants a_1, a_2 represent the rigid displacement of the shell; they will be eliminated by factorization.

Now we seek to determine those values λ, at which a branching of the solution u^0 occurs.

Setting $u = u^0 + v$, $v \in H$ in (1) we get the equation

$$v + T_2 v + T_3 v = \lambda L v, \tag{9}$$

where the operator $L: H \to H$ is linear, bounded, selfadjoint, positive and compact.

The linearization of (9) gives

$$v = \lambda L v. \tag{10}$$

Let λ_0 be an eigenvalue of (10) with multiplicity 2 and corresponding eigenfunctions

$$\phi_1 = \begin{pmatrix}\phi_{11}\\ \phi_{12}\\ \phi_{13}\end{pmatrix} = \begin{pmatrix}\varphi_n^1(x_1)\sin\left(\frac{nx_2}{R}\right)\\ \varphi_n^2(x_1)\cos\left(\frac{nx_2}{R}\right)\\ \varphi_n^3(x_1)\sin\left(\frac{nx_2}{R}\right)\end{pmatrix},\quad \phi_2 = \begin{pmatrix}\phi_{21}\\ \phi_{22}\\ \phi_{23}\end{pmatrix} = \begin{pmatrix}\varphi_n^1(x_1)\cos\left(\frac{nx_2}{R}\right)\\ -\varphi_n^2(x_1)\sin\left(\frac{nx_2}{R}\right)\\ \varphi_n^3(x_1)\cos\left(\frac{nx_2}{R}\right)\end{pmatrix}$$

with fixed $n \in \mathbb{N}$.

We define $\mathcal{N} = \mathcal{L}(\phi_1, \phi_2)$ and $\mathcal{N} \oplus \mathcal{N}^\perp = H$ and the projectors $P: H \to \mathcal{N}^\perp$ and $Q = I - P: H \to \mathcal{N}$.

Now we want to apply the methods of Keller/Langford [2] to our problem. Setting $v = \beta_1\phi_1 + \beta_2\phi_2 + w$, $w \in \mathcal{N}^\perp$, we get equivalent to (9)

$$\begin{aligned} 0 = {} & P(\beta_1\phi_1 + \beta_2\phi_2 + w) - PL(\beta_1\phi_1 + \beta_2\phi_2 + w) \\ & + PT_2(\beta_1\phi_1 + \beta_2\phi_2 + w) + PT_3(\beta_1\phi_1 + \beta_2\phi_2 + w), \end{aligned} \tag{11}$$

$$\begin{aligned} 0 = {} & Q(\beta_1\phi_1 + \beta_2\phi_2 + w) - QL(\beta_1\phi_1 + \beta_2\phi_2 + w) \\ & + QT_2(\beta_1\phi_1 + \beta_2\phi_2 + w) + QT_3(\beta_1\phi_1 + \beta_2\phi_2 + w). \end{aligned} \tag{12}$$

Contrary to [2] we get

$$\big(T_2(\phi_i, \phi_j), \phi_k\big) = 0 \quad \text{for} \quad i, j, k = 1, 2. \tag{13}$$

So we have to consider higher order terms in the bifurcation equations and for that reason we have to substitute more carefully. Setting

$$w = \sum_{i,j=0}^{2} \beta_1^i \beta_2^j U_{ij} + \overline{w}, \quad U_{ij}, \overline{w} \in \mathcal{N}^\perp \quad \text{and} \quad \tau = \lambda - \lambda_0$$

we can solve (11) uniquely and get

$$v = \beta_1\phi_1 + \beta_2\phi_2 + \beta_1^2 U_{20} + \beta_1\beta_2 U_{11} + \beta_2^2 U_{02} + \overline{w} \tag{14}$$

with

$$w = O\big((\beta_1 + \beta_2)^3 + \tau(\beta_1 + \beta_2)^2\big). \tag{15}$$

Then we get from (12) the analytic bifurcation equations

$$H_1(\tau, \beta_1, \beta_2) = 0, \tag{16}$$

$$H_2(\tau, \beta_1, \beta_2) = 0. \tag{17}$$

To calculate the coefficients we define the operator $T_c: H \to H$ for $c \in \mathbb{R}$

$$T_c u(x_1, x_2) = u(x_1, x_2 + c).$$

We can show

$$T_c(v - \lambda Lv + T_2 v + T_3 v) = T_c v - \lambda L(T_c v) + T_2(T_c v) + T_3(T_c v), \tag{18}$$

so that together with v also $T_c v$ ist a solution of (9). For the bifurcation equations we can define a corresponding transformation operator $\mathring{T}_c: \mathbb{R}^2 \to \mathbb{R}^2$ for $c \in \mathbb{R}$

$$\begin{aligned}\mathring{T}_c(\beta_1, \beta_2) &= \big(\mathring{T}_c^1(\beta_1, \beta_2), \mathring{T}_c^2(\beta_1, \beta_2)\big)\\ &= \big(\beta_1 \sin(c) + \beta_2 \cos(c), \beta_1 \cos(c) - \beta_2 \sin(c)\big)\end{aligned}$$

and corresponding to (18) the equation

$$\begin{aligned}&\mathring{T}_c\big(H_1(\tau, \beta_1, \beta_2), H_2(\tau, \beta_1, \beta_2)\big)\\ &= \big(H_1(\tau, \mathring{T}_c^1(\beta_1, \beta_2), \mathring{T}_c^2(\beta_1, \beta_2)), H_2(\tau, \mathring{T}_c^1(\beta_1, \beta_2), \mathring{T}_c^2(\beta_1, \beta_2))\big),\end{aligned} \tag{19}$$

so that $\mathring{T}_c(\beta_1, \beta_2)$ for all $c \in \mathbb{R}$ is a solution of (16), (17), if (β_1, β_2) is a solution for fixed τ.

Therefore (16), (17) must have the form

$$0 = a\beta_1(\beta_1^2 + \beta_2^2) + b\tau\beta_1 + \beta_1 O\big((\beta_1 + \beta_2)^3 + \tau(\beta_1 + \beta_2)\big) \tag{20}$$

$$0 = a\beta_2(\beta_1^2 + \beta_2^2) + b\tau\beta_2 + \beta_2 O\big((\beta_1 + \beta_2)^3 + \tau(\beta_1 + \beta_2)\big). \tag{21}$$

We can prove from (20), (21)

$$\beta_1^2 + \beta_2^2 \leqq \text{const}\,|\tau| \tag{22}$$

and therefore we can substitute

$$\beta_i = |\tau|^{1/2}\,\sigma_i, \quad i = 1, 2, \tag{23}$$

into (20), (21) and get

$$0 = G_1(\tau, \sigma_1, \sigma_2) = a\sigma_1(\sigma_1^2 + \sigma_2^2) + b\,\text{sign}(\tau)\,\sigma_1 + O(|\tau|^{1/2})\,\sigma_1 \tag{24}$$

$$0 = G_2(\tau, \sigma_1, \sigma_2) = a\sigma_2(\sigma_1^2 + \sigma_2^2) + b\,\text{sign}(\tau)\,\sigma_2 + O(|\tau|^{1/2})\,\sigma_2. \tag{25}$$

Obviously (24), (25) have the same transformation properties as (16), (17), so that with (σ_1, σ_2) also $\mathring{T}_c(\sigma_1, \sigma_2)$ is a solution of (24), (25) for fixed τ.

Therefore we can set $\sigma_2 = 0$ and get all solutions by applying the transformation $\mathring{T}_c$ on these solutions.

Equation (25) is identically fulfilled for $\sigma_2 = 0$ and from (24) we get

$$0 = G_1(\tau, \sigma_1, 0) \equiv \bar{G}_1(\tau, \sigma_1) \equiv a\sigma_1^3 + b \operatorname{sign}(\tau) \sigma_1 + O(|\tau|^{1/2}) \sigma_1 . \tag{26}$$

The algebraic bifurcation equations to (26) are

$$0 = g^+(\sigma_1) = a\sigma_1^3 + b\sigma_1 \tag{27}$$

for $\tau \geqq 0$ and

$$0 = g^-(\sigma_1) = a\sigma_1^3 - b\sigma_1 \tag{28}$$

for $\tau \leqq 0$, respectively.

The solutions of (27) and (28) are obviously

$$\mathring{\sigma}_1^1 = 0,$$

$$\mathring{\sigma}_1^{2,3} = \pm \sqrt{-\frac{b}{a}} \quad \text{for} \quad \tau > 0 \quad \text{and} \quad \mathring{\sigma}_1^{2,3} = \pm \sqrt{\frac{b}{a}} \quad \text{for} \quad \tau < 0,$$

respectively. Furthermore we get from (27), (28)

$$\frac{dg^\pm}{d\sigma_1} (\sigma_1) = 3a\sigma_1^2 \pm b$$

and therefore

$$\frac{dg^\pm}{d\sigma_1} (\mathring{\sigma}_1^j) \neq 0 \quad \text{for} \quad j = 1, 2, 3.$$

Keller/Langford [2] showed that (26) has exactly 3 analytic solutions $\sigma_1^j(\tau)$ with $\sigma_1^j(0) = \mathring{\sigma}_1^j$ for $(\tau, \sigma_1^j(\tau)) \in \mathscr{U}(0, \mathring{\sigma}_1^j)$. Because we are interested only in real solutions we have depending on the sign of b/a either for $\tau > 0$ three solutions and for $\tau < 0$ only one solution or for $\tau < 0$ three solutions and for $\tau > 0$ one solution.

Now we have to calculate the constants a and b. We obtain

$$b = (L\phi_1, \phi_2) = \int_\Omega (\phi_{13,1})^2 \, d\Omega = \frac{1}{\lambda_0}$$

and

$$a = (T_3\phi_1, \phi_1) + 2(T_2\phi_1, U_{20}) .$$

The value of a has to be computed numerically in general. If it should be 0, we have to use higher order terms in the equations. So we suppose here $a \neq 0$ and we get

for $a > 0$: for $\tau < 0$ three solutions, $\tau > 0$ one solution,

for $a < 0$: for $\tau > 0$ three solutions, $\tau < 0$ one solution.

Applying the transformation $\mathring{T}_c$ to $(\sigma_1^j(\tau), 0)$ and taking into consideration that

$$\mathring{T}_0(\sigma_1^2(\tau), 0) = \mathring{T}_\pi(\sigma_1^3(\tau), 0),$$

we get the bifurcation diagrams of the equations (16), (17) in the neighbourhood of (0, 0, 0) as presented in fig. 1 and fig. 2, respectively.

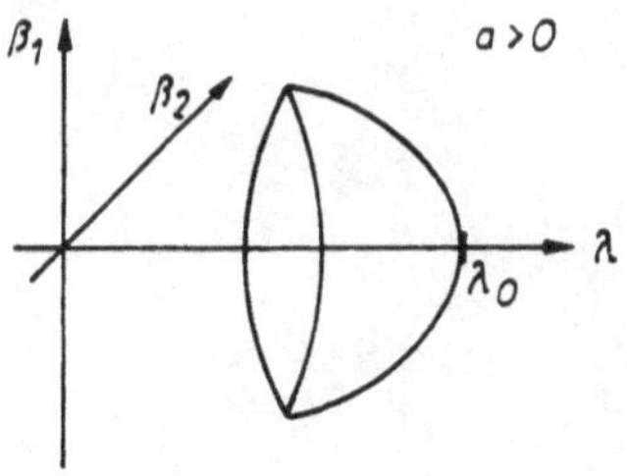

Fig. 1

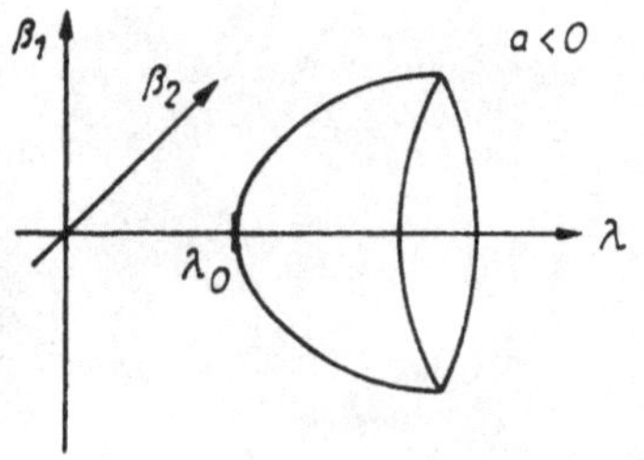

Fig. 2

Finally we want to remark that our methods also can be used to solve the buckling problem for a cylinder with other boundary conditions. But then the rotational-symmetric solution u^0 depends nonlinearly on λ and therefore the linearized problem (10) is an eigenvalue problem with nonlinear parameter. Furthermore T_2 depends analytically on λ. So the constants a and b cannot be calculated so easy. But also in this case we get equivalent bifurcation results.

References

[1] Hünlich, R., The buckling of axially compressed circular cylindrical shells with free edges. *In:* Theory of Nonlinear Operators (Proc. Internat. Summer School Berlin 1977). Abh. Akad. Wiss. DDR, Jahrg. 1978, Nr. 6 N, Akademie-Verlag, Berlin 1978, 361–363.

[2] Keller, H. B., Langford, W. F., Iterations, perturbations and multiplicities for nonlinear bifurcation problems. Arch. Rat. Mech. Anal. 48 (1972), 83–108.

Address of the author:

Humboldt-Universität zu Berlin
Sektion Mathematik
1080 Berlin
Unter den Linden 6
DDR

On difference approximations of a class of n-dimensional variational inequalities

Wilfried Weinelt, Karl-Marx-Stadt

In comparison to a preceding paper [1], this contribution deals with more general problems concerning the differential equation and the spatial dimension and gives also some new results related to the convergence of the free boundary of the coincidence set.

1. Statement of the problem

Let us consider the following elliptic obstacle problem in the bounded domain $\Omega \subset R_n$ (with boundary $\Gamma = \partial\Omega$): To find

$$u \in K = \{v \mid v \in W_2^1(\Omega),\ v \geqq \psi,\ v|_\Gamma = \mu\} \tag{1}$$

fulfilling

$$a(u, v - u) \geqq (f, v - u) \quad \text{for all} \quad v \in K, \tag{2}$$

where

$$a(v, w) = \int\limits_\Omega \left(\sum_{j=1}^{n} k(x) \frac{\partial v}{\partial x_j} \frac{\partial w}{\partial x_j} + q(x)\, vw \right) dx \tag{3}$$

and (f, w) denotes the scalar product in $L_2(\Omega)$.

In the following we restrict ourselfes to the case[1]

$$\psi(x) \equiv 0 \quad (x \in \bar{\Omega}). \tag{4}$$

Further let be

$$k(x) \geqq k_0 > 0, \quad q(x) \geqq 0, \quad f(x) \leqq 0 \quad (x \in \Omega), \quad \mu(x) > 0 \quad (x \in \Gamma). \tag{5}$$

Under relatively weak additional assumptions on the data Γ, μ, k, q and f there follow the solvability of the problem and the uniqueness of the generalized solution in the Sobolev space $W_2^1(\Omega)$, cf. [2].

Below we investigate approximations of the solution by means of the finite difference method. Of course, in order to give estimations of the speed of convergence of the approximate method, we have need of certain "regular" properties of the solution u. Therefore we shall investigate the convergence properties of the difference method

---

1 without loss of generality for sufficiently smooth ψ.

for the class of problems (1)—(5) for which

$$u \in C^{1,1}(\bar{\Omega}) \cap C^m\left(\overline{\Omega^+}\right), \quad m = 2, 3, \tag{6}$$

where

$$\Omega^+ = \Omega \setminus \Omega^-, \quad \Omega^- = \{x \mid x \in \Omega, u(x) = 0\} = \overline{\Omega^-}, \tag{7}$$

and for which the subset $\Omega^+ \subset \Omega$ has a smooth "inner boundary"

$$\Gamma_0 = \overline{\Omega^+} \cap \Omega^-.^1 \tag{8}$$

Partial results of such kind of regularity are given in [3] and [4]. From this papers one can derive the following sufficient conditions for the regularity in the sense (6)—(8) of the solution u of the problem (1)—(5) considered here:

(i) Dimension $n = 2$, Ω — strictly convex set, $k(x) \equiv 1$, $q(x) \equiv 0$.

(ii) For the solution u_0 of the problem (1)—(5) without the unilateral constraint (i.e. $\psi = -\infty$ formally) one has $\min\limits_{\bar{\Omega}} u_0(x) < 0$, u_0 — strictly convex function.

(iii) The data Γ, μ, f are, shortly speaking, sufficiently smooth.

We do not have more general results concerning regularity. But nevertheless, the sufficient conditions (i)—(iii) show that the class of regular solutions in the sense (6)—(8) is not empty, and therefore it is in our opinion a legitimate practice to investigate the approximate methods on the "good" class of regular solutions mentioned above. So we work with (6)—(8) even in the case if, for instance, Ω is a rectangle.[2]

2. Finite difference approximation

Now we approximate the given problem (1)—(5) in the rectangular domain

$$\Omega = \{x \mid x = (x_1, \ldots, x_n), \quad 0 < x_j < l_j, j = 1, \ldots, n\}.$$

We construct the uniform network

$$\mathcal{R}_h = \{x \mid x_j = i_j h_j, h_j = l_j N_j^{-1}, i_j = 0, \pm 1, \pm 2, \ldots, N_j > 0 \text{ integer}\}$$

and define

$$\omega = \Omega \cap \mathcal{R}_h, \quad \gamma = \Gamma \cap \mathcal{R}_h, \quad \bar{\omega} = \omega \cup \gamma, \quad h = \max\{h_1, \ldots, h_n\}$$

and the difference expression

$$\Lambda y = -\sum_{j=1}^{n} \left(a_j(x)\, y_{\bar{x}_j}\right)_{x_j} + q(x)\, y(x) \; (x \in \omega), \tag{9}$$

where

$$a_j(x) = k(\ldots, x_{j-1}, x_j - h_j/2, x_{j+1}, \ldots) = k(x^{-0.5j}),$$

$$y_{\bar{x}_j} = h_j^{-1}\left(y(x) - y(x^{-1j})\right), \quad y_{x_j} = h_j^{-1}\left(y(x^{+1j}) - y(x)\right).$$

---

[1] Because of (5) Ω^+ is (in general multiple) connected and $\Gamma_0 \cap \Gamma = \emptyset$. Γ_0 is called a free boundary.

[2] A further step of investigation of the numerical method may consist in considering the class of solutions with a finite number of singularities of the kind demonstrated in [5].

Problem (1)—(5) is then approximated by

$$\Lambda y \geqq f(x), \quad y \geqq 0, \quad y(x)\left(\Lambda y - f(x)\right) = 0 \quad (x \in \omega), \qquad (10)$$
$$y(x) = \mu(x) \quad (x \in \gamma).$$

It holds ([6], see also [7]):

1° *The difference scheme of inequalities* (10) *has a unique solution* y.

2° *Problem* (10) *is stable with respect to* f, *i.e. a perturbation* Δf *of* f *leads to such* Δy *of* y *that*

$$\|\Delta y\|_{(1)} \leqq C \, \|\Delta f\|_{(-1)}.$$

Here $\|\cdot\|_{(1)}$ and $\|\cdot\|_{(-1)}$ denote the discrete W_2^1-norm and the dual norm, respectively.

3° *For the error* $z(x) = y(x) - u(x)$ $(x \in \overline{\omega})$ *one has the convergence result*

$$\|z\|_{(1)} = O(h^{m/2}) \qquad (11)$$

with $m = 2, 3$ *from* (6).

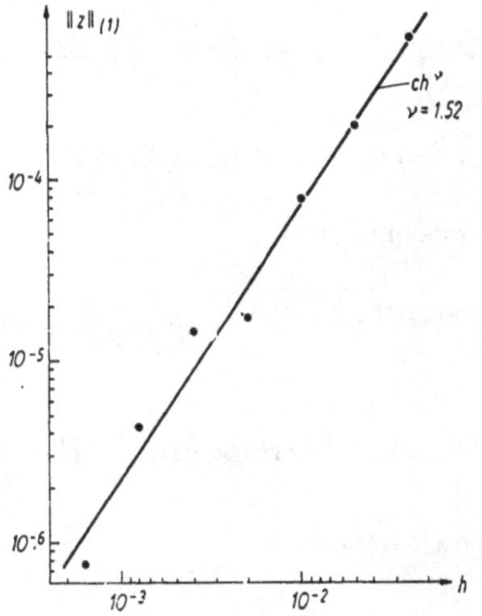

Fig. 1

Further improvements of m do not give better convergence estimations. In fig. 1 by some numerical results is demonstrated the argument that the order 3/2 of convergence will be optimal.

The convergence result (11) can be extended:

4° *If, instead of* (6), *the exact solution* u *belongs to* $W_2^2(\Omega) \cap W_2^m(\Omega^+)$ *only, then one has also*

$$\|y - \overline{u}\|_{(1)} = O(h^{m/2}), \quad m = 2, 3,$$

where the grid function $\overline{u}$ *has as components the mean values of* u *over the grid cells.*

The way of proof refers to that of 3° and to the idea of approximation of solutions $u \in W_2^m$ demonstrated in [8].

Further extensions of (11) are possible to nonlinear variational inequalities with a strongly monotone operator.

Concerning the problem of convergence of the free boundary (stated by Lions [9]) we intend to estimate the distance between the two sets Γ_0 and γ_0, the latter being the discrete free boundary

$$\gamma_0 = \omega^- \cap \overline{\omega^+}, \quad \omega^- = \{x \in \omega, \quad y(x) = 0\}, \quad \omega^+ = \omega \setminus \omega^-,$$

$$\overline{\omega^+} = \bigcup_{x \in \omega^+} S(x), \quad S(x) \text{ — the stencil of (9).}$$

One can show that the following is true: If $f(x) \leqq -f_0 < 0$, then

5° $$\operatorname{dist}(\gamma_0, \Gamma_0) = \begin{cases} O\left(h^{m/4}(\ln h^{-1})^{1/4}\right), & n = 2, \\ O(h^{(m-1)/4}), & n = 3. \end{cases}$$

The proof will be given in a forthcoming paper.

3. Numerical solution

To solve the discrete problem (10) of inequalities we have applied the following mixed penalty iteration method.

$$\begin{gathered} \Lambda y + r(y)\, y = f(x) \quad (x \in \omega), \quad y = \mu(x) \quad (x \in \gamma), \\ r(y) = \begin{cases} 0, & y > 0, \\ \varrho, & y \leqq 0, \end{cases} \quad \varrho = \text{const} > 0; \end{gathered} \tag{12}$$

$$\begin{gathered} \Lambda y^0 = f(x) \quad (x \in \omega), \quad \left(\Lambda + r(y^{m-1})\, I\right) y = f(x) \quad (x \in \omega,\ m \geqq 1), \\ y = \mu(x) \quad (x \in \gamma). \end{gathered} \tag{13}$$

Let us denote by $y_{(\varrho)}$, $y_{(\varrho)}^m$ solutions of (12) and (13), respectively. Then holds ([10], see also [11]):

6° *The problems* (12) *and* (13) *are uniquely solvable.*

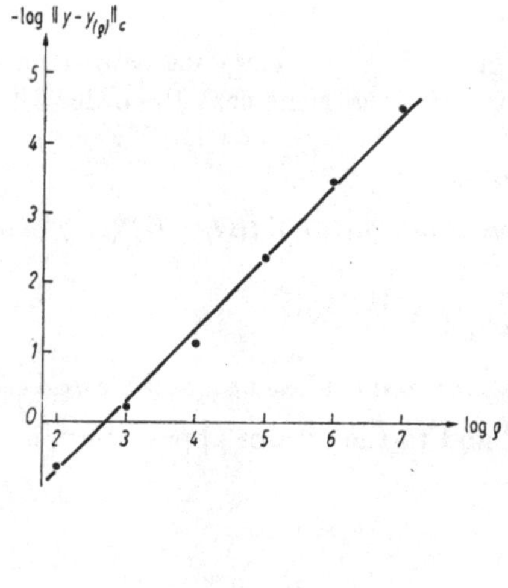

Fig. 2

7° *In* (12) *one has convergence from below, i.e.* $y_{(\varrho)}(x) \nearrow y(x)$ $(x \in \overline{\omega})$ *as* $\varrho \to \infty$, *and the estimation*

$$\max_{\overline{\omega}} \left(y(x) - y_{(\varrho)}(x)\right) \leqq C\varrho^{-1}, \quad C = \left|\min_{\overline{\Omega}} f(x)\right|. \tag{14}$$

Fig. 2 shows numerical results concerning the behaviour of the error $\|y - y_{(\varrho)}\|_C$.

8° *The iteration* (13) *converges after a finite number of steps:*

$$y_{(\varrho)}^m = y_{(\varrho)} \quad \textit{for} \quad m \geqq m_0(h). \tag{15}$$

With the help of (14), (15) it was possible to separate and to observe numerically the approximation error (11) demonstrated in fig. 1.

References

[1] W. Weinelt, On the numerical solution of a variational inequality. *In:* Theory of Nonlinear Operators (Proc. Internat. Summer School Berlin 1977). Abh. Akad. Wiss. DDR, Jahrg. 1978, Nr. 6 N, Akademie-Verlag, Berlin 1978, 405—407.

[2] J.-L. Lions, G. Stampacchia, Variational inequalities, Comm. Pure Appl. Math. **20** (1967), 493—519.

[3] D. Kinderlehrer, Variational inequalities and free boundary problems. Bull. Amer. Math. Soc. **84** (1978), 7—26.

[4] D. Kinderlehrer, L. Nirenberg, Regularity in free boundary problems. Ann. Scuola Norm. Sup. Pisa **4** (1977) 2, 373—391.

[5] D. G. Schaeffer, Some examples of singularities in a free boundary. Ann. Scuola Norm. Sup. Pisa **4** (1977) 1, 133—144.

[6] W. Weinelt, On the approximate solution of variational inequalities. To appear in "Differencialnye uravnenia" (Russian).

[7] W. Weinelt, Zur näherungsweisen Lösung von Variationsungleichungen. Tagungsberichte der 7. Tagung über Methoden und Probleme der Mathematischen Physik, Karl-Marx-Stadt 1979, 73—80.

[8] W. Weinelt, Untersuchungen zur Konvergenzgeschwindigkeit bei Differenzenverfahren. Wiss. Z. Techn. Hochsch. Karl-Marx-Stadt **20** (1978), 763—769.

[9] J.-L. Lions, On free surface problems: Methods of variational and quasivariational inequalities. Lecture Notes "Computational Mechanics", 1975, vol. 461, 129—148.

[10] W. Weinelt, G. Helmert, Iterative Lösung spezieller nichtlinearer Differenzenschemata. Beitr. Numer. Math. **7** (1979), 139—158.

[11] W. Weinelt, On the finite difference method to solve boundary value problems for differential inequalities. Žurn. vyč. mat. i mat. fis. **18** (1978), 642—652 (Russian).

Address of the author:

Technische Hochschule Karl-Marx-Stadt
Sektion Mathematik
9010 Karl-Marx-Stadt
PSF 964
DDR

On some Stefan like problem

Janina Wolska-Bochenek, Warsaw

The problem S consists in determining the functions

$$u_1 = u_1(x, t), \quad u_2 = u_2(x, t), \quad s = s(t), \quad c = u_2(s(t), t)$$

atisfying in the domain

$$\Omega = \{(x, t): 0 < x < s(t), 0 < t < T\} \tag{1}$$

the parabolic system

$$\frac{\partial u_\alpha}{\partial t} = \sum_{j=1}^{2} d_{\alpha j} \frac{\partial^2 u_j}{\partial x^2} \quad (\alpha = 1, 2) \tag{2}$$

and the boundary conditions:

$$\begin{cases} u_1(x, 0) = 1 \\ u_2(x, 0) = c_0 \end{cases} \quad 0 \leqq x \leqq s(0) \equiv s_0, \tag{3}$$

$$\sum_{j=1}^{2} d_{\alpha j} \left. \frac{\partial u_j}{\partial x} \right|_{x=0} = -a_\alpha(t), \quad 0 < t < T, \tag{4}$$

$$u_1(x, t)|_{x=s(t)} = 1, \quad 0 < t < T, \tag{5}$$

$$\sum_{j=1}^{2} d_{1j} \left. \frac{\partial u_j}{\partial x} \right|_{x=s(t)} = -\frac{ds}{dt}, \quad 0 < t < T, \tag{6}$$

$$\sum_{j=1}^{2} d_{2j} \left. \frac{\partial u_j}{\partial x} \right|_{x=s(t)} = -\frac{ds}{dt} \cdot u_2(s(t), t), \quad 0 < t < T, \quad u_2(s(0),0) \equiv c_0 > 0, \tag{7}$$

where the given square matrix $D = [d_{\alpha j}]_{2 \times 2}$ is constant, $\det D > 0$, $d_{\alpha j} > 0$ $(\alpha, j = 1, 2)$, and the given functions a_α $(\alpha = 1, 2)$ are continuous in $\langle 0, T \rangle$.

Let us denote $c(t) \equiv u_2(s(t), t)$.

The set of functions $\{u_1, u_2, s, c\}$ will be called a solution of the problem (2)—(7) if

I. s, c are positive in $(0, T)$,

II. s, c are continuously differentiable in $(0, T)$ and continuous in $\langle 0, T \rangle$, $\frac{dc}{dt}$ is Hölder continuous in $(0, T)$,

III. u_α are continuous in Ω and $\frac{\partial u_\alpha}{\partial x}$, $\frac{\partial^2 u_\alpha}{\partial x^2}$, $\frac{\partial u_\alpha}{\partial t}$ are continuous in $\bar{\Omega}$.

The problem of such a type arises in the theory of metal alloys. Concentrations of two alloy components are to be determined from the boundary data, besides one of the concentrations is prescribed on the free boundary. Conditions (6), (7) result from the law of conservation of mass [2, 4, 5].

We can easily transform the problem (2)—(7) into the following one:

$$\frac{\partial V}{\partial t} = D\frac{\partial^2 V}{\partial x^2} + \frac{dC}{dt}, \quad (x, t) \in \Omega, \tag{8}$$

$$V(x, 0) = \theta, \quad 0 \leqq x \leqq s(0) \equiv s_0, \tag{9}$$

$$D\frac{\partial V(0, t)}{\partial x} = -A(t), \quad 0 < t < T, \tag{10}$$

$$V\big(s(t), t\big) = 0, \quad 0 < t < T, \tag{11}$$

$$D\frac{\partial V\big(s(t), t\big)}{\partial x} = -\frac{ds}{dt}\cdot C(t), \quad 0 < t < T, \tag{12}$$

where

$$V(x, t) = \begin{bmatrix} v_1(x, t) \\ v_2(x, t) \end{bmatrix}, \quad C(t) = \begin{bmatrix} 1 \\ c(t) \end{bmatrix}, \quad A(t) = \begin{bmatrix} a_1(t) \\ a_2(t) \end{bmatrix}, \quad \theta = \begin{bmatrix} 0 \\ 0 \end{bmatrix}$$

and $v_1 = u_1 - 1$, $v_2 = u_2 - c$.

Consider the auxiliary problem S_0

Determine the vector V satisfying the equation (8) and the boundary conditions (9), (10), (11), assuming that $s = s(t)$, $c = c(t)$ are given, differentiable in $(0, T)$ functions and $s(0) = s_0$, $c(0) = c_0$, $\frac{dc}{dt}$ is Hölder continuous in $(0, T)$.

According to the properties of the coefficients $d_{\alpha j}$ the eigenvalues of the matrix D are real, positive and single. Thus the auxiliary problem (8)—(11) may be transformed into the problem for two separate parabolic equations with the boundary values being linear combinations of the previous ones. The following lemma holds:

Lemma 1. *The auxiliary problem* S_0 *possesses a solution* V *for* $T > 0$ *suitably chosen.*

To establish the existence of the solution of the problem S evaluate:

$$\int_0^t \int_0^{s(t)} \left(\frac{\partial V}{\partial t} - D\frac{\partial^2 V}{\partial x^2} - \frac{dC}{dt}\right) dx\, dt = 0 \tag{13}$$

with the boundary conditions (9)—(12).

We obtain

$$s(t) = s_0 + A_1(t) - \int_0^{s(t)} v_1(x, t)\, dx, \tag{14}$$

$$c(t)\, s(t) = c_0 s(t) + \int_0^t \frac{ds}{dt} c(t)\, dt + A_2(t) - \int_0^{s(t)} v_2(x, t)\, dx, \tag{15}$$

where $A_\alpha(t) = -\int_0^t a_\alpha(t)\, dt$ $(\alpha = 1, 2)$.

Thus it is possible to prove the following

Lemma 2. *If there exists the solution V of the auxiliary problem S_0 such that V, s, c satisfy the equations* (14), (15), *then V, s, c satisfy the boundary condition* (12) *i.e. form a solution of the problem* (8)—(12).

A similar problem for one parabolic equation was discussed by G. W. Evans [1] and generalized by W. T. Kyner [3].

According to their method let us define

$$s_1(t) = s_0, \quad c_1(t) = c_0$$

and by induction

$$s^{(n+1)}(t) = s_0 + A_1(t) - \int_0^{s^{(n)}(t)} v_1^{(n)}(x, t)\, dx, \tag{16}$$

$$c^{(n+1)}(t)\, s^{(n)}(t) = c_0 s^{(n)}(t) + \int_0^t \frac{ds^{(n)}}{dt} c^{(n)}(t)\, dt + A_2(t) - \int_0^{s^{(n)}(t)} v_2^{(n)}(x, t)\, dx \tag{17}$$

$(n = 1, 2, \ldots)$,

where the vector $V^{(n)}$ is the solution of the problem

$$\frac{\partial V^{(n)}}{\partial t} = \frac{\partial^2 V^{(n)}}{\partial x^2} + \frac{dC^{(n)}}{dt}, \quad (x, t) \in \{0 < x < s^{(n)}(t), 0 < t < T\},$$

$$V^{(n)}(x, 0) = 0, \quad 0 \leqq x \leqq s^{(n)}(0),$$

$$D\frac{\partial V^{(n)}(0, t)}{\partial x} = -A(t), \quad 0 < t < T,$$

$$V^{(n)}\big(s^{(n)}(t), t\big) = 0, \quad 0 < t < T.$$

The existence of the solution $s(t)$, $c(t)$ of (14, (15) being proved by the method of successive approximations (16), (17) then, in view of lemma 2 one can formulate the theorem:

Theorem. *The problem* (8)—(12) *has a solution for $T > 0$ suitably chosen.*

References

[1] G. W. Evans, Quart. Appl. Math. **9** (1951).
[2] J. S. Kirkaldy, Canad. J. Phys. **35** (1957), **36** (1958).
[3] W. T. Kyner, J. Math. Mech. **8**, 4 (1959).
[4] B. J. Lubow, W. T. Płachotnik, Fizika Metałow i Metałowiedienie **33** (1972).
[5] G. B. Szczerbedynski, J. Kondraczenko, Fizika Metałow i Metałowiedienie **30** (1970).

Address of the author:

Mathematical Institute
Technical University of Warsaw
Pl. Jednosci Robotniczej 1
00-661 Warszawa
Poland

On a system of nonlinear partial differential equations arising in the theory of heat and mass transfer in a nonlinear second-order fluid

Michael Wulst, Berlin

Abstract. The aim of this paper is to study the heat and mass transfer in a nonlinear second-order fluid within a bounded three-dimensional domain. By Schauder's Fixed Point Theorem the existence of a weak solution is shown.

The heat and mass transfer in an incompressible homogeneous fluid within a bounded domain $\Omega \subset R^3$ with boundary Γ during the time interval $(0, T)$ is governed by the following system of partial differential equations for the functions θ, p and $u = \{u_1, u_2, u_3\}$:

$$u_i' + u_j u_{i,j} - \big(S_{ij}(u)\big)_{,j} = f_i - p_{,i}{}^1 \quad (i = 1, 2, 3) \quad \text{in } Q; \tag{1}$$

$$u_{i,i} = 0 \quad \text{in } Q; \tag{2}$$

$$u = 0 \quad \text{on } \Gamma \times (0, T); \tag{3}$$

$$\theta' - k\Delta\theta + u_i\theta_{,i} = S_{ij}(u)\, D_{ij}(u) + g \quad \text{in } Q; \tag{4}$$

$$\theta = 0 \quad \text{on } \Gamma \times (0, T). \tag{5}$$

Here $S = \{S_{ij}\}_{i,j=1,2,3}$ denotes the stress tensor, p the undetermined hydrostatic pressure, u the velocity field, $D = \{D_{ij}\}_{i,j=1,2,3}$ the stretching tensor $\left(D_{ij}(u) = \frac{1}{2}(u_{i,j} + u_{j,i})\right)$, f the given volume force, g the given heat source, k $(= \text{const} > 0)$ the heat conduction coefficient and $Q = \Omega \times (0, T)$. The boundary condition (3) means that the fluid adheres to Γ. For deriving the energy equation (4) we have used the classical Fourier law for the connection between the temperature and the heat flux and a linear relationship between the temperature and the specific internal energy.

In the following we will suppose that the fluid obeys the constitutive law

$$S = -pI + \alpha(\theta, D_{\mathrm{II}})\, D + \gamma \frac{\partial}{\partial t} D \tag{6}$$

where $D_{\mathrm{II}} = \frac{1}{2} D_{ij}D_{ij}$ denotes the second main invariant of D, α the viscosity func-

---

[1] Throughout the whole paper Latin subscripts i and j take independently the values 1, 2, 3. Further, we use the convention that a repeated subscript means summation over 1, 2, 3 and that $v' = \frac{\partial v}{\partial t}$, $v_{,i} = \frac{\partial v}{\partial x_i}$.

tion depending on θ and D_{II}, γ a positive constant and I the unit tensor. Such fluids are often called second-order fluids (cf. e.g. [1, 2]).

Further, let us consider instead of the complete equation (4) the equation

$$\theta' - k\Delta\theta = 2\alpha\big(\theta, D_{II}(u)\big)\, D_{II}(u) + g \tag{7}$$

in Q where we have neglected for the sake of mathematical difficulties the heat convection term $u_i\theta_{,i}$ and the part of dissipation energy $\gamma \frac{\partial}{\partial t} D_{II}(u)$.

To formulate the main result let us briefly introduce some spaces.

Let $W_r^1(\Omega)$ $(r > 1)$ denote the usual Sobolev space. Define

$$\mathscr{V} = \{u \in [\mathscr{D}(\Omega)]^3 : u_{i,i} = 0 \text{ in } \Omega\}^1$$

and

$$V = \text{closure of } \mathscr{V} \text{ in } [W_2^1(\Omega)]^3,$$

$$H = \text{closure of } \mathscr{V} \text{ in } [L^2(\Omega)]^3.$$

Then one can introduce in the usual way scalar products in V and H. Let V^* denote the dual of V. Further, let for $q > 1$

$$W_q^{2,1}(Q) = \{v \in W_q^1(Q) : v_{,i} \in W_q^1(Q) \quad (i = 1, 2, 3),\ \delta v/\delta t \in L_q\,(Q)\}$$

Then $W_q^{2,1}(Q)$ furnished with the customary norm is a Banach space (cf. [3]).

Finally, we suppose that the viscosity function α satisfies the following conditions:

$$\alpha : R \times [0, +\infty) \to R \text{ is continuous;}$$

$$\alpha(\eta, \xi) \geqq 0 \quad \forall\ (\eta, \xi) \in R \times [0, +\infty);$$

$$\alpha(\eta, \xi^2)\, \xi \leqq c_0(1 + \xi^{r-1}) \quad \forall\ (\eta, \xi) \in R \times [0, +\infty),$$

$$c_0 = \text{const} > 0, \qquad 1 < r < 2;$$

$$\alpha(\eta, \xi_1^2)\, \xi_1 \geqq \alpha(\eta, \xi_2^2)\, \xi_2 \quad \forall\ \xi_1 \geqq \xi_2 \geqq 0, \quad \forall\ \eta \in R.$$

Let the boundary Γ of Ω be sufficiently smooth. Then we have the following

Theorem. *Let* $q = \frac{2}{r}\, l$, $\frac{r}{2} < l < 1$, $r \neq \frac{4}{3}$. *Let* $f \in L^2(0, T; V^*)$, $g \in L^q(Q)$, $u_0 \in V$, $\theta_0 \in W_q^{2-2/q}\,(\Omega)$, $\theta_0 = 0$ *on* Γ *(in the sense of trace; cf. e.g.* [4]*).*

Then there exists a pair of functions $\{u, \theta\}$ *such that*

$$u \in C([0, T]; V) \text{ with } u' \in L^2(0, T; V), \qquad \theta \in W_q^{2,1}(Q)$$

and

$$\left.\begin{aligned} &\int_\Omega u_i'(t)\, v_i\, dx + \gamma \int_\Omega D_{ij}\big(u'(t)\big)\, D_{ij}(v)\, dx \\ &+ \int_\Omega \alpha\big(\theta(t), D_{II}(u(t))\big)\, D_{ij}\big(u(t)\big)\, D_{ij}(v)\, dx \\ &+ \int_\Omega u_j(t)\, u_{i,j}(t)\, v_i\, dx = \int_\Omega f_i(t)\, v_i\, dx \\ &\text{for a.a. } t \in (0, T) \qquad \forall\ v \in V; \end{aligned}\right\} \tag{8}$$

[1] $\mathscr{D}(\Omega)$ = set of all infinitely differentiable functions having a compact support in Ω.

$$u(0) = u_0; \tag{9}$$

$$\theta' - k\Delta\theta = 2\alpha\big(\theta, D_{II}(u)\big)\, D_{II}(u) + g \quad \textit{a.e. in } Q; \tag{10}$$

$$\theta = 0 \quad \textit{a.e. on} \quad \Gamma \times (0, T); \tag{11}$$

$$\theta(0) = \theta_0. \tag{12}$$

Remarks. 1. One may easily show that each solution pair $\{u, \theta\}$ to (1)—(6) which is sufficiently differentiable and integrable also satisfies (8)—(12). Backward, if $\{u, \theta\}$ verifies (8)—(12) one may derive under appropriate regularity assumptions on u, θ, f and α that there exists a function p (uniquely determined up to an additive constant) and u, θ, p satisfy (1)—(6).

2. The proof of the theorem is based on a fixed point principle. Here one has to use a result of Wulst [5], Ladyshenskaja/Solonnikov/Uralzeva [3] and Schauder's fixed point theorem. Let us note that the difficulty of the proof arises from the nonlinearity of the viscosity function α.

References

[1] Astarita, G., and Marucci, G., Principles of non-Newtonian fluid mechanics. McGraw Hill, London 1974.

[2] Coleman, B. D., and Noll, W., An approximate theorem for functionals with applications in continuum mechanics. Arch. Rat. Mech. Anal. **6** (1960), 355—370.

[3] Ladyshenskaja, O. A., Solonnikov, V. A., and Uralzeva, N. N., Linear and quasilinear equations of parabolic type (in Russian). Moscow 1967.

[4] Nečas, J., Les méthodes directes en théorie des équations elliptiques. Prague 1967.

[5] Wulst, M., On nonlinear evolution equalities arising in the theory of second-order fluids. *In:* Studies on some nonlinear evolution equations. Seminarbericht Nr. 17, Humboldt-Universität Berlin, Juni 1979, pp. 5.1—5.37.

Address of the author:

Humboldt-Universität zu Berlin
Sektion Mathematik
1080 Berlin
Unter den Linden 6
DDR

FSC
www.fsc.org
MIX
Papier | Fördert
gute Waldnutzung
FSC® C083411

Kolibri 360 GmbH
Ferdinand-Jühlke-Straße 7
99095 Erfurt, Deutschland
produktsicherheit@kolibri360.de